T0213408

COOPERATIVE BREEDING IN BIRDS:

long-term studies of ecology and behavior

Frontispiece. Location of the studies of cooperatively breeding birds discussed in the book.

Flagstaff, Arizona (Pinyon Jays)

The Research Ranch near Elgin, Arizona (Acorn Woodpeckers)

Southwestern Research Station, near Portal, Arizona (Mexican [Gray-breasted] Jays)

Sandhills region, North Carolina (Red-cockaded Woodpeckers)

Archbold Biological Station Florida (Florida Scrub Jays)

Llanos, central Venezuela (*Campylorhynchus* wrens and Hoatzins)

Hastings Reservation, California (Acorn Woodpeckers)

Water Canyon, New Mexico (Acorn Woodpeckers)

Southwestern New Mexico (Harris' Hawks)

Galápagos Islands (Galápagos Hawks and Mockingbirds)

Guanacoste Province, Costa Rica (Groove-billed Anis)

Cambridge Botanic Gardens, England (Dunnocks)

Hatzeva Field Study Center, Israel (Arabian Babblers)

Darling Escarpment, near Perth, Australia (Splendid Fairy-wrens)

Meandarra and Laidley, Queensland, Australia (Noisy Miners)

Shakespear Park, near Auckland, N.Z. (Pukeko)

Lake Victoria, Kenya (Pied Kingfishers)

Lake Nakuru, Kenya (White-fronted Bee-eaters)

Lake Naivasha, Kenya (Green Woodhoopoes and Pied Kingfishers)

COOPERATIVE BREEDING IN BIRDS:
long-term studies of ecology and behavior

Edited by

PETER B. STACEY

Department of Biology
University of New Mexico

and

WALTER D. KOENIG

Museum of Vertebrate Zoology
University of California

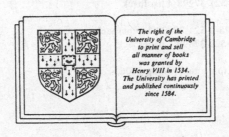

The right of the
University of Cambridge
to print and sell
all manner of books
was granted by
Henry VIII in 1534.
The University has printed
and published continuously
since 1584.

CAMBRIDGE UNIVERSITY PRESS
Cambridge   New York   Port Chester
Melbourne   Sydney

CAMBRIDGE UNIVERSITY PRESS
Cambridge, New York, Melbourne, Madrid, Cape Town, Singapore, São Paulo

Cambridge University Press
The Edinburgh Building, Cambridge CB2 2RU, UK

Published in the United States of America by Cambridge University Press, New York

www.cambridge.org
Information on this title: www.cambridge.org/9780521372985

© Cambridge University Press 1990

This publication is in copyright. Subject to statutory exception
and to the provisions of relevant collective licensing agreements,
no reproduction of any part may take place without
the written permission of Cambridge University Press.

First published 1990

A catalogue record for this publication is available from the British Library

Library of Congress Cataloguing in Publication data

Cooperative breeding in birds: long-term studies of ecology and
behaviour/edited by Peter B. Stacey and Walter D. Koenig.
     p.   cm.
Includes index.
ISBN 0 521 37298 4.–ISBN 0 521 37890 7 (paperback)
1. Birds–Behaviour. 2. Birds–Ecology. 3. Sexual behavior in
animals. I. Stacey, Peter B. II. Koenig, Walter D., 1950–
QL698.3.C66 1990
598.256–dc20   89–9773 CIP

ISBN-13 978-0-521-37298-5 hardback
ISBN-10 0-521-37298-4 hardback

ISBN-13 978-0-521-37890-1 paperback
ISBN-10 0-521-37890-7 paperback

Transferred to digital printing 2006

CONTENTS

Frontispiece   ii
Contributors   vii

Introduction   ix
P. B. STACEY AND W. D. KOENIG

1 Splendid Fairy-wrens: demonstrating the importance of longevity   1
   I. C. R. ROWLEY AND E. RUSSELL

2 Green Woodhoopoes: life history traits and sociality   31
   J. D. LIGON AND S. H. LIGON

3 Red-cockaded Woodpeckers: a 'primitive' cooperative breeder   67
   J. R. WALTERS

4 Arabian Babblers: the quest for social status in a cooperative
   breeder   103
   A. ZAHAVI

5 Hoatzins: cooperative breeding in a folivorous neotropical bird   131
   S. D. STRAHL AND A. SCHMITZ

6 *Campylorhynchus* wrens: the ecology of delayed dispersal and
   cooperation in the Venezuelan savanna   157
   K. N. RABENOLD

7 Pinyon Jays: making the best of a bad situation by helping   197
   J. M. MARZLUFF AND R. P. BALDA

8 Florida Scrub Jays: a synopsis after 18 years of study   239
   G. E. WOOLFENDEN AND J. W. FITZPATRICK

9 Mexican Jays: uncooperative breeding   267
   J. L. BROWN AND E. R. BROWN

10 Galápagos mockingbirds: territorial cooperative breeding in a
   climatically variable environment   289
   R. L. CURRY AND P. R. GRANT

11 Groove-billed Anis: joint-nesting in a tropical cuckoo   333
   R. R. KOFORD, B. S. BOWEN AND S. L. VEHRENCAMP

12 Galápagos and Harris' Hawks: divergent causes of sociality in
   two raptors   357
   J. FAABORG AND J. C. BEDNARZ

13   Pukeko: different approaches and some different answers   385
     J. L. CRAIG AND I. G. JAMIESON

14   Acorn Woodpeckers: group-living and food storage under
     contrasting ecological conditions   413
     W. D. KOENIG AND P. B. STACEY

15   Dunnocks: cooperation and conflict among males and females in
     a variable mating system   455
     N. B. DAVIES

16   White-fronted Bee-eaters: helping in a colonially nesting species   487
     S. T. EMLEN

17   Pied Kingfishers: ecological causes and reproductive con-
     sequences of cooperative breeding   527
     H.-U. REYER

18   Noisy Miners: variations on the theme of communality   559
     D. D. DOW AND M. J. WHITMORE

     Summary   593
     J. N. M. SMITH
     Index   613

CONTRIBUTORS

R. P. Balda, *Department of Biological Sciences, Northern Arizona University, Flagstaff, AZ 86011, USA*

J. Bednarz, *Department of Biology, University of New Mexico, Albuquerque, NM 87131, USA, and Hawk Mountain Sanctuary, Route 2, Kempton, PA 19529, USA*

B. S. Bowen, *Department of Biology C-016, University of California at San Diego, La Jolla, CA 92093, USA*

E. R. Brown, *Department of Biological Sciences, State University of New York, Albany, NY 12222, USA*

J. L. Brown, *Department of Biological Sciences, State University of New York, Albany, NY 12222, USA*

J. L. Craig, *Department of Zoology, University of Auckland, Private Bag, Auckland, New Zealand*

R. L. Curry, *Department of Biology, University of Michigan, Ann Arbor, MI 48109, USA*

N. B. Davies, *Department of Zoology, University of Cambridge, Cambridge CB2 3EJ, UK*

D. D. Dow, *Department of Zoology, University of Queensland, St Lucia, Queensland, Australia 4067*

S. T. Emlen, *Section of Neurobiology and Behavior, Cornell University, Ithaca, NY 14853-2702*

J. Faaborg, *Division of Biological Sciences, University of Missouri, Columbia, MO 65211, USA*

J. W. Fitzpatrick, *Department of Zoology, Field Museum of Natural History, Roosevelt Road at Lake Shore, Chicago, IL 60605-2496, USA*

P. R. Grant, *Department of Biology, Princeton University, Princeton, NJ 08544, USA*

I. G. Jamieson, *Department of Zoology, University of Auckland, Private Bag, Auckland, New Zealand*

W. D. Koenig, *Hastings Reservation, University of California, Star Route Box 80, Carmel Valley, CA 93924, USA*

R. R. Koford, *Department of Biology C-016, University of California at San Diego, La Jolla, CA 92093, USA*

J. D. Ligon, *Department of Biology, University of New Mexico, Albuquerque, NM 87131, USA*

S. H. Ligon, *Department of Biology, University of New Mexico, Albuquerque, NM 87131, USA*

J. M. Marzluff, *Department of Biological Sciences, Northern Arizona University, Flagstaff, AZ 86011, USA*

K. N. Rabenold, *Department of Biological Sciences, Purdue University, West Lafayette, IN 47907, USA*

H.-U. Reyer, *Max-Planck-Institut für Verhaltenphysiologie, D-8131 Seewiesen, West Germany*

I. C. R. Rowley, *CSIRO Wildlife and Rangelands Research, LMB 4, PO Midland, Western Australia 6056, Australia*

E. Russell, *CSIRO Wildlife and Rangelands Research, LMB 4, PO Midland, Western Australia 6056, Australia*

A. Schmitz, *Departimento Biología de Organismos, Universidad Simón Bolívar, Apartimento 89000, Caracas 1086-A, Venezuela*

J. N. M. Smith, *Department of Zoology, University of British Columbia, 6270 University Boulevard, Vancouver, BC V6T 2A9, Canada*

P. B. Stacey, *Department of Biology, University of New Mexico, Albuquerque, NM 87131, USA*

S. D. Strahl, *Wildlife Conservation International, New York Zoological Society, Bronx Zoological Park, 185th Street and Southern Boulevard, Bronx, NY 10460, USA*

S. L. Vehrencamp, *Department of Biology C-016, University of California at San Diego, La Jolla, CA 92093, USA*

J. R. Walters, *Department of Zoology, North Carolina State University, Campus Box 7617, Raleigh, NC 27695-7617, USA*

M. J. Whitmore, *Department of Zoology, University of Queensland, St Lucia, Queensland, Australia 4067*

G. E. Woolfenden, *Department of Biology, University of Southern Florida, Tampa, FL 33620, USA*

A. Zahavi, *Department of Zoology, Tel Aviv University, Ramat Aviv 69978, Tel Aviv, Israel*

INTRODUCTION

P. B. STACEY AND W. D. KOENIG

Cooperative breeding is a reproductive system in which one or more members of a social group provide care to young that are not their own offspring. The aid-givers may be non-breeding adults, in which case they are usually called 'helpers' or 'auxiliaries', or they may be co-breeders, and share reproduction with other group members of the same sex. Although the care given usually includes providing food, it may involve other parental-type behaviors as well, including territorial defense, nest or den construction, incubation and defense against predators.

Cooperative breeding is relatively rare: it is currently known to occur in about 220 of the roughly 9000 species of birds, and a smaller number of mammals and fish (for recent compilations, see Emlen 1984; Brown 1987). However, several theories of cooperative breeding suggest it should occur most frequently in tropical and warm temperate habitats, rather than in cold temperate or highly seasonal areas (see e.g. Fry 1972; Brown 1974; Stacey and Ligon 1987), and the reproductive biology of most species in these regions is presently unknown. In some habitats, such as the savannahs and woodlands of Australia, cooperative breeding is very common among some groups such as ground-foraging insectivorous birds (Ford *et al.* 1988). It also appears frequently in certain taxonomic groups, such as the New World jays (family Corvidae: Hardy 1961, Hardy *et al.* 1981), the Old World babblers (family Timaliidae: Gaston 1978), mongooses (subfamily Herpestinae: Rood 1984), and some primates such as tamarins and marmosets (family Callitrichidae: Terborgh and Goldizen 1985; Sussman and Garber 1987).

Although the existence of helpers at the nest has been known for many years (Skutch 1935), cooperative breeding has been studied in detail only over the last two decades. There were a few good early studies of cooperatively breeding birds, particularly those of Davis (1942), Robinson (1956), Selander (1964) and Rowley (1965). However, a major difficulty was the absence until 1964 of an adequate theoretical framework within which the evolution of the apparently altruistic behavior of helpers could be understood. For example, one of the most widely mentioned early examples of alloparental care involved a photograph of a Northern Cardinal (*Cardinalis cardinalis*) feeding goldfish in a small pond (see e.g. Welty 1975). It was difficult to see how such behavior could be adaptive. The most reasonable explanation was simply that the cardinal had recently lost its own nest, and because such a bird would be highly motivated to feed young, it had simply been responding to the most available stimulus (as often occurs in domestic sheep and cattle). Although there were other examples of helping that did involve members of the same species, the origin and relationships of the animals to each other were usually unknown. It seemed plausible that many of these cases also involved 'mistakes', and the phenomenon of helping was generally relegated to the back shelf of biological curiosities.

Cooperative breeding reappeared briefly in 1962 when it was discussed by Wynne-Edwards in *Animal Dispersion in Relation to Social Behaviour* (1962), as a possible example of a self-limiting reproductive behavior. He hypothesized that for the Acorn Woodpecker (*Melanerpes formicivorus*: see Chapter 14), group storage of acorns and other food constituted an 'epideictic' display, which allowed the local population of woodpeckers to measure the amount of available resources, and, if they were too low, would lead some individuals (the helpers) or entire groups to forgo breeding for that season. Since such behavior seemed to require a form of group selection for its evolution, and the mechanisms by which group selection could occur in natural populations were not well understood, cooperative breeding again disappeared from the spotlight.

It was not until Hamilton (1964) and Maynard Smith (1964) developed the theory now referred to as kin selection that there was a firm basis upon which the empirical study of cooperative breeding could begin. These authors argued that the fitness of each individual is determined by the total number of genes present in following generations that are identical by descent with its own. Copies of genes could arise either directly through the production of the individual's offspring (the usual or 'direct' component of inclusive fitness: Brown and Brown 1981) or indirectly, through the reproduction of the individual's relatives (the 'indirect' component). Kin selection was applied initially to the evolution of the sterile, or non-reproductive, worker castes in social insects, and helping at the nest among vertebrates was not mentioned in either initial article. However, it soon became apparent that helpers in birds and mammals might constitute an analog of worker castes in insects. In both cases, certain individuals appeared to act altruistically: they appeared to sacrifice their own individual reproduction to help to care for the offspring of others.

By the mid 1970s, a number of studies had been started on cooperative breeding, most dealing with birds. Although relatively rare, this breeding system presented excellent opportunities to test various hypotheses about the evolution of cooperation and competition in social behavior. It has been of wide interest therefore to both the specialist and to the general public. Several fundamental questions had to be addressed for each species. First, what was the origin of the 'extra' group members? Individual marking, usually by color banding, quickly showed that in most species (but not all) helpers are usually the previous offspring of the breeding pair that had remained in their natal group or territory. Once this had been demonstrated, the problem then became: why, on reaching physiological maturity, would the young fail to disperse and breed on their own, as they do in most other species? Are there specific ecological conditions that would lead to juvenile philopatry in some species but not in others? This failure to disperse was particularly puzzling, since by remaining at home, most helpers are not able to breed. (It had originally been suggested that helpers are physiologically immature and not capable of individual reproduction, but this has been shown to be incorrect for almost all species studied.)

The second key question is closely related to the first: given that a helper does stay at home, why would it provide care to young that are not its own offspring? For those cooperative breeders that nest in colonies rather than in group territories, such as the White-fronted Bee-eater (Chapter 16) and the Pied Kingfisher (Chapter 17), the analogous question is: why would the helper associate itself with another breeding pair within the colony, and help to raise that pair's offspring? To the extent that finding and bringing food to the young (or other forms of alloparental care) require both time and energy, these behaviors may be costly to the fitness of the helper. Is there a compensating benefit that is derived from helping and that might outweigh the apparent costs of not dispersing and attempting to breed individually? Is helping really altruistic, or is it simply an extraordinary (albeit obscure) type of selfish behavior?

It is now clear that answering these questions requires a combination of long-term observations of individual behavior and population level analyses of demography and genetic structure based either on a cross-sectional sample of many different groups observed over a short period of time, or, as more recently, on the histories of individual animals that are followed throughout their lifetimes. In addition, because this type of reproductive system has repercussions on almost all aspects of a species' biology, from physiology to social behavior and foraging ecology, studies of cooperative breeders have produced some of the most detailed portraits of free-ranging animal populations now available.

Although considerable progress has been made in our understanding of cooperative breeding, the chapters of this volume illustrate clearly that considerable controversy continues to exist, even for the basic questions posed above. There have been several general reviews of cooperative breeding (e.g. Brown 1978, 1987; Emlen and Vehrencamp 1983), as well as numerous theoretical papers on individual topics (e.g. Brown 1974; Gaston 1978; Woolfenden and Fitzpatrick 1978; Vehrencamp 1979; Koenig and Pitelka 1981; Emlen 1982a, b; Stacey 1982; Ligon 1983; Wiley and Rabenold 1984; Jamieson and Craig 1987; Stacey and Ligon 1987; Mumme *et al.* 1988; S. Zack, 1989, unpublished results). The primary literature for most species, however, is highly fragmented. Many studies have now lasted for 10, 15, or more years, and include several generations of helpers and breeders. Sufficient information is now available from many studies to begin testing the hypotheses and models of cooperative breeding using empirical data rather than simply relying on logical argument and selected examples. As editors, we felt that this was an opportune time for a volume in which researchers could summarize the central results of their studies in a single location.

Another important development is that a sufficient number of long-term studies on different species now exist to make it possible to undertake a search for common themes and patterns. To facilitate this process, we provided each contributor to the volume with a list of suggested topics to be included in their chapters whenever possible. The topics ranged from such basic ecological factors as the species' habitat requirements and foraging behavior, through patterns of social interaction and the group mating system, to various population-level parameters such as dispersal, reproductive success and survival. We also asked the authors to present specific and quantitative data whenever possible, and to provide several figures or tables that we believed would be most useful for interspecific comparisons. These included information on the distribution of group sizes throughout the study, the relationship between group size and reproductive success, and the stability of population numbers over all years of the study.

Not surprisingly, each study has focused on different aspects of cooperative breeding, and each has lasted for a different length of time. None of the chapters includes all of the suggested topics. However, most authors were themselves extremely cooperative, and provided extensive new analyses and unpublished data in their chapters.

We also suggested that each chapter should include a section in which the authors discussed what they believed to be the critical unanswered questions for their species and for the field of cooperative breeding as a whole. It is our goal that not only will this volume be a compendium of information about specific studies but that it will also serve as a source of new ideas and direction for students interested in research on cooperative breeding. The health of this field is indicated by the wealth of ideas and diversity of opinions found in these concluding sections.

There have been a large number of high-quality studies of cooperative breeding, and one of the most difficult problems we faced as editors was to decide which studies to include. It simply was not possible for the volume to contain all of them and still be of reasonable length. There are excellent studies of cooperative breeding in vertebrates other than birds, including a cichlid fish (Taborsky 1984), many mammalian carnivores (for reviews, see MacDonald 1983; Emlen 1984; Bekoff and Daniels 1984), and at least one rodent, the naked mole-rat, *Heterocephalus glaber* (Jarvis 1981). An intriguing possibility is that the social organization of many primates can be viewed as cooperative breeding (see e.g. Terborgh and Goldizen 1985; Sussman and Garber 1987). Finally, the behavior of many invertebrates, such as social spiders, might also fit the definition of cooperative breeding given above, although they also have not been considered as such traditionally.

Given the potential range of studies that could be included, we felt it necessary to limit the focus of the volume to a single group, birds. Birds have several advantages for testing hypotheses about cooperative breeding. Most species either nest in colonies or have small territories, and it is not uncommon to be able to obtain accurate data on 20 to 30 groups each year. This allows the researcher to examine the range of different social contexts that normally exists within any population, as well as to consider both the mean and variance of each demographic parameter. Birds are easily and permanently marked through banding, which makes it possible to follow individuals throughout their lives, even if they disperse from their natal groups or territories. Finally, most cooperatively breeding birds have relatively short generation times, and one or more complete generations can be included in a single 10 or 15 year study.

Because demography plays a central role in all current theories of cooperative breeding, we decided to emphasize long-term studies whenever possible, since these would be most likely to include the widest range of information about the biology of the species and would provide the most quantitative data for interspecific comparisons. This meant that we were unable to include many excellent studies, which, although of relatively short duration, have none the less made very important contributions to the field. The reader is referred to the text by Skutch (1987) and Table 2.2 of Brown (1987) for references to many of these studies.

Our second goal was to illustrate the great diversity that exists among cooperatively breeding birds. Although all species have individuals that act as helpers at the nest, it has become clear that cooperative breeding can and is expressed in an extraordinary number of different contexts. The classical picture of a single territorial pair with a small number of non-breeding helpers that are previous offspring is only the simplest. It is represented here by chapters on the Splendid Fairy-wren (Rowley and Russell), the Green Woodhoopoe (Ligon and Ligon), the Red-cockaded Woodpecker (Walters), the Arabian Babbler (Zahavi), the Hoatzin (Strahl and Schmitz), and *Campylorhynchus* wrens (Rabenold). We then include a series of chapters on New World jays, illustrating how variable cooperative breeding can be, even within the same taxonomic group. A very primitive form is represented by the Pinyon Jay (Marzluff and Balda). This species lives in large flocks and nests in loose colonies. Helping is restricted almost exclusively to yearling males, and then only in certain lineages (the most productive). The Florida Scrub Jay (Woolfenden and Fitzpatrick) is highly territorial and has single breeding pairs that may be assisted by one or a few helpers. Only males help after their first year, after which all females disperse. In the Gray-breasted or Mexican Jay (Brown and Brown) several pairs join together to defend a common territory. Each pair nests separately, and at least some helpers are believed to move between different nests. A similar system of plural nesting, but with occasional instances of two females laying eggs in the same nest, is also found in the Galápagos Mockingbirds (Curry and Grant). Joint breeding is also found in the Groove-billed Ani (Koford *et al.*), but here the females always lay their eggs together in the same nest. Since each bird feeds all of the young in the nest, one individual will be both a breeder and a helper for another's offspring at the same time. These groups also may contain unpaired birds, who act as 'typical' helpers.

Superimposed upon this diversity of spatial systems is a wide range of mating systems that can exist within the social unit. In most of the species described above, breeding is restricted to one or more monogamous pairs. Many other patterns are possible. In the Galápagos Hawk (Faaborg), for example, groups consist of one female and several unrelated males, each of whom may mate with the female and presumably share parentage of the young. Offspring are not recruited into the group, but become non-breeding floaters until a new group forms. Another raptor, the Harris' Hawk (Bednarz), also occurs in multi-male groups, and was originally believed to be polyandrous as well. However, Bednarz's recent evidence indicates that the extra males are usually non-breeding yearlings that help primarily through cooperative hunting rather than actually feeding the young. In the Pukeko of New Zealand (Craig and Jamieson) and Acorn Woodpecker (Koenig and Stacey), not only can several males mate with one female, but several females may lay eggs in the same nest. Groups also usually include one or more non-breeding helpers as well. This combination of within-group polyandry and polygyny (or polygynandry) can produce extremely mixed parentage of the young in these species, and makes determination of individual reproductive success difficult. Perhaps the extreme example of variable mating systems is the Dunnock (Davies). These small birds exhibit an incredible combination of both spatial and reproductive patterns in the same population, including monogamy, polyandry, polygyny and polygynandry. Each spring new units may form, often in a combination different from that of the previous year. The ecological reasons behind this extraordinary pattern are just beginning to be unraveled.

Most of these species live and breed within all-purpose territories. Another group of cooperative breeders are colonial nesters. These include the White-fronted Bee-eater (Emlen), where individuals may also hold feeding territories away from the nesting colony, and the Pied Kingfisher (Reyer), where they do not. Helpers in these species join already established pairs at the colony, and they may play an important role in bringing food to the nest. An intriguing and somewhat intermediate form between the colonial and territorial species, and certainly the most complex system of cooperative breeding studied to date, is the Noisy Miner of Australia (Dow and Whitmore). Here, large numbers of birds occupy a common 'territory' that is defended from all other birds, including other species. Each colony is further divided into coteries, and within each coterie, females nest individually. A female can be helped by many different males, and each male may help a number of different females. (Pinyon Jay flocks, mentioned above, may represent a primitive form of this type of organization.) Although the colonial species are ecologically very different from territorial ones, many of the same questions can be asked about the adaptive nature of cooperative breeding.

As this volume illustrates, the study of cooperative breeding is a dynamic and active field. All the puzzles posed by this type of breeding system are far from being solved. A central challenge is to understand the diversity of patterns described herein. Cooperative breeding has arisen independently in many different taxonomic lines and it can occur in a wide variety of habitats. Is there a single model that can explain the origin of this system, or is its occurrence in each species a unique event? New theories have recently been proposed, to explain both the ecological basis of philopatry among young birds and the selective nature (or lack thereof) of helping behavior itself. Our goal in this book is to present an overview of what is currently known about a diverse group of cooperatively breeding birds. We hope this will facilitate the search for common patterns among the species, and provide directions for new research in the field.

Bibliography

Bekoff, M. and Daniels, T. J. (1984). Life history patterns and the comparative social ecology of carnivores. *Ann. Rev. Ecol. Syst.* **15**: 191–232.

Brown, J. L. (1974). Alternate routes to sociality in jays – with a theory for the evolution of altruism and communal breeding. *Amer. Zool.* **14**: 63–80.

Brown, J. L. (1978). Avian communal breeding systems. *Ann. Rev. Ecol. Syst.* **9**: 123–55.

Brown, J. L. (1987). *Helping and Communal Breeding in Birds*. Princeton University Press: Princeton, NJ.

Brown, J. L. and Brown, E. R. (1981). Kin selection and individual selection in babblers. In *Natural Selection and Social Behavior: Recent Research and New Theory*, ed. R. D. Alexander and D. W. Tinkle, pp. 244–56. Chiron Press: New York.

Davis, D. E. (1942). The phylogeny of social nesting habits in the Crotophaginae. *Quart. Rev. Biol.* **17**: 115–134.

Emlen, S. T. (1982a). The evolution of helping. I. An ecological constraints model. *Amer. Nat.* **119**: 29–39.

Emlen, S. T. (1982b). The evolution of helping. II. The role of behavioral conflict. *Amer. Nat.* **119**: 40–53.

Emlen, S. T. (1984). Cooperative breeding in birds and mammals. In *Behavioural Ecology: An Evolutionary Approach*, ed. J. R. Krebs and N. B. Davies, pp. 305–39. Sinauer: Sunderland, MA.

Emlen, S. T. and Vehrencamp, S. L. (1983). Cooperative breeding strategies among birds. In *Perspectives in Ornithology*, ed. A. H. Brush and G. A. Clark, Jr, pp. 93–120. Cambridge University Press: Cambridge.

Ford, H. A., Bell, H., Nias, R. and Noske, R. (1988). The relationship between ecology and the incidence of cooperative breeding in Australian birds. *Behav. Ecol. Sociobiol.* **22**: 239–49.

Fry, C. H. (1972). The social organisation of bee-eaters (Meropidae) and cooperative breeding in hot-climate birds. *Ibis* **114**: 1–14.

Gaston, A. J. (1978). The evolution of group territorial behavior and cooperative breeding. *Amer. Nat.* **112**: 1091–100.

Hamilton, W. D. (1964). The genetical evolution of social behaviour. I and II. *J. Theoret. Biol.* **7**: 1–52.

Hardy, J. W. (1961). Studies in behavior and phylogeny of certain New World jays (Garrulinae). *Univ. Kansas Sci. Bull.* **42**: 13–149.

Hardy, J. W., Webber, T. A. and Raitt, R. J. (1981). Communal social behavior of the southern San Blas Jay. *Bull. Florida State Mus.* **26**: 203–64.

Jamieson, I. G. and Craig, J. L. (1987). Critique of helping behavior in birds: a departure from functional explanations. In *Perspectives in Ethology*, vol. 7, (ed. P. G. Bateson and P. H. Klopfer), pp. 79–98. Plenum Press, New York.

Jarvis, J. U. M. (1981). Eusociality in a mammal: cooperative breeding in naked mole-rat colonies. *Science* **212**: 571–3.

Koenig, W. D. and Pitelka, F. A. (1981). Ecological factors and kin selection in the evolution of cooperative breeding in birds. In *Natural Selection and Social Behavior: Recent Research and New Theory*, ed. R. D. Alexander and D. W. Tinkle, pp. 261–80. Chiron Press: New York.

Ligon, J. D. (1983). Cooperation and reciprocity in avian social systems. *Amer. Nat.* **121**: 366–84.

MacDonald, D. W. (1983). The ecology of carnivore social behaviour. *Nature (Lond.)* **301**: 379–84.

Maynard Smith, J. (1964). Group selection and kin selection. *Nature (Lond.)* **102**: 1145–47.

Mumme, R. L., Koenig, W. D. and Pitelka, F. A. (1988). Costs and benefits of joint nesting in the Acorn Woodpecker. *Amer. Nat.* **131**: 654–77.

Robinson, A. (1956). The annual reproductory cycle of the magpie, *Gymnorhinus dorsalis* Campbell, in south-western Australia. *Emu* **56**: 233–336.

Rood, J. P. (1984). The social system of the dwarf mongoose. In *Advances in the Study of Mammalian Behavior*, ed. J. F. Eisenberg and D. G. Kleiman, pp. 454–88. Amer. Soc. Mammalogists, Special Publ. 7.

Rowley, I. (1965). The life history of the Superb Blue Wren (*Malurus cyaneus*). *Emu* **64**: 251–97.

Selander, R. K. (1964). Speciation in wrens of the genus *Campylorhynchus*. *Univ. Calif. Publ. Zool.* **74**: 1–305.

Skutch, A. F. (1935). Helpers at the nest. *Auk* **52**: 257–73.

Skutch, A. F. (1987). *Helpers at Birds' Nests*. Iowa University Press: Iowa City.

Stacey, P. B. (1982). Female promiscuity and male reproductive success in social birds and mammals. *Amer. Nat.* **120**: 51–64.

Stacey, P. B. and Ligon, J. D. (1987). Territory quality and dispersal options in the acorn woodpecker, and a challenge to the habitat saturation model of cooperative breeding. *Amer. Nat.* **130**: 654–76.

Stacey, P. B. and Ligon, J. D. (1990). The benefits of philopatry hypothesis for the evolution of cooperative breeding: habitat variance and group size effects. *Amer. Nat.* (in press).

Sussman, R. W. and Garber, P. A. (1987). A reinterpretation of the mating system and social organization of the Callitrichidae. *Int. J. Primatol.* **8**: 73–92.

Taborsky, M. (1984). Broodcare helpers in the cichlid fish *Lamprologus brichardi*: their costs and benefits. *Anim. Behav.* **32**: 1236–52.

Terborgh, J. and Goldizen, A. W. (1985). On the mating system of the cooperatively breeding saddle-back tamarin (*Saguinus fuscicollis*). *Behav. Ecol. Sociobiol.* **16**: 293–9.

Vehrencamp, S. L. (1979). The roles of individual, kin, and group selection in the evolution of sociality. In *Handbook of Behavioral Neurobiology*, vol. 3, ed. P. Marler and J. C. Vandenbergh, pp. 351–94. Plenum Press: New York.

Welty, J. C. (1975). *The Life of Birds*. Saunders: Philadelphia.

Wiley, R. H. and Rabenold, K. N. (1984). The evolution of cooperative breeding by delayed reciprocity and queueing for favorable social positions. *Evolution* **38**: 609–21.

Woolfenden, G. E. and Fitzpatrick, J. W. (1978). The inheritance of territory in group-breeding birds. *BioScience* **28**: 104–8.

Wynne-Edwards, V. C. (1962). *Animal Dispersion in Relation to Social Behaviour*. Oliver and Boyd: Edinburgh.

1 SPLENDID FAIRY-WRENS: DEMONSTRATING THE IMPORTANCE OF LONGEVITY

1

Splendid Fairy-wrens: demonstrating the importance of longevity

I. ROWLEY AND E. RUSSELL

The Splendid Fairy-wren, *Malurus splendens*, is one of 13 species in the genus *Malurus*, which, in turn, is one of five genera in the endemic Australasian passerine family Maluridae (Schodde 1982). Only a few of the 26 malurids have been studied sufficiently closely to reveal their social organization. In seven species, individuals have been caught and color-banded so that each could be identified in the field. In each case they have been found to breed cooperatively, and it is likely that this trait may be widespread throughout the family.

The malurids are characterized by having 10 (or fewer) tail feathers, certain distinctive skull features, and a gap in the feathering between the scapulars – otherwise they are typical small passerines. They are not closely related to the wrens of Eurasia and the Americas (the Troglodytidae), despite the unfortunate similarity of their common names.

Adult birds of the genus *Malurus* are strongly sexually dimorphic, with the males being brilliantly plumaged whilst the females are usually a sedate gray-brown. Such marked sexual dimorphism is unknown in other cooperative breeders. *Malurus splendens*, or *splendens* as we shall call them from here on, are a vivid example of this dimorphism which becomes apparent in the first spring after they are hatched. Juveniles and immatures of both sexes are brown like the females, whilst the males are brilliant blue and black.

A member of the genus *Malurus* may be found in virtually every habitat in Australia, from the desert through the shrublands and the heathlands to temperate forest, tropical woodland and riverine fringing forest. *Splendens* has three recognized subspecies and occurs across the continent south of the Tropic of Capricorn, except where it is replaced by *M. cyaneus* in the south-east. In central Australia, *splendens* live in arid shrublands; further

3

east they live in woodland with a good understory of shrubs. In Western Australia, where we have studied these birds, they live in a scattered woodland of eucalypts, *Eucalyptus calophylla* and *E. wandoo*, with underneath a dense and varied heathland 1–2 m tall consisting mainly of xerophytic shrubs of the families Proteaceae and Myrtaceae.

These small birds fly only weakly as compared to most other passerines of similar size and they spend most of their time bounding over the ground in a series of hops, propelled by their strong legs, or hopping their way through the bushes. When their nest or young are threatened, the adult birds perform a fascinating *Distraction display* – the *Rodent run* – and then, instead of hopping along the ground as they usually do, they run with one leg after the other, scuttling like a small mouse.

The range of arthropods caught and eaten by *splendens* is very wide and varies from minute midges through flies, ants, spiders and beetles to quite large grasshoppers and phasmids 50 mm long. A few seeds are also eaten

Fig. 1.1. The Gooseberry Hill Study Site (near Perth, Western Australia), for the Splendid Fairy Wren.

sometimes. Most of these prey are captured by the birds pouncing on them from a standing position either on the ground or in a bush. Occasionally conspicuous and slow-moving items such as swarming termites are caught as the birds fly a brief sortie from the top of a bush, and we have watched them perched over a column of ants whilst they robbed the workers of their loads. In mid-summer, when Marri trees blossom, *splendens* may spend most of the day foraging in the canopy more than 5 m from the ground, where large numbers of insects are attracted to the flowers. Although they are therefore versatile foragers, they never search in litter by turning it over, nor do they dig in the ground with bill or feet.

Our study area covers 120 ha of the Darling Escarpment that forms the eastern margin of the city of Perth (Fig. 1.1). The topography rises steeply from 50 m at the Helena River to Gooseberry Hill at 200 m and is dissected by several winter-flowing streams. In the past, this area was regularly burned, and so parts of Gooseberry Hill were burnt in 1974, 1976, 1977, 1978 and 1981. The last two fires were very minor and hardly affected the study area at all, so that between 1978 and 1984 most of the vegetation was unburnt and responded by growing taller and denser. Then on 30 January 1985 a wildfire swept through and burnt 95% of the *splendens* territories, consuming virtually all the vegetation, which was tinder dry at that time of year (Rowley and Brooker 1987).

We have been studying a population of *splendens* on Gooseberry Hill since 1973, when we caught and color-banded the adults and their progeny in eight groups. Since then the study has grown both in area and in population density so that at its peak in 1984 we monitored 34 groups containing 115 adults that fledged 111 young that year (Fig. 1.2).

Adult birds are resident in the same place all year round and live in groups of from two to eight individuals ($\bar{x} = 3.26 \pm 1.23$), each group defending an area of from 1.6 to 9.0 ha (Fig. 1.3). Outside of the breeding season, a group of *splendens* may forage with other small passerines in mixed species flocks, handing over to, or being replaced by, the next-door neighbors as the flock moves over the territorial boundary. These territories persist in much the same location year after year. Even when a territory was vacant after a fire, the neighbors did not encroach and eventually new owners took up the old boundaries. In the early years, no new territories were established in the area under study at the time. Then, in 1979, an area previously vacant was occupied for one year only by a novice pair, which bred unsuccessfully. As the vegetation recovered from the fires of 1974–8, survival of adults was higher, and 1980 and 1981 were years of particularly high production of yearlings per group (1.66 in 1980 and 1.57 in 1981, compared with the mean over 13 years of 0.96 ± 0.45). Thus the population density increased considerably as the vegetation cover became more dense (see Fig. 1.2). In 1981, four new territories were established, where previously there were only 13 territories, and in 1982 another six new territories were established. After the population stabilized again in 1983, only one new territory was established and two lost, and in 1984 one new territory was established and one old territory was lost. After the major fire of January 1985, three territories were lost in the next season and one in 1986/7, while no new territories were established.

Fig. 1.2. Population density of adult *Malurus splendens*, 1973–86. Population size is based on censuses taken at the start of the breeding season each year. The incidence and relative magnitude of fires is indicated by arrows. The extended period without major fires (1979 to the end of 1984) was marked by increasing density of vegetation and increasing density of the *splendens* population, with increasing survival of young and the occupation of a few new territories. During the period of increasing density, the sex ratio moved closer to unity.

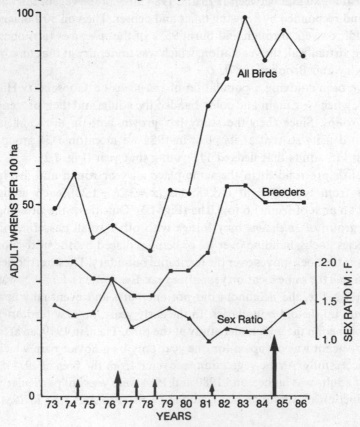

The basic unit is the socially bonded breeding pair, which remains together so long as both birds survive and, over 14 years, a third of the groups in the Gooseberry Hill population consisted of just two individuals (Fig. 1.4). However, under the prolonged absence of fire and consequent increase in population density (Fig. 1.2), some instances of polygyny were recorded. Table 1.1 shows that almost every possible combination of males and females has occurred at some time during the study, with one extra male or female being commonest. The size of any particular group waxes and wanes depending on the breeding success achieved in the previous

Fig. 1.3. Territories (capital letters) of *Malurus splendens* in the Gooseberry Hill study area in 1985. Territories are stable from year to year and contiguous, so that each is surrounded by five to seven neighboring territories.

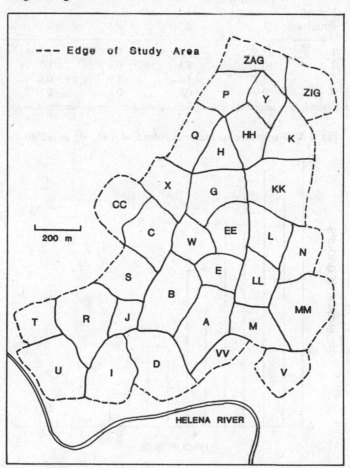

season. Certain years have been far more productive than others in terms of the number of young that survived to adulthood, which has meant that the size both of individual groups and of the population as a whole was much larger after such a year. However, there does seem to be an upper limit to group size in *Malurus* and, outside of the breeding season, groups greater than eight are rare and usually disintegrate.

Progeny tend to remain in the family group after they reach independence and many stay on for at least a year after they reach sexual

Table 1.1. *Group composition of* Malurus splendens *1973–86* (n = 256 *group-years*). *Percentage of groups with number of males and females as indicated*

No. of females	No. of males				
	1	2	3	4	5
1	33.5	19.5	8.9	2.3	0.4
2	10.5	8.6	6.6	1.2	0
3	3.1	1.6	1.9	0.8	0.4
4	0	0.8	0	0	0

Fig. 1.4. Group size in *Malurus splendens*, 1973–86 ($n = 256$).

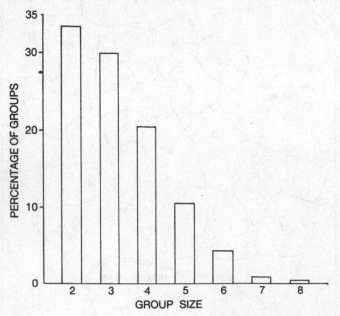

maturity (at one year old). Of 121 known males that reached one year old, 18.2% dispersed at some time in their life, 10.7% before they were one year old. Only 17.4% of males did not help at all, 40.5% helped for only one year, 33.9% helped for two years or longer, and one for as long as seven years (see ♂ 275 in group C, Fig. 1.6). Females disperse more and help less. Of 98 known females that reached one year old, 35.7% dispersed at some time in their lives, 21.4% before reaching one year old. Only 28.6% of females did not help at all, 49% helped for only one year, and 18.4% helped for two years or longer. Thus, most helping was done by birds that were one year old (Fig. 1.5).

Fig. 1.5. The percentage of helpers in *Malurus splendens* at each age, based on 176 male and 102 female helper-years. Open bars indicate males and hatched bars females.

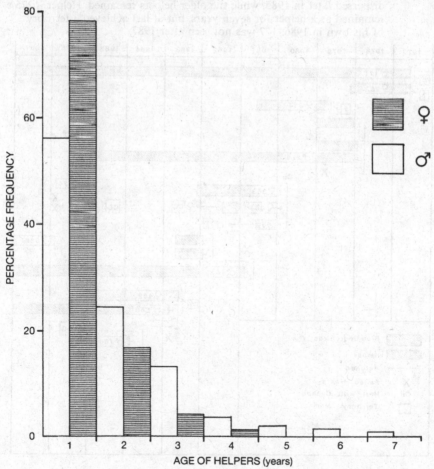

The history of group C between 1977 and 1986 is shown in Fig. 1.6. The territory had been vacant for six months when ♂177 and his new mate moved in from an adjacent territory where he had been hatched and where he had helped for one year. During 10 years, ♂177 has mated with seven different females and has been responsible for producing five offspring that have become established as breeders. In 1985, he had three surviving male

Fig. 1.6. History of one territory (territory C) of *Malurus splendens* over 10 years. Our longest surviving known-age male (♂177) was hatched in territory E in 1975. He helped there in 1976, and in 1977 took over as the breeding male in C. Since then he has been mated with a succession of seven females (including two daughters, ♀363 and ♀364), and has produced five offspring which have achieved breeding status. At the start of the breeding season in 1985, the group contained five males and one female. One of the younger males dispersed later in 1985, while the older helpers remained. Helper ♂275 remained as a helper for seven years, but at last achieved a territory of his own in 1986. 177 was not seen after 1987.

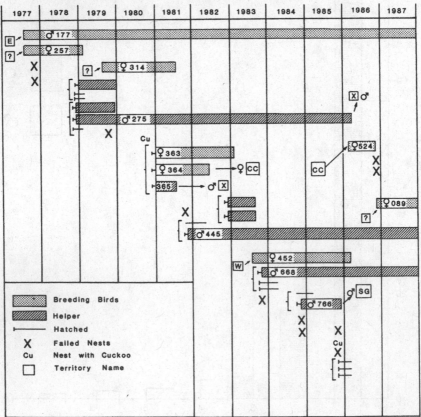

helpers, each from a different season, one of whom ($\male$ 275) was helping for his seventh year. In 1986, $\male$ 275 succeeded to the next-door territory (X) and bred. Throughout this time, territory C boundaries have remained virtually unchanged.

The overall sex ratio of the population has varied between 1.81 and 0.97 over the years (Figs. 1.2 and 1.7). Initially we thought that the presence of helpers was a consequence of greater dispersal and related mortality of females. More recently, the sex ratio of helpers has been close to unity, with even a slight surplus of females in 1981. The percentage of groups with helpers has remained stable over a wide range of helper sex ratios. Overall, there was no correlation between the population sex ratio and the percentage of groups with helpers ($r_s = -0.2$, not significant (n.s.), which does not support the hypothesis that shortage of sexual partners constrains males to remain as helpers (*contra* Emlen and Vehrencamp 1985). In 1981, when there was a slight surplus of females, the proportion of groups without helpers was higher than usual, but the availability of females was

Fig. 1.7. Relationship between the sex ratio (male:female) and the percentage of groups with helpers in *Malurus splendens*. The sex ratio is based on all adult birds. The percentage of groups with helpers has remained stable over a wide range of sex ratios, and there is no correlation between the two.

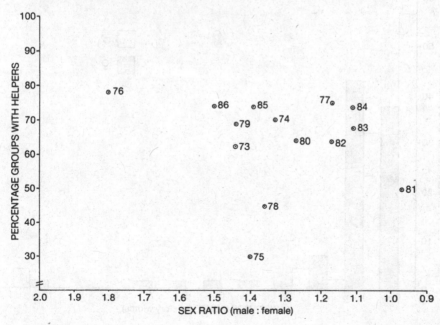

only one of many factors, since demographic and habitat changes during
this period were considerable.

Vacancies that occur in the group are filled from within the group if there
is a helper of the right sex. Of 56 known females that have become breeders
in the study area, 22 (39%) bred only in their natal territory, 15 as sole
breeders and seven breeding only as secondary female in a polygynous
group. The other 34 females dispersed. For 53 males that became
established as breeders, 30 (57%) were in their natal territory. There appear
to be no incest bars so that inbreeding (where r between breeders ≥ 0.5) is
remarkably high (20%) (Rowley *et al.* 1986, but see note added in proof). If
the female is the casualty and there is no helper to take her place, the male(s)
will leave the territory and go 'wife-hunting', returning with a new female
after being absent for several days. The arrival of a new female from
elsewhere (as opposed to from within the group) leads to a resurgence of
territorial aggression, possibly because she does not know the boundaries
of her new territory and can learn only by trespassing and being repulsed.
When a new female is installed, any male helpers that are present in the
group generally continue in that role. Female helpers usually take over the

Fig. 1.8. The dispersal distances of male and female *Malurus splendens*,
measured in territory widths. Open bars indicate males and hatched
bars females.

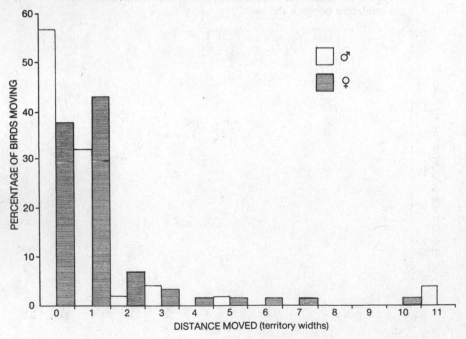

breeding role, but if they do not then they disperse rather than help the new incumbent.

Besides filling a vacancy within the natal group, the other important way of achieving a breeding situation is to occupy a vacancy in an adjoining territory (32% of known males becoming breeders and 43% of females). Together these two methods accounted for 89% of known males and 80% of known females which have become breeders (Fig. 1.8). Helpers are sexually mature and form a reservoir of potential breeders that regularly interact with their neighboring groups and so any vacancy there is detected by them before anyone else. Consequently most recruitment is either by inheritance of the natal territory or local dispersal. A few birds of both sexes have dispersed for longer distances outside the study area, but the low numbers of birds which have moved across two to five territories (and thus would have been detected within the study area) suggest that long-distance dispersal is not common. Over the length of this study, 25 unbanded males and 34 unbanded females have entered groups within the study area as breeders. Approximately 70% of these (both male and female) have entered groups on the boundary of the study area, but only one bird has moved more than two territories from the edge of the study area. There is a trend toward greater dispersal of females, with 57% of males and 38% of females remaining as breeders in their natal territories (Fig. 1.8).

Breeding system

Most *Malurus* are repeat breeders. The breeding season for *splendens* lasts from September to January, during which time a group may raise two or at most three broods of young to independence. For an experienced female with no helpers it takes a mean of 66 ($\pm$9) days from the time when the first egg is laid until the brood reaches independence and the first egg of the next clutch is laid. However, if there are helpers, the mean time between clutches is significantly shorter (55 ($\pm$10) days; Mann–Whitney U-test, $P < 0.001$). For a subgroup of females with more than two years of breeding experience and two or more helpers, re-nesting is even quicker (50 ($\pm$7) days; $P < 0.001$; this is discussed further by Russell and Rowley 1988). We suggest that in this case the helper contribution is to take over the feeding and shepherding of the fledglings, freeing the female to re-nest earlier than if she and the male alone were responsible for the fledglings. On several occasions the female has been found building her next nest within a fortnight of the previous brood fledging, although those young would still require care for at least another month – care that was provided by the helpers.

Parasitism on *splendens* by Horsfield's Bronze-cuckoo, *Chrysococcyx basalis*, varies from year to year; an average of 21% of nests per year are parasitized, with a range of 0 to 43%. The young cuckoo stays in the nest for longer than *splendens* nestlings and is fed by all group members. Nevertheless, it is dependent on its foster parents for a shorter time than is a brood of *splendens* nestlings, because the fledgling cuckoo is not fed for so long. After a group with helpers has raised a cuckoo, the time to the next clutch is even shorter than for a brood of wrens (only 46 ($\pm$ 12) days), and re-nesting after such parasitization is sufficiently quick for many groups to raise two broods of wrens in a season as well as a cuckoo (see Fig. 1.9). Group members may join in defense of the nest against cuckoos (Payne *et al.* 1985). Although one might expect larger groups to be more obvious and more subject to parasitism by cuckoos, there is no relationship between incidence of parasitism and group size.

In most groups the oldest male appears to mate-guard the breeding female before she starts incubating and if there are other males in the group they are usually kept away. Copulation in *splendens* is very brief and usually takes place in the middle of a bush. We have seen only a dozen instances and these have not always been by the oldest male. Not only is there this chance of a quick mating by a subordinate helper male in the group, but, especially early in the season, males whose females are sitting on eggs are found wandering outside their territories and one was seen to steal a mating. This philandering is often accompanied by a most spectacular display in which the intruding male carries a purple or pink flower and puffs up his feathers into what we have named the *Blue and black display*. It appears as if *Petal carrying* may be a form of appeasement display which serves as a 'flag of truce', but its origins and real function remain obscure. We cannot therefore be absolutely confident that the senior male always fathers all the brood. In fact, electrophoretic studies on proteins suggest, to date, that stolen matings do occur. However, while paternity may sometimes be in doubt, maternity never is.

Until 1981 only one female in each group built nests and laid eggs and we presumed that only one male fertilized those eggs. Now we are not so sure, for as the population has increased we have found that about 10% of groups (30 from 257 group-years) have had two concurrent nests each belonging to a different female. This state of affairs has come about in two ways. First, in one case a pair formed within the group and then budded off part of the group territory which they then defended and in which they nested. At the end of the breeding season they, along with their progeny and territory, all reverted back to become part of the old group association.

Second, in the majority of cases (29), groups included two females building nests and laying eggs at much the same time. Although their timing overlapped, there was no suggestion that the two females synchronized in any way. These 'x' and 'y' nests belonged to the old female and to a female that (in about 80% of cases) had already helped in the group for one previous season. The second female was in most cases (19/25 where the relationship was known) the daughter of the primary female. In some cases

Fig. 1.9. Summary of the breeding biology of *Malurus splendens* in a mediterranean climate on Gooseberry Hill, Western Australia. The figure illustrates a typical first nesting attempt starting in September, with 10 days of nest-building and laying, 14 days of incubation, 10 days as nestlings and 31 days from fledging to independence. When an experienced female has helpers (upper), they take over the care of the dependent fledglings, and the female may begin to build a second nest. She then has time to lay three clutches before the end of December; clutches laid in January rarely succeed. A female with no helpers does not start her second attempt until the first brood is independent, and has time for only two attempts before the end of December. If she raises a cuckoo at her second attempt, she still has no time to re-nest, whereas a female with helpers would have time after raising a cuckoo to lay a third clutch before the end of December. d, day.

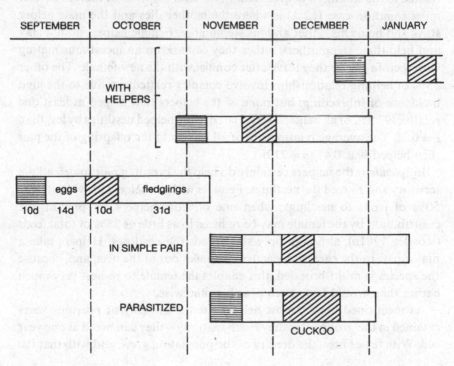

there was only one male who mated bigamously with both females, but where there was more than one adult male in the group we could not be sure who fertilized the eggs. Whoever was genetically responsible for the eggs, in groups with more than one male where a second female nested, this latter female was sometimes guarded by a particular subordinate male. In all cases the broods of young produced from these x and y nestings coalesced after they fledged and were cared for by the group as a whole. This sometimes led to groups containing 10 or more individuals by the end of the season. Most often, the secondary female was left to feed her brood by herself, and so secondary nests were seldom successful. However, if either nest failed, the whole group helped to feed at the other nest, and a primary female might help to feed at the nest of a secondary female and vice versa.

Most of the group members are related to each other, since the average group grows by the progeny remaining as helpers in the family throughout the winter and many of those that survive stay on in the group throughout the next (and other) breeding seasons as non-breeding helpers. Only four of 177 (2.3%) individuals that have helped in this study were not progeny of the group where they helped. By far the most common situation (53%) is for the helper (of either sex) to help its parents, which means that the helper is related to the siblings it helps to rear at the level of $r = 0.5$ or more. The next most common case (12%) is where the mother dies and the male helper stays and helps his father and his stepmother. Female helpers do not stay and help their stepmothers: either they embark on an incestuous mating with their father or they leave after conflict with the new female. The other 35% of helping relationships involve complex relatedness due to the high incidence of inbreeding; but none of the helpers that helped at least one relative (97.5% of all helpers) was related to the helped nestlings by less than $r = 0.25$. The average relationship of all helpers to the offspring of the pair they helped was 0.47 ($n = 210$).

In *splendens* the helpers certainly do help the breeding pair to defend the territory and to feed the nestlings. Females with no helpers provide at least 50% of feeds to nestlings; when one or more helpers are present, the contribution by the female may be reduced to as little as 25% of total feeds (Rowley 1981a), although only she broods the nestlings. Helpers take a major part in the care of dependent juveniles out of the nest, and, because the species is multi-brooded, this enables the female to re-nest very much earlier than would have been possible otherwise.

As mentioned above, most helpers are the progeny of previous years retained in the group after they reach maturity – they can breed at one year old. With fewer fires, the density of the population grew, and with that the size of the groups has increased significantly, from a mean of 2.9 ± 0.3 (1973–80) to a mean of 3.4 ± 0.2 (1981–6) (Mann–Whitney U-test, $P < 0.001$). At the same time, the number of females per group has increased significantly from 0.2 to 0.6 (Mann–Whitney U-test, $P < 0.002$). We have already mentioned that this led to there being more than one simultaneous nest in a number of groups. This retention of related females also led to an increase in inbreeding, so that overall we had a level of at least 20% inbred matings – matings where the male and female were related at $r = 0.5$ or greater. The most frequent inbreeding pairs were father–daughter, mother–son and brother–sister, combinations which arose when one or both parents died and were replaced by their offspring from within the group. Although such a high figure for inbreeding is unprecedented among vertebrates, we have noticed no deleterious effects (Rowley *et al.* 1986). Productivity of fledglings has remained high and, with high juvenile mortality occurring anyway (about 65% of fledglings die before they reach one year), the combination of a few lethal genes (if it occurs at all) becomes but part of this normal pruning process. Even allowing for the possibility of stolen matings by subordinate males or near neighbors, the average subordinate male is closely related to the breeding female, and the level of inbreeding is unlikely to be very much decreased (but see note added in proof).

The number of fledglings produced by a group in any year is affected by many things. Variation from year to year in the timing and quantity of rainfall and the occurrence and intensity of fires have a marked effect on territory quality, while the incidence of predation and the frequency of cuckoo parasitism have significant effects on offspring production. Besides these environmental and interspecific variables, offspring production is affected by the presence of helpers and by the experience of the breeding female (Fig. 1.10, Table 1.2). Larger groups of four or more birds produced significantly more fledglings per year (Mann–Whitney U-test, $P < 0.01$) than groups of two and three, which are not significantly different. A female breeding for the first time as a member of a simple pair fledges only half as many young as a female with two years of experience and no helpers (1.7 versus 3.3, Mann–Whitney U-test, $P < 0.01$). The presence of two helpers leads to increased production in groups where the female has at least two years of breeding experience. The presence of just one helper does not seem to lead to a net increase in production, even when the female has at least two years of breeding experience. After their first novice year, there is little difference in production per nest between females with different levels of help and experience. The main difference lies in the number of nests per year that females have. As shown above, the presence of helpers reduces the time to re-nesting, and this effect is greatest with two or more helpers. Older, experienced females with helpers have more nests per year, largely because they are more likely to re-nest after an earlier successful nest (Table 1.2).

The most successful groups are those with older experienced females and at least two helpers (Fig. 1.10), and it is these birds which are best able to take advantage of favourable conditions to produce two, and sometimes even three, broods in a season. The significance of two helpers being more effective than one in allowing a female to start re-nesting sooner is probably due to each fledgling being looked after by one adult. With normal broods of three, it is not until two helpers are present that there are enough adults

Fig. 1.10. Relationship between group size and young fledged per group per year in *Malurus splendens*. Production by groups of two and three is not significantly different, but groups of four or more produce significantly more than do groups of two and three. *n* is the number of groups studied.

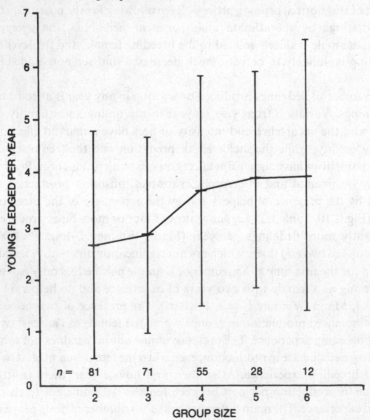

(two helpers plus the breeding male) to take care of three fledglings and release the female to re-nest. We are not yet sure that a particular adult always takes care of a particular fledgling, but our observations suggest this.

For such small birds, *splendens* are long-lived. We have constructed a survivorship (survival rate) curve based on 620 birds banded as nestlings (Fig. 1.11). Only about 35% of fledglings survive to one year old, but from then on survival is higher, with about 20% of the one-year-olds surviving for at least four years. The longest surviving bird was ♂177, who hatched in September 1975 and was last seen in December 1987 (see Fig. 1.6); two other males have also lived for more than 10 years. The longest-surviving female lived for more than eight years, and bred in eight seasons; she was responsible for producing 10% of all known-age young that have themselves become breeders up to the present (nine individuals).

About 67% of all breeding females survived from one breeding season to the next. The presence of helpers improves a female's chances of survival. Of those with helpers, 76% survived to the next breeding season, but for those with no helpers, survival was significantly lower (55%; $\chi^2 = 10.7$; $P < 0.01$). Helpers had no effect on the survival of breeding males, which was 72%. This level of mortality of breeding adults means that on average we would expect about one-third of male and female breeders to be replaced in any one year, and, over 12 years, the average has been about 0.32

Table 1.2. *Effects of breeding experience and group size on reproductive success in* Malurus splendens

Female breeding experience	Group size	No. of females	Fledglings per nesting attempt	Fledglings per year per female	% re-nest after success	Nest per female per year
Novice	2	41	1.1^a	1.7^a	15^a	1.5^a
	>2	32	1.3^a	2.1^a	13^a	1.6^a
1 year	2	16	1.5^a	3.0^b	38^b	2.0^b
	3	15	1.6^a	3.1^b	33^b	1.9^b
	>3	14	1.2^a	2.7^b	43^b	2.2^b
>2 years	2	15	1.8^a	3.3^b	40^b	1.9^b
	3	18	1.4^a	3.2^b	56^{b,c}	2.3^c
	>3	35	1.6^a	4.3^c	69^c	2.6^c

^{a,b,c} In each column values which have the same superscript are not significantly different (Mann–Whitney U tests, $P < 0.05$).

Fig. 1.11. Survivorship curve for known-age male and female *Malurus splendens* banded as nestlings from 1973 to 1984. The figure in parenthesis at each point is the number of banded birds at risk which could have survived to that age. Only 35% of fledglings survive to one year old, but survival after that is higher and relatively constant. About 20% of the birds that survive to one year go on to survive to four years old. The longest surviving male has lived for 11 years so far.

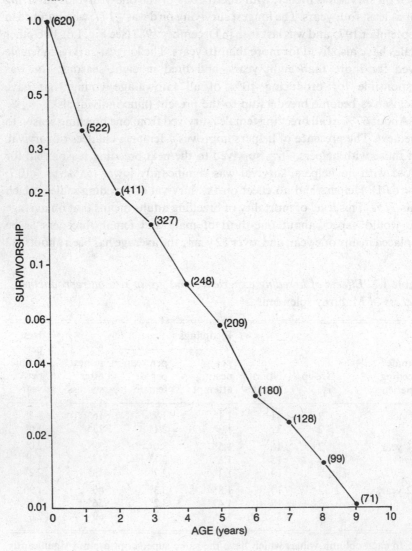

for males and 0.35 for females. Competitors for these vacancies included established helpers of one or more years' experience as well as young birds of the previous breeding season which had never helped, and on average there were three non-breeding adults available to compete for every vacancy (3.5 ($\pm$ 1.2) males and 2.1 ($\pm$ 1.9) females, based on a mean over 13 years of: non-breeders available at the start of the breeding season/number of vacancies filled during the subsequent year).

Speculations

Over the past 20 years, many people have expressed surprise at the phenomenon of cooperative breeding, largely because it is rare in the temperate northern hemisphere, where so much of ornithological theory has been generated and where simple pair breeding is so much commoner. Cooperative breeding appeared to involve altruism and to be an aberration that was hard to explain in terms of evolutionary theory. But when one looks closely at most species that breed cooperatively, one realizes that there is nothing abnormal about the formation of these groups – the offspring just stay on much as they do in our own species. The remarkable thing perhaps is that so many other species shed their young so quickly.

It will help to clarify our thinking about this form of social organization if we draw up a balance sheet of the costs and benefits both to the parents and to the offspring, with particular reference to *Malurus splendens*.

From the point of view of the parents, the costs include:
  (1) The young in the territory consume food which may be in short supply, leading to competition.
  (2) Helpers may sometimes betray the nest to cuckoos and predators as originally suggested by Skutch (1961).
  (3) When there is more than one male in a group, there may be a cost to the dominant male in shared paternity and in trying to avoid it.
  (4) A second female who builds a nest in addition to the nest of the primary female no longer acts as a helper but as a competitor to the primary female.

The benefits to the parents of having the young remain include:
  (1) They sometimes help to produce more fledglings, through the increased frequency of re-nesting, but productivity was enhanced by helpers in only 20% of all group-years.
  (2) They increase the survival of the breeding female – this not only helps her but also the breeding male, who has fewer replacement, inefficient novice females to lower his overall success. One of the most important factors contributing to a parent's lifetime fitness is

longevity (Rowley and Russell, 1989); a long life increases the probability of breeding in a good year as well as the simple more years/more nests relationship.

(3) Helpers that stay at home have an increased survival probability and a better chance of breeding. Since a one-year-old has a higher reproductive value than a potential fledgling, parents benefit by promoting the welfare of helpers even at the expense of subsequent young.

(4) Helpers provide a buffer against the extremes of between-year-variation in conditions. In poor years, when few succeed, experienced females with helpers are more likely to produce fledglings; the small cohort of recruits from such a year then has fewer competitors for the vacancies that follow. On the other hand, in the very good years, it is the experienced females with helpers who can really exploit the conditions and produce two or three broods of fledglings (Table 1.3).

(5) For *splendens*, with a high level of cuckoo parasitism, helpers may allow breeders to cope better with the effects of cuckoos by re-nesting quickly.

Table 1.3. *Yearly variation in production of fledglings by* Malurus *females with at least two years of previous breeding experience and at least two helpers (E2.2)*

Year	Mean fledglings per female per year	Total females	No. E2.2 females	No. of E2.2 which produce		
				> mean	>3 fledglings	0 fledglings
1974	1.8	9	0	0	0	0
1975	2.1	11	0	0	0	0
1976	3.5*	8	1	0	1	0
1977	3.6*	8	1	1	0	0
1978	5.2*	10	2	0	2	0
1979	2.2	13	1	0	0	0
1980	3.5*	11	2	1	1	0
1981	3.6*	22	5	0	5	0
1982	2.8	30	2	0	2	0
1983	2.6	41	3	1	2	0
1984	2.5	40	6	1	3	0
1985	2.2	38	12	2	5	1
1986	2.4	37	10	2	3	1

Mean clutch size is 3.0, and years with mean production of fledglings per female of >3 are considered to be 'good' years and are marked with an asterisk.

From the young's point of view, the costs include:
  (1) The 'sacrifice' of its own reproduction. This cost may be negligible. If a vacancy occurs, the helper takes it; meanwhile it is in a familiar place, with the advantages of group-living, such as increased vigilance against predators, access to group experience and the possibility of improved foraging. Our data on 87 known-age males and females who have become breeders shows that over 90% bred by the time they were two years old, so that most waited no more than one year. Survival from one to two years old, and thereafter, is high enough to suggest that the chances of surviving to breed for several years after succeeding at two years old are such as to outweigh the possible loss of reproduction from one year spent helping.
  (2) They do extra work in feeding and may thus suffer greater exposure to predators.
The benefits to the young helper include:
  (1) The chances of survival are better in a group on familiar ground (see point (1) for the young, above).
  (2) In *splendens* there is the potential for indirect benefit from helping to produce closely related kin, since helpers are related closely enough to the offspring of the breeders they help to derive indirect benefit (*sensu* Brown and Brown 1981). Rowley (1981a) used Vehrencamp's (1979) index of kin selection to calculate indirect fitness as a fraction of total inclusive fitness due to pursuing particular helping strategies in *splendens*. In most cases less than half the increase in inclusive fitness due to helping came from the indirect component, the rest being due to the direct component, through the increased chance of getting a breeding vacancy. Subsequent analysis with a much larger data set (Russell and Rowley, 1988) has indicated that helping leads to increased production by breeders in less than 20% of group-years, reinforcing the view that, on average, increased production of kin is only a small part of the inclusive fitness of individuals that pursue the helping strategy.
  (3) Helpers gain experience. Novice breeding females that have helped for one year produce more fledglings per year (2.5, $n = 8$, no helpers) than one-year-old novices that have never helped (1.5, $n = 10$, no helpers), although the difference is not significant.
  (4) They have a good chance of filling a breeding vacancy either at home or in a neighboring territory. Although these birds are long-

lived, vacancies do occur, and we have calculated that on average three non-breeding adults are available to compete for every vacancy. Our data show that dispersal is very local and suggest that birds that stay at home have a good chance of an opportunity near at hand. Each territory is surrounded by five or six other territories, three or four of which will have helpers, so that any vacancy that occurs is likely to be filled immediately. A dispersing bird from a distance has little chance to find and fill such a vacancy before the locals do so.

If food availability throughout the year is adequate to allow progeny to remain, then the parents' own fitness (in terms of progeny achieving breeding status) is enhanced if they allow the young to stay in the natal territory, where chances of survival are highest in an environment they know better than any other. From the point of view of the young, if the chance of surviving *and* achieving a breeding opportunity are greatest at home, then staying at home would be favoured over dispersal. Whilst this suggests why it might be better for juveniles to stay at home, it does not yet explain why they should help.'

In many birds, responsiveness to young not necessarily their own offspring is very non-specific, and they may even respond to young of another species (Shy 1982). If one supposes that responsiveness to stimuli from the nestlings is to some extent innate, then if progeny stay in the family group and encounter those stimuli to which they are already programmed to respond, since they must have mechanisms for that behavior. For the non-breeding group members not to help would require that their responsiveness to nestlings be switched off while that individual remains in the family group. If the mechanism is there, they should attack predators, feed nestlings and so on because much of that behavior is relatively non-specific as to causal stimuli, and therefore readily elicited. There are costs involved in feeding nestlings, but, for the helpers, this may be offset by the indirect benefits from increased survival of breeders and production of kin. Responsiveness to stimuli from nestlings may be enhanced by hormonal changes, as in the breeding female, but this raises a further question: what is the actual reproductive state of the helpers? Has their retention in the family group involved the evolution of some mechanism for the suppression of reproduction in subordinate helpers? Our data on the occurrence of groups with two breeding females and on broods with multiple paternity suggest that there is no such suppression in *splendens*.

Cooperative breeding in *Malurus*

Australia seems to have a disproportionate number of cooperative breeders (68/520 species (13%) of land birds). At different times, cooperative breeding has been suggested as an adaptation either to stable or to harsh, unpredictable environments. Ford *et al.* (1988) have recently pointed out that most cooperative breeders occur in forest, and that in Australia this is a relatively equable habitat. Like most other Australian habitats, it is variable from year to year, but *within-year variation*, from season to season, is less than in many other habitats. They suggest that the forest habitat has sufficient food to support both parents and progeny throughout the year, a necessary prerequisite for the evolution of cooperative breeding.

For the seven species of *Malurus* that have been looked at so far, and for *splendens* in particular, a critical factor seems to be that these small birds, weighing only 10 g, live for a remarkably long time. One reason for this appears to be the relatively mild Australian winter, and so there is little climatic pressure to migrate, and few species do so. It would seem logical that a residential life is less hazardous than that of a migrant passing through unfamiliar terrain. Evidence for this is so far indirect, but Fry (1980) and Woinarski (1985) demonstrated the relative longevity of small Australian passerines compared with those of northern hemisphere temperate zones. As a specific comparison, survival of breeders in the 10 g *Malurus splendens* is 72% (male) and 67% (female) as against 28% in the 11 g Blue Tit, *Parus caeruleus*, and 41% in the 20 g Pied Flycatcher, *Ficedula hypoleuca* (from survival rates quoted by Lack (1954)). Many cooperatively breeding species have high survival rates, as reviewed by Brown (1987), who suggests that this is an important determinant of the availability of potential helpers. In many parts of Australia, the climate does tend to be variable, with big differences between years. Once or twice during a decade, breeding may be a fruitless occupation for small insectivorous passerines, but, on the other hand, one or two years may be exceptionally productive, and long-lived birds may reap the benefit.

A penalty of being a long-lived resident is that all the available good real-estate tends to become occupied, posing problems for dispersing progeny. If the probability of successfully dispersing and finding a breeding opportunity is low, then it is reasonable to assume that survival will be better for birds that stay at home, provided there is enough for them to eat. Long-lived birds can afford to wait for a good breeding opportunity.

We are particularly interested in the way that this pattern of cooperative breeding recurs throughout a genus that occurs in a great variety of habitats throughout Australia from desert to rainforest fringes. Although our main study has been on *Malurus splendens* near Perth, we have also studied *M. cyaneus* in south-east Australia, *M. elegans* that lives in high-rainfall *Eucalyptus* forest in south-west Australia, and *M. coronatus* that inhabits the riverine fringing forests of the tropics in northern Australia. Table 1.4 shows that all these species are regular cooperative breeders.

Data for *M. cyaneus* (Rowley 1965) showed a clear helper effect on productivity, groups with helpers producing more fledglings than did groups with no helpers, but age and experience of females were not distinguished. The rapid turnover of females suggests that many of the females with no helpers were novice females, which in *M. splendens* are significantly less productive than experienced females with or without helpers. Earlier comparisons for *M. splendens* (Rowley 1981a) showed no significant increase in production with group size, but in subsequent years the proportion of larger groups has increased, and we have been able to separate females with no, one, and two or more years of breeding experience and to demonstrate that two or more helpers do lead to

Table 1.4. *Comparison of group composition between four species of* Malurus

Malurus	Birds in group[a]		Total groups	Total birds	Sex ratio M:F	Annual survival of breeders (%)	
	Simple pair	>2 birds				M	F
cyaneus[b,c]	29 (67.4)	14 (32.6)	43	107	1.38	64	63
splendens[c]	78 (34.5)	148 (65.5)	226	733	1.23	72	67
elegans[c]	26 (20.5)	101 (79.5)	127	485	1.17	78	76
coronatus[c,d]	77 (83.7)	15 (16.3)	92	202	1.23	71	66

M, male; F, female.
[a] Figures in parentheses are percentages.
[b] Rowley (1965).
[c] Rainfall in study areas: *cyaneus*: 622 mm over all months; *splendens*: 880 mm in winter; *elegans*: 1000 mm in winter; *coronatus*: 1300 mm in summer.
[d] Rowley (1987).

increased productivity by older experienced females. In *M. elegans*, we have found a similar effect (Rowley *et al.* 1988). There is thus evidence that some indirect benefit through increased production of kin is attributable to helpers.

For these species, we would suggest that probably the most important extrinsic constraint on the presence of helpers is the demographic one of longevity. The highest incidence of groups with helpers and the largest group sizes occur in *M. elegans*, the species with the highest annual survival rate of breeders. The available habitat has not always been entirely saturated for *splendens*, since new territories were established in some years as the habitat changed, but there were clearly more non-breeders than vacancies. Lack of marginal habitat, suggested by Koenig and Pitelka (1981) as an environmental factor predisposing to cooperative breeding does not seem to us to be a significant factor in a species with large expanses of suitable habitat and very limited dispersal.

The sex ratio constraint, a shortage of females, which means that for a male breeding is not an alternative to helping, is clearly not important for *splendens* (Fig. 1.7). In *M. elegans*, the larger groups have almost as many surplus males as females. Although data for *cyaneus* (corrected from Emlen (1978) and replotted by Rowley (1981*a*) and Emlen and Vehrencamp (1985)) show a clear relationship between sex ratio and the number of groups with helpers, in the light of our knowledge of other species of *Malurus* the number of groups and years are too few for us to accept this as an adequate explanation of delayed breeding (*contra* Emlen and Vehrencamp 1985; Brown 1987). We feel that the most crucial aspect is that the species all tend to be long lived and therefore breeding vacancies occur rarely, and the important thing for a malurid is to make sure that your offspring survive long enough to be around when that opportunity to breed does arrive. What better opportunity could there be for a wren than to inherit the territory he knows so well? And where better to await such a chance than at home, if, as Ford *et al.* (1988) suggest, it is able to support them and their parents?

We feel that there are several important and exciting avenues for further research:

(1) What is the hormonal status of different individuals in a group during the breeding season. Are the helpers in fact able to breed, or are they suppressed by the dominant birds?

(2) What is the availability of food in territories and how does it change throughout the year, for cooperative species versus non-cooperative species? Is it a prerequisite for cooperative breeding

that a territory is able to provide enough food for breeders and helpers throughout the year?

(3) To what extent is group size a matter of past 'luck' and not of present conditions. Is an optimum group size selected for?

(4) By comparing related species which live (e.g. *Malurus* spp.) in very different habitats but are all cooperative breeders, can we identify common ecological or demographic features? Much has been written about the ecological attributes of habitats which may promote cooperative breeding. Comparisons of similar species in different habitats has much to contribute to this debate.

(5) Paternity: is the actual difference in fitness between dominant males and others as great as it appears, or do helpers or outsiders achieve a significant number of stolen copulations? Do the helpers really 'give up' their chance of breeding? Obviously *splendens* have complicated social lives and to try to work out the genetics of the situation is hopeless without the aid of modern technology in the form of electrophoretic analysis. Such analysis has been started.

In the course of a long-term study, many people have provided help in the field and in the analysis of data. We are most grateful to Craig Bradley, Graeme Chapman, Michael, Lesley and Belinda Brooker, Joe Leone, and Bob and Laura Payne. The manuscript was read at various stages by Michael Brooker, Bob Payne and Graeme Smith.

Bibliography

Brooker, M. G., Brooker, L. C. and Rowley, I. C. R. (1988). Egg deposition by Bronze-Cuckoos, *Chrysococcyx basalis* and *C. lucidus. Emu* **88**: 107–9.

Brown, J. L. (1987). *Helping and Communal Breeding in Birds: Ecology and Evolution.* Princeton University Press: Princeton, NJ.

Brown, J. C. and Brown, E. R. (1981). Kin selection and individual selection in babblers. In *Natural Selection and Social Behaviour: Recent Research and New theory*, ed. R. D. Alexander and D. W. Tinkle, pp. 244–56. Chiron Press: New York.

Emlen, S. T. (1978). The evolution of cooperative breeding in birds. In *Behavioural Ecology: an Evolutionary Approach*, ed. J. P. Krebs and N. B. Davies, pp. 245–81. Blackwell: Oxford.

Emlen, S. T. and Vehrencamp, S. L. (1985). Cooperative breeding strategies among birds. In *Experimental Behavioural Ecology and Sociobiology*, ed. B. Holldobler and M. Lindauer, pp. 359–74. Fisher-Verlag: Stuttgart.

Ford, H. A., Bell, H. L., Nias, R. and Noske, R. (1988). Cooperative breeding in Australian birds: is it a feature of stable or unpredictable environments or both or neither? *Behav. Ecol. Sociobiol.* **22**: 239–49.

Fry, C. H. (1980). Survival and longevity among tropical land birds. In *Proc. Pan-African Ornitholog. Congr. 1976* **4**: 333–43.

Koenig, W. D. and Pitelka, F. A. (1981). Ecological factors and kin selection in the evolution of cooperative breeding in birds. In *Natural Selection and Social Behaviour:*

Recent Research and New Theory, ed. R. D. Alexander and D. Tinkle, pp. 261–80. Chiron Press: New York.

Lack, D. (1954). *The Natural Regulation of Animal Numbers*. Oxford University Press: London.

Payne, R. B., Payne, L. L. and Rowley, I. C. R. (1985). Splendid wren *Malurus splendens* response to cuckoos: an experimental test of social organization in a communal bird. *Behaviour* **94**: 108–27.

Payne, R. B., Payne, L. L. and Rowley, I. C. R. (1988a). Kinship and nest defence in cooperative birds: Splendid Fairy-wrens *Malurus splendens. Anim. Behav.* **36**: 939–41.

Payne, R. B., Payne, L. L. and Rowley, I. C. R. (1988b). Kin and social relationships in Splendid Fairy-wrens: recognition by song in a cooperative bird. *Anim. Behav.* **36**: 1341–51.

Rowley, I. C. R. (1957). Co-operative feeding of young by Superb Blue Wrens. *Emu* **57**: 356–7.

Rowley, I. C. R. (1962). 'Rodent-run' distraction display by a passerine, the Superb Blue Wren *Malurus cyaneus* (L). *Behaviour* **19**: 170–6.

Rowley, I. C. R. (1963). The reaction of the Superb Blue Wren, *Malurus cyaneus*, to models of the same and closely related species. *Emu* **63**: 207–14.

Rowley, I. C. R. (1965). The life history of the superb blue wren. *Emu* **64**: 251–97.

Rowley, I. C. R. (1968). Communal species of Australian birds. *Bonner Zoologische Beiträge* **3/4**: 362–8.

Rowley, I. C. R. (1974). Co-operative breeding in Australian Birds. *Proc. 16th Intern. Ornith. Congr.*: 657–66.

Rowley, I. C. R. (1975). *Bird Life*. Collins: Sydney, NSW.

Rowley, I. C. R. (1978). Communal activities among White-winged Choughs *Corcorax melanorhamphus. Ibis* **120**: 178–97.

Rowley, I. C. R. (1981a). The communal way of life in the Splendid Wren, *Malurus splendens. Z. Tierpsychol.* **55**: 228–67.

Rowley, I. C. R. (1981b). A relic population of Blue-breasted Wrens *Malurus pulcherrimus*, in the central wheatbelt. *West. Aust. Nat.* **15**: 1–8.

Rowley, I. C. R. (1983a). Re-mating in birds. In *Mate Choice*, ed. P. Bateson, pp. 331–60. Cambridge University Press: Cambridge.

Rowley, I. C. R. (1983b). Commentary on co-operative breeding strategies among birds. In *Perspectives in Ornithology. Essays presented for the Centennial of the American Ornithologists' Union*, ed. A. H. Brush and G. A. Clark, Jr, pp. 127–33. Cambridge University Press: Cambridge.

Rowley, I. C. R. (1987). *Conservation of the Purple-crowned wren* Malurus coronatus *in Northern Australia*. Final Report on Project 47 to World Wildlife Fund (Australia).

Rowley, I. C. R. and Brooker, M. G. (1987). The response of a small insectivorous bird to fire in heathlands. In *Nature Conservation: the Role of Remnants of Native Vegetation*, ed. D. A. Saunders, G. W. Arnold, A. A. Burbridge and A. J. M. Hopkins, pp. 211–18. Surrey Beatty and Sons: Sydney, NSW.

Rowley, I. C. R., Emlen, S. T., Gaston, A. J. and Woolfenden, G. E. (1979). A definition of 'group'. *Ibis* **121**: 231.

Rowley, I. C. R. and Russell, E. M. (1989). Lifetime reproductive success in *Malurus splendens*, a cooperative breeder. *Proc. 18th Intern. Ornith. Congr.*: 866–75.

Rowley, I. C. R., Russell, E. M. and Brooker, M. G. (1986). Inbreeding: benefits may outweigh costs. *Anim. Behav.* **34**: 939–41.

Rowley, I. C. R., Russell, E. M., Brown, R. and Brown, M. (1988). The ecology and breeding biology of the Red-winged Fairy-wren *Malurus elegans. Emu* **88**: 161–76.

Russell, E. M. and Rowley, I. C. R. (1988). Helper contributions to reproductive success in the splendid fairy-wren *Malurus splendens. Behav. Ecol. Sociobiol.* **22**: 131–40.

Schodde, R. (1982). *The Fairy Wrens. A Monograph of the Maluridae.* Lansdowne: Melbourne, SA.

Shy, M. M. (1982). Interspecific feeding among birds: a review. *J. Field Ornithol.* **53**: 370–93.

Skutch, A. F. (1961). Helpers among birds. *Condor* **63**: 198–226.

Vehrencamp, S. L. (1979). The rules of individual, kin and group selection in the evolution of sociality. In *Handbook of Behavioral Neurobiology,* ed. P. Marler and J. G. Vandenbergh, vol. 3, pp. 351–94. Plenum Press: New York.

Woinarski, J. C. Z. (1985). Breeding biology and life history of small insectivorous birds in Australian forests: response to a stable environment? *Proc. Ecol. Soc. Aust.* **14**: 159–68.

Postscript added in proof

Since this chapter was written, further testing of paternity likelihood in *Malurus splendens* by allozyme electrophoresis has changed our thinking about mating systems, relatedness and inbreeding. A total of 91 offspring of 24 mothers and 37 putative fathers was typed at 10 polymorphic loci. All young were compatible with their mothers but at least 65% were not fathered by any of the males in their group. The promiscuous mating system demonstrated in this work would reduce the level of inbreeding in the group but still allow individuals the security of group-living in a stable year-round territory. It would also reduce the relatedness of helpers to the nestlings that they fed, so that indirect fitness benefits would be even less than previous estimates suggested.

Reference

Brooker, M. G., Rowley, I. C. R., Adams, M. and Baverstock, P. R. (1989). Promiscuity: an inbreeding avoidance mechanism in a socially monogamous species? *Behav. Ecol. Sociobiol.* (in press).

2 GREEN WOODHOOPOES: LIFE HISTORY
TRAITS AND SOCIALITY

2

Green Woodhoopoes: life history traits and sociality

J. D. LIGON AND S. H. LIGON

The woodhoopoes and scimitar-bills form a small, homogeneous coraciform family, Phoeniculidae, restricted to sub-Saharan Africa. Of the eight species of phoeniculids recognized by Ligon and Davidson (1988), the Green Woodhoopoe (*Phoeniculus purpureus*) is the most common and widespread, occurring from East to West Africa, and south to the tip of South Africa.

As suggested by the species' broad geographical and ecological range, Green Woodhoopoes are not particularly habitat specific. They occur in savannah, open woodland, palm groves, riverine forest in arid thorn bush, and in wooded gardens, from near sea level to over 2000 m. They are, however, absent from dense forest, such as the rainforests of West Africa. In general, their habitat requirements seem to be open woodland, with at least some trees large enough to provide nest and roost cavities.

Unlike several other cooperative breeders, Green Woodhoopoes are strongly sexually dimorphic in weight and in bill length, and, in addition, all vocalizations of adults are sexually diagnostic. We suspect that the larger size of males is related to intense male–male competition for territories and females. Whatever its adaptive basis, large body size also appears to carry a cost to males of all ages, in the form of greater annual mortality relative to females (see below).

Foraging woodhoopoes move as group as they search tree trunks, branches and twigs for arthropods. Although the tail is long and flexible, it is used for support as the birds hang beneath or on the sides of branches. Male and female woodhoopoes tend to forage in different parts of the trees (our unpublished data), and this difference appears to be related to differences between the sexes in body size and bill length. Females forage more on small terminal branches and twigs, while foraging males are more

33

often seen on the main trunk and large branches, as well as on the ground at the bases of trees.

At our study area near Lake Naivasha, in the central Rift Valley of Kenya, woodhoopoes occupy open woodland consisting almost entirely of one tree species, the yellow-barked acacia, *Acacia xanthophloea* (Fig. 2.1). These trees, growing on a former lake bed, form a belt of varying width that surrounds the lake. Our primary study site was Morendat Farm, with 18 marked flocks (not all present at the same time) in contiguous territories. We also studied woodhoopoes along the south side of Lake Naivasha, near Crescent Island, where seven social units were banded, and between the Morendat River and North Lake Road along the Naivasha–Nakuru highway, where we banded eight flocks (Fig. 2.2).

Our study of Green Woodhoopoes began in mid 1975 and the main body

Fig. 2.1. Prime habitat of Green Woodhoopoes on Morendat Farm, near Lake Naivasha, Kenya. Trees are yellow-barked acacias (*Acacia xanthophloea*). This area is grazed by cattle and seven native species of artiodactyls.

of the project ended in January 1982. J.D.L. returned to the study site in mid 1984. We ultimately marked a total of 386 woodhoopoes in 33 flocks for individual recognition; 269 birds were banded on, or just adjacent to, Morendat Farm. By January 1982, 93% of the Green Woodhoopoes on Morendat Farm were of known parentage. This was possible both because

Fig. 2.2. Map of Lake Naivasha, Kenya. Hatched areas indicate sites where all social units of Green Woodhoopoes were individually marked and studied. Numbers indicate minimum and maximum numbers of flocks present at one point in time over the period 1975–84 (see the text). The block of 13–16 territories represents Morendat Farm, the primary study site.

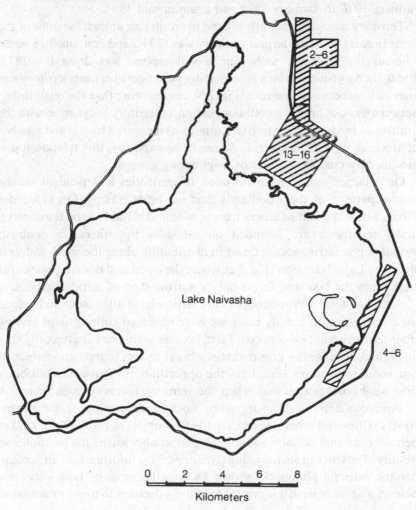

2–6

13–16

Lake Naivasha

4–6

0   2   4   6   8
Kilometers

mortality was high and because most woodhoopoes of each sex are extremely philopatric.

Spacing system

Green Woodhoopoes are highly territorial and each established social unit holds a year-round territory. As compared to most other territorial cooperative breeders, woodhoopoe territories are large. Because the woodhoopoes engage in noisy confrontations with neighboring groups on a daily basis, and because the conspicuous displays occur at territorial boundaries, it is not difficult to measure territory sizes. We measured territories on Morendat Farm at approximately six-month intervals from January 1976 to January 1982 and again in mid 1984.

Territory size is significantly related to group size, at least for some of the time. In early 1976 the largest territory was 133 ha and contained 15 birds, whereas the smallest, with four woodhoopoes, was 26 ha ($r = 0.813$, $P < 0.01$). As group size on a given territory changes over time, territory size also may increase or decrease (Fig. 2.3), emphasizing that the relationship between a social unit of woodhoopoes and its territory is dynamic, with the number of birds affecting territory quality as measured by size and number of trees, as well as vice versa. As might be expected, this relationship is modified by changes in sizes of neighboring groups.

On a larger scale, the distribution of territories is dependent on the spatial pattern of the woodlands that the birds occupy. On Morendat Farm, where the belt of acacia trees is wide and suitable habitat extensive, many territories are bounded on all sides by others. In contrast, woodhoopoe territories are linear in distribution along the south and west shores of Lake Naivasha (Fig. 2.2), where the woodland does not extend far back from the lake and forms only a narrow strip of suitable habitat.

Although Green Woodhoopoes are territorial at all seasons and defend their territories on a daily basis, we have observed entire groups leaving their territories in two contexts. First, because territories are typically very large, and because the group almost always moves together, members of one social unit on occasion have the opportunity to move surreptitiously into neighboring territories when the territory owners are elsewhere. A conspicuous activity of an intruding flock is exploration of the owners' roost cavities and other tree holes in the territory. As a result of this kind of behavior, an individual woodhoopoe presumably learns the location and quality of cavities in surrounding territories. This information can account for the extreme philopatry shown by woodhoopoes of both sexes (see below), and we believe it is critical to a bird's decision to move or not move into a nearby territory when the opportunity arises (i.e. when a bird or birds of its sex die, creating a breeding vacancy). As will be considered below, the quality and number of available and potential roost sites within a given territory is probably the key factor determining the overall lifespan and reproductive success of the birds occupying it.

Second, on three occasions, we followed entire social units as they flew over open country far from their territories. Two of these events took place in January 1979, when the birds appeared to be food-stressed, on the basis of our unpublished foraging data. In both cases the woodhoopoes flew to areas where they encountered no other birds and where they foraged intensively (i.e. an isolated grove of trees) before returning to their territories later in the day. Thus it appears that short-term movements beyond the territorial boundaries may occur for very different reasons,

Fig. 2.3. Changes over time in sizes of three territories of Green Woodhoopoes on Morendat Farm. Numbers along each line represent flock size in June of each year, prior to the onset of breeding for that year. The solid line represents the DD territory; the dashed line represents the AD territory; and the dash–dot line represents the BRF territory, which came into existence in 1977, primarily in space earlier held by the DD flock.

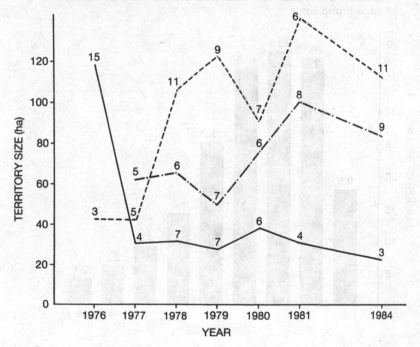

either (1) to evaluate quality of neighboring territories, or (2) to forage temporarily in unoccupied areas where arthropod prey is more plentiful.

Social system

Most social units are composed of a breeding pair, plus one to three helpers (Fig. 2.4). Average group size in a given year often reflects reproductive output the previous year; i.e. groups tend to be largest in the year following a year especially favorable for reproduction (cf. Table 2.2 and Fig. 2.5). The smallest stable social units are simple pairs; these can usually persist and breed only in relatively isolated areas where pressure from neighboring groups is low. In areas where most territories are surrounded by others, one to three helpers make a vast difference in the ability of a pair to control territorial space and to breed.

Within a social unit of Green Woodhoopoes, agonistic interactions are rarely seen except among recently fledged juveniles. As is also the case for some other cooperatively breeding birds, it appears that status relationships are established at this time, and that following establishment of a dominant–subordinate relationship, continued expression of overt

Fig. 2.4. Distributions of group sizes in Green Woodhoopoes. The number above each bar indicates percentage of total represented by each group size.

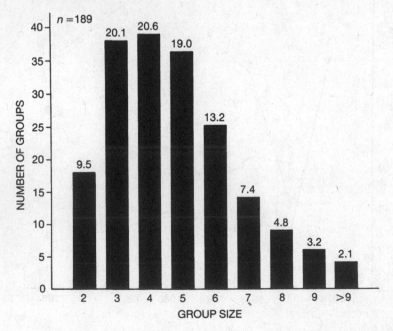

aggression is normally detrimental to the interests of individual group members, with two exceptions. (1) Conflict among same-sex former nestmates reappears in the context of competition for breeding status (see below). (2) Among older birds, aggression occurs uncommonly at the entrance to the roost cavity. This takes place only if cavities are in short supply, and cavities are too small to hold all the woodhoopoes attempting to gain entry.

Woodhoopoes are monogamous, and thus the sex ratio of breeders is 1:1. The sex ratio (male:female) of the study population as a whole ranged over eight years from 0.67:1 to 0.94:1 ($\bar{x} = 0.74:1$, s.d. $= \pm 0.13$). However, the sex ratio within a given social unit can be highly biased over a long period. One social unit, BRF, which originated in 1977, had sex ratios at the onset of each breeding period ranging from 1:3 (1977) to 1:6 (1979 and 1980). In mid 1984, it was 2:6. A few other flocks showed similar female-biased sex ratios over time.

New social units develop in a variety of ways. The most common is for a group of two or more related males to join two or more females in the latter's territory, or vice versa. This unisexual 'team' movement from one territory to a nearby one occurs typically when an opportunity for immigration has been created by the disappearance of most or all members of one sex or the other. Such emigrant teams are composed of birds of unequal age, e.g. father–son, uncle–nephew, older and younger brothers. Once we recorded three generations moving together – a grandfather, two sons and a grandson – to a new territory. The three younger birds occupied a territory adjacent to that of the old male. Following the deaths of the breeding females in both flocks, all four moved to a territory containing two females.

We have observed male same-age former nestmates emigrate together on three occasions; in each instance the dominant bird behaved so aggressively to the subordinate that the latter soon returned to its previous (natal) territory. In striking contrast, immigrant males of unequal age from the same flock exhibit no visible agonistic behavior regardless of their genetic ties (Ligon and Ligon 1983). Whatever the explanation for the inability of same-age siblings to form a stable founder unit, it appears that the social bonds developed earlier between older and younger individuals usually eliminate overt aggression from their relationship and make possible the mutually beneficial acquisition and retention of a new territory.

As described earlier, an individual Green Woodhoopoe exists in a social environment composed of cooperating individuals, and with neighboring groups competing with one another. The outcome of most intergroup interactions is dependent on the relative sizes of the competing groups. Larger groups tend to dominate smaller ones, and very small groups (two to three individuals) sometimes cannot successfully defend any territorial space if neighboring groups are numerically superior. Thus alliances are a critical component of an individual woodhoopoe's existence. This is seen most clearly in the group emigration of two or more same-sex individuals to a breeding vacancy in a nearby territory. Team emigration greatly increases the chances that the members will be able to acquire and retain possession of the new territory in the face of competition and aggression by neighboring groups. Present-day woodhoopoes occupy a social environment where cooperative alliances are critically important. Typically, unisexual groups of allies are relatives (Ligon and Ligon 1978a, 1982, 1983).

However, the high mortality (see below) not infrequently leads to a situation where an individual loses its same-sex relatives, or in the case of females, a bird might emigrate alone to a new area. Such single birds are at an overwhelming competitive disadvantage, and it appears that allying with an unrelated individual is a sound and commonly employed strategy. In short, formation of teams by unrelated birds, particularly females, is not a rare event (Table 2.1), although we have recorded such alliances only among birds acquiring low-quality territories. The more productive territories tend to be passed from parent to offspring (see below). Formation of teams composed of unrelated birds, therefore, is one kind of strategy for procurement of breeding space and status.

An interesting aspect of team formation by unrelated birds is the dominant–subordinate relationship. When the age of each bird was known, as a result of banding, the younger was always subordinate and later served as a helper if it had not disappeared during the period between acquisition of a territory and breeding, a common occurrence. When we did not know with certainty the ages of the participating birds, morphological traits associated with age, especially bill color, also suggested that the dominant

Table 2.1. *Composition of unisexual 'teams' of two or more Green Woodhoopoes immigrating to a territory, 1975–81*

	Kin	Non-kin	?
Male	14	4	2
Female	8	14	6

Summary: 18 of 40 were composed of non-relatives.

bird was always older than the subordinate. Thus the same relationship between age, dominance status and breeding status was present in both related and unrelated teams.

Breeding biology
Mating system

In each flock of Green Woodhoopoes, regardless of its size, only one breeding pair exists. Prior to nesting, the dominant male guards his mate, usually by interposing himself between the female and any approaching adult male. The alpha male may then groom the other male or initiate a vocal rally (Ligon and Ligon 1978a) with it. Anthropomorphically, it appears that the alpha male attempts to distract the male that it thwarted or supplanted by 'changing the subject'. This seemingly benign form of mate guarding serves not only to ensure paternity but apparently also not to discourage subordinate males from bringing food to the incubating female and later to the alpha male's offspring. Although the breeding pair is functionally monogamous, male helpers sometimes attempt to copulate with the breeding female, and rarely they do mount the female, but we have recorded such mountings only *after* the eggs were laid.

Green Woodhoopoe pairs copulate conspicuously and comparatively frequently. An unusual aspect of the breeding biology of this bird is the duration of copulation. We have timed copulations that lasted as long as $2\frac{1}{2}$ min. The extended period apparently required for transfer of sperm may be of significance in that 'stolen' copulations resulting in insemination must be rare, since the female must cooperate by remaining still for an extended period of time and since the alpha male almost certainly would detect and break up a mating attempt. We believe, therefore, that confidence of paternity in Green Woodhoopoes is high, both because the alpha male assiduously guards his mate before the eggs are laid and because the duration of copulation makes a quick, 'stolen' copulation unlikely.

Confidence of maternity also is probably high. Clutch size normally ranges from two to four eggs. Four eggs is a common clutch size in all groups, including those containing only one female. More than one female probably laid eggs in no more than three of 73 nests for which we have appropriate information. (Five eggs were laid in two of these nests, eight in the other.) In each of these cases two or more adult females had recently joined or been joined by new males. Another line of evidence supporting the suggestion that only one female normally contributes to the clutch is the pattern of incubation. The breeding female is the sole incubator of eggs. Of 51 nests watched for extended periods of time during incubation, two females appeared to incubate in only one. (Clutch size was four.) These observations lead us to conclude that the alpha female, like the alpha male, is normally the sole breeder of its sex.

Timing of breeding

Breeding in each year normally begins in late May or early June, following the 'long rains' of March–June, and generally terminates in December or earlier. The number of successful nesting efforts per year appears to be controlled largely by food availability, and this, in turn, is determined largely by patterns of rainfall (Ligon and Ligon 1982). Two broods during the period June–December are common and, if environmental conditions are extremely favorable, up to three broods can be produced by one breeding pair during this interval. If conditions are poor, however, breeding activity is limited, and only one, or even no, nesting attempt may occur. In such years, there is considerable territory-to-territory variation in the time of breeding as well as in the number of nests attempted. A major effect of unfavorable rainfall patterns, such as occurred in 1979 (see next section and Table 2.2), is the decreased number of nesting attempts per season. We have complete data on this point for the years 1979–81 (Table 2.3). In 1979 nesting attempts per flock were far fewer than in 1980 and 1981. This is a primary reason that mean annual reproductive success for 1979 was so low compared to the latter two years, since the mean number of nestlings produced per successful nest was about the same in all three years (Table 2.3).

Table 2.2. *Precipitation patterns and annual reproductive success in the Green Woodhoopoe*

Year	Dry season's precipitation (cm)	Long rains' precipitation (cm)	Mean no. young produced/ flock[a]	Mean flock size
1975	0.41	18.06	1.5	5.8 (22)[b]
1976	2.29	14.00	1.9	5.6 (22)
1977	10.06	43.15	2.4	4.8 (22)
1978	17.22	45.72	0.7	6.3 (25)
1979	27.21	17.70	0.5	5.3 (24)
1980	5.67	36.70	1.2	4.0 (24)
1981	2.11	36.05	2.5	3.9 (21)

[a] Number of offspring surviving to the end of the calendar year.
[b] Number of flocks.

The relative effects of helpers and precipitation patterns on annual reproductive success

Although most social units of Green Woodhoopoes include non-breeding helpers (Figs. 2.4 and 2.5), mean number of young produced per year does not differ significantly between pairs and groups with helpers ($\bar{x}$ pairs $= 0.94$, $n = 17$, $\bar{x}$ groups $= 1.36$, $N = 155$. Kruskal–Wallace $\chi^2 = 1.53$, d.f. $= 1$, $P > 0.20$); nor does the probability that no young will be produced/year differ between pairs and groups ($n = 155$ groups, 17 pairs; Kruskal–Wallis test $\chi^2 = 0.846$, $P > 0.36$). However, the overall correlation between group size and number of young produced is significant ($r_s = 0.16$, $P < 0.05$, $N = 172$), but group size explains little of the variance in reproductive success (see below).

The year-by-year data (Tables 2.2 and 2.3) suggest that the pattern of rainfall (timing and amount) might account for much of the variation recorded for each group size (Fig. 2.6). Our analyses revealed that, perhaps surprisingly, the quantity of precipitation during the 'long rains', which occur March or April–May or June, just prior to the onset of breeding, does not significantly influence production of young woodhoopoes on an annual basis ($r = 0.063$, $P > 0.42$, $n = 171$). Rather, *dry season* (January–March) rainfall is inversely associated with reproductive success ($r = 0.335$, $P < 0.0001$, $n = 171$). This counter-intuitive finding is explained by the biology of the many species of moth larvae which comprise the vast

Table 2.3. *The relationship between patterns of rainfall and predation to reproductive success in the Green Woodhoopoe*

Year	Timing and quantity of rainfall	Mean no. nesting attempts per flock	% nesting attempts producing at least one fledgling	Mean no. fledglings produced per successful nest
1979	Unfavorable[a]	1.00 (21, 0–2)[b]	48[c] (21)[d]	1.67 (9, 1–2)[e]
1980	Favorable	2.40 (20, 1–5)	44 (48)	1.65 (20, 1–4)
1981	Favorable	2.00 (19, 1–3)	82 (38)	1.81 (31, 1–4)

[a] See Table 2.3.
[b] Number of flocks, range in number of nest attempts.
[c] Low percentage of successful nests reflects high rate of predation.
[d] Number of nests.
[e] Number of successful nests; range in number of fledglings produced per successful nest.

majority of food items brought to nestlings and eaten by both adult and young birds during the potential six or seven month nesting season (Ligon and Ligon 1982). Numbers of these caterpillars through the year are apparently related most directly to the amount of precipitation that falls during the putative dry season. During this period, the moth caterpillars pupate in the ground and survive well only if the soil remains dry. If extensive rain falls during the 'dry' season, mortality of the pupae is very high (M. Clifton, National Museum of Kenya, personal communication), with few adults emerging to give rise to the generation of caterpillars that appears following the long rains.

A multiple regression analysis considering the long rains, dry season precipitation, and number of helpers, provides an indication of the relative importance of each of these factors in the annual production of young woodhoopoes. Dry season rainfall had by far the greatest effect ($r^2 = 0.112$, $n = 171$, $P < 0.0001$), while the effect of precipitation during the long rains is negligible (multiple $r^2 = 0.01$, $P > 0.16$). When the effect of precipitation during the dry season is controlled, the relationship between number of helpers and number of young woodhoopoes produced per year is still significant (multiple $r^2 = 0.024$, $P < 0.03$). What all of this means with

Fig. 2.5. Annual variation in total numbers of Green Woodhoopoes on Morendat Farm for the years 1975–81 and 1984 (open boxes). Censuses were taken in June in each year prior to breeding. Numbers above boxes indicate number of social units. Triangles indicate the total number of breeders in each year (two per social unit).

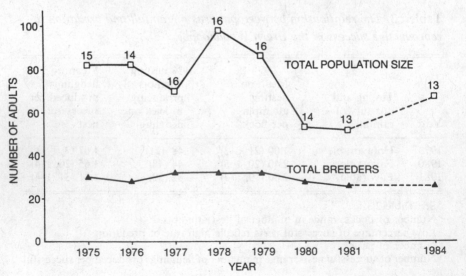

regard to the evolution of helping behaviour is problematic. Although helpers have the potential to increase the production of young birds, this potential is rarely achieved, as a result of overriding environmental factors.

Nest predation

Another important factor influencing year-to-year variation in breeding success is predation. Nest predation was high in both 1979 and 1980, with fewer than half of all nests producing any fledglings. Known predators of nestlings include driver ants (tribe Dorylini), Harrier Hawks (*Polyboroides radiatus*) and Pearl-spotted Owlets (*Glaucidium perlatum*). In addition, we suspect that an arboreal agamid lizard ate eggs at a few nests. In contrast, in 1981, when weather factors were favorable and nest predation was low, over 80% of all nests produced one or more fledglings. The combination of increased nesting attempts and decreased predation in 1981 resulted in the largest average number of young birds produced per flock over a seven year period, even though the average number of helpers per flock was at its lowest (Table 2.2).

Fig. 2.6. Relationship between group size at the onset of breeding of Green Woodhoopoes in each year and the number of young surviving to the end of the calendar year. Circles indicate means, vertical lines show one standard error. Numbers above points indicate sample size for each group size.

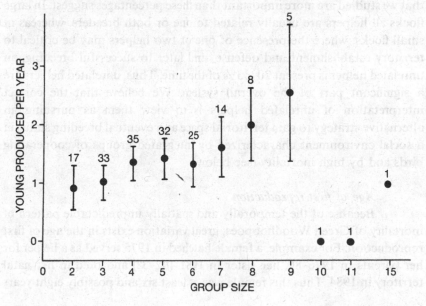

Number of helpers and relatedness of helpers and nestlings

Genetic or kinship ties within social units of Green Woodhoopoes are highly variable. Most helpers are either offspring of one or both members of the breeding pair or are siblings of one breeder. Both the number of helpers per social unit and the mean relatedness of helpers to the breeders vary over time. These values are high in the year following a year of unusually high reproduction and both tend to decrease in succeeding years of average or poor reproduction. For example, reproductive success was great in 1977, and in 1978 the mean number of helpers was 4.3 (Table 2.2). By 1981, following two years of below-average annual reproduction, this value had decreased to 1.9. Similarly, mean relatedness of helpers to nestlings was 0.35 in 1978 and by 1981 this had fallen to 0.29. The pattern of decreasing relatedness over time is based primarily on mortality of breeders. One-year-old helpers are more likely to be helpers for both parents (to help full siblings) than are those same birds at two years of age, simply because the probability is high that by that time one or both parents will be dead.

Helpers known not to be at all closely related to the breeders, 'unrelated', also occur regularly. An unrelated helper was present in 20–30% of all social units on Morendat Farm over the four year period 1978–81. The percentage of unrelated helpers within the helper population as a whole ranged from 5.7 to 10.7. Unrelated helpers in the woodhoopoe population that we studied are more important than these percentages suggest. In large flocks all helpers are usually related to one or both breeders, whereas in small flocks, where the presence of one or two helpers may be critical to territory establishment and defense, and later to successful breeding, an unrelated helper is present 20–33% of the time. Thus, unrelated helpers are a significant part of the overall system. We believe that the correct interpretation of unrelated helpers is to view them as pursuing an alternative strategy to gain territorial space and eventual breeding status in a social environment characterized by integrated groups of cooperating birds and by high mortality (see below).

Age of first reproduction

Because of the temporally and spatially unpredictable pattern of mortality of Green Woodhoopoes, great variation exists in the age of first reproduction. For example, a female hatched in 1975 served as a helper for her parents in 1976–80, her sister in 1981–(82–3?) and bred in her natal territory in 1984. Thus this female was at least six and possibly eight years old when she first bred. At the other extreme, a female hatched in late 1975 quickly inherited breeding status and nested in her natal territory in June 1977, when she was 20 months of age. (Late in incubation, this bird became sick and soon disappeared.) Mean age of first reproduction for 15 females was 3.9 years (range 1.6–8) and for 19 males the comparable value was 3.3 years (range 1.5–5.0).

Duration of breeding status

Duration of breeding status likewise is highly variable. Many birds survive as breeders for less than a year (Ligon 1981*a*), whereas others breed for several years. Several females that held breeding status in 1975, at the beginning of the study, bred for five to seven years thereafter. One male that first bred in 1975 survived for at least seven years, and two others that first bred in 1978 were still alive and breeding in 1984.

Philopatry and inbreeding

Green Woodhoopoes of both sexes tend to be extremely philopatric. Eighteen of 38 females banded as nestlings or fledglings, and that we later recorded as breeding, did so in their natal territories and 14 did so in an adjacent territory. Three of the remaining six birds bred only two or three territories away from their natal territory (mean number ($\pm$ SEM) of territories females moved = 1.24 ± 0.39). We have similar data for 33 males: nine attained breeding status in their natal territories, 20 bred in an adjacent territory, and four bred two territories from their birth sites (mean number ($\pm$ SEM) of territories males moved = 0.85 ± 0.11). The conservative dispersal pattern of male woodhoopoes is similar to that of males of several other cooperative breeders; however, female woodhoopoes differ from the usual pattern in that about half bred in their natal territories and 85% bred in their hatching site or one territory away from it. Female Green Woodhoopoes thus employ the full range of dispersal possibilities, from no movement to long-distance (*ca* 13 territories) emigration. Although we undoubtedly failed to relocate some dispersers, despite systematic searches for banded birds well beyond the boundaries of our study sites, we have no reason to believe that many birds of either sex simply moved beyond our areas of activity.

An inevitable outcome of extreme philopatry by both sexes is the presence of neighborhoods of territories that contain close relatives. For example, in 1981 and 1984, 0.5 genetic relationships among *breeding* birds in different flocks were common in our study population (see Fig. 1 of Ligon and Ligon 1988). This means, of course, that many young birds produced in this neighborhood were minimally related at the first cousin ($r = 0.125$) level. The relatedness of young birds in neighboring flocks, together with the conservative pattern of dispersal, suggests that relatives from different flocks sometimes pair and breed. (In contrast, we have never recorded a pair bond between close relatives in the same flock, such as siblings or parents and offspring.)

To investigate matings between relatives, J.D.L. next visited the study site in mid 1984, by which time a few of the birds hatched in 1980 and 1981 should have attained breeding status. During the $2\frac{1}{2}$ year interval January 1982–June 1984, survival of breeders had been relatively high, and only two marked 1980-hatched woodhoopoes and no 1981-hatched birds had attained breeder status. Relatedness among breeders in neighboring territories again was high, and in one flock the breeding pair was nephew–aunt ($\bar{r} = 0.25$), while in another, the relatedness of the breeders was somewhat less, 0.188. In first case, the male simply moved 'next door' into his father's natal territory, whereas in the second case the movement pattern was more circuitous and thus more interesting. Male B/WR was hatched in the CT3 territory in early 1980; in early 1981 he and his father deserted female BR (B/WR's mother) in the CT3 territory and moved into the ST territory, one of the best territories on the study area (see below). Male B/WR was still a non-breeding helper in the ST territory in January 1982. By June 1984, he was a successful breeder in the BF territory, and was mated to female G/WR (a half-aunt plus second cousin, $\bar{r} = 0.188$; see Fig. 2 of Ligon and Ligon 1988).

Given the conservative, short-distance movement by both males and females, the fact that close relatives occasionally form pair bonds is not surprising. What is surprising is that we have recorded it so infrequently. Our limited data do not suggest that relatives more distantly related than parents and offspring, or siblings, avoid mating with each other. Rather, the rarity of such matings appears to be due to the high annual rate of mortality, together with the generally long period of time between hatching and first breeding. Most close relatives in different flocks do not have the opportunity to breed with each other simply because mortality typically intervenes. When this is not the case, close relatives (e.g. $\bar{r} = 0.25$) may form pair bonds and breed, as described above.

The woodhoopoe groups that we marked and studied are part of a semi-isolated population that occupies the acacia woodland ringing Lake Naivasha. On the basis of suitable habitat available around the lake and knowledge of territory sizes, we estimate that in 1975 about 105 social units made up the Lake Naivasha population of Green Woodhoopoes. Both before and since 1975, the number of birds undoubtedly has declined coincident with the destruction by man of acacia woodland. Thus the 33 groups that we marked represent a sizable fraction of the current population of woodhoopoes at Lake Naivasha.

Certain empirical data suggest that inbreeding depression may exist in the Green Woodhoopoes that we studied. One result of inbreeding in both wild and domestic birds is reduced hatchability (Romanoff 1972; Van Noordwijk and Scharloo 1981). In the woodhoopoes, the percentage of unhatched eggs is high and strikingly similar regardless of clutch size or year. One or more eggs did not hatch in 82% (49/60) of the nests for which we know both original clutch size and number of eggs hatched. Of the 32 unhatched eggs that we examined from 22 nests, over half contained obvious embryos that had died at various stages of development up to the point of hatching; thus many of the unexamined eggs that did not hatch also may have been fertile. Separation of the data by year reveals a similar pattern: about one-third of all eggs laid do not hatch. We know of no other species for which hatching failure, apart from predation, is so consistently high.

In part, we interpret the philopatry and hatching failure in a way similar to that of Bengtsson (1978). The decision to disperse and avoid inbreeding, or to be philopatric and risk inbreeding depression, should be dependent on the relative costs of inbreeding versus dispersal. For the woodhoopoes, we suggest that the costs of long-distance dispersal are so great that the risk of inbreeding below the level of parent and offspring or siblings is accepted. Moreover, although there may be a cost of inbreeding in the form of reduced hatchability of eggs, it is not apparent that overall this cost is particularly important. A single pair of birds may rear up to three broods during the potential breeding period June–December (Ligon and Ligon 1982). As a result of the high survivorship (survival rate) of nestlings (Ligon and Ligon 1988) and the multi-brooded reproductive pattern of woodhoopoes, loss of offspring via inbreeding depression may exert little effect on the lifetime fitness of breeders. That is, other factors in a sense compensate for the mortality of inbred embryos (also see Van Noordwijk and Scharloo 1981).

Basic demography
Reproduction
As described earlier, reproductive output in Green Woodhoopoes is strongly affected by environmental factors, such as density of predators, and especially by quantity of dry-season rainfall, which determines the

abundance of moth larvae, the main food brought to nestlings. These usually override the relationship between number of helpers and number of young fledged either per nest or per year. A regression of number of young produced per year on flock size for each of the years 1975–81 revealed a significant correlation for only 1975 and 1981 (Table 2.4). Thus, although helpers appear to reduce the burden of breeders (Ligon and Ligon 1978*a*), no consistent simple relationship exists between the number of helpers and the number of young woodhoopoes produced per year (Table 2.4) or overall (Fig. 2.6).

However, this is not to say that the presence of helpers is irrelevant to the reproductive output of breeders; to the contrary, helpers are typically critical to the potential success of would-be breeders, in that unassisted pairs are rarely able to breed at all (Ligon 1983*a*, *b*). Unassisted pairs apparently breed successfully only under a combination of unusual conditions: (1) in years of abundant food; (2) in territories isolated or almost isolated from neighboring flocks; and (3) when the overall woodhoopoe population is low, further decreasing conflict with neighboring groups. In addition, breeders with a greater than average number of helpers may breed more times per year than those with few helpers. It appears, therefore, that helpers are critical to most potential breeders in that: (1) an unassisted pair has an extremely low probability of breeding at all, due to the difficulty of gaining or holding a territory; and (2) under favorable environmental conditions helpers allow the breeders to re-nest more frequently.

One case illustrates that, under the right conditions, an unaided pair can

Table 2.4. *Values by year for relationship between group size and number of young woodhoopoes produced per annum*

Year	Multiple r	p	n
1975	0.55[a]	0.02	18
1976	0.18	0.41	23
1977	0.13	0.60	20
1978	0.11	0.61	25
1979	0.13	0.54	24
1980	0.09	0.69	24
1981	0.49	0.02	21
All years combined	0.10	0.21	155

[a] A single outlying point led to significance here.

produce well above the average number of young per annum. In 1981 the spatially isolated NL6 territory held a pair that produced a single female fledgling during the first nesting period and four males during the second, for a total of five young. This pair occupied an 'island' (acacia forest surrounded by grassland) of high-quality habitat in a year when food was abundant and neighboring woodhoopoes were absent, and they suffered no losses of young to predators. However, as mentioned above, most woodhoopoes must contend with daily aggression from adjacent flocks; under these circumstances helpers appear to be essential for successful breeding.

Mortality
Unlike most cooperative breeders for which long-term data are available, Green Woodhoopoes exhibit high annual rates of mortality, about 40% per year for males and 30% per year for females (Ligon 1981*a*; Ligon and Ligon 1988). Woodhoopoes without exception use cavities or other cover for roosting, and a large cost is apparently associated with this behaviour. All of our evidence suggests that most predation on these birds occurs at the roost holes.

Most of the 386 woodhoopoes that we banded failed to live for many years. At the time of our last period of field work in mid 1984, eight known-age female breeders averaged 6.1 years of age and six known-age male breeders averaged 4.8 years. Two other male breeders still alive in 1984 were banded in January 1977 and June 1978 and were probably nine and eight years old, respectively. Our longevity record is of a female hatched in 1975 and seen by us as a breeder in her natal territory in 1984 and by H. Dinkeloo (personal communication) in 1985.

In most birds, immatures survive far less well than adults. In contrast, in the woodhoopoes rates of disappearance of birds of different age and/or social status are almost identical within each sex (Table 2.5). We believe that this provides additional, indirect evidence for the importance of nocturnal roost-site predation; that is, increased age and experience provide no benefit if predation occurs primarily while the birds are roosting. For example, since males of all ages frequently roost together, all are vulnerable when in the roost cavity. We have occasionally recorded the overnight disappearance of one or two birds, at the same time that their roost partners lost their tails. In such cases we assume that the predator was unable to reach all birds in the cavity and pulled the tail out of the fortunate survivor that was just out of reach. (When woodhoopoes in cavities are threatened, they point the tail toward the entrance, elevate their rump feathers and produce a strong odor from a substance released from the oil gland (Ligon and Ligon 1978a).)

Thus the cavity roosting habit appears to be risky. Our evidence suggests that major nocturnal predators are driver ants and Large-spotted Genets (*Genetta tigrina*). Birds occupying cavities in weak wood, or cavities that are shallow and/or with large entrance holes, are vulnerable to genets and possibly to wild and feral cats (*Felis libyca* and *F. domesticus*). All cavities are vulnerable to invasion by raiding driver ants.

It seems paradoxical that Green Woodhoopoes should inevitably roost in cavities where they are so susceptible to capture. Earlier we suggested that perhaps these birds are unable to remain endothermic in the absence of insulation afforded by the roost cavities (Ligon and Ligon 1978a). However, we have only recently been able to test this suggestion (Ligon *et al.* 1988). We measured metabolic rates and body temperatures of three Green Woodhoopoes and discovered that the birds did indeed appear to be incapable of maintaining normal body temperatures at the relatively moderate ambient temperature of 19°C, and that their behavioral responses to such nocturnal ambient temperatures were inappropriate; i.e. although the birds increased their activity as ambient temperature decreased, they nevertheless became hypothermic. At 24:00 h and at an ambient temperature of 19 °C, the body temperatures of the three woodhoopoes had dropped from a 'typical' avian daytime level of 40–42 °C to 33–36 °C.

Variation in territory stability and quality

During the 10 years of this study, four of the 13 original established territories on Morendat Farm disappeared. During this same interval, five

Table 2.5. *Mean percentage minimum annual[a] survival in from 22 to 27 flocks of individually marked Green Woodhoopoes*

	Males			Females		
	Breeders	Helpers	Immatures[b]	Breeders	Helpers[c]	Immatures[b]
% survival	62	60	61	70	67	73
Range	50–82	30–81	50–79	57–86	53–78	60–78

[a] Based on six years, January 1976–January 1982.
[b] Birds from about one to six months of age at the beginning of their first calendar year. The survivorship data are based on the first full year (January–January) of life, thus nestlings and many individuals in juvenile plumage are not included.
[c] A few female helpers undoubtedly dispersed and were not relocated.

new territories were initiated, one of which lasted only a few months. Thus the number of territories at the end of the study was the same as at the beginning.

Because territories are restricted in number, one possible aspect of territoriality is that such behavior serves to reserve a future breeding site for offspring of the current breeders, i.e. some young birds that remain in their natal territories will inherit breeding status there. Green Woodhoopoes of either sex may breed within their natal territories. In this species, therefore, inheritance of breeding space and status does occur.

However, at first glance, our data do not appear to support strongly territory inheritance over the long term. In 10 of the 14 territories present on Morendat Farm in 1975–6, all group members and all of their descendants disappeared in 10 years or less. In other words, in 71% of the territories, inheritance of breeding status did not occur at all, or did so for only one generation before complete turnover of the genetic lineage took place.

Because the territories on Morendat Farm show a lot of variation in productivity and survival of occupants over time, we attempted to categorize territories with regard to these ultimate aspects of quality. Meaningful proximate measures of territory quality are usually difficult to obtain. Although one can measure various vegetational attributes of a territory, just what the measured features mean to the animals occupying them is often in doubt. Similarly, assessments of food quality and quantity, especially for arboreal insectivorous species may be unreliable. Therefore, we have not emphasized these proximate measures, but instead focus on the ultimate effects of teritory quality: natality (N) and mortality (M), making the reasonable assumption that in the most productive territories $N/M > 1$: i.e. those territories in which the number of woodhoopoes produced is greater than the number dying are the best ones, from the perspectives of the breeding birds occupying them. Note that we are not simply separating the more productive and the less productive; rather, we use an independent criterion ($N/M > 1$ or $N/M < 1$) to designate two categories.

To provide an independent measure of territory quality, we compared various vegetational characteristics of 11 territories measured in 1975–6 by use of a modified point-quarter method. Although boundaries have changed over the following decade, the core areas of the surviving territories have remained more or less constant. Three of these territories have N/M ratios greater than one. We compared tree sizes by DBH measurements and found that average tree sizes in these three territories were almost significantly larger (Mann–Whitney U-test, $0.05 < P < 0.1$)

than in the other eight territories. We believe that tree size is a very important aspect of territory quality for Green Woodhoopoes, for two reasons. First, large trees provide better foraging substrate and more food than do smaller ones (our unpublished data). Second, and even more critical is the relationship between tree size and cavities: bigger trees have a greater likelihood of holding cavities suitable either for roosting or nesting by the woodhoopoes. In some cases loss of a few or even only one key cavity tree can lead to the disappearance of an entire social unit (Ligon and Ligon 1978a). Thus a causal relationship may exist between tree sizes on a given territory and rates of natality, and, especially, mortality.

We measured annual natality as the number of juvenile birds surviving to the end of the year in the territory in which they were hatched. Mortality is based on marked birds lost in each year. Because long-distance emigration occurs in mature female helpers, unaccounted for disappearances of such birds are not included in our tabulations. This information alone does not tell us specifically how one territory is superior to another, but, since the number and quality of cavities are critical both to reproduction and survival, we believe that these data reflect in large part the cavity situation in a given territory, relative to other territories.

Only five of 16 territories had an N/M ratio greater than one. Because similar ratios can be obtained with very different values, we compared these five territories with the other 11. The five (high-quality $=$ HQ) territories had both significantly higher annual natality and significantly lower annual mortality than did the 11 low-quality (LQ) territories (Mann–Whitney U-test, $P < 0.0025$ and $P < 0.005$, respectively). In all five of the HQ territories, offspring of one or both of the original breeders have inherited breeding status. In fact, in 1984, a *daughter* of the original breeder or breeders held reproductive status in four of these five territories. (In nearly all other cooperatively breeding species, territories are inherited largely or exclusively by males.) In the fifth, the MSG territory, a son inherited breeding status and bred successfully in late 1979 and 1980. However, after losing his mate late in 1980, this male moved with four of his sons to the then male-less AD territory. Desertion of an HQ territory by several birds, leaving it temporarily unoccupied, was a unique event during our study.

To look more closely at the fates of individual breeders, we calculated the mean number of new breeders per year for each territory and found that the five HQ territories had a significantly lower turnover of breeders (Mann–Whitney U-test, $P < 0.001$) than did the remaining 11 territories; i.e. on average, individuals persist longer as breeders in the HQ territories. We also compared the mean number of young woodhoopoes produced per individual breeder in the two categories and found that each breeding bird in an HQ territory produced on average significantly more offspring over its lifetime than did breeders in the LQ territories (Mann–Whitney U-test, $P < 0.001$).

To contrast further the differences in potential fitness of breeders in the HQ territories in 1975–6 with those in the LQ territories, we tallied the number of known emigrants to another territory, the number of those that attained breeding status, and the number that actually bred successfully, producing at least one surviving offspring (Table 2.6). As with the other comparisons, breeders in HQ territories produced more total emigrants, more emigrants that became breeders, and more that bred successfully. Table 2.6 makes one additional, important point: an emigrant from a LQ territory can, and sometimes does, attain breeding status in a HQ territory. What this means is that a chance exists that a breeder in a LQ territory will produce an offspring that ultimately breeds in a HQ territory; i.e. LQ territories can produce 'big winners'.

Our data indicate that neither flock size nor territory area at a single

Table 2.6. *Relative fitness parameters of Green Woodhoopoe breeders occupying specific territories*

Territory	No. known emigrants	No. emigrants becoming breeders	No. emigrants breeding successfully
$N/M > 1$			
MSG	13	10	7
BRF	4	1	1
ST	4	3	3
BF	8	6	4
AD	5	4	4
$N/M < 1$			
CO	1	1	1 (in BRF)
LH	2	1	0
DD	8	6	3 (in Lh, KG, H5)
WWH	3	2	1
KG	4	1	1
H5	3	1	1
SS	2	1	1 (in AD)
Cr	0	—	—
CR5	0	—	—
3St	0	—	—
CT3	1	1	1 (in BF)

point in time may be a sound indicator of the relative fitnesses of the breeders or the lineages occupying a particular territory. The three largest groups present in 1975 (DD, KG, CR) were notably lacking in long-term success (Table 2.7). Possibly of greatest significance is the striking dichotomy among groups in number of descendants within the study area after 10 years. The original breeders of 1975 either have a good many descendants (12–18), or none (seven pairs with none, one with one), with no intermediate values (Fig. 2.7). All of the lineages occupying high-quality territories have been very successful, as have two individuals hatched in low-quality territories. Both of these latter birds (one male, one female) gained success as a result of obtaining breeding status in a high-quality territory. The success of the two birds emigrating from low- to high-quality territories suggest that quality of the territory, rather than quality of the birds is indeed the primary factor determining reproductive success and

Fig. 2.7. The number of known living descendants in 1984 of Green Woodhoopoe breeders occupying territories on Morendat Farm in 1975 or soon thereafter. The BR territory, indicated by the hatched bar, came into existence in 1977; since that time it has been highly productive. The success indicated for the SS and CO territories is based on the movement of one individual hatched in each of those flocks to breeding status in a high-quality territory (see the text).

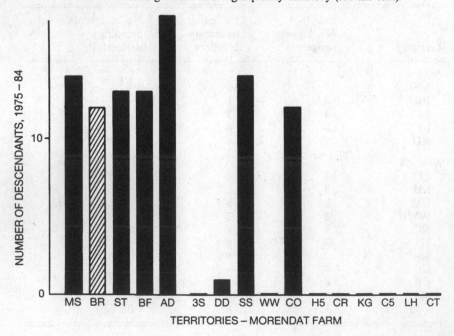

survival of offspring over time. This suggests in turn that high-quality territories, in particular, should be inherited by offspring from their parents, and, as previously mentioned, this has been the case.

Discussion

Many critical life history features of Green Woodhoopoes appear to be related to a physiological limitation. These birds apparently lack the physiological means to contend with even moderately low ambient temperatures. To avoid the low nocturnal temperatures (7–10 °C) characteristic of our study site at Lake Naivasha in Kenya, the woodhoopoes almost invariably roost in cavities in trees. Often, several birds roost together.

The necessity of cavity-roosting appears to be responsible for the comparatively high rate of predation on woodhoopoes. Because of predation pressure and competition with other species for cavities (e.g. honey bees, acacia rats, dormice, other kinds of birds), cavity quality (i.e. size of entrance, size of cavity, strength of wood surrounding the cavity) and number of potential roost holes are probably the major factors determining reproductive success and survival of birds occupying a particular territory.

Table 2.7 *Initial group size of Green Woodhoopoes and descendants a decade later*

Territory	Maximum group size 1975–6	No. breeding emigrants	No. known living descendants of original breeding pair as of 1984
N/M > 1			
MSG	6	10	14
ST	8	3	13
BF	7	6	13
AD	4	4	18
N/M < 1			
CO	5	1 (to BRF)	12
LH	4	1	0
DD	16	6	1
WWH	6	2	0
KG	11	1	0
H5	4	1	0
SS	5	1 (to AD)	14
CR	10	0	0
CR5	5	0	0
3St	5	0	0

In addition to its effect on mortality rate, the dependence of Green Woodhoopoes on holes in trees for roosting, together with the scarcity of high-quality cavities, has affected dispersal patterns and this, in turn, has affected the birds' reproductive biology in several ways, possibly including the evolution of cooperative breeding. The best evidence that cavity numbers and/or distribution strongly influenced the social system of Green Woodhoopoes comes from a study in South Africa. In coastal riverine forest where cavities appeared to be plentiful (night-time temperatures also may have been warmer), over 38% of the social units consisted of only a pair of birds (average: 3.4 per social unit), while inland, at a higher and much drier site, group size averaged 4.7 (du Plessis 1989). This difference was significant and appeared to be related to availability of cavities for nests and roosts. The picture that emerges is that territory quality, in terms of characteristics of the trees – especially cavities – is the primary determinant of lifetime fitness in Green Woodhoopoes (Fig. 2.8).

Scenario for the development of reciprocity among non-kin

In Green Woodhoopoes a demographic feature – high mortality – appears to have set the stage for the first and most critical step in the hypothetical evolutionary development of (1) cooperative breeding, and later (2) reciprocity (*sensu* Alexander 1974) among non-kin (Ligon 1983; Ligon and Ligon 1978b, 1983). Because most mortality occurs at roost cavities and because safe roost sites are scarce, mated pairs of woodhoopoes allowed their grown offspring to remain in the parents' territory and to use roost sites therein. This retention of matured offspring, beneficial to both the breeders and the grown young birds, had at least two critical consequences for the breeders. (1) With more than two adults per group, opportunities for territorial expansion were increased relative to neighboring pairs without allies. This means an increase in territory quality. Moreover, territorial defense became less costly on an individual basis. (2) Additional adults in the group represented potential help to the breeders: potential defenders of the territory during nesting activities and potential feeders and protectors of the nestlings and fledglings. The probable reduction of the cost of nesting to the breeders, as a result of the presence of the additional woodhoopoes, meant greater potential annual reproduction. In addition, when opportunities arose to occupy a new territory, older adult non-breeders in a social unit now had allies composed of younger, subordinate flock mates, whose assistance increased the emigrants' chances of both obtaining and retaining ownership of a new territory. This clearly benefits the original breeders because one or more (in sequence) of their offspring gains breeding status as a result of unisexual sibling-group emigration. Each individual emigrant also stands to gain by virtue of group movement, as described below.

Once interdependence became established among related individuals for territorial enlargement and defense, reproduction, and later, acquisition of new territorial space, the unaided pair, or single individual, was placed at an overwhelming competitive disadvantage. However, lone woodhoopoes without supportive kin do occur frequently because of the high mortality and the subsequent displacement of a surviving singleton from a territory by a group, and because of dispersal in the case of some individual females. Thus, because of the group-based population structure, the best option open to most lone birds, if they are to gain territory ownership and eventually to breed, is to procure subordinate allies of their sex, as well as a mate. This they can sometimes do by allowing younger floaters of their sex to join them. These younger birds then provide critical aid in establishing and sometimes expanding territorial boundaries. The young birds will also later feed the nestlings of the dominant birds. By such cooperation they in

Fig. 2.8. Diagram showing the factors leading to among-territory variation in quality. Cavities for roosting are probably the single most important factor influencing territory-to-territory variation. The variation in turn is reflected by the Green Woodhoopoe breeders' lifetime production of offspring that themselves survive to attain breeding status.

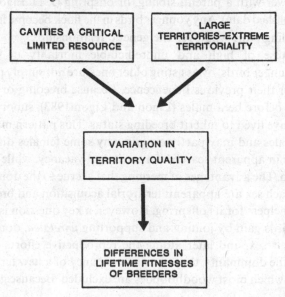

turn will eventually obtain the benefits described above, provided that they live long enough.

Because non-breeding flock members generally are active helpers whether or not they are related to the nestlings, it has been argued elsewhere that for this species kinship ties alone are insufficient to account for helping behavior (Ligon 1981a; Ligon and Ligon 1983). If so, what are the critical personal returns on the helper's investment? The older woodhoopoe forms personal bonds with younger birds that it may later use to gain breeding status for itself. As described earlier, breeding vacancies are usually filled by a former helper accompanied by one or more younger birds of its sex that it has helped to rear. Members of these unisexual teams may be full siblings, half-siblings, uncle–nephew, or unrelated. In contrast, full siblings that are the same age do not form harmonious, permanent, emigrant teams. Thus age patterns among groups of emigrant woodhoopoes predict more accurately than do kinship ties whether an individual will stay permanently in the new territory or return home. Because the oldest bird in an immigrant group becomes the initial breeder, with the next oldest in line to breed next (depending on the pattern of mortality), and because each bird has helped to rear those younger than itself, the presence of the younger birds and their alliance and assistance to the oldest, next oldest, etc., can be viewed as part of the 'repayment' for the earlier helping behaviour of older birds. Similarly, many helpers never emigrate from the parental territory. Rather, many of them become breeders on the natal territory and mate with an immigrant but never with a parent, sibling, or offspring (J. D. Ligon and S. H. Ligon, unpublished data). Any younger birds in the flock become helpers for the new breeding pair, whatever the genetic relationships.

However, because of high and unpredictable mortality in Green woodhoopoes, younger birds, by assisting older ones, are not simply paying back the latter for their previous beneficence. Because breeding or alpha males usually die before beta males (Ligon and Ligon 1983), subordinate males generally have lived to inherit breeding status. This pattern may not hold true for females and may partly explain why some females disperse, wandering widely in apparent search of a territorial vacancy, while males almost never do so. The advantages of merging that accrue to the dominant woodhoopoe of each sex are apparent: territorial acquisition and breeding status and, later, helpers for its offspring. However, a key question is: what do subordinate birds gain by joining and supporting *unrelated*, dominant individuals of their sex, and later aiding the reproductive efforts of the latter? First, like the dominants, they obtain occupancy of a new territory, living space from which most woodhoopoes are excluded. Because gaining and successfully holding a territory is dependent in large part on numbers, and because flock size affects territory size, and thus at least some aspects of territory quality, each individual gains by the presence of flockmates. Within the territory each bird has access to known, dependable roost sites and foraging areas. Second, a subordinate woodhoopoe will advance to breeding status in the same territory if the dominant bird(s) of its sex dies, as is frequently the case, or gain an ally for emigration if a vacancy occurs in a nearby territory. In both cases it often 'inherits' younger birds in the flock as helpers for its own reproductive efforts (Ligon and Ligon 1978*b*, 1983).

In brief, a single woodhoopoe of either sex is far less likely to attain and retain territory ownership and breeding status than are two or more individuals of the same sex. This sets the stage for the evolution of cooperative interactions between birds that may or may not be genetically related.

General issues

As described in this chapter and elsewhere, Green Woodhoopoes exhibit traits related to two general evolutionary problems.

The significance of helpers. Unlike some of the other species considered in this volume, the relationship between number of helpers and number of young birds produced per annum is not strong. Environmental factors typically swamp any potential positive effect of helpers. Moreover, unlike two other cooperatively breeding coraciiform relatives of Green Woodhoopoes also studied in Kenya, the White-fronted Bee-eater and the Pied Kingfisher (see Chapters 16 and 17), death of nestling woodhoopoes by starvation is exceedingly rare (< 5%), regardless of group size (Ligon and Ligon 1988). These two related points – weak effect of number of helpers and rarity of nestling starvation – suggest that adaptive explanations for helping behavior in Green Woodhoopoes should rely primarily on direct rather than on indirect benefits. We have argued above that reciprocity can account for the benefits gained by serving as a helper. Although the benefits obtained via helping are not easy to quantify meaningfully (e.g. what numerical value does one put on procurement of breeding status by a former helper, obtained with the aid of one or more younger flockmates?), they clearly are large and are not obviously affected by environmental variation.

Both Trivers (1971) and Wilkinson (1988) have pointed out that reciprocity or reciprocal altruism can occur between kin and 'generate a substantial selective force independent of kin selection even when performed among related animals' (Wilkinson 1988, p. 98). Although a distinction between reciprocity and kin selection may not be possible (Rothstein 1980), we have attempted to separate these two selective pathways to cooperation in Green Woodhoopoes by looking separately at social context and kinship ties. First, both related and unrelated helpers appear to incur the same costs of helping and both have the opportunity to obtain most of the same personal benefits (e.g. allies for later attaining breeding status). Unrelated helpers have an additional one. In one case a young immigrant female woodhoopoe helped to rear a single unrelated male nestling. Three years later, in the same territory, she mated with this male and bred successfully. Thus her participation in flock activities was eventually rewarded by (1) breeding status in the territory in which she helped, and (2) mating with the bird she had earlier provisioned and protected.

Second, the form of social interactions among same-sex adults likewise suggests that social context rather than kinship dictates the relationship between two such individuals (Fig. 2.9). Two woodhoopoes of the same sex normally cooperate when they live in the same flock, and compete aggressively when they occupy adjacent territories, regardless of kinship ties. In short, the social relationship predicts the form that personal interactions will take, whereas the genetic relationship does not. Recall, too, that relative age of cooperating emigrants (i.e. same age versus unequal age), rather than the genetic relationship, determines the outcome of

Fig. 2.9. Diagram showing the kinds of interaction exhibited by two woodhoopoes of the same sex, either kin or non-kin and in either the same or neighboring social units. Birds in the same social unit cooperate while those in adjacent groups compete, as at territory boundaries, regardless of kinship ties.

INTERACTION BETWEEN:

	KIN	NON-KIN
SOCIAL UNIT: SAME	COOPERATION	COOPERATION
SOCIAL UNIT: NEIGHBORING	AGGRESSION	AGGRESSION

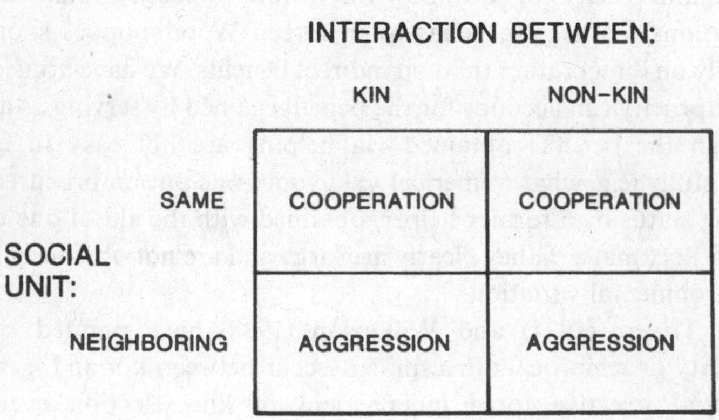

emigration to a new territory by a unisexual team of two or more individuals. For all of these reasons we have concluded that for Green Woodhoopoes, helping behaviour is not critically tied to kinship ties (and, by extension, that aid provided by unrelated birds is not an accident or mistake). However, we add the caveat that this interpretation of helping in woodhoopoes, if correct, may indicate little or nothing about the bases of helping behavior in other kinds of birds. The significance of helping behavior as a phenomenon cannot be accounted for with a single adaptive explanation. Moreover, the act of feeding nestlings in some cases may not be adaptive *per se* (see, e.g. Woolfenden and Fitzpatrick 1984; Jamieson 1986; Jamieson and Craig 1987). We believe that the evidence indicates that helping behavior is adaptive in Green Woodhoopoes, particularly for the helpers themselves, whether or not they are related to the young birds they assist.

Physiology, phylogeny and sociality. Recently Ligon *et al.* (1988) suggested a causal relationship between the apparent inability of woodhoopoes to tolerate what appear to be moderately low night-time ambient temperatures and the development of group-living. The requirement of roost sites in tree holes in order to avoid low ambient temperatures, combined with the scarcity of available cavities, has apparently led to the retention of young birds in their natal territories, setting the stage for the eventual evolution of the suite of behaviors that characterize cooperative breeding. The physiological traits of Green Woodhoopoes, and possibly some of their coraciiform relatives, may account, at least in large part, for the fact that all members of this diverse order are cavity nesters and that many or most also roost in cavities. If this physiological limitation of woodhoopoes is viewed as a 'phylogenetic constraint', the usage of cavities to ameliorate this problem can be viewed as a behavioral bypass around the constraint. We suggest that study of physiological problems such as this, and available solutions to them, may yield valuable new insights into the diversity of selective factors leading to social behavior, including avian cooperative breeding.

Our field studies of Green Woodhoopoes in Kenya were made possible via the support of R. E. Leakey and G. R. Cunningham-van Someren of the National Museums of Kenya. E. K. Ruchiami, of the Office of the President of Kenya, kindly granted research permits. S. Zack, D. C. Schmitt and J. C. Bednarz assisted with the field work, and W. and R. Hillyar and R. and B. Terry provided logistical support. This project was generously supported by the National Science Foundation, the

National Geographic Society, the American Museum of Natural History, the National Fish and Wildlife Laboratory, and the University of New Mexico. We thank all of these individuals and institutions.

Bibliography

Alexander, R. D. (1974). The evolution of social behavior. *Ann. Rev. Ecol. System* 5: 325–83.

Bengtsson, B. O. (1978). Avoiding inbreeding: at what cost? *J. Theoret. Biol.* 73: 439–44.

du Plessis, M. (1989). The influence of roost-cavity availability on flock size in the Redbilled Woodhoopoe, *Phoeniculus purpureus*. *Ostrich Monogr.* 1.

Jamieson, I. G. (1986). A functional approach to behavior: is it useful? *Amer. Nat.* 127: 195–208.

Jamieson, I. G. and Craig, J. L. (1987). Critique of helping behavior in birds: a departure from functional explanations. In *Perspectives in Ethology*, ed. P. P. G. Bateson and P. H. Klopfer, vol. 7, pp. 79–98. Plenum Press: New York.

Ligon, J. D. (1980). Communal breeding in birds: a test of kinship theory. *Proc. 17th Intern. Ornith. Congr.*: 857–61.

Ligon, J. D. (1981a). Demographic patterns and communal breeding in the Green Woodhoopoe (*Phoeniculus purpureus*). In *Natural Selection and Social Behavior*, ed. R. D. Alexander and D. W. Tinkle, pp. 231–43. Chiron Press: New York.

Ligon, J. D. (1981b). Sociobiology is for the birds. *Auk* 98: 409–12.

Ligon, J. D. (1983a). Commentary on cooperative breeding strategies in birds. In *Perspectives in Ornithology. Sponsored by the American Ornithologists' Union*, ed. A. H. Brush and G. A. Clark Jr, pp. 120–7. Cambridge University Press: Cambridge.

Ligon, J. D. (1983b). Cooperation and reciprocity in avian social systems. *Amer. Nat.* 121: 336–84.

Ligon, J. D. (1984). Communality in the Green Woodhoopoe (*Phoeniculus purpureus*). In *National Geographic Society Research Reports*, ed. P. H. Oehser, J. S. Lea and N. L. Powars, pp. 451–61. National Geographic Society: Washington, DC.

Ligon, J. D. (1985). Woodhoopoe. In *A Dictionary of Birds*, ed. B. Campbell and E. Lack, pp. 658–9. British Ornithologists' Union, Buteo Books: Vermillion, SD.

Ligon, J. D. (1988). The question of parent–offspring conflict in cooperatively breeding birds. *Proc. 19th Intern. Ornith. Congr.*: 1231–43.

Ligon, J. D., Carey, C. and Ligon, S. H. (1988). Cavity roosting, philopatry and cooperative breeding in the Green Woodhoopoe may reflect a physiological trait. *Auk* 105: 123–7.

Ligon, J. D. and Davidson, N. K. (1988). Order Coraciiformes, Phoeniculidae, Woodhoopoes . In *The Birds of Africa*, vol. III., ed. E. K. Urban, C. H. Fry and S. Keith, pp. 356–70. Academic Press: London.

Ligon, J. D. and Ligon, S. H. (1978a). The communal social system of the Green Woodhoopoe in Kenya. *Living Bird* 17: 159–97.

Ligon, J. D. and Ligon, S. H. (1978b). Communal breeding in Green Woodhoopoes as a case for reciprocity. *Nature* (Lond.) 276: 496–8.

Ligon, J. D. and Ligon, S. H. (1982). The cooperative breeding behavior of the Green Woodhoopoe. *Sci. Amer.* 247: 126–34.

Ligon, J. D. and Ligon, S. H. (1983). Reciprocity in the Green Woodhoopoe (*Phoeniculus purpureus*). *Anim. Behav.* 31: 480–9.

Ligon, J. D. and Ligon, S. H. (1988). Territory quality: key determinant of fitness in the group-living Green Woodhoopoe. In *The Ecology of Social Behavior*, ed. C. N. Slobodhikoff, pp. 229–53. Academic Press: San Diego, CA.

Romanoff, A. I. (1972). *Pathogenesis of the Avian Embryo*. John Wiley and Sons: New York.

Rothstein, S. I. (1980). Reciprocal altruism and kin selection are not clearly separable phenomena. *J. Theoret. Biol.* **87**: 255–61.

Trivers, R. L. (1971). The evolution of reciprocal altruism. *Quart. Rev. Biol.* **46**: 35–57.

Van Noordwijk, A. J. and Scharloo, W. (1981). Inbreeding in an island population of Great Tits. *Evolution* **35**: 674–88.

Wilkinson, G. S. (1988). Reciprocal altruism in bats and other mammals. *Ethol. Sociobiol.* **9**: 85–100.

Woolfenden, G. E. and Fitzpatrick, J. W. (1984). *The Florida Scrub Jay. Demography of a Cooperative-breeding Bird*. Princetown University Press: Princeton, NJ.

3 RED-COCKADED WOODPECKERS: A 'PRIMITIVE' COOPERATIVE BREEDER

3

Red-cockaded Woodpeckers: a 'primitive' cooperative breeder

J. R. WALTERS

The Red-cockaded Woodpecker (*Picoides borealis*) is a monotypic species endemic to the south-eastern United States, similar in appearance to its more familiar congeners, the Downy (*P. pubescens*) and Hairy (*P. villosus*) Woodpeckers. It is isolated taxonomically from its relatives (Short 1982), and ecologically is even more distinct, exhibiting an unusual set of life-history features that includes cooperative breeding.

The ecology of an endangered cooperative breeder

The Red-cockaded Woodpecker was placed on the federal list of endangered species in 1968. It was once abundant in the Piedmont and coastal plain of the south-east, ranging north to New Jersey, west to Texas, and inland to Kentucky, Tennessee and Missouri (Jackson 1971). Now it is virtually extirpated north of North Carolina, and in all interior states but Arkansas. Most remaining populations are isolated and small, and many continue to decline. Only four populations of 300 groups or more exist. The total population is still reasonably large, however, numbering perhaps as many as 10 000 individuals.

The conservation of the species is a major controversial issue in the south-east (US Fish and Wildlife Service 1985; Jackson 1986; Ligon *et al.* 1986), affecting land use practices over vast areas. In particular, there is conflict between protecting the bird and managing timber. This interest in conservation was the impetus for most studies that provide information on basic biology. The ecology of the species, almost unknown in 1970, is now well researched.

Habitat requirements

The species is closely tied to mature pine savanna, particularly longleaf pine (*Pinus palustris*). In the past two centuries, this habitat, once

69

abundant, has been greatly reduced by forest clearing, logging and fire suppression (Jackson 1971; Lennartz *et al.* 1983). Fire suppression alters the habitat by allowing hardwood understory to develop. Logging practices often result in conversion of longleaf to other pine types, or alter the habitat by reducing or eliminating the older age classes of trees. Clearly, the decline of the species is due to habitat loss (US Fish and Wildlife Service 1985; Jackson, 1987; Ligon *et al.* 1986).

Within this habitat, Red-cockaded Woodpeckers select old, living pines for construction of cavities for nesting and roosting. Use of living pines for cavity construction is highly unusual among woodpeckers. This habit may be an adaptation to vulnerability or low density of snags in the fire-maintained ecosystems that the species inhabits (Jackson 1986), or it may be related to resource partitioning (R. Lancia, personal communication). The birds appear to select trees infected with red-heart fungus (*Phellinus pini*) for cavity construction, presumably because excavation is easier if the heartwood is softened. However, the birds apparently do not inoculate the trees with fungus (Connor and Locke 1982), as was once suspected (Jackson 1977). Old trees are selected because fungus infection is more prevalent among them, and because they contain sufficient heartwood for excavation. Recent evidence indicates that they may select trees that have undergone a period of suppressed growth (Conner and O'Halloran 1987).

Cavity trees generally average 80–120 years in age, depending on the species of pine (Jackson *et al.* 1979; US Fish and Wildlife Service 1985). However, the birds typically have a limited population of old trees from which to select trees for excavation because of logging. In virgin pine savanna in Oklahoma, cavity trees averaged 149 years of age (Wood, 1983), and in our study area in the Sandhills of North Carolina, where many old trees occur, some cavity trees are over 200 years old.

Cavity trees are conspicuous because the birds chip away at the sapwood around the cavity to form resin wells, from which sap flows. Trees used for extended periods become caked with sap, and plates of exposed heartwood develop around the cavity. The sap appears to deter predators, especially snakes (Jackson 1974).

Foraging

Red-cockaded Woodpeckers are highly specialized in their foraging behavior. Because of their implications for management, foraging and ranging behavior are the most studied aspects of the species' biology. Most foraging time (usually over 90%) is spent scaling bark from, and occasionally pecking on, living pines to capture invertebrates (Skorupa and

McFarlane 1976; Ramey 1980; Hooper and Lennartz 1981; DeLotelle *et al.* 1983, 1987; Repasky 1984; Porter and Labisky 1986). The remaining fraction of foraging effort may comprise a diversity of methods, depending on location and season. These include foraging on dead pines and hardwoods, taking insect larvae from cones, feeding on mast and berries, and even taking insect larvae from corn (Baker 1971). What is eaten is not as well documented as where foraging occurs, but ants, termites and beetle larvae appear to be common foods (Short 1982).

Within pine habitat, there is considerable evidence that birds select larger trees for foraging (Hooper and Lennartz 1981; DeLotelle *et al.* 1983; Porter and Labisky 1986). There is some indication of preference for older trees as well, but an age effect independent of size has yet to be clearly demonstrated. Variation between studies in available ages and sizes of trees clouds attempts to define optimum foraging habitat. Clearly, small young trees (e.g. < 30 years for longleaf) are seldom used if older trees are available, but preferences beyond this level are less clear, and how preferences relate to foraging success is unknown.

Interestingly, the species exhibits one of the more striking sexual dimorphisms in foraging behavior known among birds. Although both sexes forage on the upper regions of the trunk, only females regularly forage low on the trunk, whereas only males regularly forage on limbs and twigs (Ligon 1968; Skorupa 1979; Ramey 1980; Hooper and Lennartz 1981; Repasky 1984). This is not a subtle trend that appears only with appropriate statistical analysis, but a behavioral difference that is immediately conspicuous. The foraging difference cannot be related to any morphological dimorphism (Mengel and Jackson 1977; Short 1982). In fact the only documented morphological dimorphism among adults is that females lack a red cockade, the tiny spot of red on the white cheek patch that gives the bird its common name.

The social system

Initially, the social system of the species did not receive the same degree of attention that was afforded to habitat and foraging requirements. Currently, however, the social system is the focus of three long-term studies, Lennartz's in South Carolina (Lennartz *et al.* 1987), Jackson's in Mississippi (J. A. Jackson, unpublished results), and ours in North Carolina. In North Carolina, P. Doerr, J. Carter and I have been studying a large population (225 groups from a total population of 450 groups) in the Sandhills region since 1979 (Carter *et al.* 1983; Walters *et al.* 1988). The habitat is second-growth longleaf pine with scattered old-growth trees, understory of scrub oaks (*Quercus* spp.), and ground cover of wiregrass (*Aristida stricta*; Fig. 3.1). Open savanna occurs in some areas, but, as a result of fire exclusion, understory is dense in others. Pond pines (*P. serotina*) occur along drains and streams, and loblolly pine (*P. taeda*) occurs as second growth on some old field sites. We are studying also a small population of 12 groups in the coastal plain at the Sunny Point Military Ocean Terminal. This area contains three primary habitat types: (1) longleaf pine and scrub oak forest similar to that in the Sandhills; (2) pine flatwoods with longleaf overstory, open understory and dense ground cover; and (3) pocosins, wet areas of dense understory with scattered pond pines. In 1986 we began studying another population of 30 groups in similar habitat at Camp LeJeune. Lennartz studied 30 groups in coastal plain habitat for seven years in the Francis Marion National Forest (Lennartz *et al.* 1987), and has been studying a population of similar size in the Piedmont in recent years. Habitat in the latter region is characterized by more understory and

Fig. 3.1. Longleaf pine savanna in the Sandhills study area, North Carolina.

loblolly pine than the Sandhills and coastal plain sites. Jackson's study area in Mississippi is also in Piedmont habitat.

Two terms may create confusion. First, social units of Red-cockaded Woodpeckers are traditionally called clans, but these units do not meet the definition of clan used elsewhere in the cooperative breeding literature (*e.g.* for White-fronted Bee-eaters, see Chapter 16). Therefore I do not use the term clan here, but instead use group. Second, the set of cavity trees inhabited by a group is traditionally called a colony. These trees are often clustered in space, leading early observers, unaware of the species' group-living habit, to assume that the birds were colonial. The species is not colonial as the term is defined in the avian literature, so I use the term cluster instead of colony. Bear in mind, however, that cluster is defined by use of trees rather than their spatial arrangement; therefore trees forming a cluster are not necessarily clumped spatially.

Red-cockaded Woodpecker groups typically consist of a single female and one to four males. Copulations are frequent and conspicuous, but those observed have been restricted to the (behaviorally) dominant male within the group. The breeding system therefore has been assumed to be monogamous, and the subordinate males to be non-breeding helpers (Lennartz *et al.* 1987). That males sometimes remain as helpers for extended periods (see below), often are unrelated to the breeding female and pair with the breeding female following the dominant male's demise makes one suspicious that some helpers may copulate with the breeding female, but there is as yet no evidence that this occurs. If subordinate males do copulate with the female, the behavior involved must differ greatly from that associated with copulation with the dominant male.

Usually fewer than 50% of the groups in a population contain male helpers, and only a few contain more than one male helper (Fig. 3.2). Female helpers are rare, occurring in only 1·0% of groups in the Sandhills (13 cases) and 2% in the Francis Marion Forest (2 cases, Lennartz *et al.* 1987). However, 8% of the groups in the small, isolated population at Sunny Point have included female helpers. Groups larger than three individuals are uncommon (Fig. 3.3).

Spacing behavior

Red-cockaded Woodpeckers are non-migratory and territorial. A group usually travels together within their home range, giving frequent contact calls. Groups are least cohesive during incubation and when nestlings are being fed, and in late summer when individuals often leave the group to excavate cavities. Intruders are met with characteristic vocalizations, wing-spreading displays, and physical attack. Drumming, however, is quiet and is performed usually on the trunk or limbs of a living pine. Presumably drumming does not function in long-range territorial defense in this species, unlike in many other woodpeckers (Short 1982).

All group members participate in territorial defense, including fledglings, but not necessarily with equal vigor. Typically, adult males are the primary defenders against intruder males, and adult females against female intruders, but during the breeding season both sexes may respond to a single intruder. The birds are highly aggressive toward intruders and neighboring groups during breeding. Outside the breeding season, however, after an initial bout of displays and vocalizations, the resident group will sometimes forage peaceably with an intruder within the territory, or near a neighboring group along the territorial boundary. At this time of year, the birds often forage in interspecific flocks with species such as nuthatches, parids, kinglets, Pine Warblers (*Dendroica pinus*), and other woodpeckers. In our study area these flocks may be enormous, and contain as many as six woodpecker species.

Territory size varies with quality of habitat as defined by density and age of pines. In the best remaining habitat, territories are 50–70 ha, but they may be twice as large in poor habitat (Hooper *et al.* 1982; Porter and Labisky 1986; DeLotelle *et al.* 1987). Also, a group may use from almost

Fig. 3.2. Proportion of groups of various compositions in three populations of Red-cockaded Woodpecker. The North Carolina coastal plain population is the Sunny Point population (see the text). The North Carolina Sandhills data are for the years 1981–6. Lower shaded area, solitary males: clear area, pairs: cross-hatched area, pairs with one male helper: and upper shaded area, pairs with two or more male helpers. Sample sizes are 1335, 93 and 60 group-years respectively.

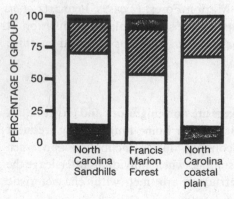

none up to several hundred additional hectares of extra-territorial home range. The amount of extra-territorial range varies enormously among sites, again as an apparent function of habitat quality, being greatest where habitat is poor. There is also considerable variability in territory and home range sizes within study areas. This, too, may be related to variation in habitat quality, but it may also be related in part to the dynamics of the breeding system (see below). It is also possible that populations are below carrying capacity in many areas, and thus that variation in population density contributes to variation in territory and home range size (Hooper *et al.* 1982).

The significance of this extra-territorial range is unclear. Its extent is difficult to estimate, as the estimate is strongly influenced by how often and for how long a particular group is followed, and by the method of estimation. Groups range outside their territory most often in late summer, when fledglings are present, and in mid-winter, when foraging conditions are presumably poorest (Skorupa and McFarlane 1976; Wood 1983; Blue 1985; but see Porter and Labisky 1986). This suggests that extra-territorial range may be critical to survival. On the other hand, some extra-territorial range consists of portions of territories of other groups (DeLotelle *et al.* 1987), or areas devoid of foraging habitat, suggesting that extra-territorial forays may have a social function, or no adaptive function at all.

Fig. 3.3. Distribution of group sizes of Red-cockaded Woodpecker in the North Carolina Sandhills during the years 1981–6 ($n = 1335$ group-years).

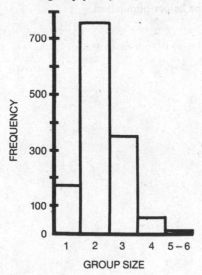

Origin of helpers

Male helpers usually, but not always, help on their natal territories. Most commonly, they are previous offspring of both breeders or of the breeding male (Fig. 3.4). Some helpers are more distant relatives of the breeding male, typically a full or half sibling. Male helpers frequently are unrelated to the breeding female, but almost never (two cases in Sandhills, one in Francis Marion) help a related female and an unrelated male, due to the pattern of territory inheritance (see below). A few (5%, Fig. 3.4) helpers are unrelated to either breeder in the Sandhills. Most of these (14 of 18 males, 6 of 6 females) immigrated from another group. The frequency of unrelated helpers is slightly higher in South Carolina (11%), but the sample size is much smaller ($n = 16$).

Of the 24 female helpers reported, 15 have been on their natal territories and eight were unrelated to either breeder. Some or all of the latter might be better viewed as competing for breeding status than as helping (see below).

Fig. 3.4. Relatedness of Red-cockaded Woodpecker helpers to young that they help to raise in the North Carolina Sandhills. Lower clear portion, helpers that are full siblings of young: lower shaded area, half-siblings: lower cross-hatched area, more distantly related than half-siblings: upper clear portion, related but of unknown degree: upper shaded area, of unknown relatedness: and upper cross-hatched area, unrelated. Sample sizes are indicated above each bar. The status (breeder or helper) of most (66–91%) males was known in 1983–6, but not 1980–2 (19–51%). The year 1985 exemplifies a year following high reproductive output, so that an unusually large number of one year old males are available to enter the helper population.

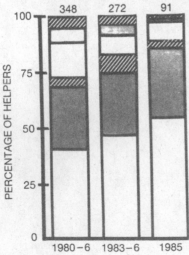

Interestingly, when female helpers help only one parent, it is as often their mother (three cases) as their father (three cases).

Acquisition of breeding status by helpers

Helpers become breeders both by inheriting territories and by dispersing to nearby territories. In the North Carolina Sandhills, about 30% of helpers per annum become breeders, roughly half by each method (Fig. 3.5). Helpers have not been observed to return once leaving a group, nor have birds been observed to revert to helping status once becoming breeders.

Floaters and solitary males

An interesting aspect of the breeding system of this species is the presence of a significant number of floaters of both sexes. Also, in North Carolina (Fig. 3.2) and elsewhere (J. A. Jackson personal communication) a significant fraction of 'groups' consist of solitary males. Jackson (personal communication) has observed males occupying a cluster alone for as long as six years in South Carolina. In North Carolina, many of these males remain in the same territory for several breeding seasons, and are paired in some years and solitary in others (Fig. 3.6; see also Walters *et al.* 1988). Sometimes males become solitary by losing their mate, whereas other solitary males are birds that occupy a cluster but fail to attract a mate.

Fig. 3.5. Transition probabilities for Red-cockaded Woodpecker helper males, measured from one breeding season to the next, based on a sample of 273 bird-years from the North Carolina Sandhills (1980–6). Mortality is not corrected for emigration from the study area. We estimate that such a correction would lower mortality rate to 18% and raise dispersal to 17%.

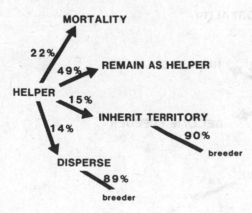

We have interpreted these lone males as potential breeders that simply lack a mate. Consistent with this, in the Sandhills, groups containing multiple males sometimes lack a female during a particular breeding season (12 cases). However, there is a gradient in site fidelity and defense between individuals that are truly solitary males and those that are truly floaters; that is, wander widely and do not defend or affiliate with a particular area. Clear cases of both exist, but those that are intermediate make any classification arbitrary, and difficult to apply. Thus, differences in methods may account for some of the disparity in frequency of solitary males between the North Carolina Sandhills and the Francis Marion Forest in South Carolina (Fig. 3.2). In North Carolina many intermediate individuals are classified as solitary males, whereas in South Carolina they are classified as floaters. However, much of the difference likely is attributable to a real difference in availability of dispersing females. In the Sandhills, solitary males are characteristic of areas and periods in which production of female fledglings is low (see below). Elsewhere, solitary males are most frequent in declining, small, low-density populations where dispersing females likely are few. Intermediate individuals may account for the mobility and high mortality of solitary males compared to breeding males (Fig. 3.6).

There may also be a gradient, in terms of relationship with group members, between floaters and immigrant helpers. Floaters sometimes

Fig. 3.6. Transition probabilities for solitary male Red-cockaded Woodpeckers, measured from one breeding season to the next, based on a sample of 131 bird-years from the North Carolina Sandhills (1980–6). Mortality is not corrected for emigration from the study area. We estimate that such a correction would lower mortality rate to 34% and raise dispersal probability to 22%.

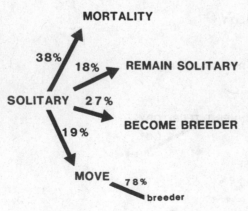

affiliate with particular groups, intruding in their territory regularly, roosting in their cluster, and even travelling with them, despite aggression from group members. If association with a group becomes consistent and aggression infrequent, such individuals might be classified as group members. Among females, such situations are rarely stable over consecutive breeding seasons (one case). This suggests that immigrants are competing for breeding status with the resident female, and should not be called helpers. Perhaps females retained on their natal territories should not be classified as helpers either, until data on female helper participation in brood care are available. We have watched only seven related and three unrelated female helpers tending fledglings. Four related, and one of the unrelated helpers, fed the young.

We observe male floaters associating with particular groups less often than we observe females doing this. Some of these associations also appear to be cases of competition for breeding status, but more often such males integrate themselves into groups and coexist with resident males for several years. These males do assist in feeding young, and therefore the term helper is appropriate. Thus a gradient between affiliated floaters and immigrant helpers is less evident with males than with females.

Dispersal behavior
Helpers
When helpers disperse, they almost invariably move only a short distance. More than 60% of the observed dispersals have been to an adjacent territory (Table 3.1), and only one has exceeded 4 km. In contrast, the median dispersal distance for fledgling males is nearly 4 km (see below).

Table 3.1. *Dispersal distances of Red-cockaded Woodpeckers* (km)[a]

Category of disperser[b]	Mean	Median	Maximum	% to adjunct territory	N[c]
Fledgling female	4.8	3.2	31.5	27	217
Breeding female	2.1	1.3	15.0	61	106
Fledgling male	5.1	3.9	21.1	31	88
Helper male	1.8	1.0	17.1	61	41
Solitary male	2.3	1.3	8.5	43	23

[a] Includes both natal and breeding dispersal.
[b] At the beginning of the one year interval, measured from one breeding season to the next, in which the dispersal occurred.
[c] All data from the North Carolina Sandhills.

These data are consistent with the idea that the strategy of helper males is to await a breeding vacancy in the vicinity of (or on) the territory in which they help.

Breeders and solitary males

In Red-cockaded Woodpeckers, male breeders are site-faithful. Paired males have been observed to desert a mate only twice in the Sandhills. We also have observed three cases in which breeding males were aggressively displaced by an immigrant male. In two cases the displaced male behaved as a floater in the vicinity of its old cluster until it perished, and the other occupied an adjacent cluster until it died. In contrast, solitary males often abandon their clusters and move to new ones (Fig. 3.6). Movements of up to 8·5 km have been recorded, although most cover only short distances (Table 3.1). Some, but not all, of these movements may be attributed to birds that were behaving more like floaters than like territorial males (see above). There are other indications that being unpaired reduces male site fidelity. Most of the movements of breeding males that we have recorded are cases in which a male lost its mate and then moved from its original cluster to another nearby (12 cases in 838 bird-years). However, some solitary males remain in the same cluster, unpaired, for years (see above).

In contrast to breeding males, breeding females regularly desert intact groups. In North Carolina, roughly 17% of adult females surviving from one breeding season to the next switch groups ($n = 912$ bird-years). Correcting for estimated dispersal out of the study area raises this figure to 19%. Most of these movements cover short distances (61% to an adjacent territory), but a few are much longer (Table 3.1). The maximum distance recorded within the North Carolina Sandhills is 15 km, but one individual moved over 90 km from a site in the Piedmont to one in the Sandhills.

Movement of breeders is not unique to Red-cockaded Woodpeckers. The rate of 'divorce', i.e. splitting of pairs surviving from one breeding season to the next, is 5%, which is similar to that observed in Florida Scrub Jays (7%, see Chapter 8). However, in Scrub Jays, divorce results from movement by both sexes, whereas in Red-cockaded Woodpeckers almost all movement (32 of 34 cases of divorce) is by females.

Females may be compelled to move by several factors. Some movements appear to be related to incest avoidance. When a breeding male dies leaving a female and her helper son occupying a territory, the male inherits breeding status and the female leaves. Such cases account for only 25% of 93 cases of female movement in the North Carolina Sandhills. If a helper unrelated to the female was present when her mate died, the female remained and paired with him (seven cases). If no helpers were present when her mate died, the female left 50% of the time ($n = 76$). Movement in this circumstance may be a response to lack of a replacement male, or may represent some form of mate choice. Females that remained paired with an immigrant replacement male.

In 34% of female movements, females abandoned their previous mate to join another group. Of these movements, 59% ($n = 32$) followed successful reproduction. There is no trend for the abandoned group to have consistently lower reproductive success or smaller group size than the group joined (Walters *et al.* 1988), suggesting that these movements do not function to improve reproduction. Perhaps conflicts within groups or with outsiders cause females to leave in such cases. The possibility that immigrants may force females from their groups was discussed above. Female movement following territory inheritance by sons is indicative of within-group reproductive competition, and there is evidence from South Carolina that this may be manifested even before the death of the breeding male (see below).

Fledglings

Fledglings appear to practice one of two life-history strategies: (1) remain on the natal territory as a helper until a breeding vacancy arises nearby, or (2) disperse to wander in search of a breeding vacancy. Nearly all females practice the latter strategy, although female helpers behaving as described in (1) occur rarely (see above). In the North Carolina Sandhills, substantial numbers of males practice each of the two strategies (Fig. 3.7). For both sexes, dispersal during the first year is spread over a considerable period. Some individuals disperse as early as July of the year in which they fledge, whereas others remain with the natal group until just before the next breeding season.

One might conclude that for males the distinction between the helper and disperser strategies is artificial, reflecting a difference in timing rather than in strategy of dispersal, were it not for the distribution of dispersal distances. Simply, birds dispersing after serving as helpers move only short distances, whereas those dispersing in their first year move much farther (Table 3.1). Furthermore, nearly all male floaters are aged one year, and they are often far from their natal territories. This suggests that male floaters are individuals adopting the disperser strategy that fail to acquire a territory by age one year. Some of these floaters acquire a territory at age two or three. Helpers have never been observed to become floaters. This suggests that helpers do not wander in search of breeding vacancies, but instead monitor neighboring groups.

Inbreeding

That nearly all females disperse precludes incestuous matings of females on their natal territories. Females rarely remain on their natal territories as breeders even if their mother disappears, and, in those few instances where they have remained and bred, they did so with unrelated males. Avoidance of incestuous matings between females and their male offspring was described above. Another circumstance in which inbreeding may occur is when relatives of opposite sex disperse to the same territory. Although individuals of both sexes sometimes disperse long distances, a large proportion of each disperses to neighboring territories. We have observed only three incestuous matings of this type. In these cases, the individuals involved (full or half siblings) had no previous contact with one another, not even as nestlings, and only one group nested. It is not yet clear whether a mechanism to avoid this type of incestuous mating exists. In any case, incestuous matings are extremely rare in Red-cockaded Woodpeckers.

Reproduction

A group of Red-cockaded Woodpeckers produces a single nest in the spring, usually located in the cavity of the breeding male. Resin wells on the nest tree are worked vigorously. Clutch size is two to four, occasionally

Fig. 3.7. Transition probabilities for Red-cockaded Woodpecker fledgling males, measured from time of fledging to the next breeding season, based on a sample of 616 individuals from the North Carolina Sandhills. Mortality is not corrected for emigration from the study area. We estimate that such a correction would lower mortality rate to 50%, and raise dispersal to 19%.

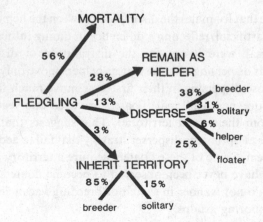

one or five, averaging just over three throughout the species' range (3.0–3.3; Ligon 1970; Beckett 1971; Carter *et al.* 1983; Lennartz *et al.* 1987). Eggs hatch 10–11 days after clutch completion, and nestlings fledge 24–9 days after hatching. Incubation may begin before the clutch is complete, and brood reduction involving the loss of the last-hatching, smallest chick(s) is common (Ligon 1971). In North Carolina, four young are sometimes fledged, and five young were fledged from one nest in 1986, but two or three fledglings are typical. This appears to be representative of the species throughout its range (Lennartz *et al.* 1987; J. A. Jackson personal communication). Although some studies report smaller broods of fledglings (but not clutch sizes) and hence more brood reduction (see, e.g. Ligon 1970), the variation involved is no more than that observed among years in the Sandhills. The average number of fledglings produced per breeding group (those containing females) has ranged from 1.2 to 1.7 over six years (*n* = 165–85), largely due to differences in brood reduction and the number of groups nesting.

Red-cockaded Woodpecker nests appear to suffer little predation. Recall the suggested role of resin wells in preventing predation, discussed above. In North Carolina at least one chick has fledged from 77.5% of the nests detected (*n* = 1117). Nesting success is slightly overestimated because some nests undoubtedly failed prior to detection, but nevertheless surely is comparable to, or even greater than, other temperate cavity nesters (Ricklefs 1969).

Even when whole-brood loss occurs, factors such as abandonment and conflicts with conspecific intruders or other cavity-nesting species are often implicated as the cause, rather than predation. In North Carolina, only 25.1% of nest failures occur after young are banded (age 4–10 days), although the intervals from laying to banding and banding to fledging are roughly equal. This pattern implicates desertion rather than predation as the primary mortality factor (Ricklefs 1969). Despite low predation rates, because of brood reduction the proportion of eggs that become fledglings is not large for a cavity nester (Ricklefs 1969). In South Carolina, over 50% of eggs produced result in fledged young, and 65% of nestling mortality occurs as partial brood loss, largely through brood reduction (Lennartz *et al.* 1987). In our Sandhills study area, 69.7% of eggs in successful nests in which clutch size was known (*n* = 1909 eggs from 580 clutches) produced young that reached banding age, whereas 91.0% of banded nestlings survived to fledging (*n* = 1331). The mortality prior to banding includes eggs that fail to hatch as well as loss of young nestlings through brood reduction. In 1986, the year for which we have the best data, 32.8% of this partial brood loss was due to inviable eggs (*n* = 122 losses).

We have detected relatively little annual variation in components of reproductive success. Survival of banded nestlings in successful nests has ranged from 84.5% to 94.5% over seven years, and survival to banding age from 64.5% to 71.9%. Nest success rates are more variable, ranging from 71.9% to 87.0% in the Sandhills, and to 92.0% at Camp LeJeune in 1986 ($n = 25$).

Re-nesting may occur if eggs or nestlings are lost, but not if fledglings are lost. Therefore only one brood is produced per year. In North Carolina, re-nesting is common in some years, with young being fledged as late as early August. In other years, there is little re-nesting and the last young are fledged by the first week of July.

The fledglings travel with the group until they disperse. Fledglings are fed regularly for an extended period, well beyond when they begin to forage for themselves. It is not unusual to see young being fed two months after fledging, and young are occasionally seen begging as late as the winter following fledging (Ligon 1970).

Age effects

Reproductive success improves dramatically with age in Red-cockaded Woodpeckers. Among females, one year old breeders fare much worse at reproduction than older females, and two-year-olds fare slightly

Fig. 3.8. Number of Red-cockaded Woodpecker fledglings produced per individual per year as a function of age, for breeding females (■), breeding males (▲), and all males (★). Males of undetermined status are excluded. Sample sizes all exceed 75 through age 3 years, range from 24 to 62 for ages 4 and 5 years, and range from 10 to 12 for age 6 years.

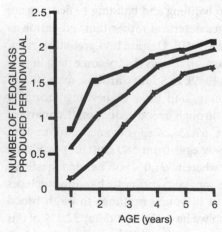

worse (Fig. 3.8). In addition, those few females that are helpers or floaters instead of breeders are nearly all one-year-olds (lowering mean annual reproductive success of one-year-olds to 0.73 fledglings per female), and the remainder are two- or three-year-olds. Reproductive success appears to increase slowly beyond age two years.

An age effect is even more apparent among males (Fig. 3.8). One-year-old breeders fare very poorly, and thereafter reproductive success improves steadily through at least age six years (Fig. 3.8). Of course, young males contribute even less to the population's reproductive output than the data in Fig. 3.8 indicate, because many males do not acquire breeding status until a late age. Only 16% of males acquire both a territory and mate at age one year, whereas 69% have both at age three years (Fig. 3.9). Younger birds are more often helpers, solitary males and floaters.

The effect of age on reproductive success is not a result of older breeders being more often assisted by helpers. Analysis of variance reveals that breeder age, presence of helpers and year all affect the number of young fledged (P. Manor, unpublished results). In fact, age and presence of helpers are not highly correlated among female breeders: 27% of one year old females are assisted by helpers ($n = 182$), and this increases only to 40% for females age four years and older ($n = 35$). Confounding of these two factors is much greater among male breeders. The corresponding figures are 3% and 45%, and the effect of helpers is only marginally significant ($P = 0.052$).

The poor performance of one year old male breeders, coupled with their unusually high mortality (see below), suggests that breeding is difficult for

Fig. 3.9. The proportion of Red-cockaded Woodpecker males that are breeders (lower shaded portion), solitary (lower clear portion), helpers (cross-hatched portion), and floaters (upper clear portion) as a function of age. Sample sizes are indicated above the bars. Males of undetermined status are excluded from the analysis.

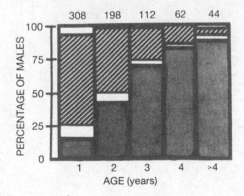

these males, but the reasons are unclear. Pairs including a one year old male often fail to nest (36%, $n = 50$), have a high rate of nest failure among pairs that do nest (34%, $n = 32$), and fledge few young from successful nests mean = 1.52, $n = 21$).

Effect of helpers

Helpers participate in many breeding activities, including territory defense, construction and maintenance of nest cavities, incubation (helpers have brood patches), brooding and feeding nestlings, and feeding and tending fledglings (Lennartz and Harlow 1979). However, their role has yet to be well quantified: the absolute contributions of helpers, their contribution relative to breeders, and the variability between individuals in their contribution are not well documented for any of the activities listed, although some data exist (Ligon 1970; Lennartz and Harlow 1979).

However, there are substantial data on the effect of helpers on reproductive success. Lennartz *et al.* (1987) found that, in South Carolina, groups with helpers fledged 0.65 more young per year than pairs unassisted by helpers ($n = 93$ group-years), due primarily to reduced loss of whole broods during the nestling stage. These data suggest that helpers reduce predation or increase resistance to disturbance rather than improve provisioning of young. Brood reduction occurred in groups with helpers as well as those lacking helpers. Our data from North Carolina are remarkably similar to those from South Carolina. From 1981 through 1985, groups with helpers produced 0.62 more fledglings per year than groups lacking helpers ($n = 1052$ group-years) (Fig. 3.10). Mean annual

Fig. 3.10. The relationship between group size and reproductive success, measured as Red-cockaded Woodpecker fledglings produced per year. Vertical bars indicate $\pm$ one standard error. Data from the North Carolina Sandhills (1980–5).

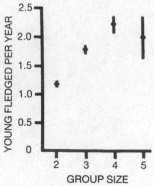

reproductive success is positively correlated with group size over the small range of group sizes (two to four) for which sufficient data are available (Fig. 3.10). One problem with these analyses is that the presence of helpers may be confounded by other factors that affect reproductive success, notably territory quality and age of breeders. In the South Carolina data, group size is positively correlated with measures of territory quality, and reproductive success also is positively correlated with territory quality (Lennartz *et al.* 1987). In addition, the ages of many breeders are unknown. In our data, helpers improve reproductive success significantly, even when effects of year and breeder age are taken into account (see above).

However, the increment attributable to helpers is less than in the unadjusted data. Comparing same-aged females, those assisted by helpers produce 0.51 more fledglings per year than unassisted females (averaging means of four age classes, $n = 35$–182 per class; averaging over all females, the value is 0.46, $n = 381$). For males, the corresponding value is only 0.28 (averaging means of four age classes, $n = 15$–69 per class), a small effect compared to that of male age. One year old breeding males are almost never assisted by helpers, and the probability of helper assistance is perfectly correlated with breeder age among the five age classes in our sample. Thus, females assisted by helpers are more likely to be paired with an older male than are unassisted females. This suggests that part of the helper effect on female reproductive success is attributable to mate's age, and that the effect observed among males is closer to the true value. However, the latter still may be subject to being confounded by territory quality. Comparing pairs breeding on the same territory with and without helpers ($n = 38$ pairs), we found no effect of helpers on number of young fledged per year (P. Manor unpublished results). In fact, these pairs did better without helpers (by 0.31 fledglings), although not significantly so, despite the fact that they were generally older in the year they were assisted by helpers than in the year they were unassisted.

Lennartz *et al.* (1987; Gowaty and Lennartz 1985) report an odd interaction between effect of helpers and status of breeding females that may relate to this last result. The presence of helpers dramatically increased the reproductive success of groups with new (to the group) females ($n = 14$), but did not improve reproductive success of groups including females that had bred previously in the same group ($n = 34$). Many of the helpers in the latter group likely were offspring of the breeding female, whereas those in the former group could not be. This is intriguing, given the potential for reproductive conflict between females and their helper sons (see above). However, in our much larger dataset in which relatedness between helpers and breeders is known, whether helpers are related to the breeding female does not influence their effect on reproductive success (P. Manor, unpublished results). Still, questions about the effect of helpers on reproduction, and its interaction with breeder age, relatedness between breeders and helpers, and territory quality, need to be investigated further.

Sex ratio of young

The Red-cockaded Woodpecker is the only cooperative breeder for which a biased sex ratio among nestlings and fledglings has been reported. Gowaty and Lennartz (1985) found that the broods they examined in the Francis Marion National Forest contained 59% males. In this species, older nestlings and fledglings can be easily sexed according to the presence (male) or absence (female) of a red crown patch. Gowaty and Lennartz (1985) relate the imbalance among young to the interaction between effects of helping and female experience within the group on reproductive success described above. They found that experienced females assisted by helpers did not produce more sons than daughters, whereas other females did. They propose that failure to produce a biased sex ratio is a tactic that favors retention of females in the face of intersexual reproductive competition between them and their helper sons. One might also view other classes of females (new females with helpers, females without helpers) as overproducing potential helpers. Accordingly, Emlen *et al.* (1986) propose that males are less costly to produce because they repay a portion of their cost through helping behavior, selecting for a biased sex ratio (see also Lessells and Avery 1987). This model is similar in approach to well-known models of sex ratio, specifically the local resource competition model, and could apply to many cooperative breeders.

An overall sex ratio bias does not occur among fledglings in North Carolina (*n* = 984 fledglings, 49.6% male). The North Carolina data have not yet been subdivided according to group composition to determine if sex ratio varies among groups in the same manner as in South Carolina.

Mortality

Data from North Carolina, although not yet sufficient to assess fully the effects of age, provide considerable information about survivorship (survival rate). The mortality pattern is one common to many cooperative breeders, and includes (1) high mortality during the first year, (2) low mortality among breeders and (3) low mortality among helpers (Table 3.2). Sex differences in mortality and high mortality rates among floaters and solitary males suggest a link between movement and increased mortality. Correcting for estimated dispersal off the study area (for method, see Walters *et al.* 1988) leaves relative rates of mortality among status classes unchanged, but the values for both fledgling classes are significantly altered (Table 3.2). Surprisingly, we have found no significant annual variation in (uncorrected) mortality rates (Walters *et al.* 1988).

The limited data available on known-age birds indicate that survivorship may be independent of age among adults generally (Table 3.3). However, mortality of one-year-olds is higher than that of older birds among breeding males ($P < 0.05$, χ^2-test). This would suggest that attempting to breed or even to hold a territory at age one year has a mortality cost for males. This may explain the poor reproductive performance of one year old males, especially their frequent failure to nest. No effect of age on mortality was evident among breeding females.

We have measured mortality only for ages one to five years. Mortality among breeding males at these ages averaged slightly less than that for all breeding males (0.18 versus 0.24), a group which presumably includes older birds among those of unknown age, suggesting that mortality may rise at some older age. This may also be true for breeding females (0.30 versus 0.32), but in this case the values are much closer. Data from older birds are needed to confirm this possibility.

Table 3.2. *Mortality rates of Red-cockaded Woodpeckers*

Status class[a]	Uncorrected annual mortality (disappearance)	Corrected for dispersal out of the study area	Sample size[b]
Females			
Fledgling	0.66	0.57	595
Breeding	0.32	0.30	912
Floater[c]	0.45	—	20
Males			
Fledgling	0.56	0.49	616
Solitary	0.38	0.34	131
Helper	0.22	0.18	273
Breeding	0.24	0.24	838
Floater[c]	0.38	—	29

[a] At the beginning of the one year interval, measured from one breeding season to the next, for which the probability of mortality is calculated.
[b] Data from the North Carolina Sandhills.
[c] Includes only those observed at the beginning of the interval, and not those inferred to be alive from subsequent data.

Population dynamics

Red-cockaded Woodpeckers appear to compete over existing territories rather than to colonize new ones. Territorial inheritance by males contributes to this, but, even when birds disperse, they join existing groups or at least occupy existing clusters of cavity trees rather than excavate cavities in unoccupied areas. Excluding our study, only one successful colonization has been observed, and this newly created cluster was abandoned after two years (Jackson 1987). All populations studied are either stable in size or declining, so perhaps colonization would be more frequent in increasing populations. Another possibility is that new groups form more often by territorial budding (Woolfenden and Fitzpatrick 1978) rather than by colonization. In North Carolina, we have observed six cases of territorial budding, but no cases of colonization. In the budding cases, a portion of a cluster was defended by a former group member (three cases) or a floater (three cases), who then attracted a mate and bred.

There are other indications that territorial budding may be a common process by which new groups form in this species. In our study area there are many abandoned clusters of cavity trees. In 25 cases, abandoned clusters were 'captured' by a group that already occupied another cluster, and thereby expanded its territory. Captured clusters also arise by one group's occupation of a second cluster that formerly housed another group (67 cases). Thirty-one captured sites were reoccupied in a subsequent year

Table 3.3. *Mortality rates of known-aged Red-cockaded Woodpeckers[a]*

	Status class[b]				
Age (years)	Breeding females	Breeding males	Helper males	Solitary males	All males[c]
1	0.33 (220)	0.27 (51)	0.21 (206)	0.47 (32)	0.26 (322)
2	0.27 (143)	0.16 (86)	0.23 (97)	0.29 (14)	0.21 (204)
3	0.28 (78)	0.14 (78)	0.31 (26)	—	0.21 (114)
4	0.30 (44)	0.14 (52)	—	—	0.18 (63)
5	0.17 (18)	0.20 (30)	—	—	0.21 (33)

[a] Data from the North Carolina Sandhills, 1980–6.
[b] At the beginning of the one year interval, measured from one breeding season to the next, for which the probability of mortality is calculated. Sample sizes in parentheses.
[c] Excludes floaters.

by birds that defended the captured cluster, splitting what for a period was one territory into two. Fifteen of these new groups have nested successfully. This suggests that budding could occur, given a territory of sufficient size and the proper distribution of cavity trees. We also have observed 22 new groups form by reoccupying abandoned clusters of cavity trees. Eleven of these groups have nested successfully. Still, our population's dynamics revolve around continued occupation of existing clusters. Of the 1191 group-years recorded (one group-year is one group observed during one breeding season), 95.3% were cases in which the group occupied a cluster that had also housed a group the previous year.

Population dynamics that are based on competition over existing territories ought to result in a population that is relatively stable in size. At first glance Red-cockaded Woodpeckers do not appear to fit this pattern because many populations studied have declined, but decline can usually be attributed to habitat deterioration. The Florida population studied by Baker (1983) is an important exception. In the North Carolina Sandhills, a population averaging 451 adults varied by only 37 individuals (8%) over five years (1982–6: Fig. 3.11), and similar stability has been observed in the Francis Marion National Forest in South Carolina (Lennartz *et al.* 1987). These observations suggest that, in good habitat, populations are remarkably stable, like those of many other cooperative breeders.

Our preliminary data suggest that those population fluctuations that occur may be related to reproductive output rather than mortality. There is

Fig. 3.11. The number of Red-cockaded Woodpecker adults (▲) and groups (●) recorded during the breeding season in the North Carolina Sandhills. Includes data only from those clusters monitored continuously since 1981. Data from additional clusters added to the sample in later years are not included in the figure.

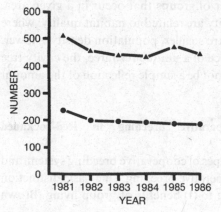

no annual variation in mortality rates among either adults or fledglings, whereas reproductive output varies greatly (see above). Furthermore, population variation can usually be related directly to prior variation in reproduction. For example, large group size in 1985 followed high reproductive output in 1984 rather than improved survival of fledglings in the winter of 1984–5. The preponderance of solitary males in one portion of the study area in 1984 followed low production of female fledglings in that same area in 1982 and 1983, rather than any increase in female mortality.

The relationship between reproduction and population size has important implications for interspecific interactions. Other woodpeckers, various other cavity-nesting birds, flying squirrels (*Glaucomys volans*), and a few other animals use cavities excavated by Red-cockaded Woodpeckers. Often active cavities are usurped by other species, and occasionally a group is thereby prevented from nesting. This has generally been called interspecific competition, but as yet no evidence that such interactions have population-level effects exists. Use of the term competition, therefore, is premature. However, it follows from the view of population dynamics presented here that interfering with nesting would affect population size, and therefore that interspecific interactions may constitute competition.

The above discussion addresses processes that determine the number of individuals in the population, not the number of groups. The number of individuals may of course affect the number of groups, for example by influencing the likelihood of territorial budding or abandonment. However, much of the variation in population size is manifested in changes in group size rather than group number. For example, the North Carolina population increased by 8% between 1984 and 1985, but the number of groups declined by 3% (Fig. 3.11). In fact, the number of groups in our population has declined continuously. Currently we can only speculate about what determines the number of groups that occur in a given area. Certainly variations in group density are related to habitat quality: where pine stands are less dense or trees are smaller, population density is lower. However, because of the importance of a unique resource, the cavity tree cluster, the number of groups may not be a simple reflection of the amount of foraging habitat (see below).

The evolution of cooperative breeding in Red-cockaded Woodpeckers

Clearly there are several types of cooperative breeding system, and several evolutionary routes to cooperative breeding. One basic distinction is between systems that evolve due to (1) benefits of group living (Brown

1982), and (2) ecological constraints that select for retention of young (Emlen 1982). Members of Red-cockaded Woodpecker groups travel and forage together, and cooperate in territory defense and raising young, so several potential benefits of group-living exist. The data necessary to determine if these benefits are sufficient to account for the evolution of cooperative breeding are not yet available. Still, ecological constraints models appear more promising for this species. Effects of group size on reproductive success do not appear sufficiently large to result in increased per capita reproductive success with increased group size (Fig. 3.10). Also, there is evidence that limited opportunities for dispersal and difficulty in reproducing at a young age, the two types of constraints identified by Emlen (1982), apply to Red-cockaded Woodpeckers. We are currently testing the dispersal model of Woolfenden and Fitzpatrick (1984) on the Sandhills population. This model is a formalization of the concept of habitat saturation.

Habitat saturation and cavity trees

There is an appealing correspondence between several aspects of the natural history of Red-cockaded Woodpeckers and the habitat saturation model. Habitat saturation refers to a condition in which all suitable breeding territories are nearly always filled, so that a young bird has a low probability of successful dispersal. In this demographic regime, there is selection for delaying dispersal until a suitable opening occurs, and helpers arise as retained young. In the case of the Red-cockaded Woodpecker, the alternative life-history strategies of the model ((1) remaining on the natal territory until a breeding vacancy arises nearby, and (2) dispersing after fledging to search for a territory) are not merely theoretical abstractions. They could in fact be derived as descriptions of behavior. It is possible, therefore, to determine directly whether the lifetime reproductive success of males practicing the retention option is sufficient, relative to those practicing the dispersal option, to account for retention of young. We are currently completing these calculations.

However, a positive result of our pending test will not be conclusive proof of habitat saturation. The demographic formulation of the habitat saturation model is simply an accounting of fitness, as defined by lifetime reproductive success, using survival and reproductive schedules under alternative life-history tracks. Other models (or general models that include all alternatives) use similar accounting, and therefore have a similar demographic formulation (Emlen 1982; Stacey 1982; Brown 1985). However, these possibilities differ in which aspects of demography are critical in creating the appropriate conditions for retention, and in which ecological factors drive the demography. To account fully for the evolution of cooperative breeding by habitat saturation, one must also explain what produced the shortage of breeding vacancies upon which the demographic model is predicated. The demographic model thus depends on an ecological model. The simple model of Koenig and Pitelka (1981) is an attempt to fill this void. They propose that a sharp decline in reproductive success with decreasing territory quality favors competing for high-quality territories over accepting available low-quality ones, leading to saturation of high-quality territories. Theirs clearly is a habitat quality model, and therefore it is ironic that habitat saturation is usually viewed as a habitat quantity argument, which it is not. The key is appropriate variation in habitat quality, not amount of habitat.

A resource whose presence or absence dramatically affects reproductive success could produce appropriate variation in habitat quality. If there is a sudden, sharp drop in reproductive success between the poorest of the territories that contain the resource and the best of those that do not, then a clear distinction between high-quality and low-quality territories, dependent on the resource, exists. One possible example is the acorn storage tree in Acorn Woodpeckers. If territories that contain a storage tree are of much higher quality than those that do not, given the effort required to construct one, it may pay to compete for territories with existing storage trees rather than accept those without them (Koenig and Pitelka 1981). Acorns stores have a large positive effect on reproductive success, and virtually everyone who has studied Acorn Woodpeckers has noted the link between cooperative breeding and acorn storage (see Chapter 14). This is especially evident in comparisons between populations in which cooperative breeding is the rule and those in which it is absent or less common (see, for example, Stacey and Bock 1978).

Similarly, both Lennartz *et al.* (1987) and Ligon (1970) postulated a link between cooperative breeding and cavity trees in Red-cockaded Woodpeckers. In terms of the habitat saturation model, cavity trees could function like acorn storage trees in producing a break in the habitat quality–reproductive success curve. Certainly an enormous investment of time and energy is necessary to produce the resource. Because Red-cockaded Woodpeckers so rarely nest other than in cavities in living pines, it is not possible to determine the effect of the resource on reproductive success, but a positive effect seems likely. Cavities in living pines are probably less vulnerable to fire than those excavated in snags, as well as less vulnerable to predation (see above).

That Red-cockaded Woodpeckers compete over existing territories rather than create new ones is consistent with the above interpretation. If this interpretation is correct, destruction of cavity tree clusters should lead to elimination of a group without replacement, which indeed seems to be the case (but see Hooper 1983). Also, one should be able to induce birds to settle in unoccupied areas by constructing cavities, a prediction we plan to test.

The proposed role of cavity trees in the evolution of cooperative breeding in Red-cockaded Woodpeckers provides an answer to a nagging question about habitat saturation, namely why territories cannot be compressed to accommodate more breeding units. A general solution is that territories are despotic (Fretwell and Lucas 1969), or that there is a non-linear relationship between territory size and the number of birds that can be supported. For example, the area required for breeding may be so much greater than that required for subsistence that only one nest is possible in an area that can support many adults. In the case of the Red-cockaded Woodpecker, one need not resort to these explanations. If cavity tree clusters define territories, the number of territories in an area may depend on the distribution of these clusters rather than bird density and foraging needs. Over longer time scales, the distribution of territories would depend on the distribution of potential clusters (i.e. sets of trees suitable for excavation), which is still a property of the environment rather than of the birds. This may explain in part why territory and home range size vary so greatly in this species.

Other nagging questions about habitat saturation and Red-cockaded Woodpeckers remain to be addressed by future research. First, how can both life history options (disperser and helper) be maintained in the population? Are the payoffs sufficiently similar that their relative value can change with dominance, population density, habitat or other factors, and do these factors determine which option a given male chooses? Alternatively, is one option being selected against, or is there frequency-dependent selection?

Second, how can the habitat be saturated for males but not females? Do the demographies of the sexes differ sufficiently to account for this difference in behaviour? Female mortality is higher than male mortality, and females have less chance to inherit territories because they always lose in reproductive competition with male relatives. Both differences lower the payoff of the helper option for females relative to males. Females do remain as helpers occasionally, and usually in situations where the opportunity for successful dispersal appears to be reduced (e.g. in isolated populations or when fledgling cohorts are unusually large). One would expect female retention to be favored occasionally if the relative payoffs of the two options for females are not too different from those for males.

Third, the presence of abandoned territories is inconsistent with the habitat saturation model (Koenig and Pitelka 1981). One might postulate that populations in which abandoned territories occur are temporarily below saturation levels; that is, that population levels become so low due to poor reproduction over a prolonged period that there are too few birds to occupy all suitabl e territories. This seems unlikely given the continued presence of a pool of helpers to replace breeders in these populations. It is more reasonable to speculate that abandoned territories are no longer of high quality. One might predict, then, that abandoned territories no longer contain suitable cavities. For example, encroachment of understory vegetation on cavities may render them unsuitable (Jackson 1987), or cavities may be destroyed by Pileated Woodpecker (*Dryocopus pileatus*) excavation (J. R. Walters, personal observation). Also, we have observed the location of a group's cluster to shift gradually over time as new cavity trees are added and old ones abandoned, so that eventually what appear to be two clusters occur in an area that only ever contained one group. Thus, some abandoned clusters may not be associated with some minimum foraging area in the vicinity of the cluster, and are therefore not usable. If such factors cannot account for all but temporarily abandoned clusters, the habitat saturation model is invalid for this species.

Finally, some may view a species with a formerly wide distribution and a limited capacity to colonize as a paradox. This is less of a dilemma that it seems at first. Territorial budding, if it occurs, provides a capacity for range expansion within what was once continuous habitat, albeit at a slow rate. It is also possible that juvenile retention evolved after the species' period of range expansion. In this case, it is no more a paradox than the presence of flightless birds on isolated islands.

A primitive cooperative system

There are important differences between the breeding system of Red-cockaded Woodpeckers and those of other cooperative breeders to which the habitat saturation model has been applied. A few individuals seem to behave as dispersers in virtually all cooperative breeders, but not as many as in Red-cockaded Woodpeckers. This suggests that, in terms of the evolutionary models, the Red-cockaded Woodpecker exhibits a 'primitive' stage of cooperative breeding. In this view, many differences between Red-cockaded Woodpeckers and other species may be interpreted as secondary adaptations in the latter. That is, these other species exhibit behaviors that are advantageous within the context of a cooperative breeding system, and evolved secondarily in that context. Egg tossing and cooperation among same-sex siblings in competing for breeding vacancies in Acorn Woodpeckers (Mumme *et al.* 1983; Hannon *et al.* 1985) are examples.

Conclusion

The Red-cockaded Woodpecker is proving to be an interesting cooperative breeder, different in some important ways from others. Much basic work remains to be done on this species, as several fundamental characteristics of the breeding system have yet to be fully described, for example helper contributions (and variation therein) to breeding activities. The species is especially well suited to population studies, due to the relative ease with which one can monitor large numbers of individuals. It is therefore a prime subject for investigating questions related to demography. On the other hand, that it is declining and endangered poses special problems. Many experimental manipulations that might be instructive would violate the statutes that protect this bird. Also, it is sometimes difficult to interpret results from populations that are declining, as these presumably are subject to selective pressures different from those in which cooperative breeding evolved. The absence of territorial budding and colonization are good examples. Is this characteristic of the species, or an artifact of its recent decline? Intensive research on this species continues. Additional data are being collected from previously studied populations, and new populations are being added to the sample. In five years many outstanding questions should be answered, and the Red-cockaded Woodpecker may emerge as the most thoroughly studied cooperative breeder of all.

This paper is dedicated to Dr Frank Pitelka, a kind mentor for too short a period, on the occasion of his seventieth birthday.

My research on Red-cockaded Woodpeckers has been done in collaboration with Dr Phillip Doerr. I thank him for inviting me to join him in this work, and for his continuing contributions to the research and ideas described here. I also thank P. Stangel and Drs R. Conner, P. Gowaty, R. Hooper, R. Lancia, M. Lennartz and L. Maguire for sharing their insights about Red-cockaded Woodpeckers with me. Research in the North Carolina Sandhills has been funded by the National Science Foundation (BSR-8307090), the US Fish and Wildlife Service via Section 6 of the Endangered Species Act of 1973 administered through the North Carolina Wildlife Resources Commission, the North Carolina Agricultural Research Service at North Carolina State University, and donations from individuals and conservation organizations. I thank private landowners, personnel from the Fort Bragg Military Reservation, and personnel from the North Carolina Wildlife Resources Commission for providing access to research areas, and assistance in the field. R. Blue, J. H. Carter III, S. Everhart, J. Harrison, M. LaBranche, J. Lape, P. Manor, R. Repasky, M. Reed, P. Robinson, T. Stamps and several undergraduate interns assisted in collecting data. I am especially grateful to Jay Carter for his tireless efforts over many years. J. Carter, M. Reed, the editors, and Drs P. Doerr, J. Jackson, R. Lancia and M. Lennartz provided helpful comments on previous drafts of the manuscript. This is paper 11299 of the Journal Series of the North Carolina Agricultural Research Service.

Bibliography

Alexander, R. D. (1974). The evolution of social behavior. *Ann. Rev. Ecol. Syst.* **5**: 325–83.

Baker, W. W. (1971). Observations on the food habits of the Red-cockaded Woodpecker. In *The Ecology and Management of the Red-cockaded Woodpecker*, ed. R. L. Thompson, pp. 100–7. Bureau of Sport Fisheries and Wildlife and Tall Timbers Research Station: Tallahassee, FL.

Baker, W. W. (1983). Decline and extirpation of a population of Red-cockaded Woodpeckers in northwest Florida. In *Proceedings of Red-cockaded Woodpecker Symposium II*, ed. D. A. Wood, pp. 44–5. Florida Game and Fresh Water Fish Commission and U.S. Fish and Wildlife Service.

Beckett, T. (1971). A summary of Red-cockaded Woodpecker observations in South Carolina. In *The Ecology and Management of the Red-cockaded Woodpecker* ed. R. L. Thompson, pp. 87–95. Bureau of Sport Fisheries and Wildlife and Tall Timbers Research Station: Tallahassee, FL.

Blue, R. J. (1985). Home range and territory of Red-cockaded Woodpeckers utilizing residential habitat in North Carolina. M.Sc. thesis, North Carolina State University. NC.

Brown, J. L. (1982). Optimal group size in territorial animals. *J. Theoret. Biol.* **95**: 793–810.

Brown, J. L. (1985). The evolution of helping behavior – an ontogenetic and comparative perspective. In *The Comparative Development of Adaptive Skills: Evolutionary Implications,* ed. E. Gollin, pp. 137–71. Erlbaum: Hillsdale, NJ.

Carter, J. H. III. (1974). Habitat utilization and population status of the Red-cockaded Woodpecker in south-central North Carolina. M. Sc. thesis, North Carolina State University.

Carter, J. H. III, Stamps, R. T. and Doerr, P. D. (1983a). Red-cockaded Woodpecker distribution in North Carolina. In *Proceedings of Red-cockaded Woodpecker Symposium II*, ed. D. A. Wood, pp. 20–3. Florida Game and Fresh Water Fish Commission and U.S. Fish and Wildlife Service.

Carter, J. H. III, Stamps, R. T. and Doerr, P. D. (1983b). Status of the Red-cockaded Woodpecker in the North Carolina Sandhills. In *Proceedings of Red-cockaded Woodpecker Symposium II*, ed. D. A. Wood, pp. 24–9. Florida Game and Fresh Water Fish Commission and U.S. Fish and Wildlife Service.

Carter, J. H. III, Walters, J. R., Everhart, S. H. and Doerr, P. D. (1989). Restrictors for red-cockaded woodpecker cavities. *Wildl. Soc. Bull.* **17**: 68–72.

Conner, R. N. and Locke, B. A. (1982). Fungi and Red-cockaded Woodpecker cavity trees. *Wilson Bull.* **94**: 64–70.

Conner, R. N. and O'Halloran, K. A. (1987). Cavity-tree selection by Red-cockaded

Woodpeckers as related to growth dynamics of southern pines. *Wilson Bull.* **99**: 398–412.

DeLotelle, R. S., Epting, R. J. and Newman, J. R. (1987). Habitat use and territory characteristics of Red-cockaded Woodpeckers in central Florida. *Wilson Bull.* **99**: 202–17.

DeLotelle, R. S., Newman, J. R. and Jerauld, A. E. (1983). Habitat use by Red-cockaded Woodpeckers in central Florida. In *Proceedings of Red-cockaded Woodpecker Symposium II*, ed. D. A. Wood, pp. 59–67. Florida Game and Fresh Water Fish Commission and U.S. Fish and Wildlife Service.

Emlen, S. T. (1982). The evolution of helping. I. An ecological constraints model. *Amer. Nat.* **119**: 29–39.

Emlen, S. T., Emlen, J. M. and Levin, S. A. (1986). Sex-ratio selection in species with helpers-at-the-nest. *Amer. Nat.* **127**: 1–8.

Everhart, S. H. (1986). Interspecific utilization of Red-cockaded Woodpecker cavities. Ph.D. thesis, North Carolina State University.

Fretwell, S. D. and Lucas, H. L., Jr (1969). On territorial behavior and other factors influencing habitat distribution in birds. I. Theoretical development. *Acta Biotheoret.* **19**: 16–36.

Gowaty, P. A. and Lennartz, M. R. (1985). Sex ratios of nestling and fledgling Red-cockaded Woodpeckers (*Picoides borealis*) favor males. *Amer. Nat.* **126**: 347–53.

Hagan, J. M. and Reed, J. M. (1988). Red color bands reduce fledging success in red-cockaded woodpeckers. *Auk* **105**: 498–503.

Hannon, S. J., Mumme, R. L., Koenig, W. D. and Pitelka, F. A. (1985). Replacement of breeders and within-group conflict in the cooperatively breeding acorn woodpecker. *Behav. Ecol. Sociobiol.* **17**: 303–12.

Hooper, R. G. (1983). Colony formation by Red-cockaded Woodpeckers: hypotheses and management implications. In *Proceedings of Red-cockaded Woodpecker Symposium II*, ed. D. A. Wood, pp. 72–7. Florida Game and Fresh Water Fish Commission and U.S. Fish and Wildlife Service.

Hooper, R. G. and Lennartz, M. R. (1981). Foraging behavior of the Red-cockaded Woodpecker in South Carolina. *Auk* **98**: 321–34.

Hooper, R. G., Niles, L. V., Harlow, R. F. and Wood, G. A. (1982). Home ranges of Red-cockaded Woodpeckers in coastal South Carolina. *Auk* **99**: 675–82.

Jackson, J. A. (1971). The evolution, taxonomy, distribution, past populations and current status of the Red-cockaded Woodpecker. In *Proceedings of Red-cockaded Woodpecker Symposium II*, D. A. Wood, pp. 4–29. Florida Game and Fresh Water Fish Commission and U.S. Fish and Wildlife Service.

Jackson, J. A. (1974). Gray rat snakes versus Red-cockaded Woodpeckers: predator–prey adaptations. *Auk* **91**: 342–7.

Jackson, J. A. (1977). Red-cockaded Woodpeckers and pine red heart disease. *Auk* **94**: 160–3.

Jackson, J. A. (1986). Biopolitics, management of federal land, and the conservation of the Red-cockaded Woodpecker. *Amer. Birds* **40**: 1162–8.

Jackson, J. A. (1987). The Red-cockaded Woodpecker. In *Audubon Wildlife Report 1987*, pp. 479–93. The National Audubon Society: New York.

Jackson, J. A., Lennartz, M. R. and Hooper, R. G. (1979). Tree age and cavity initiation by Red-cockaded Woodpeckers. *J. Forest* **77**: 102–103.

Koenig, W. D. and Pitelka, F. A. (1981). Ecological factors and kin selection in the evolution of cooperative breeding in birds. In *Natural Selection and Social Behavior*, ed. R. D. Alexander and D. W. Tinkle, pp. 261–80. Chiron Press: New York.

Lennartz, M. R. and Harlow, R. F. (1979). The role of parent and helper red-cockaded woodpeckers at the nest. *Wilson Bull.* **91**: 331–5.

Lennartz, M. R., Hooper, R. G. and Harlow, R. F. (1987). Sociality and cooperative breeding of Red-cockaded Woodpeckers, *Picoides borealis. Behav. Ecol. Sociobiol.* **20**: 77–88.

Lennartz, M. R., Knight, H. A., McClure, J. P. and Rudis, V. A. (1983). Status of Red-cockaded Woodpecker nesting habitat in the South. In *Proceedings of Red-cockaded Woodpecker Symposium II*, D. A. Wood, pp. 13–19. Florida Game and Fresh Water Fish Commission and U.S. Fish and Wildlife Service.

Lessells, C. M. and Avery, M. I. (1987). Sex-ratio selection in species with helpers at the nest: some extensions of the repayment model. *Amer. Nat.* **129**: 610–20.

Ligon, J. D. (1968). Sexual differences in foraging behavior in two species of *Dendrocopos* woodpeckers. *Auk* **85**: 203–15.

Ligon, J. D. (1970). Behavior and breeding biology of the Red-cockaded Woodpecker. *Auk* **87**: 255–78.

Ligon, J. D. (1971). Some factors influencing numbers of the Red-cockaded Woodpecker. In *The Ecology and Management of the Red-cockaded Woodpecker*, ed. R. L. Thompson, pp. 30–43. Bureau of Sport Fisheries and Wildlife and Tall Timbers Research Station: Tallahassee, FL.

Ligon, J. D., Stacey, P. B., Conner, R. N., Bock, C. E. and Adkisson, C. S. (1986). Report of the American Ornithologists' Union Committee for the Conservation of the Red-cockaded Woodpecker. *Auk* **103**: 848–55.

Mengel, R. M. and Jackson, J. A. (1977). Geographic variation in the Red-cockaded Woodpecker. *Condor* **79**: 349–55.

Mumme, R. L., Koenig, W. D. and Pitelka, F. A. (1983). Reproductive competition in the communal Acorn Woodpecker: sisters destroy each other's eggs. *Nature (Lond.)* **306**: 583–4.

Porter, M. L. and Labisky, R. F. (1986). Home range and foraging habitat of Red-cockaded Woodpeckers in northern Florida. *J. Wildl. Manag.* **50**: 239–47.

Ramey, P. (1980). Seasonal, sexual, and geographic variation in the foraging ecology of the Red-cockaded Woodpecker (*Picoides borealis*). M. Sc. thesis, Mississippi State University.

Reed, J. M., Doerr, P. D. and Walters, J. R. (1986). Determining minimum population sizes for birds and mammals. *Wildl. Soc. Bull.* **14**: 255–61.

Reed, J. M., Carter, J. H. III, Walters, J. R. and Doerr, P. D. (1988*a*). Indices of red-cockaded woodpecker populations: an independent test of the circular scale technique and a new index. *Wildl. Soc. Bull.* **16**: 406–10.

Reed, J. M., Doerr, P. D. and Walters, J. R. (1988*b*). Minimum viable population size of the Red-cockaded Woodpecker. *J. Wildl. Manag.* **52**: 385–91.

Repasky, R. R. (1984). Utilization of home range and foraging substrates by Red-cockaded Woodpeckers. M. Sc. thesis, North Carolina State University.

Repasky, R. R., Stamps, R. T., Carter, J. H. III, Blue, R. J. and Doerr, P. D. (1983). A computerized filing system for Red-cockaded Woodpecker cavity tree data. In *Proceedings of Red-cockaded Woodpecker Symposium II*, ed. D. A. Wood, pp. 111–12. Florida Game and Fresh Water Fish Commission and U.S. Fish and Wildlife Service.

Ricklefs, R. E. (1969). An analysis of nesting mortality in birds. *Smithson. Contrib. Zool.* no. 9.

Rothstein, S. I. (1980). Reciprocal altruism and kin selection are not clearly separable phenomena. *J. Theoret. Biol.* **87**: 255–61.

Short, L. L. (1982). *Woodpeckers of the World*. Delaware Museum of Natural History: Greenville, DE.

Skorupa, J. P. (1979). Foraging ecology of the Red-cockaded Woodpecker in South Carolina. M. Sc. thesis, University of California at Davis.

Skorupa, J. P. and McFarlane, R. W. (1976). Seasonal variation in foraging territory of Red-cockaded Woodpeckers. *Wilson Bull.* **88**: 662–5.

Stacey, P. B. (1982). Female promiscuity and male reproductive success in social birds and mammals. *Amer.* Nat. **120**: 51–64.

Stacey, P. B. and Bock, C. E. (1978). Social plasticity in the Acorn Woodpecker. *Science* **202**: 1297–1300.

Stamps, R. T., Carter, J. H. III, Sharpe, T. L., Doerr, P. D. and Lantz, N.J. (1983). Effects of prescribed burning on Red-cockaded Woodpecker colonies during the breeding season in North Carolina. In *Proceedings of Red-cockaded Woodpecker Symposium II*, ed. D. A. Wood, pp. 78–70. Florida Game and Fresh Water Fish Commission and U.S. Fish and Wildlife Service.

US Fish and Wildlife Service (1985). *Red-cockaded Woodpecker Recovery Plan.* U.S. Fish and Wildlife Service: Atlanta, GA.

Van Balen, J. B. and Doerr, P. D. (1978). The relationship of understory to Red-cockaded Woodpecker activity. *Proc. Ann. Conf. Southeast. Assoc. Fish and Wildl. Agencies* **32**: 82–92.

Walters, J. R. (1989). The Red-cockaded Woodpecker. In *Endangered, Threatened and Rare Fauna of North Carolina.* Part 3 *Birds*, ed. D. S. Lee and J. F. Parnell, Occasional Papers of the North Carolina Biological Survey (in press).

Walters, J. R., Doerr, P. D. and Carter, J. H. III (1988). The cooperative breeding system of the Red-cockaded Woodpecker. *Ethology* **78**: 275–305.

Walters, J. R., Hansen, S. K., Carter, J. H. III, Manor, P. D. and Blue, R. J. (1988). Long-distance dispersal of an adult Red-cockaded Woodpecker. *Wilson Bull.* **100**: 494–6.

Wood, D. A. (1983). Foraging and colony habitat characteristics of the Red-cockaded Woodpecker in Oklahoma. In *Proceedings of Red-cockaded Woodpecker Symposium II*, ed. D. A. Wood. pp. 51–8. Florida Game and Fresh Water Fish Commission and U.S. Fish and Wildlife Service.

Woolfenden, G. E. and Fitzpatrick, J. W. (1978). The inheritance of territory in group-breeding birds. *Bioscience* **28**: 104–8.

Woolfenden, G. E. and Fitzpatrick, J. W. (1984). *The Florida Scrub Jay.* Princeton University Press: Princeton, NJ.

4 ARABIAN BABBLERS: THE QUEST FOR SOCIAL STATUS IN A COOPERATIVE BREEDER

4

Arabian Babblers: the quest for social status in a cooperative breeder

A. ZAHAVI

The bird and the study area

The Arabian Babbler (*Turdoides squamiceps*) is a member of the. Paleotropic family Timallidae. Only a few forms of the genus *Turdoides* have penetrated into the Palaearctic arid zone north of the tropics (Meinertzhagen 1954). These Palaearctic forms are distributed over hot deserts from India to Morocco and south to the arid deserts of East Africa. *T. squamiceps* occurs in the Arabian and Sinai peninsulas, extending into the hot deserts of Israel. In Israel it is common along the Rift Valley, north to Jerico; it is also found west of the rift in several of the large wadis.

The Arabian Babbler weighs 65–85 g; it is about 280 mm long, of which over half (145–55 mm) is tail. Its color is a dull gray-brown and very cryptic against the desert background. Young birds have dull gray irises that change in color during the first year to pale yellow in most males and dark brown in females. Babblers are quite terrestrial and hop and walk more than they fly. When they fly, they do so slowly and consequently are vulnerable to predation when they cross open terrain. This may be the reason why they usually stay near bushes and trees, where they may escape into cover, with their strong legs and long tail enabling them to dodge and outmaneuver any predator. They usually fly near the ground, low-flying birds being more difficult to locate and also a more difficult target for a swift potential predator.

The diet of the Arabian Babbler consists mainly of small animals, mostly arthropods, found on the ground or within the vegetation. They also spend much of their time digging for food in the ground and under the bark of trees, as well as preying on small lizards, geckoes and snakes. They have been observed to kill small song birds but are unable to eat them as their blunt beak is not efficient at cutting up prey. They feed on berries (e.g.

Ochradenus, *Lycium* and *Nitraria*), when these are available, nectar of *Loranthus* and sometimes flowers (primarily Compositae). They also feed at garbage dumps, when these are present in their territories.

Our study site extends over 25 km² in the Rift Valley around Hatzeva Field Study Centre, 30 km south of the Dead Sea (Fig. 4.1). The study was initiated in 1971, when we color-banded 125 babblers, in 20 groups living at the site, as well as birds in the surrounding area. These original 20 groups have been followed since 1971. Since 1974, we have been feeding the babblers with bits of bread, and by 1976 we had succeeded in taming most individuals in the focal study groups, which allowed us to move freely among them and observe their behavior patterns at close range. We do not, however, feed them while recording their behavior, unless such feeding is a part of an experiment.

Because part of the study area borders on farms, about half of the 20 groups have access to supplementary food resources and also suffer from man-induced mortality from pesticides and automobiles. However, five to six groups that inhabit territories distant from these farming areas (the southern groups), three to four other groups living to the north of the study area and some 20 untamed groups outside the study area and also away from farming, provided us with a population of individuals living in undisturbed areas for comparison with groups influenced by human settlements.

Spacing system

Groups of babblers defend year-round territories, the boundaries of which may be crossed occasionally while foraging, but are respected immediately upon the appearance of territorial owners. The available habitat is filled with territories throughout most years. Non-territorial birds, about 5–10% of the population, are mostly birds that were chased away from their territories. These non-territorial birds are continuously chased by residents whenever they are detected and are referred to here as 'refugees' because of their origin and behavior. They do not breed, but some may survive as refugees for more than a year and eventually succeed in joining a breeding group to breed or to form, with other refugees, a new resident group on a vacant territory. Because our study area is disturbed by agriculture, we do not have a good indication of the number of new territories which may form in natural areas. In the southern area inhabited by the five undisturbed groups, which we have observed for 16 years, we saw only one case in which a group was eliminated and one case in which a new group tried to breed (but was unsuccessful and disbanded). Apart from

Fig. 4.1. Arabian Babbler study area at Hatzeva Study Centre, Israel.

Table 4.1. *Life history of the group of Arabian Babblers from 1972 to 1987*

Babbler	Sex	Y72	Y73	Y74	Y75	Y76	Y77	Y78	Y79	Y80	Y81	Y82	Y83	Y84	Y85	Y86	Y87
aatz	M	*	*	* *	* X												
mtmt	M	*	*	* *	* *	*	*	*	*	*	*	*	*	*	* D		
actz	M		= = = X														
cctz	M		= = =	= * * X													
aztz	F		+ + + X														
tvcc	F		= = =	= = = = X													
catz	F		= >															
avtz	F		= >															
zhvt	M	N	= = =	= = = =	* * * X												
zvvt	F	N	= = = X															
cact	F			J + X														
ahzt	M			J *	* * X													
atva	M			N = = =	= = =	= =	*	* X									
htmc	M			N = = =	= = =	= =	*	* X									
mtmz	M			N = = =	= = =	= =	=	= =	= X								
tamh	F				JX													
tltz	F				J + + X													
halt	F				J + + X													
mmtm	F				J +	+ +	+ +	+ +	+ +	+ +	+ +	+ +	+ +	+ + >			
zmht	F					NX												
cltz	L					NX												
hlct	L					N =	= =	= =	= =	= =	= = > B						
zzat	M								N =	= =	= = = > B						
hlst	M								N =	= =	= = = > B						
tsch	M								N =	= =	= = = =	= = =	= = =	= > B			
tsmv	M									N =	= = = X						
aatz	M									N =	= = = X						
ctlc	M									N =	= = X							
hhtc	F									N =	= = = =	= = =	= = =	= = > B			
ztaz	M										NX							
ltac	L										NX							
thll	L										NX							
tmlm	L										NX							
tzmz	L										NX							

atsc	F	N===========*******P
sath	M	N=============**X
svtl	L	N=X
zsvt	M	N=========>B
msvt	M	NX
stla	F	N==========>
vtls	F	NX
shlt	M	N============>B
vstm	M	N============>B
ltvl	F	N============>B
thzc	L	N============>B
vthl	L	N============>B
ctmc	L	N=======>
lztv	L	N==N=====>
vzct	L	N==N=====
vvtl	L	N==N=====
llmt	F	N=N====>B
zmct	L	NX
vmct	F	N=X
tvls	F	J++>B
ztxx	F	N====X
avtc	M	N==========***
tbzl	F	N==========P
mthv	M	M=========***P
lvtl	L	NX
cllt	M	N==========**P
vtlv	F	J+++++++P
ztc	L	N=X
lltv	M	N====P

= = =, present as non breeder; * * * *, present as breeding male; + + + +, present as breeding female.

First column gives the babblers name. M, male; F, female; L, fledgling; N, nestling or fledgling; J, joined the group as a nestling or fledgling; B, succeeded in gaining a breeding slot after dispersal; >, dispersed; P, present in parental territory; X, disappeared from the group – most probably died; >, dispersed but probably died later.

these, we have observed about 20 cases in which new groups were formed in the disturbed area, a few of which succeeded in breeding. Most vacancies for new groups occur following periods of drought, as babblers do not breed during dry years. After the dry years of 1977 and 1978, the number of groups declined to 15.

Large territories may be over 1 km², or stretch over several kilometers along a dry river bed, while small territories, still suitable for the production of young, may be as small as 0.2 km². The differences are due partly to the large variation in vegetation cover and partly to the variation in productivity of different parts of the study area. Territories located on shallow sand or in areas where ground water is near the surface provide good conditions for vegetation, and thus more food for the babblers. However, food may not be the only factor determining territory size. For example, we have observed two instances in which groups annexed the territories of neighbors and successfully held them against the refugees. Barren areas, hills or plains between wadis, which are open to predators, often serve as permanent territorial boundaries; in many cases these boundaries are the same today as they were in 1971.

Social system

Babbler groups occupy the same territories over many years. In 1987, three of the five territories in the southern area were held by the same male line that occupied them in 1971. The other two territories were taken over by males originating from neighboring groups. A male babbler (rarely a female) may spend all its life and breed within its parental territory. Hence, successful groups may pass their territories through the male line over many generations. Table 4.1 presents the history of a group which has survived since 1971 within the same territory.

Groups range in size from 2 to 22 individuals and vary considerably in their membership. Until 1981 there was no group with more than 12 birds. Table 4.2 provides information on the size and composition of breeding groups in 1986. Emigration and immigration are uncommon (Table 4.1). Individuals may stay in the same group for years, often for their entire lifetime, although they may disperse from their natal groups to breed. Other individuals are forced to leave because of intragroup fights or when neighbors or refugees displace them from their group. Birds that disperse on their own occasionally return following an unsuccessful attempt to join another group as a breeder.

Groups that breed successfully grow into large groups that are able to defend their territories over many years. Such 'stable' groups provide a good base for the survival of their offspring, to the age of two to six years, and consequently for successful dispersal. In contrast, small groups of two to four individuals are continuously threatened by their neighbors, or by non-territorial groups. Even stable groups sometimes pass through a phase in which they are reduced in size and are vulnerable to intruders (see, for example, Table 4.1, 1979). Offspring of groups that lose their territories have fewer options for finding a breeding slot in the population than offspring of resident groups.

The dispersal pattern is strongly influenced by incest taboo. Babblers do not mate with group members born within their group while they were there. We observed only two cases in which the incest taboo was broken: one was a four-year old son that mated with his mother, more than a year after the death of his father; the other was a brother–sister pair, and also constituted the only observed case in which two nests occurred simultaneously within a territory. The lack of sexual interest within the taboo group is not a matter of dominance by the breeders, as birds that previously showed no interest in copulation may copulate within minutes after an alien bird joins the group. Because of male philopatry and the

Table 4.2. *Group composition of Arabian Babblers in 1986*

Name	Size	Breeders	Helpers	Fledglings
bms	13	2	4	7
bar	16	2	6	8
k30	6	2	2	2
pol	13	2	6	5
mtr	6	2	2	2
stv	10	2	5	3
zeh	9	2	5	2
bre	4	2	0	2
sth	4	2	0	2
mzr	17	3	3	11
bts	15	3	6	6
tmr	5	3	1	1
tok	14	4	4	6
bok	11	3	4	4
zmg	5	4	0	1
zom	5	3	0	2
sva	7	4	2	1
bbs	9	5	0	4
meh	6	4	2	0

Breeders are adult birds which have a potential breeding partner.
Helpers are one-year-old birds or adults without a potential mate.
Fledglings are birds three to twelve months old.

incest taboo, females, that are subordinate to males, must disperse in order to breed. The only opportunity for a female to breed within her natal territory is when an unrelated male, or males, take over the breeding role in the group.

Babblers are only accepted into a resident breeding group if a reproductive vacancy exists, but alien fledglings may occasionally be accepted. New groups form on vacant territories or in marginal areas. They may consist of two to eight individuals with all possible sex combinations. A newly forming group may continue to accept new members until it succeeds in producing offspring (five cases).

Babblers maintain a strict dominance hierarchy. Older birds always dominate younger birds as long as they remain members of the same group. Breeding males dominate breeding females. Siblings of the same brood fight with one another to establish dominance within several days of fledging (Carlisle and Zahavi 1986). Fights among older group members of the same sex always result in the expulsion of the loser from the group. Brothers usually succeed in attaining, with time, a higher rank than their older sisters. Once females are more than one year old, they tend to avoid intragroup aggression. Hence, we often do not know if all males of a certain age dominate their sisters. Most females disperse from their parental group at the age of one to five years, to breed in alien groups. All males dominate all alien females that join the group.

The rank of young babblers may be easily determined by observation of aggressive acts. Several displays by young Babblers, such as the flip callnote, allopreening on the back and allofeeding, are performed in a way that reflects the rank of the interacting individuals, e.g. the flip call is often used to greet an approaching dominant. A subordinate male never feeds a more dominant male; a dominant male is more likely to allopreen a subordinate on the back. However, rank among adult males is difficult to determine through aggressive interactions (Gaston 1977), but they display their rank by the same displays used by young babblers. Babblers attain the status of the alpha breeders in accord with their social rank. Subordinate males are either excluded from breeding or, if they do breed, they copulate less with the breeding females.

Breeding system
About 50% of the groups are 'simple groups' composed of a pair of breeding birds and their offspring (Table 4.2). Babblers mature when they are two years old, but very few of them succeed in finding a breeding slot at that age.

The loss of a breeding female is followed by the acceptance of one or more new females into the group. Alien females often appear in groups of two to eight individuals, and often create a breeding slot by killing or chasing away a breeding female and usually also chasing away all other females of the group. Subsequently, all adult male babblers have the option to participate in breeding and up to three have been observed to copulate with a breeding female ('complex groups'). Two, and sometimes up to four, of the females joining such a group may lay eggs in the same nest and succeed in fledging offspring. The success of these joint clutches is not greater per nest, and is usually less, than simple clutches. A clutch is referred to as a joint clutch when we observe more than one female laying in the nest, when two eggs appear on the same day or when there are more than five eggs at the nest and more than one potential breeding female. I refer to a male as a breeding male when that male is an adult bird and alien to the breeding female. In most cases, these males have been observed copulating with the females (see below).

As a rule, multiple clutches are laid only until the first brood of the female coalition is raised. Subsequently, one of these females, usually the former dominant female, becomes the sole breeding female in the group. Other females are either chased from the group or remain as non-breeding helpers. The alpha male usually copulates much more frequently than do the other males and most probably at the best time for fertilizing the eggs (Zahavi 1989). Subordinate males sometimes do not copulate at all.

A multi-male group, in which more than one male copulates with the breeding female, may persist over several broods and years, until gradually one of the males drives away or kills the other males and it once again reverts to being a simple group of a pair of breeders and their offspring. Complex groups may also develop when a group of males chases the male line from a territory. These males may then breed with the female of the former group and its adult daughters. Eggs are often broken in joint clutches with several breeding females and sometimes in nests attended by several breeding males. In at least three instances we observed the babblers taking an egg out of the nest. We also have indirect evidence that the babblers destroyed their eggs, but in most cases, it is not possible to distinguish the destruction of nests by babblers from destruction by other predators.

Simple groups may split up when birds of either the male or female lineage find an opportunity to breed. The highest ranking son is always among the dispersing males, while the second son, the third in the hierarchy, usually stays in the natal territory. The third son tends to disperse and the fourth to stay. A similar situation was described by Ligon and Ligon (Chapter 2) for Green Woodhoopoes.

Females also disperse in groups. However, as females have much less opportunity to breed within their natal territory, it is not surprising that when an opportunity does arise, sometimes all mature daughters (often two to eight) leave together ($n = 16$; $\bar{x} = 3.5$). In contrast, the largest group of males to disperse consisted of three individuals ($n = 14$; $\bar{x} = 2.1$). Males and females that are not, for whatever reason, satisfied with their new group after a split, sometimes return to their parental group (10 cases observed).

Birds, both dominant and subordinate, chased from their group, form the substantial population of refugees discussed earlier. Such individuals continue to contest breeding opportunities in established groups and search for vacant territories. Five breeding males (two of them over nine years old) that were chased from their groups succeeded in breeding elsewhere in the population. Females are much less successful in having a second chance to breed after they have bred as an alpha in a territory: I only observed one case during the study.

In rare circumstances, fledglings are 'adopted' by a group, either by themselves (three cases) or along with their elders (about six cases). In two cases, fledglings which dispersed with their elders eventually succeeded in attaining a breeding status in the group that adopted them.

Females disperse further than males and only rarely inherit breeding status in their parental groups (Table 4.3). However, most babblers still find

Table 4.3. *Dispersal of breeding Arabian Babblers in and out of the southern groups*

Distance	Males	Females
0	9	1
1	15	19
2	5	5
3	0	1
6	0	2
?	1	2

Distance is measured in territories crossed by dispersing birds: 0 distance, the bird stays at home; 1, in neighboring territory; ?, birds of unknown origin.
Data for the five southern groups over 16 years for birds which settled as breeders irrespective of their success.

their mates among the territories which border on their natal group. This may sometimes result in a high degree of relatedness among babblers in adjacent groups. One male supplanted his brother and bred with one of his nieces. The displaced male succeeded in breeding in an adjacent territory and his sons in that territory moved and bred with his daughters (their half-sisters) that had remained in the former territory. On the other hand, babblers that attend to a fledgling that is not related to them avoid breeding with it (several observations). All group members older than a few months help at the nest. It is not possible to distinguish breeders from non-breeders from their 'helping' behavior. However, non-breeders are usually excluded from the nest in its early stages until the clutch is completed.

Helping at the nest reflects the social rank and status of the helper (Carlisle and Zahavi 1986). The mere presence of a dominant at the nest may inhibit the subordinate from coming to the nest or even from landing near it (Fig 4.2(a) and (b)). Babblers compete in helping, and invest in preventing other group members from helping. Most of this competition is done without aggression. However, subtle intention movements of the dominant, such as walking towards the subordinate or around it, allopreening on the back, or allofeeding, replace threat displays. Babblers tend to interfere more with the help provided by babblers one rank below them than with helpers that are lower than that in the hierarchy.

Competition to copulate among male breeders is also mostly done without aggression. The dominant male, apparently, inhibits the attempts of the subordinate males by its mere presence near the female. Competition at that time among males is displayed through sentinel activity and allofeeding. Babblers do not copulate if another babbler is within sight, even if that babbler is a fledgling some 20 m away. Consequently, the male must lead the female away from the group in order to copulate. I interpret this 'shyness' of babblers to copulate in the presence of other babblers as a demand by the female to force the courting male to display its ability to dominate its competitors. The reasoning for this is as follows: it is more difficult to inhibit a group member from following than it is to control interference from a nearby rival. So, being able to stop the group from following the courting pair displays a higher level of status than the ability to intimidate a rival located within pecking distance.

If a babbler persists in following the courting male, the latter usually approaches the follower and expresses its disapproval by vocalizations and displays, including allopreening on the back or walking around the follower. This usually stops the follower. My preliminary observations suggest that a male that disturbs the breeding pair by persistently following

116   *A. Zahavi*

Fig. 4.2. Relationship of nest-visiting rates to dominance (α through
ω) for immature Arabian Babbler helpers. Visits by individual helpers
are plotted as the percentage of the sample mean for each set of
yearling and juvenile helpers visiting a particular nest. The points for
each set of helpers are uniformly distributed along the x-axis between
the value for the most dominant (α) individual, plotted on the left of
both graphs, and that for the most subordinate (ω) individual, plotted
on the right. The various sets of points on either graph are not
directly comparable in a statistical sense. Lines plotted on the graphs
are therefore not true regression lines, but are intended merely to
emphasize trends which were confirmed by statistical analysis (see the
text). (a) Visits to attended nests. This graph summarizes 187
observations from three nests visited by three to nine yearling helpers.
In all observations, the incoming helper relaced another bird on the
nest. The solid indicates that rate of nest visits declined with
decreasing social rank of the helper. (b) Visits to unattended nests.
This graph summarizes 598 observations from four nests visited by
three to nine yearling helpers. Incoming helpers came to nests which
were not attended by another bird in their group. There was no
apparent relationship between rate of nest-visits and the helper's rank.
(From Carlisle and Zahavi 1986.)

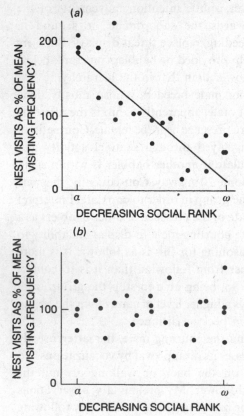

it is more likely to copulate with the female during the fertile days. In several cases the breeding female actively sought out the follower in order to copulate with him. Apparently, a female cannot be forced to copulate; she merely needs to vocalize, as is often the case, in order to attract the attention of the group and stop any potential suitor. Even if a dominant finds a subordinate courting the breeding female, a resolution generally occurs without overt aggression and often by allofeeding and allopreening. In most cases, its aggression is turned towards the female rather than the male.

Basic demography
Eggs are laid usually from February to July; up to three broods (four on one occasion) may be produced by a single female within a season. Clutch sizes range from three eggs, common at the beginning and end of the breeding season, up to five eggs, occurring only at the height of the season. Joint clutches, however, included up to 13 eggs. The largest number of fledglings produced from any nest was six (one clutch only). One of the fledglings died immediately after it fledged.

Joint clutches of two or more females do not fledge more young than do nests with a single female. Consequently, the fitness of females that nest jointly is lower than single-nesting females. The fitness consequences of joint nesting are decreased because breeding attempts are often aborted and eggs are missing or broken as a result of the conflict among the breeding females.

Annual variation in reproductive success is the result of several factors. Primary among these is rainfall; in dry years, which occur relatively frequently (1973, 1977, 1978, 1984, 1987), there is little food and no breeding, except for groups with access to garbage dumps and gardens. Under more natural conditions, groups may breed even in dry years only if they have access to lush vegetation around a spring.

Predation, especially by snakes and less frequently by shrikes, appears to be a major source of nest destruction. However, it is difficult to quantify its importance because of destruction by babblers (see p. 113). Such intragroup nest destruction is known to occur in complex groups and in groups containing several males.

On days prior to laying, the breeding female often appears to lead the group during the morning into a confrontation with a neighboring group. I believe that these fights provide her with information concerning the potential of her group to defend the brood. These fights and other intrusions by larger groups into territories of small neighboring groups interfere with the breeding of the small groups in various ways, including delayed breeding destruction of nests, and also in the killing of fledglings during intergroup fights (observed three times). Adults are usually not killed under these circumstances. Such killing happens when one adult attempts to supplant another in the territory. We also recorded the loss of whole broods of grown fledglings that we suspect were killed by rival groups. Fledglings of large stable groups that do not suffer from interference survive better than fledglings of small groups that border on strong neighbors. Such evidence suggests that at least some losses are attributable to intergroup conflict rather than to predation.

The relationship between group size and breeding success is complex (Zahavi 1974; Fig. 4.3 and 4.4). In some years, small groups of two to five birds are less successful than larger groups (Fig. 4.3) while in other years, small groups breed as well as the larger ones (Fig. 4.4). In all years, groups of 5–12 individuals do not differ in their reproductive success, I believe that it is unlikely that helping at the nest in this species has evolved to increase the reproductive success of the group (see below). The lower success of small groups is probably a combination of several factors: first, as described earlier, small groups experience greater interference from neighbors than do large groups. Second, small groups are often new and consist of individuals living on territories of marginal quality. Thirdly, small new groups are often composed of individuals that have not yet sorted out the terms for their cooperation, i.e. the degree of compromise on reproduction. They still do not agree on how they will share in the reproduction.

Fig. 4.3. Breeding success in relation to group size for Arabian Babblers in 1981. Data values are the number of groups for each size and number of young fledged.

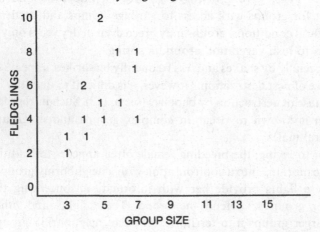

There is high mortality among fledglings and young birds during their first few months of life (Fig. 4.5) Subsequent mortality is much lower. We do not yet know whether the death rate is the same for young and old birds or how it varies with the sex and status of the individual concerned.

Some birds, both males and females, bred successfully for nine years and were still breeding when 12 or 13 years old; the oldest babbler in the study was 14 years old. Breeders that live a long time not only fledge many

Fig. 4.4. Breeding success in relation to group size in Arabian Babblers in 1982. Data values as for Fig. 4.3.

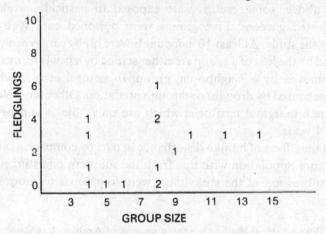

Fig. 4.5. Survival of 103 babbler fledglings born during 1981 to July 1987.

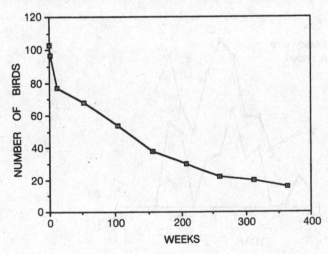

offspring but succeed in helping to displace the breeding birds in neighboring territories. One male that bred for 12 years, nine of which were good breeding years, succeeded in fledging 20 birds to the age of one year; 10 of these eventually succeeded in attaining breeding status (Table 4.1).

The number of territories in the study area fluctuated between 16 and 24, while the population size varied from 65 to 220 individuals. The low point in the population occurred in 1979, before the breeding season, after two consecutive years in which there was no breeding. The high point occurred in 1982, after the breeding season, following a string of four good years (Fig. 4.6).

As discussed above, some groups were exposed to pesticides while feeding in cultivated gardens. Five groups were poisoned entirely by pesticides during the study. At least 10 individuals were hit by cars. Vacant territories created by the loss of a group are either settled by a newly formed group or are annexed by a neighboring group. In natural areas, such instability may be caused by drought or through predation. Other unstable areas include small marginal territories which are only able to support breeding in good years.

We can assess the effects of human disturbance in part by comparing the data from the entire population with that from the southern undisturbed groups. In this subsection of the study there were four to six territories,

Fig. 4.6. Population in the five southern groups of Arabian Babblers. Top line is total population at the end of the year. Bottom line is number of birds ringed that year.

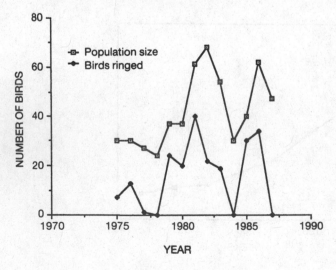

including one which has been observed only since 1979, and the population fluctuated between 20 and 75 individuals. These demographics are similar to those shown for the population as a whole and suggest that the overall effect of human disturbance was not significant. Only a single newly formed group was observed in the southern area. It successfully fledged young, but the fledglings disappeared following a fight with a large neighboring group. There were more new breeding groups which were formed within the disturbed areas (15 groups). These areas had vacant habitat and a rich supply of food.

Some special activities
I describe below some of the more peculiar social behaviors of babblers.

(1) *Play*. Babblers play in a way similar to that of young mammals. As far as I know, the only other bird species in which such elaborate play has been described is the Australian magpie, *Gymnorhina tibicen*, another cooperatively breeding species (Pellis 1981). Babblers mock fight, rolling on the ground like animal cubs. They mock chase, trying to replace one another from particular locations, and playtug twigs with one another (Posis 1984).

(2) *Night roost*. Babblers clump with one another at their night roosts, usually lining up along one branch. Roosting trees are traditional and are used over long periods sometimes exceeding several years. However, nocturnal disturbances such as predation may cause frequent changes in roost location. The alpha male always roosts on the edge of the clump, with the beta male, or less frequently with the gamma male, generally on the other side. In groups with only a single adult male, the breeding female usually roosts on the opposite side. The clumping at the nocturnal roost is often accompanied by special vocalizations, gaping and a special display of neck quivering.

(3) *The morning dance*. This is a special social ritual that usually occurs before sunrise. It takes place on the ground about 50–100 cm from a bush, and is initiated by one group member, usually a subordinate. Sometimes it occurs on a branch inside a bush. The dancers, which may include the entire group, clump, changing during the dance back and forth from a row into a tight ball. They then attempt repeatedly to go into the middle of the row or the ball, sometimes continuing the dance for up to 30 min.

(4) *The water dance*. When babblers have access to open water they drink and bathe. Wet birds then clump and perform a ritual which is similar in many ways to the morning dance. This behavior is unlikely to be an adaptation for getting dry, as the dance probably delays drying and even dry birds that have not bathed may join the dance once it has begun.

(5) *Clumping*. Babblers, like many other species of group-living birds, tend to clump with one another in a variety of circumstances; when roosting at night, while resting during the day, when confronting other groups, when trying to appease other group members, when dancing, playing, and while allopreening. In all, birds may sometimes spend several hours a day, especially during the hot hours of the day, clumped with one another.

(6) *Allopreening*. Babblers may allopreen over long periods, especially during the hot hours of the day or before or after roosting. Allopreening has a social significance and many of the details of the pattern of allopreening are correlated to the social rank of participants (Gaston 1977).

(7) *Egg coloration*. Arabian babbler eggs are glossy green, without any spots. This absence of spots, which is relatively common in social rather than non-social jays (Brown 1963), may also be a social adaptation. I interpret the lack of spotting in the eggs of this species as an adaptation that makes egg recognition difficult and thus enables several females to collaborate in laying at the same nest without 'cheating' one another. I believe that the emerald green, glossy color of eggs, which probably makes them more conspicuous to predators, may be a strategy to extract more care for the vulnerable 'showy' eggs from other breeders in the group. A similar suggestion has been made for the tinamou (Bohl 1970).

(8) *Sentinel activity*. All group members act as sentinels. An individual may perch as a sentinel for more than 30 min at one time and for more than an hour a day. I regard sentinel activity as a display, since it reveals the rank and the status of the babbler (see below). The alpha male perches as a sentinel more than other males and is often the first sentinel of the day. Babblers compete to act as sentinels. Subordinates usually refrain from replacing a dominant, but if they do so they perch lower or on a more distant tree. Dominant males, on the other hand, may replace the subordinate on the same perch and often relieve him with allofeeding. When two babblers chance to ascend as sentinels at the same time, the subordinate usually descends within 1–2 min (Zahavi *et al.* unpublished results).

(9) *Allofeeding*. Adult babblers feed one another. This activity may be regarded as a display, since it reflects the social rank and status of the bird. A subordinate adult male never feeds a dominant. The same is generally true for females. Although females are subordinate to males, a female courting a male may attempt to feed it. The male then often attacks the female. The same may happen among first-year birds. A dominant that is fed by a subordinate usually responds with aggression. The alpha male often actively searches to feed the beta male, especially during the breeding season when both compete to copulate.

Speculations
Why live in groups?
Arabian Babblers may first breed when two years old, lay two to three clutches per year, and live for 12–14 years. In most years, all suitable habitat for survival is occupied by breeding groups. There is a constant surplus of adult birds ready to occupy vacancies, and thus young birds have little chance of establishing themselves as breeders. Under these circumstances, parents that let their offspring remain with them on their natal territory for several years as non-breeding helpers, while they wait for an opportunity to find a vacancy, increase their fitness over parents that force their offspring to disperse (Zahavi 1976; Koenig and Pitelka 1981). However, there are other species with a surplus of non-breeding adults that do not form breeding cooperations. Hence the saturated habitat is not sufficient to explain why a particular species forms groups. In the following, I discuss additional factors which may enhance or retard the formation of groups.

Babblers do occasionally form new breeding groups with unrelated individuals. Why do small groups in this case not accept new members to help in defending their territory against their large neighbors and thus ensure more successful reproduction? I believe the answer lies in the high cost of group-living. By joining with additional birds, babblers invite reproductive rivals to live with them. Not only may such individuals be forced to share in future breeding attempts but they would most likely suffer an increased risk of being chased away from the territory or being killed by the additional group member.

An adult babbler living with both its parents cannot reproduce owing to the incest taboo. Some of these offspring are six years old and still do not breed. Offspring may usurp territories once one of the parents dies. We have witnessed five cases in which an offspring killed its parent of the same sex, following the demise of its other parent. We have not, as yet, ever observed offspring replacing a parent when both parents are alive. We have observed one member of a pair helping its mate to fight a rival from another group. It is possible that an offspring may find it difficult to fight both its parents and thus have to wait until one of them dies before it attempts a coup. Fighting between an offspring and its parent may be a very quick event. There are no early warning signs until the offspring starts the fight, which may be over in less than an hour. The defeated parent, if not killed, usually escapes into a mouse hole or thick vegetation. It may return after it recovers and may succeed in regaining its former status. Once such a fight begins, rivals never compromise. The loser is either killed or chased away from the territory.

Species no doubt differ in the advantages gained from retaining offspring at home. One important factor making such retention more likely is if offspring are better able to search for, and succeed in filling, vacancies from the safety of their natal territory. Babblers that stay within their natal group may use the help of their group to establish themselves as breeders elsewhere, as has also been described in the Florida Scrub Jay (Woolfenden & Fitzpatrick 1984, and Chapter 8). Hence, living with parents and brothers increases the chances of a babbler acquiring a breeding territory. More than half of the breeding babblers breed in a territory adjacent to their natal territory (Table 4.3). A second factor is the benefit to parents of retaining their offspring. In babblers, parents gain considerable benefits from having at least some of their offspring remain, as they may lose their territory to intruders if all their offspring disperse. We observed four cases in which the breeding male lost its territory when he remained the only adult male of the group.

The cost to parents of having offspring remain in the natal territory may be reduced when territory size is not determined by food for adult birds. Territories of Arabian Babblers support large groups, even when food shortage does not permit breeding. A possibly similar reduction in the cost of group life occurs in other cooperatively breeding species, in which a nesting hole or storage place for acorns, rather than the carrying capacity of the territory *per se*, determines the extent of the territory.

I do not suggest that a saturated habitat, territories with surplus food and a reasonable gain in the defense of the territory from living in groups are the only conditions that may select for cooperative breeding. Any reasonable gain from collaboration, such as acquiring mates, defense against predators, and so on, may do so. Collaborations among breeders may form if reproduction is worse without them. However, a slight gain in reproductive success as a consequence of collaboration may be difficult to ensure, given the high risks involved with group living.

Why help if a babbler may stay and not help?

Although babblers benefit from staying within their natal territories, this does not necessarily force them to help their group. Dominants never force subordinates to provide any service to the group; on the contrary, breeders interfere with the attempts of helpers to help (Figs. 4.2(*a*) and (*b*)). Babblers do not exploit the tendency of other babblers to act as altruists. No model of group selection, kin selection, or reciprocity can explain why babblers do not exploit the tendency of others in the groups to help, and certainly does not explain the reason for investing in hindering other babblers from acting as altruists.

Woolfenden & Fitzpatrick (1984) suggested that helping at the nest has been selected for because large groups more effectively aid their offspring in acquiring breeding status. However, this model for the evolution of group living may not be stable; a bird that does not help but rather lets others do so, benefits more than other members of its group, as already suggested by Maynard-Smith (1964) and Trivers (1971). The potential to evolve such social parasitism was one of the main reasons for rejecting the theory of group selection. The fact that a helper may benefit from helping its kin does not decrease the likelihood that social parasites will evolve to exploit their kin and, therefore, kin selection may not explain the evolution of helping behavior.

Reciprocity may also be rejected as an explanation for altruism in babblers. If a helper helps because of an expected benefit in the future, there are two possibilities. Either the benefit it (or its offspring) will obtain is in the interest of the benefactor, in which case the benefactor will help regardless of whether it was helped earlier or not, or, alternatively, the help from the expected reciprocation act would not benefit the benefactor, in which case it should simply not reciprocate. Our observations in no way suggest that there is any mechanism or expectation that a babbler will reciprocate help given to it or its parents at an earlier time. For example, a dominant babbler acting as a sentinel may display aggression toward a subordinate that comes to replace it (reciprocate), or a dominant (among first-year birds) may hit a subordinate which feeds it.

Dominant babblers help at the nest more than do subordinates of the same age (Carlisle and Zahavi 1986). They also invest more in other 'altruistic' activities such as sentinel behavior (Zahavi *et al.*, unpublished results). The pattern we observe in these altruistic adaptations suggests that babblers compete in order to act as altruists and that they compete especially with individuals which immediately follow them in rank.

I interpret all the apparently 'altruistic' activities of babblers as selfish adaptations for displaying social status (Zahavi 1976, 1977a, b, 1981). This hypothesis requires neither reciprocity, which we do not observe in babblers, nor relatedness among group members. By social status I do not simply imply social rank. The rank of each individual is well known to all group members, and usually stays the same throughout the life of the individual within a particular group. Hence there is no point investing in advertising it. Social status as defined here, however, may change from one day to the next according to the performance of the individual.

The nature of social status may be explained by comparing differences among various male coalitions. We have observed hundreds of copulations (unpublished data). In one group, the alpha male may have all the copulations, while in another it shares a high proportion of copulations with the beta male. The former case is often associated with an old bird (usually a father and son), while the latter case may be of two males, often siblings, that do not differ greatly in age. In neither case is there ambiguity as to who dominates whom. None the less, babblers that are closer in age to one another dare to interfere with each other more than those where there is a large age difference.

I suggest that the difference between the two combinations of male babblers is a consequence of difference in status. Social status may be considered as the degree of dominance (or control) a higher-ranking individual has over a subordinate. In this sense, social status is the ability of one individual to control a second, as it is perceived by the second individual. Hence, it is reasonable to assume that babblers will pay attention to actions correlated with the 'claim' of other group members to their social status in order to assess their own options to share in reproduction. It is also reasonable to assume that babblers would invest considerable time and energy and risk their lives in advertising their claim for social status.

Our experience with babblers suggests that they are very attentive to the actions of other babblers. A babbler that feeds another advertises his approach by means of a special vocalization which attracts the attention of other group members. I suggest that these additional birds come to observe the interaction and to learn the details of its pattern for the same reason that they would gather around an aggressive interaction. Thus, the pattern of allofeeding reflects the social status of the interacting birds. A subordinate female often rejects food offered by a male, after which the dominant male may show overt aggression toward the subordinate female (Carlisle and Zahavi 1986). I believe that the reason for this is that acceptance of food *decreases* the status of the recipient. Younger subordinate birds are more likely to beg and to accept the food when the dominant offers food than a subordinate bird, similar in age to the dominant, which may reject the offer. Our evidence suggests that babblers that reject insects offered by a dominant bird are, in fact, hungry at the time, since such individuals will eagerly take a dry piece of bread offered to them by investigators. The subordinate bird is, apparently, prepared to be attacked by the dominant rather than accept food which lowers its social status.

I suggest that altruistic activities such as helping at the nest, sentinel activity, allofeeding, and risks undertaken to defend the territory or in mobbing a predator contribute directly to the status of the performer and are, therefore, *not* altruistic. Instead, like a peacock's tail, they function as an advertisement (Zahavi 1977a). Their effect on the recipients is a consequence of their value in advertising social status.

This model, which is based on simple individual selection, explains all details of the patterns of 'altruistic activities' exhibited by Arabian Babblers. I further speculate that this model applies to a wide variety of species, including man. For example, the Kwakiutl North-west Coast Indians have the potlatch custom to 'shame' their enemies by giving them presents (Benedict 1946).

Altruism is not the only way to acquire status. Aggression and waste may be two additional means by which individuals may raise their social status. I suggest that 'altruism' has evolved in babblers, as well as other cooperative breeders, for two reasons. First, adult babblers display no aggression within the group, and thus may have no mechanism to threaten one another within their group. The reason for this reduced aggression is that within a cooperative society such aggression is more costly than towards outsiders. Any display of aggression which is not successful is witnessed by all group members, and any injury or weakness may be exploited by rivals that are constantly present within the group, waiting for an opportunity to change their rank. The second reason for the apparent altruism in cooperative societies is related to the importance of attracting collaborators. Threats and fights are not appropriate strategies for attracting individuals to cooperate. Alternatively, an individual investing in the welfare of its group is more likely to attract other individuals as collaborators. Its ability to invest also functions as a measure of its potential as a good-quality bird deserving high status.

Group-living animals require collaborators. Since individuals may stay to collaborate or may abandon the group, dominants must advertise their advantages as collaborators. Advertising for collaborators in group-living is essentially not different from advertising for a mate. To sum up: the advertisement of the qualities of the dominant may deter rival groups, attract mates, and convince male collaborators to stay with a dominant who will be a good defender of the common territory. Advertisement may also function to inform collaborators about its willingness to share in reproduction, while deterring subordinates from challenging its rank.

It may be argued that babblers refrain from attacking collaborators because wounding a collaborator is bad. However, this suggestion is based on what I believe to be an unstable model. If the dominant is programmed not to fight with a subordinate (so that the group will not weaken), some individuals could exploit this pattern to their own selfish advantage. The same argument is true for the investment in a group member. These suggestions are based on a subtle group selection argument.

The evolution of social mechanisms designed to invest in other individuals is not unique to group breeders, but exists in all social activities. Individuals opting to have a partner for social cooperation should be able to display honestly that they deserve what they wish to obtain from the partner, and that they are going to let that partner receive his share from the collaboration. I believe that a great deal of social display, including much that is often interpreted as altruistic, has evolved in order to yield this information.

Testing the bond

Babblers and many other animals display several social behaviors believed to strengthen the social bond. Among them are allopreening (analogous to allogrooming in mammals), clumping, and social dancing. All these activities share one thing in common: they impose a cost on the social partner to the activity on top of the cost to the initiator (Zahavi 1977a). The theory that these signals reinforce the social bond originates from the observation that they take place only among group members. This theory does not explain why these activities should impose such a cost on the individuals performing them. When a babbler clumps with a group member it restricts its options to move. This is particularly apparent during the morning dance, which occurs only at dawn. Indeed, early morning would seem to be the absolute worst time to perform such a display, as it is the best time to feed and the time when, because of poor light, birds may be most vulnerable to predation. I suggest that babblers dance in the morning, and generally in the open away from cover, because it is *difficult* to dance at that time; indeed, I believe it is this difficulty which provdes the very reason to dance in the first place.

There is no reason why a display should strengthen a social bond unless that display provides information, not available earlier, concerning the desire of the individuals involved to bond with one another. Furthermore, I suggest that such information should be tested by the partners, because important decisions may result from that information. It is likely that members of social units that are willing to sustain a cost imposed on them by other group members do so because they have a desire to be group members (Zahavi 1977b). It is this readiness to sustain a cost which may distinguish between an individual that is ready to invest in the social bond and one that is not. Following this logic, an activity which tests the social bond should impose a burden on the individual participants. The social bond itself may be strengthened by the cost imposed on participants, as by the very act of participating they honestly reassure one another concerning their motivation to bond.

These activities do not strengthen the bond between individuals that do not intend to bond. Such individuals do not participate in the costly displays and, consequently, are eliminated from the social group.

Cooperative breeding and social behavior

I plan to continue the study of Arabian Babblers for as long as possible. There is no end to the kinds of question that arise as a result of watching these fascinating birds. Although at the start of the study I was intrigued by the problems which seemed to me, at the time, to be unique to cooperative breeders, I now realize that analogous behavior is exhibited by non-cooperative breeders as well. The interpretation of behavioral mechanisms among the more 'normal' breeders is already loaded with interpretations. It is easier to examine the basis of social behavior by looking at a system which is comparatively unknown, one in which behavioral mechanisms are expressed in different dimensions, than those occurring in more conventional breeding systems.

My ideas about testing the social bond, as well as those concerning the interpretation of mobbing or warning calls (Zahavi 1977b, 1981), came from watching babblers. There has not yet been a year in which the babblers have not presented me with interesting problems. Watching them is a continuous intellectual exercise. They are tame and easy to observe in their open habitat. The social bond formed between these long-lived birds and the observer makes their study a delight.

I thank the Society for the Protection of Nature in Israel which helped the study by providing living space and various other facilities at the field study center of Hatzeva. Grants from the Israel Academy of Science and the Binational Science

Fund provided financial support for a number of years. The manuscript and many of the ideas presented in it are consequences of discussions with my wife (Dr A. Kadman-Zahavi). The manuscript benefited in its presentation by Mrs V. Ely, Dr R. Payne, Dr W. Koenig and Dr P. Stacey. Thanks are also due to the many students and volunteers who participated in following the babblers, taming them, and in the collection of the data.

Bibliography

Benedict, R. (1946). *Patterns of culture*. Mentor Books: New York.

Bohl, W. H. (1970). A study of the crested tinamou of Argentina. *U.S. Dept. Int. Bur. Sport Fish. Wildl. Spec. Sci. Rep. Wildl.* **131**.

Brown, J. L. (1963). Social organization and behavior of the Mexican Jay. Condor **65**: 126–53.

Carlisle, R. T. and Zahavi, A. (1986). Helping at the nest, allofeeding and social status in immature Arabian Babblers. *Behav. Ecol. Sociobiol.* **18**: 339–51.

Gaston, A. J. (1977). Social behaviour within groups of Jungle Babblers. *Anim. Behav.* **15**: 828–48.

Koenig, W. D. and Pitelka, F. A. (1981). Ecological factors and kin selection in the evolution of cooperative breeding in birds. In *Natural Selection and Social Behavior:Recent Research and New Theory*, ed. R. D. Alexander and D. W. Tinkle, pp. 261–80. Chiron Press: New York.

Ligon, D. J. and Ligon, S. H. (1982). The cooperative breeding behavior of the Green Woodhoopoe. *Sci. Amer.* **247**: 126–34.

Maynard-Smith, J. (1964). Group selection and kin selection. *Nature (Lond.)* **201**: 1145–7.

Meinertzhagen, J. (1954). *Birds of Arabia*. Oliver & Boyd: London.

Pellis, S. M. (1981). Exploration and play in the behavioural development of the Australian Magpie *Gymnorhina tibicen*. *Bird Behav.* **3**: 37–49.

Posis, O. (1984). Play in babblers. M.Sc. thesis. Tel-Aviv University.

Trivers, R. (1971). The evolution of reciprocal altruism. *Quart. Rev. Biol.* **46**: 35–57.

Woolfenden, E. G. and Fitzpatrick J. (1984). *The Florida Scrub Jay: Demography of a Cooperative Breeding Bird*. Princeton University Press: Princeton, NJ.

Zahavi, A. (1974). Communal nesting by the Arabian Babbler: a case of individual selection. *Ibis* **116**: 84–7.

Zahavi, A. (1976). Co-operative nesting in Eurasian birds. *Proc. 16th Int. Ornith. Congr.* 685–93.

Zahavi, A. (1977a). Reliability in communication systems and the evolution of altruism. In *Evolutionary Ecology*, ed. B. Stonehouse and C. Perrins, pp. 253–259. University Park Press: Baltimore, MD.

Zahavi, A. (1977b). The testing of a bond. *Anim. Behav.* **25**: 246–7.

Zahavi, A. (1981). Natural selection, sexual selection and the selection of signals. *Proc. 2nd Intern. Congr. Syst. Evol.*: 133–8.

Zahavi, A. (1989). Mate guarding in Arabian Babbler, a group-living songbird. *Proc. 19th Intern. Ornith. Congr.*: 420–7.

5. HOATZINS: COOPERATIVE BREEDING IN A FOLIVOROUS NEOTROPICAL BIRD

5

Hoatzins: cooperative breeding in a folivorous neotropical bird

S. D. STRAHL* AND A. SCHMITZ†

Study species

The Hoatzin (*Opisthocomus hoazin*) is a relatively large bird (700–900 g, 70 cm overall length) which inhabits riparian forests and swamps in the Amazon and Orinoco river basins (Figure 5.1). As a folivorous, semi-ruminant, cooperatively breeding bird, the Hoatzin is among the most unusual of avian species. The swimming and climbing abilities and functional wing claws of the young are well known among ornithologists (e.g. Beebe 1909). The latter have remained the most recognized 'primitive' feature of the bird, prompting early speculation on the reptilian affinities of *Opisthocomus*. One author (Penard 1908, p. 309) went so far as to suggest that the Hoatzin 'formed the transition between the birds and the creeping animals', though numerous avian groups possess wing claws at some stage of development (Heilmann 1927). In no other group are these claws as well developed as in young Hoatzins, nor are they used for climbing or predator escape in other species.

The earliest taxonomic references to the Hoatzin are by Hernandez (1651), who presented an accurate anatomical description but confused its habitats with those of other species, erroneously stating that:

> The bird subsists upon snakes. It has a powerful voice which resembles a howling or wailing sound. It is heard in the autumn and held inauspicious by the natives. The bones of this bird relieve the pain of wounds in any part of the human body; the odor of the plumage restores hope to those who, from any disease, are steadily

* Author to whom correspondence should be sent at Universidad Simon Bolivar, Dpto. Biologia de Organismos, Apartado 89000, Caracas 1086-A, Venezuela.

† Present address: Department of Wildlife and Range Science, 118 Newins-Ziegler Hall, University of Florida, Gainesville, FL 32611, USA.

wasting away. The ashes of the feathers when devoured relieve the Gallic sickness, acting in a wonderful manner...

More accurate descriptions of the Hoatzin were not available until the nineteenth century, when the species became a subject of great interest as a possible 'missing link' between birds and reptiles.

The taxonomic status of *Opisthocomus* has remained a subject of controversy over the past several centuries. During this period, it has been placed into four orders by various authors, including the monotypic Opisthocomiformes, Galliformes (as originally described, *Phasianus hoazin*, by Muller 1776), Gruiformes, and Cuculiformes (for a review, see Sibley and Ahlquist 1973). None of these classifications is universally accepted by systematists, and the true status of the species is a bone of contention among many. The most recent work on egg-white protein and

Fig. 5.1. A distributional map of *Opisthocomus hoazin* in South America.

DNA hybridization show affinities between it and the Crotophaginae (Sibley *et al.* 1988), as does the scleral ossicle work of de Queiroz and Good (1988).

The diet and digestive physiology of the Hoatzin also represent aberrant life history traits. It is one of the world's few obligate avian folivores, and is certainly the only species in this category that can fly. In Venezuela, we have observed Hoatzins eating the leaves, flowers and fruits of over 50 species of plants, although 10–12 make up 90% of the diet on an annual basis. Roughly 82% of the annual diet is green leaves, while flowers comprise 10% and fruits 8%. Both fruits and flowers are eaten mainly on an opportunistic basis, which depends on seasonal phenology of the food plants.

The Hoatzin's anatomical adaptations to leaf-eating are unique, and include an enlarged muscular crop, a reduced sternal carina, and an epidermal callosity on the tip of the sternum, on which the bird rests its weight when roosting. The crop, along with the esophagus anterior to the proventriculus and gizzard, constitutes over 70% of the weight of the entire digestive tract when empty and over 25% of the overall weight of the bird when full. This organ acts much like the rumen of a cow in the fermentation of the leaves on which the Hoatzin feeds, and has therefore replaced the proventriculus and gizzard as the primary site of digestion (Strahl *et al.*, unpublished work; see also Gadow 1891).

The digestive physiology of the Hoatzin is the subject of an ongoing study by A. Grajal. Needless to say, the energetics of this species seem to be quite different from those of other birds, and the ecological and evolutionary constraints that folivory places on *Opisthocomus* affect all phases of its life history. In particular, the slow growth rate which a low energy food imposes on young birds has subjected the Hoatzin to heavy mortality in the nestling/fledgling periods. The escape behavior of the young, along with their wing claws, has probably evolved to compensate for this.

Study area

The work described here is part of a long-term investigation into the behavioral ecology, diet and social system of the Hoatzin in the central plains *llanos*) of Venezuela. The study site for this research is Fundo Pecuario Masaguaral, a 10 000 ha cattle ranch bordering the Guárico river, at 67° 35' W, 8° 34' N, at an altitude of roughly 65 m above sea level (see also Chapter 6). The vegetation on Masaguaral ranges from open Copernicia palm savanna with isolated woodland and 'islands' in the west to thick gallery forest in the east along the Guárico river and Caracol Creek (Fig. 5.2). Much of the savanna regions and portions of the forest are flooded during the wet season. Hoatzin populations were studied along the Caracol Creek and an overgrown Drainage Canal in the eastern sections of the ranch, and in an open swamp (Headquarters (HQ) population) in the west. Fig. 5.3(a)–(c) depicts these typical Hoatzin-infested habitats.

The climate of the Venezuelan *llanos* is strongly seasonal, with a six-month rainy season (May–October), a four-month dry period (December–March), and two transition months. Average annual rainfall over the past 25 years was between 1400 and 1500 mm (range 890–2300 mm). Average monthly rainfall patterns for the period 1982–6 are shown in Fig. 5.4. Virtually no rain falls during December–March, and much of the study site drys out completely, which is a dramatic change from the seasonal flooding during the wet season.

Spacing and social system

Although the Hoatzin is found throughout much of the Amazon and Orinoco river drainages in South America, early field research was confined almost entirely to Guyanan populations along the Berbice and Abary rivers (Quelch 1888, 1890; Beebe 1909; Grimmer 1962). Despite

Fig. 5.2. A map of Hato Masaguaral showing 1988 boundaries. Watercourses and swamps are indicated as by solid lines as follows: 1, Headquarters (HQ) area; 2, Drainage Canal region; 3, Caracol Creek.

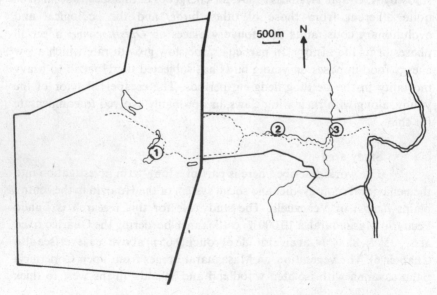

several less extensive studies in Venezuela and Brazil, quantitative data on the diet, social system and habitat preferences of *Opisthocomus* were virtually non-existent until recently.

Beebe (1909) suggested that the Hoatzin nested colonially in Guyana and Venezuela, while 50 years later, at the same site in Guyana, Grimmer (1962) observed cooperative breeding. Sick (in Thompson 1964) described Hoatzins as being promiscuous or polygamous in their social habits, but did not discuss breeding dispersion. McClure (1921), however, found the birds to be monogamous in Guyana.

Out study has documented a seasonally territorial, cooperative breeding system in the Hoatzin in the *llanos* of Venezuela (Strahl 1985, 1988). Here, the birds defend small, linear territories along watercourses and the flooded forest margins of swamps. Territory size, as measured along the length of defended watercourse, averaged 43.2 m, with a range 10–95 m. Fig. 5.5 shows the typical dispersion of Hoatzin territories in the wet season along a section of the Caracol Creek. Hoatzin territories are all-purpose, and all diurnal activities take place within them. Daily use of surface area is usually confined to a 1000–3000 m^2 region. Due to the small size of the territories and the birds' affinity for water, it is relatively easy to monitor the size and reproductive success of many units in a given year.

Hoatzin territories are usually along a watercourse, and one or more boundaries on the side away from the water may not border on other social units. The birds are partially nocturnal, and during moonlit nights sometimes move outside of the diurnal activity area in the direction of an undefended boundary to forage. These nocturnal bouts can take them up to several hundred meters into the interior of the forest.

We have also documented changes in territoriality linked with seasonal desiccation of certain areas during the dry season (Schmitz 1987; Schmitz and Strahl 1989). In this period, some social units migrate up to 2 km from their wet season territories. These birds depart from their territories at the onset of the dry season, moving along streams and flooded areas over the course of a week or more to areas which remain wet during the dry season. At these sites, migrating social units often join to form flocks of 25–50 (maximum 125–150) individuals. Members of each social unit associate closely with each other within these large dry season flocks.

The presence of permanent water and green leaves appear to be the proximate environmental factor influencing these migrations. In seasonally dry areas, most trees shed their leaves during the period between January and April. As a 'ruminant' leaf-eater, the Hoatzin requires fresh green leaves and a constant water supply. Migrations to sections of the creek

Fig. 5.3. Typical Hoatzin habitat in the (a) Headquarters (HQ), (b) Drainage Canal and (c) Caracol Creek habitats.

from the seasonally dry drainage canal are in response to the lack of suitable forage and water in the latter area during the January–April period. The dry period is extremely stressful for Hoatzins in terms of the effects of drought on the quality and quantity of available food, and its length and severity greatly influence the survival of adults and young.

In areas of permanent water, many Hoatzin units retain their territories

throughout the year, even in the presence of elevated intruder pressure from neighboring flocks of migrant birds. Other resident units in these areas are 'swamped' by the heavy influx of migrant birds during the early dry season. These units become part of the dry season flocks, and re-establish their territories at the start of the subsequent rainy season when the migrant birds depart. Year-round territorial units are usually those that have high annual reproductive success associated with predator-free island habitat, indicating a long-term cost-benefit relationship between this and dry-season territoriality. They adopt the high cost of territoriality in the face of high intruder pressure perhaps due to the tremendous benefits of nesting on islands.

Social unit size and composition

Hoatzin social units on our study site range in size from two to eight at the onset of the breeding season (Fig. 5.6). Two is the most common unit size, and units larger than five are relatively rare (< 5%). Mean social unit size is 2.99 (S.E.M. = ±0.1, n = 590 unit-years). Annual means have varied between 2.54 and 3.22 over the past six years (n = 57–117 annual unit-years), and is strongly linked with survivorship (survival rate) during

Fig. 5.4. Mean monthly rainfall at the study site for the period 1982–7. Vertical bars indicate one standard error above and below the mean value.

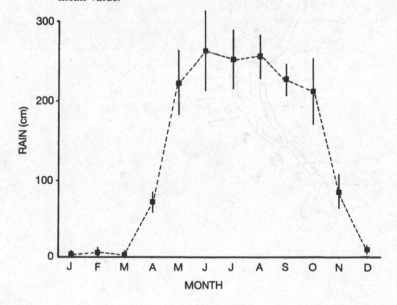

Fig. 5.5. Distribution of Hoatzin territories along the lower Caracol Creek in 1987. Dashed lines indicate territorial boundaries, and numbers represent unit sizes.

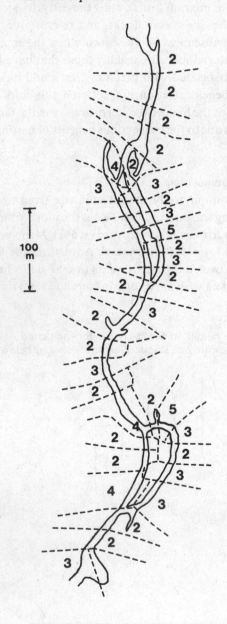

the dry season. Hoatzin social units are usually composed of a monogamous breeding pair of adults (see below) and a variable number (up to six) of non-breeding 'helpers at the nest' (*sensu* Skutch 1961; Brown 1975).

Helpers engage in all reproductive activities except fertilization and laying of eggs (Strahl 1985). Non-breeders vary in the amount of help that they perform. While some individuals are more active than the breeding adults in territorial defense and aid provided to young, others provide little or no aid and only rarely participate in interactions. Male helpers are more active than females. When helpers are present, they significantly reduce the amount of effort provided by breeding females. Immigrant non-breeders usually help very little, but more often in territorial defense than in rearing young.

A total of 329 individuals has been banded from 1981 to 1986. Young are captured by shaking them from the trees prior to attaining full flight capabilities (45–55 days post-hatching). Only 22 adult birds have been captured or banded during the study. For this reason, many study flocks are composed of unbanded adult birds. Reliable identifications are made on units under study from blinds using patterns of wear on the white terminal bands of the tail feathers and the presence of immature tail feathers. These identifications are updated periodically and allow individual unit members to be followed through an entire season. Sex is determined behaviorally from copulations, which are frequent as part of Hoatzin territorial displays.

Fig. 5.6. A frequency histogram of Hoatzin unit sizes for 590 combined unit-years during the 1981–7 period.

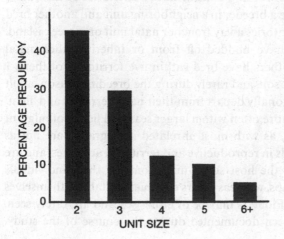

The composition of Hoatzin social units is stable during the breeding season. Units occupy the same territories from year to year, and even those that migrate during the dry season return to within a few meters of their previous year's boundaries. In the early wet season prior to the initiation of the first nest, new territories occasionally 'bud' (*sensu* Woolfenden and Fitzpatrick 1978) off existing ones.

Non-breeder unit members are usually offspring retained from a previous year ($n = 262$ of 297 helper-years), though unrelated immigrants were sometimes found in this category. On rare occasions a non-breeder dispersed with a sibling from a previous year in a 'budding' incident ($n = 7$). The known sex ratio among banded non-breeders was 45:26 (males:females), which differs significantly from 50:50 (binomial test: $P < 0.05$). Assuming that this sex ratio reflects that of the overall non-breeding population, and that a third of the Masaguaral Hoatzin population was composed of non-breeders (Fig. 5.6), the sex ratio of the population was roughly 57:43 (male:females).

Of yearling Hoatzins, 89% remain in their natal territory as non-breeders ($n = 171$ banded birds). Over half of second-year birds and a third of third-year birds stay in their natal units, but by the fourth and fifth years these proportions drop to 20% and $< 10\%$, respectively. By the fifth year, over 90% of individuals still living have attained breeding status.

Female Hoatzins tend to disperse sooner and over larger distances than males. No females have been found among non-breeders of known sex which remained in their natal territory for more than two years ($n = 18$). Of sexed birds that dispersed in their first two years, 88% were females. Females also disperse further than males. Only two females have established themselves in breeding units within 10 territories of their natal unit. One of these became a breeder in a neighboring unit and another bred in her third year four territories away from her natal unit on a creek island. By contrast, 13 males have budded off from or inherited their natal territories, and several others have bred within five territories of them.

During the late dry season, and rarely during the breeding season, adult non-breeders may occasionally depart from their natal territory and 'float' between other units or more often within larger seasonal flocks of migrant individuals. These birds, as with most unrelated immigrants, are not as active as other individuals in reproductive and territorial activities, and are not always accepted by the host social unit. Some of these individuals return to their natal groups, whereas others eventually establish themselves elsewhere or disappear. Floaters may be of either sex and are rarely seen. Only 11 floaters have been documented during the course of the study.

However, given the short periods during which they depart from their natal unit and their lack of tenure in the groups they visit, we feel that floaters may be more common than our observations can detect. Floating prior to the breeding season may be a mechanism by which individuals assess opportunities for acquiring a territory and mate. The significance of these 'floaters' will be discussed later.

Enforced dispersal has also been documented of birds whose natal units have gone extinct due to the death of one or both of the breeders or pressure from neighboring groups. These 'refugee' individuals are usually found as floaters, often for extended periods. Refugees have been documented as dispersing up to 4 km along the water courses, and as with floaters, refugees have difficulties establishing themselves as breeders.

Breeding system

Hoatzins exhibit a wide array of breeding systems. They are normally monogamous, but can rarely be polygynous (see below). Of all social units, 45% were composed of a pair without helpers, while 48% are single monogamous pairs with helpers. Seven per cent of Hoatzin units have more than one breeding pair, which may nest both jointly (in the same nest) and separately (in different nests within the same territory), even within the same breeding season.

In Peru, Torres (1987) has partially documented coloniality for *Opisthocomus*, wherein a non-territorial group of 28 individuals nested colonially in a single tree. We have not observed colonial nesting in our study population, although other authors have also suggested it for hoatzins in Guyana.

Unlike most birds, monogamy in the Hoatzin cannot be judged by the presence or absence of extra-pair copulatory events, as copulations also fulfill a display function in territorial defense in this species (Strahl 1985, 1988). Non-breeders (especially females) sometimes participate in territorial display copulations, but never in reproductive copulations. There is a marked difference between these two. Display copulations are shorter in duration, often include enhanced aggressive post-copulatory displays are directed towards (or illicited by) neighbouring groups, and are usually incomplete (i.e. they lack certain components of full copulations and often do not include cloacal contact). Male non-breeders usually elicit highly aggressive responses from breeding males when attempting display copulations. Display copulations take place throughout the year. Reproductive copulations are longer in duration, include all components of complete copulatory behavior, and take place just prior to egg-laying.

Reproductive copulations can be used to define the presence or absence of polygyny, as non-breeders (by definition) do not engage in them.

While most Hoatzin units include only one monogamous pair of breeders, some larger units contain two pairs of breeding adults ($n = 9$ of 102 units monitored during 1982–6). Within all such units observed to date, each pair is monogamous, with no observed extra-pair copulations (except display copulations). In over 590 unit-years of data, we have fully documented only one case of polygyny in a trio of two females which laid four eggs in the same nest with one male, and suspect one other in a group of six birds on the basis of behavioral interactions in a unit of seven individuals. Polygyny may be somewhat more common than our data indicate, but it is certainly rare in our population.

Hoatzins breed at two years or more of age, although they may remain as non-breeders in the natal territory for over five years. We have documented only one case of a yearling breeding, which produced one infertile egg in a territory budded from the parental unit. Most young remain as helpers in their first year ($>90\%$), and 'dispersing' yearling birds are often those that are left behind in the dry season home ranges by migrating units (44%). Only 10% of banded birds have bred in their second year, although an additional 20% have dispersed from the natal flock. Third-year birds show a roughly equal percentage between breeding and helping. The majority ($>90\%$) of banded individuals have established or inherited a territory and are breeding by their fifth year.

In summary, Hoatzins are territorial during the breeding season, but some units migrate to form non-territorial flocks in the dry season. During the breeding season, they demonstrate nearly all of the possible combinations of breeding systems: monogamy with and without helpers, polygyny with and without helpers, cooperative breeding with joint and plural nesting, and perhaps (in Peru) coloniality.

Reproductive rates and effects of unit size

Hoatzins showed low overall reproductive success in Venezuela from 1982 to 1987, averaging 27% for individual nests ($n = 813$) and 44% annual success for social units ($n = 590$ unit-years). Units often re-nested when the first nest was not successful, although this was not always the case (especially in years of low rainfall). As many as six re-nestings have been observed by one unit in one season. Larger units produce first nests earlier in the season, are able to re-nest more times in a given season, and have an increased production of young. The presence of helpers also significantly decreases the time and effort spent by the breeding female in incubation, brooding and feeding of the young (Strahl 1985).

Annual success rates are greatly influenced by predation pressure, and units nesting on predator-free islands have significantly higher reproductive success than groups nesting elsewhere. Island units demonstrated 89% annual success from 1982 to 1986 ($n = 95$ unit-years). During 1982–5, non-island units had a 29.9% success rate ($n = 291$ unit-years). However, units with helpers in each habitat have higher overall reproductive success, and there is a significant correlation between unit size and the number of young produced per year (Fig. 5.7: Spearman rank correlation = 0.428, $P < 0.01$). This is very marked among successful units, where the correlation is even stronger (Spearman rank correlation = 0.448, $P < 0.01$).

The mechanism by which helpers influence reproductive success lies in the growth rate and development of the young. Hoatzin units are unable to defend their nest or young against their main predator at the site, the wedge-capped capuchin monkey (*Cebus olivaceus*), regardless of the size of the unit. As a result, the rapidity with which the young leave the nest is extremely important, as the nest is a visual and olfactory focal point for predators (Strahl 1985). The benefits of departing the nest early are also reflected on an evolutionary scale by the elaborate escape behavior and physical attributes of the young, such as the retention of functional wing claws and swimming ability. Hoatzin fledglings depart the nest when still

Fig. 5.7. The mean number of Hoatzin young raised to independence for social units of different sizes on the study site. Vertical bars represent two standard errors above and below the mean. Data are only for units with a single breeding female.

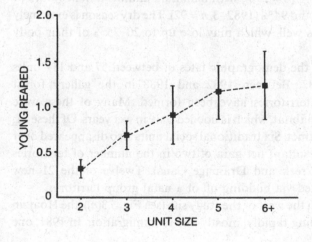

downy at an age of 14–25 days, and are not able to fly for an additional 40–50 days thereafter. Young in units with helpers depart the nest and develop flight capabilities more quickly than young attended by pairs without helpers. Helpers advance the nest departure date by one week.

Survivorship and population stability

As adults, Hoatzins demonstrate relatively high survivorship values, which range from 60% to 80% depending on the year and the severity of the preceding dry season. The average value for the first three years of adult life (ages 1–3: the data are biased by undocumented long-distance dispersals) is 70.3%. Once established in a territory as a breeder, adult survivorship approaches 90% (86.9% in the 61 bird-years of banded individuals that have attained breeding status within the population). Tenure as a breeder is unknown, as the majority of banded birds that have become breeders are still living. One unbanded bird, recognizable by a deformed foot, has been the male breeder in a migratory Drainage Canal unit since 1981 (8+ years tenure). The risks of dispersal and acquiring a breeding position, as illustrated by high apparent mortality among floating birds, may account for lower survivorship in earlier years.

Survival of young and yearlings from the age of independence until the subsequent breeding season is highly dependent on rainfall patterns and the length of the dry season. In years when the rains persist into December and the following rainy season beings early, young may have up to 90% survival from independence through their first year. Survivorship over the first dry season is markedly reduced if the rains end early, or commence late in the following year. An average value for survivorship of young over this critical period is 67.8% ($n = 276$), with minimum and maximum values of 44% (1986–7, $n = 43$) and 94% (1982–3, $n = 92$). The dry season is extremely stressful for adults as well, which may lose up to 20–25% of their body weight.

We have followed the demographic fates of between 57 and 117 units each year since 1982. Between 1982 and 1988 in the gallery forest population, 21 new territories have been formed. Many of these were formed in 'marginal' habitat, which is flooded only in wet years. Of these 21, 13 have since gone extinct. Six traditional social units also disappeared over this period, with a resultant net gain of two in the number of territories along the Caracol Creek and Drainage Canal. Twelve of the 21 new territories were formed via budding off of a natal group territory.

In the HQ region in the west of the study site (see Fig. 5.3(a)), the Hoatzin population is expanding rapidly, mostly due to immigration. In 1981, one unit of seven individuals was present in this region. In 1987, 55 individuals were resident in 17 social units, with over 100 birds congregated in the area during the dry months. The source of immigrant birds is unknown. They are not from the gallery forest population, as no birds banded in that region have appeared in the HQ area. Over 70 young have banded in this region alone, where nesting success is somewhat higher than in the gallery forest.

Overall group sizes and composition have been affected by annual survivorship trends, especially over the past several years. Fig. 5.8 shows the composition of 46 social units in the population for which we have data over a six year period. The dry seasons following the 1984–6 breeding periods have been relatively severe, and average unit size has been reduced from a high of 3.2 in 1984 to a low of 2.6 in 1987. In addition, the 1985–7 breeding seasons were very poor, due to sporadic rains and a late start to the breeding season. This greatly reduced recruitment during these years. Long-term climatologically dependent trends in reproductive success and survival are apparently quite severe, and drought years have particularly strong effects on demography.

Fig. 5.8. Overall structure and production of young in 46 Hoatzin social units monitored during the period 1982–7. Numbers of breeders (shaded area) and non-breeders (white area: includes helpers and other non-breeding birds) are for the onset of the breeding season each year. Number of young produced (hatched area) represents the total number of young to reach banding age (50 days) for the combined units in that year.

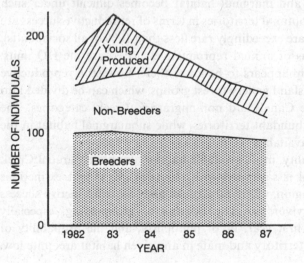

Discussion and speculation: Evolution of cooperative breeding in the Hoatzin

Demographic conditions

Over the past two decades, various demographic models have been put forward regarding the evolution of cooperative breeding in birds. The most generally accepted of these is that of 'habitat saturation' in its various forms, as developed from the mid 1960s by Selander (1964), Brown (1969, 1974), Emlen (1982) and others. More recently, Stacey and Ligon (1987) emphasized the benefits of group-living and the importance of territory quality as being causal factors in the evolution of cooperative systems in birds. When habitat saturation is defined as the occupation of all medium- to high-quality territories, these two theories appear to be similar.

The demography of Hoatzins is complicated by two main factors. First, there is a wide variation in territory quality. Island units have nearly 90% reproductive success annually. Non-island units in the gallery forest have a far lower success rate (< 30%). The HQ region has intermediate success in an area where seasonal drying may affect long-term survival. There also are territories which are infrequently occupied, and which on an annual basis appear to be submarginal habitat. Secondly, many Hoatzin units are migratory in the dry season. Migrations are apparently stressful and are a frequent source of mortality among young birds. Thus, territory quality can also be measured in terms of seasonal availability. On the other hand, year-round territories, such as those on islands, have the added costs of defense during the dry season.

Defining optimal and marginal habitat becomes difficult under such conditions. Clearly, optimal territories in terms of reproductive success are on islands, yet these are exceedingly rare (less than 10% of all social units). Second best in terms of annual reproductive success are the HQ units, although survivorship appears to be lower there. Lowest in reproductive success are the non-island gallery forest groups, which can be divided into migratory (Drainage Canal) and non-migratory (Creek) categories. The latter are the most abundant territories, while submarginal habitat in the former is the most available.

Regardless of quality, most available habitat along the Caracol Creek and Drainage Canal is saturated with Hoatzins. Most unit extinctions occur in the latter region, where migrations and low reproductive success decrease group survivorship. The benefits of group-living, especially survivorship in the first year, favor philopatry, and the probability of acquiring a suitable territory and mate in any given habitat are quite low.

The complete saturation of suitable territories, along with the infrequent occupation of submarginal areas, suggests that Hoatzins fit Brown's (1987) version of the habitat saturation model (Strahl 1988).

Folivory and cooperative breeding in the Hoatzin

In few cooperative avian species is dietary ecology such an obvious factor in the evolution of sociality as it is in the Hoatzin. Leaf-eating is one of the most obviously aberrant factors of the life history of this species. The Hoatzin, at approximately 750 g, is far below the theoretical minimum body weight predicted for homeotherms with ruminant-like digestive systems, based on energy extraction rates from forage, metabolic rates of the animal, and passage rates of digesta (Dement and van Soest 1983). The physiological constraints that folivory and foregut fermentation place on the Hoatzin are reflected in their slow growth rate, behavioral thermo-regulation, generally sedentary habits, and perhaps delayed maturation (Strahl 1985, 1988).

The morphological adaptations of this species to accommodate their oversized, heavy crop, the long flightless period after leaving the nest, and the predator-escape behavior and morphology of the young are all linked closely with folivory. Folivory also may dictate the affinities of *Opisthocomus* to watercourses, which further facilitate predator avoidance in the young. The primary effect of helpers is found in the enhanced growth rate and survival of young, with a secondary effect of reducing the energetic costs of reproduction in female breeders. The former increases the number of young reared when a unit has a successful nest. The latter effect permits more nests per year by allowing the breeders to produce a first nest earlier, and by reducing the inter-nest interval in subsequent nestings. This enables groups with helpers to have a higher probability of at least one successful nest in a given season. The contributions of helpers are extremely important in a species, such as the Hoatzin, that is energetically stressed by its foraging mode.

Fitness components of cooperative breeding

The discussion of the genetic evolution of cooperative breeding systems is plagued by the annoying fact that what we observe in the present may or may not have any bearing on the conditions under which the system evolved. The genetic benefits that accrue to helpers by their actions and to breeders as a result of the presence of helpers are not necessarily those that have driven the system, yet they are all that we can arm ourselves with in speculating about evolutionary pathways. These genetic benefits and losses are (handily) additive (Hamilton 1963, 1964), so, in sorting through the data, we may weight each direct and indirect component.

In analyzing cooperative breeding system, it is essential that we do not discard the rule for the exception. Individual life history anecdotes should not be employed as a main means of discarding alternative hypotheses. For instance, in our population, OB-BB inherited his natal territory (future direct selection) after remaining (future direct selection) as a helper (indirect selection) for five full breeding seasons. Actually, OB-BB was the only helper to remain for such an extended period, and although inheritance of territory is present in the Hoatzin, it is also somewhat rare.

The 'decision' of a helper to provide aid must not be isolated as the only reference point for the evolution of cooperative systems. Instead, two decisions must be viewed as complementary processes: the decision to *Stay* in the natal unit, and the decision to *Help*. The fact that a potential helper has (in the usual case) remained in the natal unit is perhaps a more important issue which often reflects the demographic conditions of the population to which it belongs (Selander 1964; Brown 1969, 1978; Emlen 1982; Stacey and Ligon 1987).

In many species, the indirect genetic benefit from helping is far lower than the expected direct genetic benefit from breeding *if* a territory and mate can be obtained (Brown 1978). The Hoatzin does not appear to be an exception to this generalization. However, the probability of obtaining a territory and mate in *Opisthocomus* is so low that dispersal, especially in the first year, is not viable and/or may drastically reduce the chances for survival of the individual. The initial decision to *Stay* also reflects the low survivorship of young from independence to yearling stage. The energetic demands on yearling Hoatzins also seem to preclude breeding in the first year. In any case, both present and future direct fitness components favor the *Stay* decision.

In most species, the decision to *Help*, once remaining on the natal territory, is associated with increased production of young by the breeders (Brown 1987). In the Hoatzin, helpers significantly increase reproductive success (Strahl 1985; Strahl and Schmitz 1989). The extent to which they do so is influenced by environmental conditions, microhabitat and other factors, but a consistent trend is present. The behavior of helpers has an important effect on the energetic expenditures of the breeding pair, and also on the growth rate, fledging age and survival of the young. Once a potential helper has delayed dispersal and remained in the natal unit, the decision to *Help* produces higher inclusive fitness than the decision *Not Help*, principally by adding a significant proportion of indirect fitness to the equation.

However, other long-term direct benefits may also accrue to helpers. The decision to *Stay* may ensure (through enhanced survivorship) increased direct fitness in the long term. Also, in at least 13 units, male former helpers 'budded' territories off the natal territory, indicating that territorial inheritance may be important for males (Woolfenden and Fitzpatrick 1978). In five such instances, they were accompanied by one or more other unit members (usually siblings), suggesting reciprocity or delayed mutualism (Ligon 1983). Although this suggests that future direct fitness is a strong component in the evolution of cooperative breeding in the Hoatzin, several additional factors must be considered. Relatively few of the above instances of social dispersal included helpers and former recipients of their aid. In four of the budding events, the new breeders were accompanied by members of their unit which *had not* been reared by them, suggesting that alloparenting is not a prerequisite for receiving future aid. In the vast majority of dispersal events (including budding), recipients did not disperse along with their former alloparents or other unit members. Furthermore, in only one case did a female bud a territory (in conjunction with a neighboring male), and females were never observed to inherit breeding status within their natal unit.

These data indicates that future direct fitness benefits, via reciprocity or delayed mutualism, while sometimes present for males are not especially common even within that sex in the Hoatzin. Alloparenting is apparently not a prerequisite for future help in acquiring a territory, nor does it ensure that the recipient will provide future aid to the donor. We therefore feel that while the decision to *Stay* may influence future direct fitness, the decision to *Help* is more related to production of young (indirect fitness) than to the future direct component.

However, indirect selection is not solely or even mostly responsible for the evolution of what we observe as the social system (including both staying as a non-breeder and helping) in the Hoatzin. On the basis of the observed benefits and losses in direct fitness, one would expect that the average Hoatzin should disperse and breed if it has a relatively high chance of obtaining a territory and mate, as the overall losses in potential direct fitness attributable to the *Stay* and *Help* decisions far outweight the benefits accrued from indirect fitness. Clearly, this is the case in our population, such that obtaining a breeding position is of paramount importance. In the absence of breeding opportunities, the option of staying as a non-breeder seems to maximize direct fitness (higher survivorship via the benefits of a territory and group-living) and the choice of the helping option maximizes indirect fitness.

The risks of choosing not to *Stay* are illustrated by the presence of 'floaters' in our Hoatzin population. Floaters may be assessing the availability of both mates and breeding opportunities in the population at large. In some cases floaters remain in the units which they visit, most often as breeders. In others, they are harassed by breeders and helpers alike, and usually depart. Some floaters return to their natal unit after a period of several days, weeks, or even months. Two floaters have been found dead after visiting three or more units over the course of several weeks, and a number of others have disappeared, which seems to indicate that high risks are involved in dispersal.

Our future research will illuminate the importance of assessing breeding opportunities and will concentrate on the probability of finding a territory and mate when dispersing. This probability, F (as used by Brown 1978, 1987), is likely to be the most important component of the cooperative breeding equation for Hoatzins. We feel that F may vary with both age, sex and habitat. Comparisons between populations and sexes will shed light on the relative weightings of the various components of fitness for this species. Further research on habitat-dependent survivorship and reproductive success, including annual variation and lifetime fitness, will, we hope, allow us to distinguish better among the existing demographic models of the evolution of cooperative breeding.

The authors thank Dr Tomas Blohm for his limitless logistic support on Hato Masaguaral over eight field seasons, and for his enthusiasm in aiding our work. L. Elliot, M. Butler, E. Vander Werf, M. F. Rodriguez, A. Grajal, M. Withiam, D. Lemmon and P. Zahler were of assistance in data collection. We also thank J. Brown, C. Bosque and the editors, for comments. Wildlife Conservation International (a division of the New York Zoological Society) funded the majority of the field work discussed in this text. S.D.S. received additional support from Chapman Fund of AMNH and Sigma Xi. A.S. was partially supported by a SEPET scholarship during her field work.

Bibliography

Beebe, C. W. (1909). A contribution to the ecology of the adult Hoatzin. *Zoologica* 1: 45–67.

Brown, J. L. (1969). Territorial behavior and population regulation in birds. *Wilson Bull.* 81: 293–329.

Brown, J. L. (1974). Alternate routes to sociality in jays with a theory for the evolution of altruism and communal breeding. *Amer. Zool.* 14: 63–80.

Brown, J. L. (1975). *The Evolution of Behavior*. W. W. Norton: New York.

Brown, J. L. (1978). Avian communal breeding systems. *Ann. Rev. Ecol. Syst.* 9: 123–55.

Brown, J. L. (1987). *Helping and Communal Breeding in Birds: Ecology and Evolution*. Princeton University Press: Princeton, NJ.

de Queiroz, K. and Good, D. A. (1988). The scleral ossicles of *Opisthocomus* and their phylogenetic significance. *Auk* 105: 29–35.

Dement, M. W. and van Soest, P. J. (1983). *Body Size, Digestive Capacity, and Feeding Strategies of Herbivores*. Winrock International: Morilton, AR.

Emlen, S. T. (1982). The evolution of helping, I. An ecological constraints model. *Amer. Nat.* **119**: 29–39.

Gadow, H. (1891). The crop and sternum of *Opisthocomus cristatus*. *Proc. Roy. Irish Acad.*, 3rd series, **2**: 147–53.

Grimmer, J. L. (1962). Strange little world of the Hoatzin. *Nat. Geogr.* **122**: 391–401.

Hamilton, W. D. (1963). The evolution of altruistic behavior. *Amer. Nat.* **97**: 354–6.

Hamilton, W. D. (1964). The genetical evolution of social behaviour. I, II. *J. Theor. Biol.* **7**: 1–52.

Heilmann, G. (1927). *The Origin of Birds*. D. Appleton & Co.: New York.

Hernandez (1651). *Nova Planetarum Animalium et Mineralium Mexicanorum Historia*, p. 320.

Ligon, J. D. (1983). Cooperation and reciprocity in avian social systems. *Amer. Nat.* **121**: 366–84.

McClure, D. (1921). Four legged birds that climb trees: the Hoatzin – a missing link between birds and reptiles found in South America. *Sci. Amer.*: 126–8.

Muller, P. L. S. (1776). *Des Ritters Carl von Linne ... Vollstandigen Natursystems Supplements- und Register-band uber alle sechs Theile oder Classen des Theirreichs*. G. N. Raspe: Nürnberg.

Penard, P. (1908). *Die Vogels van Guyana*, pp. 307–9.

Quelch, J. J. (1888). Notes on the breeding of the Hoatzin. *Ibis* **5**: 378.

Quelch, J. J. (1890). On the habits of the Hoatzin. *Ibis* **6**: 327–35.

Schmitz, A. (1987). [Some aspects of the diet and social system of the hoatzin (*Opisthocomus hoazin*) during the dry season in the Venezuelan Llanos.] In Spanish. Licenciatura thesis, Universidad Simon Bolivar.

Schmitz, A. and Strahl, S. D. (1989). [Implications of seasonal variation in social system on reproductive success in a cooperative bird, the hoatzin (*Opisthocomus hoazin*).]. In Spanish. *Proc. IIIrd Neotrop. Ornith. Congr.* Universidad del Valle, Cali, Colombia (in press).

Selander, R. K. (1964). Speciation in wrens of the genus *Campylorhynchus*. *Univ. Calif. Publ. Zool.* **74**: 1–305.

Sibley, C. G. and Ahlquist, J. E. (1973). The relationships of the Hoatzin. *Auk* **90**: 1–13.

Sibley, C. G., Ahlquist, J. E. and Monroe, B. L., Jr (1988). A classification of the living birds of the world based on DNA–DNA hybridization studies. *Auk* **105**: 409–23.

Skutch, A. F. (1961). Helpers among birds. *Condor* **63**: 198–226.

Stacey, P. B. and Ligon, J. D. (1987). Territory quality and dispersal options in the Acorn Woodpecker, and a challenge to the habitat-saturation model of cooperative breeding. *Amer. Nat.* **130**: 654–76.

Strahl, S. D. (1985). The behavior and socio-ecology of the Hoatzin, *Opisthocomus hoazin*, in the *llanos* of Venezuela. Ph.D. thesis, State University of New York at Albany.

Strahl, S. D. (1988). The social organization and behaviour of the hoatzin *Opisthocomus hoazin* in central Venezuela. *Ibis* **130**: 483–502.

Strahl, S. D. and Schmitz, A. (1989). [Demography of a cooperative bird, the hoatzin (Aves: Opisthocomidae), in relation to territory quality.] In Spanish. *Proc. 3rd Neotrop. Ornith. Congr.*, Universidad del Valle, Cali, Colmbia (in press).

Thompson, A. L. (1964). *A New Dictionary of Birds*. McGraw-Hill: New York.

Torres, B. (1987). The ecology of the Hoatzin (*Opisthocomus hoazin*) in Peru. M.Sc. thesis, Ohio State University.

Woolfenden, G. E. and Fitzpatrick, J. W. (1978). The inheritance of territory in group-breeding birds. *BioScience* **28**: 104–8.

6 *CAMPYLORHYNCHUS* WRENS: THE ECOLOGY OF DELAYED DISPERSAL AND COOPERATION IN THE VENEZUELAN SAVANNA

6

Campylorhynchus *wrens: the ecology of delayed dispersal and cooperation in the Venezuelan savanna*

K. N. RABENOLD

Social groups of birds and mammals are normally built by delayed dispersal of young. For many species of cooperative breeders, delayed dispersal is associated with delayed reproduction, creating a non-reproductive class embedded in a family structure. When these individuals collaborate in the breeding efforts of others during their tenure as non-reproductive adults, they pose an apparent dilemma for biologists: why should they help to rear others' young instead of dispersing to attempt at least to rear their own (Brown 1983, 1987)? This paper reviews a system of cooperative breeding in which few adults produce offspring of their own and in which helping is so effective in improving the reproductive success of breeders that remaining on the natal territory to aid in rearing siblings is an unusually productive alternative method of gene replication compared to dispersing to attempt breeding.

Stripe-backed Wrens (Troglodytidae: *Campylorhynchus nuchalis*) live in family groups in the savannas of Colombia and Venezuela, where they often co-occur with a congener *C. griseus*, the Bicolored Wren. A third species, the Fasciated Wren (*C. fasciatus*) lives in the semiarid scrub and valley woodlands of the coasts of Ecuador and Peru. These three species normally live in groups with (apparently) non-reproductive helpers that remain on their natal territories past physiological maturity. Other species in this genus are known to be cooperative breeders (Selander 1964), including *C. turdinus* in the Amazon basin (R. H. Wiley, unpublished results) and *C. rufinucha* in Central America (F. Joyce, personal communication). The northernmost representative of the genus is the Cactus Wren, *C. brunneicapillus*, for which cooperative breeding is not usual, at least in Arizona (Anderson and Anderson 1973). Members of the genus inhabit a wide variety of habitats, mainly topical, from desert to

159

Fig. 6.1 *Campylorhynchus* wren habitat in the Venezuelan savanna. The large leguminous tree (*Albizia* sp.) is favored nesting habitat for Stripe-backed Wrens; they often use nests constructed of sticks by thornbirds, as silhouetted below the canopy here. Bicolored Wrens nest in palms like those at the base of the *Albizia*.

mature lowland rainforest. The study described here has included the Bicolored and Fasciated Wrens, but I will focus particularly on features of the social organization of the Stripe-backed Wren.

Study area

The low savanna (*llanos*) of central Venezuela is alternately inundated by rains from May to November and baked to powder in the dry season of December to April. Stripe-backed Wrens occupy open woodlands dominated by leguminous trees up to 20 m tall. They are very active insectivores, mainly arboreal, that use a repertoire of acrobatic foraging techniques to capture arthropod prey, including lepidopteran larvae, orthopterans, and arachnids, from leaf and branch surfaces. They roost and breed in the large legumes (especially *Albizia* and *Enterolobium* spp.) in either nests constructed of sticks by Thornbirds (*Phacellodomus rufifrons*) or nests that the wrens themselves construct of grasses, often in clumps of parasitic mistletoe (Fig. 6.1). Bicolored Wrens nest and forage preferentially in palms (*Copernicia tectorum*).

Hato Masaguaral is a large wildlife reserve and cattle ranch in the Venezuelan *llanos* (45 km south of Calabozo in the state of Guárico) where owner Tomás Blohm has preserved the natural habitat and created a biological station. The 60 km² of Hato Masaguaral encompasses a mosaic of vegetation types from thick forest along the Guárico river to the open woodland favored by the Stripe-backed Wrens to still more open savannas dominated by herbaceous vegetation, shrubs and palms, where Bicolored Wrens are most abundant. More than 300 species of birds have been recorded on the reserve and at least 10, from Hoatzins (*Opisthocomus hoazin*) to Flycatchers (*Myiozetetes inornata*), have been shown to breed cooperatively (Thomas 1979, S. Strahl, unpublished results, and Chapter 5).

This reserve provides a rare opportunity to study large-scale patterns of vertebrate population dynamics; we have been able to study cooperative breeding on a scale large enough to detect nearly all dispersal and to understand interpopulation variation. In 10 km², we have been able accurately to measure demographic parameters, population trends and dispersal among nearly 100 territories each of Stripe-backed and Bicolored Wrens. For the block of 30 territories of each species upon which study was initiated in 1977, we have been able to record virtually all dispersal originating there, including rare movements over several kilometers. The sedentary nature of the wrens and the scale of study made possible by the extent of the reserve combine to make understanding of the key phenomenon of dispersal possible.

The original study area was named the Samán population after an emblematic *llanos* tree. Part of this population was first studied by R. Haven Wiley in 1974 (Wiley and Wiley 1977). With his collaboration, I began studying the wren groups of both species in the 4 km² of this area in 1977 and this work has continued with the help of Carla Christensen, Patricia Parker Rabenold, Steven Austad, Steve Zack and Joseph Haydock. Recruits to this population are marked with unique colored leg bands each year at the end of the breeding season, and detailed censuses are taken at least every six months. S. Austad (since 1982) and J. Haydock (since 1985) have studied Bicolored Wrens in this and a separate area of open palm savanna. Outlying areas known as the Caro and Guácimo populations, along with others, have been censused biannually for dispersers from the Samán population. Ted Stevens studied a separate population of Stripe-backed Wrens in 1983–5. These 'populations' are distinct clusters of territories, since suitable habitat is patchy on a scale of square kilometers, but occasional long-distance dispersal has been observed between them. S. Zack and J. Haydock banded the Guácimo population for experimentation (30 territories) in 1985, and have been studying dispersal in both species. By 1987, 85 groups of Stripe-backed Wrens in three areas (400 individuals) and 87 groups of Bicolored Wrens in two areas (200 individuals) were 95% color-banded (Fig. 6.2). Over 10 years, we have banded 856 and 527, respectively, of these two species.

The Samán population occurs in a patchwork of distinct vegetation types that are associated with different soil types and drainage regimes in the wet season. Dry sand ridges thought to be Pleistocene beaches rise a few meters above their surroundings and are scattered throughout the area (Fig. 6.3). They remain dry through the rainy season and generally support open vegetation without groves of large trees. Near these ridges are areas that drain slowly during the peak rains and therefore present plants with a variable regime of inundation in the wet season. These sites often support well-developed woodland dominated by trees of the legume family including the genera *Albizia*, *Pithecellobium*, *Enterolobium* and *Cassia*, in addition to figs (*Ficus* spp.) and palms. More shrubby woodland where the above trees are much less abundant occurs on sites that are further from the sandridge systems and are more continuously flooded. Some areas are flooded with more than a meter of water in the wet season; these are dominated by sedges and varying densities of palms, with very few large trees. Stripe-backed Wrens are not common in these last two vegetation types, as are Bicolored Wrens. The Samán area is centered on a series of sandridges and patches of woodland. It is bordered on all sides by areas

Fig. 6.2. The spatial arrangement of the four main study populations
of marked Stripe-backed Wrens and two populations of marked
Bicolored Wrens on Hato Masaguaral in the Venezuelan *llanos*.
The Samán area is characterized by leguminous woodland; it was
intensively studied in 1977–82 and monitored since. The Guácimo
area is more open woodland and was the site of principal-removal
experiments in 1985–8. The Palm area is open palm savanna. Stripe-
backed Wrens are rare in unshaded areas, except north of the Caro
area. Bicolored Wrens occur throughout, except in a large treeless
area between the LS and MS areas. Dark areas are dry-season ponds.

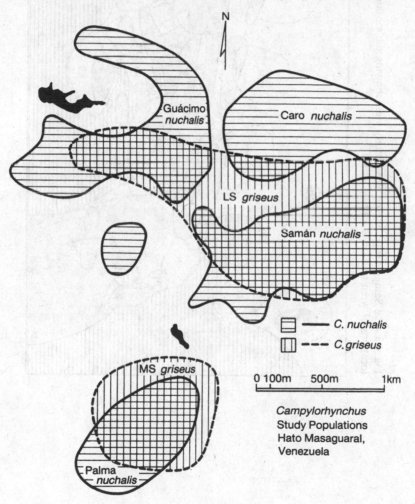

Fig. 6.3. The Samán Stripe-backed Wren study area, showing the mosaic of vegetation types and locations of wren territories (capital letters).

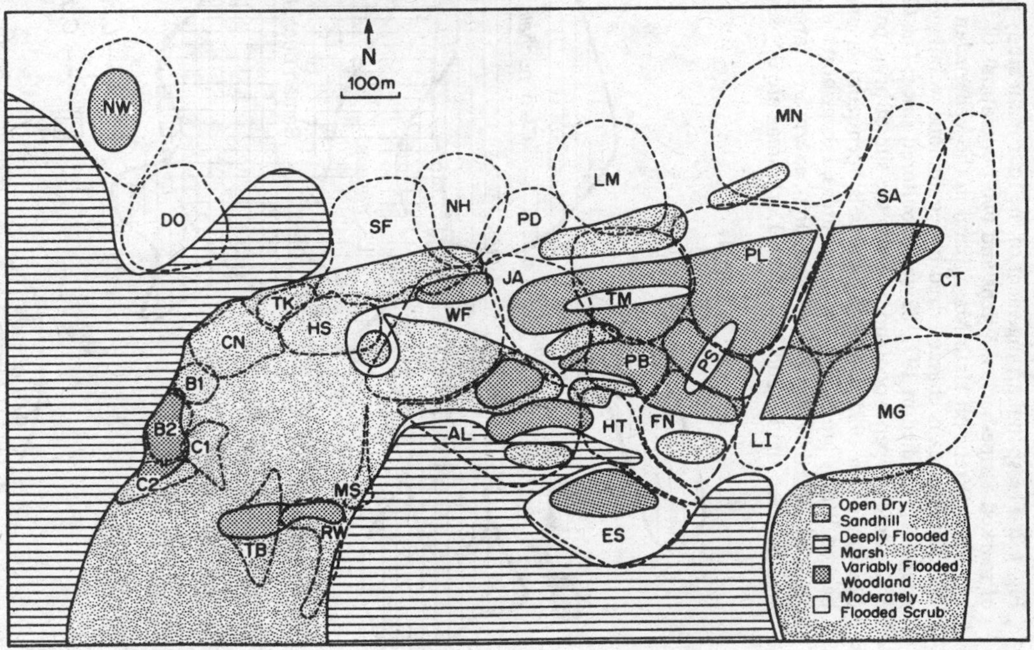

uninhabited by Stripe-backed Wrens: on the south and west by open marsh and sandridge and on the east and north by open shrubby vegetation.

Social organization

Composition of groups and territoriality

Stripe-backed Wren groups consist on average of four to five adults during the breeding season; however, small groups of two or three adults and large groups of six to ten adults are common (see group size distribution in Fig. 6.7). Extensive observational data suggest that only the two behavioral dominants, or principals, breed. 'Auxiliaries' – non-breeders – then outnumber breeders in the population, and they are almost always birds that have fledged on the territory where they are helpers (Fig. 6.4). In a few rare cases (10 of 151 auxiliaries of known parentage – 7%), dependent juveniles have been adopted from neighboring territories or immigrants from other territories have assumed helping roles. Auxiliary birds participate fully in defending the year-round communal territory, in constructing and defending the communal roosting nest, and in caring for young produced in the group. Male auxiliaries slightly outnumber females in most years (see Fig. 6.8).

Social interactions among group members are varied and occur constantly throughout the day. The species song, used to defend territories and display affiliation, is a duet of noisy notes contributed in rapid coordination by a male and female. This is our principal means of sexing individuals of the nearly completely monomorphic Stripe-backed Wrens, since same-sex partners do not perform duets, and rarely vocalize with each other at all. Duets are often performed side-by-side or while flying in tandem; the latter is a high-intensity display most common during boundary conflicts and incursions. Several contact and warning calls are also used frequently as groups move through the trees in fairly close-knit flocks. Allopreening occurs frequently and is generally reciprocal. Dominant individuals, generally the principals and older birds, are more often the donors than recipients and are more likely to preen forcefully. At times allopreening bouts are like dominance interactions, since recipients often adopt a solicitation posture resembling juvenile begging and the posture adopted when attacked. Soliciting birds erect their feathers and often squat low with spread wings. Dominant birds will sometimes peck hard at a soliciting bird, and reciprocal allopreening is often very asymmetrical with respect to the vigor with which the recipient's plumage is probed with the donor's beak.

Populations of the Stripe-backed Wren have not differed substantially in the average composition of groups, but two study populations of the Bicolored Wren have differed dramatically. In the more dense of the two populations, the majority (46–57%) of groups contained one to three auxiliaries of either sex (up to seven adults) and group size has averaged three adults. The less dense population has consisted mainly of pairs (83–91%), with the rest aided by a single male auxiliary (Fig. 6.5). In the former population, overall population density is much higher as well, since the larger groups occupy smaller territories (averaging 1.3 versus 4.3 ha).

Fig. 6.4. Genealogies and compositions for two neighboring groups of Stripe-backed Wrens. Dispersers from and into the groups are shown with arrows at the margins, along with the identities of the other groups. The number of young these dispersers produced in their new positions is shown in parentheses. D and W in the chronology refer to dry and wet seasons. Tenure of principal pairs is tracked by solid vertical lines and their offspring are shown on horizontal solid lines. The two groups exported more principal females than they imported, and they exchanged principal females in 1980. Note the long wait in auxiliary status of two males that ultimately gained principal status, marked by asterisks at the top.

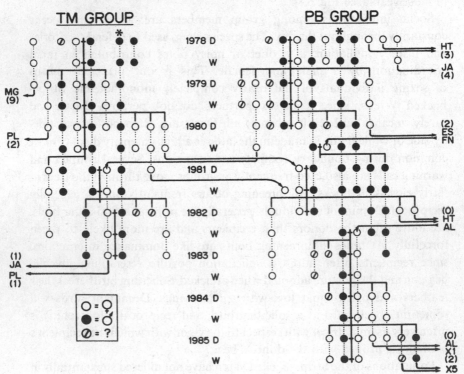

Average group sizes and population densities have been stable in both populations. The dense population occupies an open palm savanna in contrast to the mixed woodland habitat of the less dense population in the Samán area (Austad and Rabenold 1985, 1986).

Both species occupy year-round all-purpose territories with closely defended and highly traditional boundaries. Boundaries between two particular trees, from which fights between the groups are staged, can remain constant for years. Even when groups disappear, the activity boundaries of their neighbors shift very slowly to fill the gap; sometimes the gap remains incompletely filled for years. Territory sizes for Stripe-backed Wrens average just a few hectares, depending on the population density and vegetation. Defended areas ranged from 0.3 ha to 3.6 ha in the Samán population in one year. Generally, larger groups defended larger territories (2.5 ha on average for groups of four or larger versus 1.1 ha for smaller groups; data from 1978, $n = 24$, $p < 0.05$, t-test). When neighboring groups have diverged in size over a period of years, the larger group has gradually gained ground from the smaller. Auxiliaries in large groups are often the main instigators of boundary conflict. Stripe-backed Wren groups with large territories, especially those with varied vegetation, expand their use of their territory in the dry season to include a more even representation of

Fig. 6.5. Bicolored Wren group-size distributions in two populations. MS is the dense population in palm savanna and LS is the less dense population in the Samán area (see Fig. 1; 1982–3 data, Austad and Rabenold 1986).

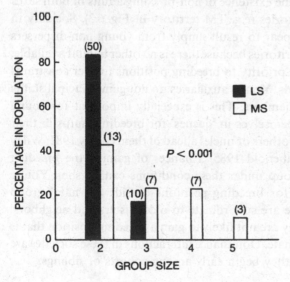

vegetation types in the territory while foraging (including less-preferred palm–fig clusters).

Dispersal and dominance

Several factors contribute to the difficulty of dispersal for young Stripe-backed Wrens, thereby favoring philopatry and the accumulation of helpers in family groups. Most fundamentally, much of the apparently preferred vegetation is already occupied by established groups. However, there have always been unoccupied areas at the edges of the Samán population that have contained vegetation indistinguishable from that in successful territories. Furthermore, there have always been vacant areas within the population that have been proven suitable by the successful reproduction of previous occupants. It is clear, especially in years of low population density, that suitable habitat is always available for colonization. However, it appears that habitat in which to breed is not by itself sufficient for establishment of breeders.

Unaided breeding pairs are very unlikely to produce young, and they seldom colonize an area to attempt breeding with an unaided mate. The limiting resource for fitness is a set of helpers, not simply habitat. It is rare for pairs to develop colonizing associations with auxiliaries. These alliances (sometimes sisters with their mother and an unrelated male breeder) are not usually stable enough to result in successful reproduction. As a result, even areas that have supported some of the most successful breeding over the decade of study in the Samán population have gone uncolonized for years when vacant, in spite of the existence of non-breeding adults of both sexes in the neighboring territories (e.g. TM territory in Fig. 6.3). Sociality in these wrens does not appear to result simply from young non-dispersers accumulating in natal territories because there is no other habitat available.

A seniority system for priority to breeding positions further constrains the options of young birds. Young auxiliaries do not gain principal status ahead of older family members. This is especially important for young males, who often find themselves in 'queues' for breeding status in their natal group with older brothers or uncles ahead of them (Wiley 1981; Wiley and Rabenold 1984; Rabenold 1985). Chances of gaining the breeding position in the natal group under these conditions can be slim. Young females in competition for breeding positions outside the natal group (explored below) likewise are subordinate to older sisters and neighbors; yearling females generally are not likely to gain a breeding position that is contested by an older female. Dominance interactions in these societies are usually very subtle, but they begin early among cohorts of siblings.

I have followed the development of social dominance among sets of same-age brothers and sisters from several Stripe-backed Wren groups. Early social interactions among siblings, before they are well integrated into adult social behavior like duetting, consist of roughhousing that can only be called play fighting. Still unsteady on their feet, they tackle and ambush one another with juvenile ferocity, breaking off and reversing roles at seemingly arbitrary times. As they grow older, the dominant siblings become more active in duetting and preening with the adults, especially the principals. They also become more active in territorial boundary hostilities and in nesting duties, including feeding young. Seldom have same-age siblings survived long enough for one to gain principal status while the others are still auxiliaries, because the wait is often several years; few tests exist of the hypothesis that early social dominance confers priority to breeding status in same-age birds. However, in the few examples where substantial observations were made while the siblings were juveniles, each time it was the bird that showed early dominance that eventually bred first ($n = 6$).

Breeding biology and helping behavior
Parentage
Available evidence strongly indicates that only one mated pair reproduces in groups of Stripe-backed Wrens (Rabenold 1985). The behavioral dominance expressed by one pair in a group over others in territorial defense and in intragroup interactions such as duetting and preening suggests that these principals could suppress breeding in others and are in fact the only breeders. The higher level of participation of the principal pair in duetting, preening, and territorial defense, and their priority in any conflicts over partners, make them easily distinguishable, even in the non-breeding season. Overt aggressive interactions are generally uncommon, but the principal pair in a group stands out clearly from the rest in that their bouts of duetting and preening are rarely interrupted by others but they often intervene in these activities when performed by others. High-intensity flying duets, in response to intrusion by other groups or experimental playbacks of recorded vocalizations from other groups, are performed much more often by the principal pair.

The wrens' behavior during egg-laying provides more direct evidence that only one pair normally reproduces in a group. During the two weeks before egg-laying, the principal male prepares a nesting chamber in which he and the principal female rendezvous several times immediately preceding egg deposition in the chamber. Close association of the pair just before and during egg-laying suggests mate guarding. No other individuals enter the nest during this time; the principal male often perches at the entrance for hours. The principal female is the only member of a group that we have seen enter the breeding chamber before the eggs hatch, aside from her mate's brief visits with her there, suggesting that other females are not laying eggs.

Even when other members of the group begin feeding nestlings, only the principal female broods the young and sleeps with them overnight. A second female with a brood patch has been seen only once in groups during the breeding season, but her history and fate were unknown. While auxiliaries may be physiologically mature, our behavioral evidence suggests that only the principal pair actually contributes to the clutch. The possibility remains of occasional furtive copulations by the principal female with other males; this is suggested by very close behavioral relations between breeding females and two males in a few scattered cases. However, no hint of courtship or copulation by auxiliaries has been witnessed, and the exclusive occurrence of copulation in the nest controlled by the principal male makes matings by auxiliaries improbable. T. Stevens has shown (unpublished data), using allozyme electrophoresis, that fertilizations outside the principal pair can occur, although it is not yet possible to estimate their likely frequency. Although we are continuing to study this issue, available evidence, including the electrophoretic study, indicates that principals are normally the only breeders in groups.

Breeding normally begins with the first substantial rain and flush of vegetation in April. Stripe-backed Wren groups often use multiple nests or several compartments in one of the large stick nests constructed by Thornbirds; the breeding chamber in which nestlings are produced and housed is always separate from chambers used for overnight roosting by adults in the group ('dormitories'). Eggs are incubated by the principal female for 19 days and nestlings spend another 17 days in the nest before fledging. Most sets of emerging fledglings have been twos, threes and fours. Clutch size has seldom been determined directly because the closed nests are generally inaccessible, being at the ends of high, thin branches. For this reason, we cannot distinguish partial predation of a nest from reduced clutch size or brood reduction by competition among nestlings.

Most nesting failure for both Stripe-backed and Bicolored Wrens is caused by predation. Almost all nest failures are sudden and do not suggest starvation of nestlings; failed nests are not characterized by lower feeding rates (Rabenold 1984; Austad and Rabenold 1985). Parasitism by Shiny Cowbirds (*Molothrus bonariensis*) affects fewer than 10% of nesting attempts, mainly those of small groups and those in grass nests; entry to the nest is probably easier for Cowbirds when fewer wrens are vigilant and with a more elastic nest. It is not uncommon for groups to attempt breeding three times in a season (normally six months) and recycling after a nesting failure often occurs within two weeks; however, only large groups attempt further breeding after fledging young.

Genetic relatedness and helping effort

Because auxiliaries nearly always help in their natal groups, they normally rear siblings. Since the fifth year of the study, the parentage of nearly all helpers has been known (assuming principals are the only breeders). Helpers' tenure is normally just two to three years before death or dispersal, and in that time only one of their breeding parents is typically replaced. Two-thirds of first-year helpers rear full siblings and another 20% rear half-siblings, so that only 13% rear less-related offspring (Rabenold 1985). Some of these relationships are still fairly close, often involving the young of an uncle. Older helpers (aged two years or more), because of parental mortality, are more likely to rear young more distantly related than full or half-siblings (44%). Yearling Bicolored Wrens rear 85% full siblings and 15% half-siblings (Austad and Rabenold 1986).

While helpers often return the care of their parents by helping to raise siblings, they seldom have the opportunity to reciprocate aid given to them by non-parents. Only one-quarter of the helpers in the Samán Stripe-backed Wren population has given aid to a non-parent principal that had previously helped to rear the helper (Rabenold 1985). Most of these cases involve helping to rear the young of older brothers and uncles. From the other side of this potential reciprocity, principals are not often aided by individuals raised with the aid of the principal when it was a helper (17% of principals, both sexes), although males in their first season as principals will often be aided by young that they had helped to rear. Reciprocity in this population is not restricted to previous benefactors, but rather is a non-specific relationship between cohorts, in which breeders reliably receive aid from younger non-breeders.

Helpers show little discrimination in providing aid in the rearing of young, regardless of variable relatedness and tendency to reciprocate (Rabenold 1985). Enough variation in helping exists to ask whether helpers reduce their level of effort when rearing young of lesser relatedness than full siblings. Limiting this analysis to those helpers whose efforts were well quantified and whose parents were known, no tendency exists for full-sibling helpers to contribute more than helpers of lesser relatedness

(Fig. 6.6). This pattern remains when sex and age are controlled. In 22 cases, young auxiliaries have been incorporated into non-kin groups either as the result of adoption as dependent young from dissolved groups or of scrambled membership in groups dislocated by a fire. In none of these cases was there any indication of levels of helping below the range of those normal for auxiliaries in their natal groups.

Because male Stripe-backed Wrens tend to breed eventually in their natal groups (60% of male principals) and females normally disperse to breed (97%), male helpers are much more likely to be rearing future helpers for themselves. In spite of this, males are not more energetic helpers than females, even when age and relatedness are controlled (Fig. 6.6). Comparing sisters and brothers helping in the same natal group, females

Fig. 6.6. Contributions to feeding nestlings by Stripe-backed Wren helpers, by sex and relatedness to nestlings fed. The measure of contribution would equal 1.0 for all group members if all contributed equal numbers of feedings (proportion of an even share $= f(n/F)$ when f is the number of feedings by subject, n is the number of group members, and F is the total feedings by all members (Rabenold 1985). Shaded bars indicate means; vertical lines indicate standard deviations.

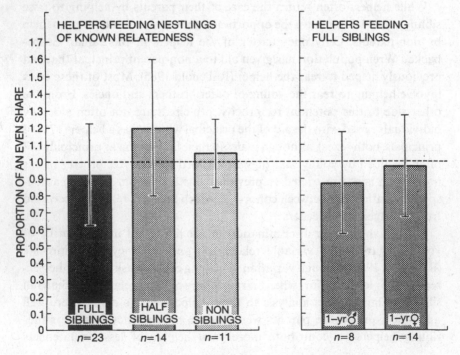

help as much as males ($n = 15$ pairs). Substantial helping by females suggests that inheritance and reciprocity are not necessary conditions for helping to occur.

Some tendency exists, however, for older birds to help more than younger ones: of 11 males whose helping was quantified in successive years, 10 of them helped more (37% more on average) when they were older (Rabenold 1985). An older male helper is both more likely to inherit a breeding position in the near future and less likely to be closely related to young being reared in the group. Overall, helping by auxiliaries in this population is insensitive to variation in likely return on the investment; it appears to be a 'hard-wired' behavior. It is probably significant that such automatic behavior occurs in the context of lasting monogamous pair bonds, stable group membership, and reliable effectiveness of aid; these factors stabilize the rewards of helping for auxiliaries.

Reproductive success: the group size effect
Helpers greatly enhance productivity
The most important constraint on breeding by young Stripe-backed Wrens is the very low reproductive success of pairs breeding unaided or with only one helper. A 'critical mass' of four adults for successful breeding limits productive breeding opportunities to taking over breeding positions vacated by deceased principals in established groups with at least two helpers. Over the ten years, the great majority of pairs and trios failed in their breeding attempts (101 of 128 group-years: 79% failure) while the great majority of groups of four or more adults produce at least one young that survived to the end of the breeding season (70 of 103 group-years: 68% success) (Fig. 6.7). Even the few breeders in pairs and trios that produced some young produced on average one fewer juvenile per year than successful breeders in larger groups (1.41 v. 2.55 juveniles surviving to December; $n = 96$, $P < 0.01$, Mann–Whitney U-test).

Detailed information is available, for the years 1978–81, on success rates of nesting attempts of Stripe-backed Wrens and on numbers of fledglings and one month old (independent) juveniles produced (Rabenold 1984). Groups with two or more auxiliary adults produced on average six times as many independent juveniles per pair-season as did unaided pairs or breeders with only one helper (2.4 versus 0.4, $n = 96$, $P < 0.0001$, Mann–Whitney U-test) (Rabenold 1984). Groups of four or more adults often produce two broods of fledglings in a single season (12 of 49 group-years), for a maximum production of eight young. Smaller groups do not successfully rear two sets of young; this accounts in large part for the disparity in reproductive success when only successful groups are considered.

The relationship between young produced and group size is a sharp step function, since the largest proportional increment comes with the second helper. However, groups of six or more adults produced on average two more independent juveniles per season than did quartets or quintets (3.7 versus 1.9, $n = 49$, 1978–81). Even if the number of independent juveniles per adult, instead of the number per breeding pair, is used for comparison, four or more adults do significantly better than smaller groups (0.50 versus 0.16, $n = 96$, $P < 0.001$, Mann–Whitney U- test). This measure of per capita productivity does not increase in groups larger than four. For pairs and trios, only 13% of clutches produced any fledglings (46 attempts) but 62% did so in larger groups (55 attempts). This magnitude of group size effect on nesting success has been consistent over subsequent years as well.

Fig. 6.7. Relationship between group size and reproductive success for Stripe-backed Wrens in the Samán area for two five-year periods. Each point represents the number of December juveniles produced by a particular group in one year; height of bar indicates mean.

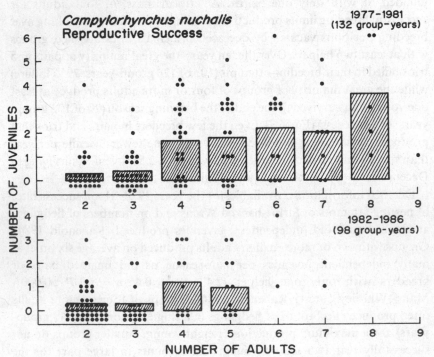

Bicolored Wren helpers produce a similarly strong enhancement of reproductive success, although the 'critical mass' is lower at three adults (Austad and Rabenold 1985). Unaided pairs produce an average of 0.4 independent juveniles per year, while breeders with helpers produce 1.3 juveniles ($n = 123$ group-years, 1982–6; $P < 0.0001$, G-test; data from S. Austad and J. Haydock; see also Austad and Rabenold 1985, 1986). Several helpers do not enhance reproductive success significantly beyond the effect produced by one. This pattern has been consistent over 10 years and in both of the two Bicolored Wren populations mentioned earlier. Pairs produced independent young from only 20% of their nesting attempts, while breeders with helpers successfully fledged young from 47% of their nests. Group size is probably held below its optimum (3) in one habitat by high juvenile mortality, but low juvenile mortality and habitat saturation in the other population combine to encourage philopatry and produce groups with more than the optimum number of helpers (Austad and Rabenold 1986). Fasciated Wrens in Peru also experience greater reproductive success with increasing numbers of helpers, but the magnitude of the effect in that species is not as well substantiated as that of the other two.

Experimental evidence of mechanism

Experimental evidence suggests that deterrence of predators by large groups is the key to reproductive success, for both Bicolored and Stripe-backed Wrens. Both species aggressively defend the vicinities of their nests against many other species, particularly hawks, snakes, cowbirds, and competitors for nests and nesting material. Wrens often loudly scold intruders and attract other species of birds that join in physically harassing an intruder. Large groups of wrens are probably more effective in mobbing predators than small groups, but the principal advantage of large group size could be the ability to cause sufficient disturbance around a predator to draw other species. The wrens seem to function as the main alarm in a predator-mobbing guild of species that share nesting habitats.

Experiments in which a live caged predator is presented beside an active nest with Stripe-backed Wren nestlings show that there is a threshold of responsiveness that corresponds to the threshold group size producing reproductive success. When groups of four or more adults are challenged, they react vigorously, with loud scolding vocalizations and close physical approach to the predator; they scold more continuously than pairs and trios ($n = 15$, $P < 0.02$, Mann–Whitney U-test). These larger groups are very successful in attracting many individuals of other bird species that are better able, because of larger size, to harass predators physically. They attract as many as 30 individuals of other species, and the increase in the average number of other birds in the nest tree, compared to pre-presentation levels, is significantly greater than the increase for pairs and trios (4.3 additional birds in the average 30 s interval versus 0.5, $P < 0.01$, Mann–Whitney U-test). Large groups are also better able to compensate in feeding rates for the interruption caused by the presence of a predator (75% increase over pre-presentation levels after removal of the predator versus 36%, $P < 0.05$, Mann–Whitney U-test) (K. N. Rabenold and P. P. Rabenold, unpublished results).

Pairs and trios normally adopt a passive response of simply ceasing nestling feeding and avoiding the vicinity of the nest. Their desultory scolding is seldom effective at attracting other species. It is possible that individuals in these small groups incur greater risk by announcing their position to a predator, and that the group is less able to repel the other species attracted to scolding once the danger is past and allies become potential competitors for nests and material. For whatever reason, the reaction of small groups to the challenge of a predator at their nest differs greatly from the noisy, aggressive reaction of groups of four or more adults.

Bicolored Wrens without helpers, whose nests were protected experimentally from climbing predators, with greased metal sleeves affixed to their nest palms by S. Austad, enjoyed the same level of breeding success as that of aided pairs. Experimentally protected pairs produced an average of 1.67 juveniles for the season while unprotected controls produced 0.44 juveniles on average ($n = 20$, $P < 0.05$, Mann–Whitney U-test). The experimental devices produced an effect of the same magnitude as one helper in improving reproductive success (Austad and Rabenold 1985).

Territory quality

In neither Stripe-backed nor Bicolored Wrens is breeder experience important, and territory quality is a secondary determinant of reproductive success greatly overshadowed by the direct beneficial effects of helpers (Rabenold 1984; Austad and Rabenold 1985). Breeding in successive years on the same territory varies as much with number of helpers as across the population in both species. However, summaries of reproductive success in each group over the 10-year study of the Samán population of Stripe-backed Wrens allows an analysis of the long-term effects of vegetation type on wren production.

Of 32 territories occupied for at least two years, average group sizes ranged from 2.0 to 5.7 (June adults) and average number of young produced per year ranged from 0 to 2.4 (December juveniles). Territories that rank in the top half by yearly production of juveniles are mostly those in the central, more wooded section of the population (WF, AL, JA, TM, PL, PB in Fig. 6.3) and most of the unsuccessful territories are peripheral in more open shrubby habitat. The top-ranked territories averaged 1.4 juveniles per year compared to 0.4 for the unsuccessful half and 4.4 adults compared to 2.8. Successful territories also had greater nesting success (60% versus 16%) and longer tenure (7.7 years occupied versus 6.5). We (C. Christensen and I) collected detailed data on the taxonomic and structural composition of the vegetation on 28 territories, using transects radiating at regular angular intervals from the centers. Successful territories averaged somewhat greater plant species diversity than unsuccessful ones (10.4 versus 8.8 using the inverse Simpson's diversity index).

If territories are divided into those composed mainly of woodland and those composed mainly of more open shrubland at the margins of the woodlands, the woodland category contains more successful groups: 4.2 versus 2.9 average June adults; 1.2 versus 0.5 juveniles per year; and 63% versus 16% nesting success ($P < 0.05$, Mann–Whitney U-tests and χ^2 test). Woodland territories averaged 10.9 for plant species diversity compared to 8.4 for shrubland; 17 plant species accounted for 90% of intercepts for the woodland samples, while 14 species were sufficient for the shrubland samples. Structural diversity and percentage cover did not differ between the two categories. Long-term average breeding success is associated with diverse woodland vegetation in the Samán population, but available measures of territory quality do not correlate as well with reproductive success in any one year because of the much stronger effect of number of helpers.

Is group enhancement consistent?

The effectiveness of helpers in enhancing reproductive success has been consistently strong throughout the study (Fig. 6.7). To analyze the group size effect for the second half of the study, it was necessary to use juveniles surviving to December as the measure of reproductive success, since detailed observations of nesting have not been made in all years. In the first five years (1977–81), pairs and trios in the Samán Stripe-backed Wren population produced 0.36 juveniles per year on average, while larger groups produced over five times as many: 1.95 juveniles per year. In the second part of the study (1982–6), although overall productivity was less than half that of the earlier years, groups of four or more adults still produced nearly five times as many juveniles per year as did small groups

(1.17 versus 0.25; $P < 0.001$, Mann–Whitney U-tests). This later difference was due mainly to differences in nesting success rather than to the frequency of multiple successful clutches by large groups. Over the 10 years, groups of six or more produced more young (2.3 per year, $n = 31$ group-years) than quartets and quintets (1.5, $n = 71$; $P < 0.05$), so that help in excess of the 'critical mass' of two helpers seems to be effective in further enhancing reproductive success.

There is some indication that the threshold for reproductive success is higher in the Guácimo population and that the magnitude of the effect of helpers is somewhat lower in that population over three years (Zack and Rabenold, 1989). Over 10 years in the Samán population, breeders in large groups (with two or more helpers) produced 176 of 214 juveniles (82%), while accounting for only 44% of the breeding population ($n = 230$ group-years) (Table 6.1). It seems clear that young would-be breeders, given the low probability overall of breeding successfully, would be selected to compete vigorously for principal positions in large groups.

Demography

Density of the Samán population of Stripe-backed Wrens has varied considerably over the decade of study, but average group composition has changed less (Fig. 6.8). In 1978, density of the population was nearly one individual per hectare (180 wrens in 200 ha). By 1987, densities had fallen to one third of the previous high. This decline was due mainly to reduced reproductive success in the 1980s. An average breeder produced fewer offspring per year in the second half of the study (0.52

Table 6.1. *Numbers of juvenile Stripe-backed Wrens, recorded at the end of the breeding season, produced in groups of different sizes, 1977–87*

	No. of juveniles	
	Observed	Expected[a]
Large groups (≥ 4 adults) $n = 102$	176	95
Small groups (2–3 adults) $n = 128$	38	119

[a] Expected values generated from hypothesis that juvenile production is uninfluenced by group size.
χ^2-test $= 66.0$, $P < 0.00001$.

juveniles in December per pair per year, versus 1.23); the cause is unclear. Production of juveniles was highest when our study of breeding in the population was most intense – 1978–9 – so that observer interference with breeding is not likely to be important. The Caro population to the north experienced a similar decline, although it has only been censused for dispersers since 1979. The density of the Guácimo population has remained fairly constant over the decade, and average group sizes were consistently above five.

Determinants of survival

Patterns of survival can provide clues to the ecological constraints most important in the life-histories of individuals. Survival has been variable over the years, but consistent patterns remain among status classes. Adult survival over the decade among Samán Stripe-backed Wrens has been 62% annually (based on 1265 cases of individuals starting a six-month census period). Male and female principals have not differed in their survivorship (survival rate), both averaging 63% ($n = 710$). Auxiliary males survive significantly better than auxiliary females (76% versus 55%

Fig. 6.8. Stripe-backed Wren population structure in the Samán area over 10 years. n is the number of groups under study; $\bar{S}$ is the average group size. Tops of the bars indicate total population size.

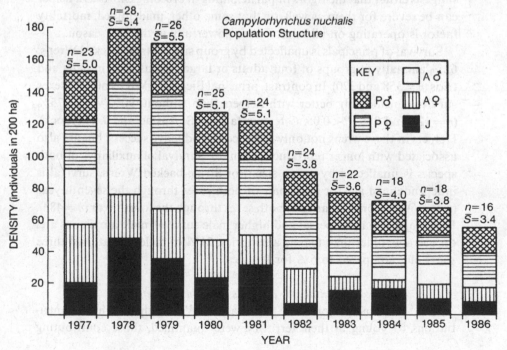

annually over the decade, $n = 257$ and 245; $P < 0.002$, χ^2-test). This asymmetry presumably reflects the cost of dispersal for females, although of course it is impossible to know whether a female has died in the act of attempting dispersal. It is not likely that dispersing females are eluding our censuses; we have searched far enough outside the Samán population to detect a number of banded emigrants that equals the number of unbanded immigrants (for both sexes). It is more likely that female auxiliary mortality is high because of the risks involved in the dispersal process; this issue is explored further below.

Mortality of principals is probably higher in the rainy breeding season than in the dry non-breeding season (21% semi-annually versus 16%, $n = 362$ and 366, but $P = 0.09$, χ^2-test). From 1983 to 1987, during the population decline, principal mortality in the wet season was twice that of the dry season (25% versus 11%, $n = 128$ and 125, $P = 0.004$, χ^2-test) and auxiliaries showed a similar pattern (26% versus 15%, $n = 54$ and 54, but $P = 0.15$, χ^2-test). High wet-season mortality presumably reflects the energetic demands and risks of the breeding effort, in spite of the fact that insect food should be harder to obtain in the unproductive dry season, when many of the trees drop their leaves. In addition, risk of predation is probably greater in the dry season without the cover of foliage. The fact that the dry/wet season comparison is similar for non-breeding helpers suggests either that the rigors of participating in breeding efforts as a helper can be severe for them as well, or that some other, unidentified, mortality factor is operating on all adults more powerfully in the wet season.

Survival of principals is unaffected by group size in Stripe-backed Wrens: 62% annually in groups of four adults or larger versus 65% in pairs and trios ($n = 358$ and 370). In contrast, principal Bicolored Wrens of both sexes survive considerably better with helpers than without: 90% versus 75% ($n = 120$ and 246, $P < 0.05$, G-test; data from S. Austad and J. Haydock). Helpers in these wrens not only enhance reproductive success, but are also associated with longer reproductive tenure. Survival of auxiliaries of both species is unaffected by group size. For Stripe-backed Wrens, survival is independent of age for principals of both sexes through the seventh year ($n = 716$) and for auxiliaries of both sexes through the fourth year ($n = 488$). Because of female dispersal and higher male survival, male auxiliaries are, on average, older than female auxiliaries: 21% of males are at least three years old, compared to 6% for females.

Survival of auxiliaries depends upon helping effort

In a sample of 100 Stripe-backed Wren helpers whose contributions to young in their territory were quantified, those contributing above the median level to the feeding of young may have had lower survival over the following year than those contributing less than median levels to the feeding of nestlings (67% versus 78%, but $P > 0.05$, χ^2-test). In another analysis, 30 pairs of helpers, matched for sex and from the same group in the same year, but differing in levels of contribution to feeding nestlings, were compared for survival. This analysis should control for mortality factors, not controlled in the previous analysis, that depend on sex or local conditions. In 16 of 20 cases when we knew which member of a pair died first, the more energetic helper was outlived by the poorer helper ($P < 0.006$, sign-test); survival was 47% to the next breeding season for those helping more, compared to 83% for those helping less ($n = 30$; $P < 0.01$, χ^2-test). Although dispersal tendencies could complicate this analysis, 22 of the 30 pairs were males (males are usually non-dispersive) and the results stand when females are removed from the analysis. These patterns argue against the possibility that participating as a helper in the breeding efforts of others is of negligible energetic or survival cost to the helpers. It is then reasonable to expect that some benefits to the fitness of helpers should compensate this reduction in survivorship.

Adult mortality and population decline

Declines in survival were marked in years when the Samán Stripe-backed Wren population dropped sharply, suggesting that lower breeding success was not solely responsible for lower density. For instance, the population density declined sharply between the breeding seasons of 1979 and 1980 (Fig. 6.8). Survivorship in this year was substantially lower than average: 49% annually for all adults, with no difference between principals and auxiliaries, compared to the long-term average of 62%. This reduction in survivorship was concentrated in the dry season of that year: 31% mortality over the six months including the dry season (twice the normal dry-season mortality). Mortality was especially high in small groups (26% annual survival in pairs and trios v. 55% in groups of four or more; $n = 249$, $P = 0.04$, χ^2-test). Multiple auxiliaries are then associated with longer tenure for principals in Stripe-backed Wrens, at least in 'bottleneck' years; this effect also appears to be strong outside of the breeding season, perhaps because of enhanced territorial defense or defense against predation.

No simple explanation is apparent for the poor survival in 1979–80, but aberrant weather is implicated. March is normally a very dry month, when fewer than 10 mm of rain fall on average.In March 1979, 100 mm of rain fell in one storm (more than has fallen in March in 35 years at a nearby weather station). This rainfall precipitated breeding in the wrens much earlier than is usual, although many attempts failed. It is possible that such an early start, when a flush of vegetation was not forthcoming owing to the resumption of dry conditions, stressed breeders and helpers alike. Reproduction in the 1979 season was eventually very high; it is also possible that the early start combined with high investment in reproduction left the adults weakened going into the following dry season, especially those in small groups with little help. As shown above, it is unusual for mortality to be higher in the dry season than in the wet season. Equally unusually, early rain fell in April of 1980. Whatever the cause, the increased mortality of 1979–80, coupled with the generally low levels of reproductive success in ensuing years, led to the decline of the population. Population fluctuations of this magnitude could have profound influences on the selective pressures on individuals in cooperative societies.

Lifetime reproductive success

Most young Stripe-backed Wrens delay breeding for at least one year, and 52% of birds reaching adulthood (age one year) never gain breeding positions. Few wrens that survive for one year manage to breed in their first potential breeding season (19%, $n = 181$), although a few do breed successfully at one year, arguing that there is behavioral and physiological potential for early breeding. Helping through at least one wet season is nearly universal among adults in the Samán population. It appears that competition for good breeding opportunities constrains young would-be breeders. Most of the principals in the population have gained this status by the age of three years, although males reach breeding status at a somewhat later age than females (Table 6.2). A few males have been non-breeding helpers for more than five years before attaining a breeding position (see Fig. 6.4). Delayed breeding is clearly the norm in all three *Campylorhynchus* spp. that we have studied.

Table 6.2. *Age of first breeding position for Stripe-backed Wrens in the Samán population*

	Age (years)							
	1	2	3	4	5	6	7	Mean
Females	19	26	17	4				2.1
Males	10	16	10	7	5	1	1	2.8

For females, 23 of 66 individuals' ages were uncertain because they were banded as adults. Half were assigned the minimum known age and half the next older category. For males, 17 of 50 were treated the same.

Lifetime reproductive success has been highly skewed in the Samán population of Stripe-backed Wrens, as is probably the case for many cooperative breeders. Life histories of cohorts of birds fledged in 1977–80 ($n = 131$) show that only 28% of those individuals contributed descendants to the next generation. While the great majority (72%) produced no offspring, four (3%) produced 10 or more. Fewer than 10% became grandparents. Attempting breeding clearly yields high variable rewards in this population, and success is determined mainly by acquisition of a breeding position in a large group. For males, this occurs mainly by birthright – inheritance of a set of helpers on the natal territory by older brothers – but females must compete outside the natal territory and appear willing to take considerable risks to attain productive positions.

Dispersal: the encapsulation of selective pressures

Competition for breeding positions

Competition can be fierce for breeding status among females. Routine netting of large groups of wrens for banding requires haste because of the likelihood that auxiliary females from the neighborhood will invade in response to the temporary removal of the principal female. This suggests substantial pressure for large-group breeding positions, but no such problems are encountered when netting small groups.

One naturally occurring vacancy in the TM Stripe-backed Wren group in 1980 resulted in competition among at least seven females that lasted at least two weeks. This group consisted of seven adults at the time and had been one of the most successful territories in the population. All the non-breeding females from the four adjacent territories contested the position, in addition to four birds from two or more territories distance. The four females from non-adjacent territories were in evidence for only a few hours early in the contest. Of the three females from neighboring territories, two were two years old and the third was a one year old sister of one of the others. This younger neighbor participated in the contest for only the first two days. In 31 h of observations, we (P. P. Rabenold and I) recorded 303 duets in which participants included one of the contestants for the principal female position: 271 of these involved one of the two neighboring two-year-olds with one of the three males of the TM group, usually the principal male. Overt fighting was infrequent in this contest, since the competing females established well-defined spheres of influence and did not often meet. The two major contestants divided the territory so that one held a few trees (trees were tagged and mapped) surrounding the nest recently used by the TM group to rear young. The other contestant held an area containing the nest most recently used for roosting (50 m from the other). The males flew between these subterritories at regular (10–15 min) intervals to duet and preen with the females; their attention was equally divided between the two. After 17 days (or longer – observations were interrupted), one contestant disappeared and was not seen again and the remaining female became the principal for several years.

Other contests for female breeding positions in large groups have been considerably more violent, but the advantage of age and proximity in such contests was consistently suggested. Of 25 females responding to four naturally occurring vacancies in large groups for which detailed observations exist, 14 (56%) were from non-adjacent territories (Fig. 6.9). For 57 cases of successful female dispersal in this population, only 32% have come from non-adjacent territories, suggesting that an advantage to neighboring females in competition might enforce the short-distance natal dispersal that is characteristic of this population. It was also suggested by these observations that females attempting dispersal to large groups could be operating in neighborhood hierarchies based on age. We designed a series of field experiments to test these suggestions under more controlled conditions.

Field experiments

Competition among Stripe-backed Wren females for experiment-ally created breeding vacancies is much more intense for positions in groups with two or more helpers than for positions in smaller groups (Zack

Fig. 6.9. Origins of female Stripe-backed Wrens contesting four naturally occurring principal vacancies (proximity of home group in territory widths). Compare to distribution of distances from origin for successful dispers in Fig. 6.12.

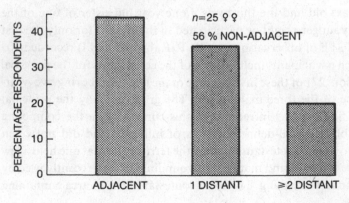

and Rabenold, 1989). We have removed breeding females in 19 groups over two years in the Guácimo population north-west of the Samán population (see Fig. 6.12). An opening in a group of four or more adults attracts more females ($P < 0.05$, Spearman Rank-test), who fight vigorously, often over many days, for breeding position, compared to smaller groups. In one removal of a principal female from a group of five, 10 auxiliary females responded to the vacancy, all banded, some within an hour of the capture of the breeding female. Four of these respondents were from immediately adjacent territories, but several were from territories up to 1 km away and must have been 'scouting' in the neighborhood at the time. Fights occurred almost continuously in the morning hours of the next two days. Over the 19 removals, a significantly greater proportion of older (more than one year) auxiliary females competed for large-group vacancies than for small groups (pairs and trios) (16 of 25 possible – 64% – within the normal dispersal distance of two territories, compared to 4 of 28 or 14% ($P < 0.005$,

Fig. 6.10. Responses to experimental removals of principal female Stripe-backed Wrens. Bars depict the percentages of auxiliary females, within a two-territory radius of the opening, responding to openings in large (more than three adults) and small groups (pairs and trios). Asterisks indicate significant differences ($P < 0.05$, χ^2-tests). n, sample size; n.s. not significant.

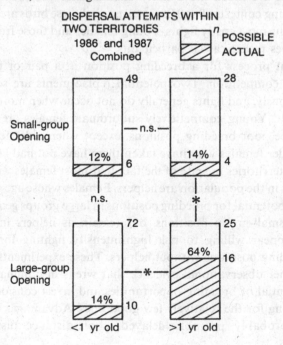

χ^2-test) (Fig. 6.10). Young females (one year) competed much less often overall and did not discriminate between large- and small-group openings.

Levels of aggression in contests for breeding positions in large groups can exceed 100 agonistic encounters per contestant per hour. Fights range from chases and displacements to mid-air tackles and pecking bouts in which contestants sometimes tumble to the ground with feet locked while delivering blows to the head that can draw blood. Fighting generally centers around the main nest of the experimental group, and females compete for duetting position with the principal male. These contests are often a cacophony of duets, alarm calls, contact calls, scolding and other wren vocalizations. Although participants are quite variable in their levels of effort for a given position, combatants are often heedless of their conspicuousness; they sometimes fall, locked together, at the feet of the experimenters or into the water. It is not difficult to imagine that exposure to predators and exhaustion increase risk of mortality, contributing to the low survivorship reported for auxiliary females.

Some large-group vacancies are filled without contest, but these are relatively isolated groups, with few auxiliary females in the neighborhood (two-territory radius). All openings in large groups that attracted more than one contestant ($n = 8$) elicited fighting. Older females and those from adjacent territories are much more likely to win these protracted and energetically draining contests ($P < 0.05$, χ^2-tests), and these birds are more likely to enage in the most costly fights. Young females and those from far away are sometimes rather casual participants.

The replacement process for a breeding position in a pair or trio is normally without competition. Two potential replacements are seldom present simultaneously, and fights generally do not occur when more than one bird is present. Young competitively subordinate females are more likely to take these poor breeding positions, except when older females compete. Most older females who have taken them have not had helping positions in natal territories in spite of the fact that most females without principal positions in the population are helpers. Females whose age grants them competitive potential for breeding positions in large groups generally do not take the small-group positions, but remain as helpers instead. Females rarely appear willing to risk high-intensity fighting for low-productivity breeding positions without helpers. These experiments have substantiated earlier observational evidence that wrens discriminate the reproductive potential of breeding opportunities and invest considerable energy in competing for the relatively few good ones. Advantages of age and proximity probably promote delayed, short-distance dispersal characteristic of this and other cooperative species (Zack, unpublished results).

Natural natal dispersal patterns

Natal dispersal in Stripe-backed Wrens is both spatially restricted and female-biased (Fig. 6.11). Most males (60%) that gain breeding status do so on their natal territories after substantial delays in age-determined queues (see Table 6.2). Male succession to breeding status generally occurs without overt fighting, since dominance relationships are well established among family members. Natal dispersal, movement from the natal territory to the site of acquisition of breeding status, most often occurs within 0.5 km for dispersers of both sexes, traversing normally fewer than two territories. Females rarely breed on their natal territories (3%); exceptions involve male dispersal into the group. Breeding dispersal, movement from one

Fig. 6.11. Sites of succession to principal status by Stripe-backed Wrens. Distances of natal dispersal are expressed both in territories traversed and in absolute distance from natal group.

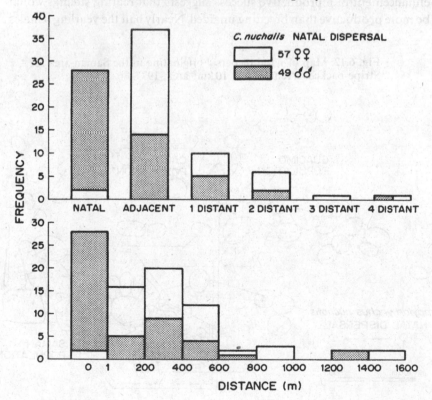

breeding position to another, often covers greater distances than does natal dispersal. Except for females without helpers, principals generally move only when they have lost a mate and a replacement does not exist, as when no male auxiliaries occur in a group where a female has lost her mate. In some cases, incest avoidance appears to motivate moves by principal females in groups where the only available males are sons. Females occasionally disperse in pairs (double arrows in Fig. 6.12). In these cases, the pairs are usually known to be sisters or mother–daughter combinations, and one assumes the principal role while the other becomes a helper. These helpers often later move on or return to the natal group.

Natal dispersal in Bicolored Wrens is less female-biased than in Stripe-backed Wrens and is variable between populations. In the more dense study area, where groups are larger, dispersal is as spatially restricted as in Stripe-backed Wrens. In the less dense population, where juvenile mortality is higher (0.76 versus 0.43, Austad and Rabenold 1986), early dispersal for longer distances is the rule (more than three territory widths on average compared to two). This early dispersal occurs in spite of the fact that group enhancement of reproductive success suggests that rearing siblings would be more productive than breeding unaided. Nearly half the yearling females

Fig. 6.12. Map of natal dispersal originating in the Samán area for Stripe-backed Wrens over a 10 km² area, 1977–86.

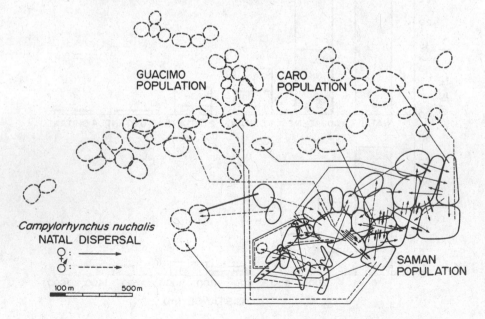

in the dense population are philopatric, compared to the universally dispersive females of the other population. Demographic circumstances in the dense population are such that breeding positions become vacant less frequently (adult survival improved by helpers) and competition for them is more keen (more and longer-lived juveniles).

Although Stripe-backed Wren dispersal is generally spatially restricted, enough exchange exists among the study 'populations' to suggest that they are not genetically isolated. Because of male sedentariness, territories represent male lineages, and these differ considerably in the amount of genetic export to the rest of the population in the form of dispersers. The WF group, for example, has contributed six females and four males as breeders to other groups while receiving only two breeding females from the outside (see also Fig. 6.4). Other groups have imported several breeders while exporting none.

Male natal dispersal is uncommon and constrained among Stripe-backed Wrens. Only two of 21 male dispersers have gained breeding positions in groups above the success threshold of four adults (males in established queues can apparently keep outsiders out), and none has successfully reproduced. Males making moves so unlikely to result in breeding success come disproportionately from trios (more than would be expected if dispersers were drawn at random from the population; $P < 0.01$, χ^2-test) (Table 6.3). Male helpers in trios ('small' groups in Table 6.3) are ineffective (unaided pairs do as well) and they can expect little reproductive success there themselves, since they are not likely to inherit helpers when they gain breeding status. In large groups of six or more, helpers are

Table 6.3. *Natal dispersal by male Stripe-backed Wrens in the Samán population*

| Origin | Destination | | | |
	Small	Medium	Large	Total
Small	9	0	0	9 [2]
Medium	1	0	0	1 [6]
Large	8	2	0	10 [12]
Total	18 [10]	2 [5]	0 [5]	20

'Small' groups are pairs and trios; 'medium' is 4–5; 'large' is 6 or more adults. Expected values, based on representation in the population of auxiliary males by group size and distribution of group sizes, are in square brackets. 'Origin' refers to natal group and 'destination' refers to group where principal status attained.

somewhat superfluous, since groups that large could afford to lose a helper without much decline in likely success. Breeding in such a group would be productive, but there is often a long queue for breeding status. In medium-sized groups of four or five, helpers contribute significantly to production of young; leaving would put the group at or below the threshold for success so that the disperser would incur a cost in sibling production. In addition, there is a short queue for breeding with a high probability of success. The statistical 'reluctance' of males to leave medium-sized groups suggests that these advantages of staying have molded the tendency to philopatry.

Female dispersal represents a clearer potential tradeoff between helping and breeding, since females only rarely gain breeding status in the natal group. As for males, dispersing females originating in natal groups of four or five are underrepresented compared to expectation. Dispersing females are statistically 'eager' to leave trios and large groups of six or more adults, where their contributions in the role of helper would be unproductive (insufficient or superfluous). Departure is rarer than expected where it would reduce sibling production in the natal group (9 of 54 dispersing females originating in groups of four or five, compared to an expected 21, on the basis of the representation in the population of auxiliary females in groups of four and five, $P < 0.01$, χ^2-test).

Breeding in a large group would result in direct fitness gains that could balance the loss of indirect gains of sibling production, but competition for these prime positions is intense and often there is none available. The few females leaving groups of four or five tend to take breeding positions primarily in large groups (eight of nine dispersers) so that the transition from helping in a group just above the success threshold to breeding in a group below the threshold is an especially rare one. If auxiliary females took breeding positions as they arose, regardless of the state of the natal group, then dispersal from groups of four and five would neither be as rare nor as limited to large-group positions as it appears to be. In contrast, dispersers from trios and groups larger than five, where helping is relatively unproductive, do appear to take breeding positions in groups of different sizes in proportion to their availability (50% to small and large groups, as expected; $n = 54$ for comparison, $P < 0.02$, χ^2-test). Female dispersers from quartets and quintets also tend to be less related to young being raised in the natal group than other dispersers; almost all had lost both parents. In short, the differences between these two classes of dispersers suggest that wrens making this critical life history transition behave as if weighing the indirect component of fitness (sibling production) when vying for breeding positions.

For male and female non-reproductives alike, the potential inclusive fitness benefits of helping decline with age because of parental mortality and erosion of relatedness with young being reared in the natal group. The potential benefits of dispersal and breeding increase as competitive ability increases with age and the possibility (at least for females) of gaining a breeding position in a large group increases. Evidence in this population suggests that the ontogenetic timing of dispersal optimizes the tradeoff between cessation of helping and attempted breeding. In addition to the above-described dispersal tendencies, the modal breeding age of two years for females marks the age when rising competitive ability for breeding positions balances falling relatedness in the natal group to produce a reversal in the helping/breeding tradeoff, favoring breeding. Calculations for analyzing inclusive fitness gains accrued by different strategies at various ages should include reproductive potential weighted by competitive ability and the effects of helping on the reproductive potential of kin weighted by relatedness (Rabenold 1985). Helpers can affect survival of breeders (as in Bicolored Wrens and in Stripe-backed Wrens in bottleneck years) as well as improve future fecundity by producing more helpers for future breeding attempts.

Discussion

Campylorhynchus wrens have proven to be fruitful subjects in attempting to answer questions about how sociality in general, and cooperative breeding in particular, might have evolved. The broader question of ecological, genetic and demographic conditions favoring delayed dispersal has required a broad survey of population structure and life histories. The more incisive questions about fine tuning in life history 'decision-making' have required detailed analysis of individuals' behavior. The populations under study at Hato Masaguaral in the Venezuelan *llanos* have provided a rare opportunity to ask questions on both levels and especially to begin to address both observationally and experimentally the central question: are key parameters in individual life histories, such as helping behavior and dispersal, graded and timed to optimize contributions to the production of non-descendant and descendant kin? This 'kin-selection' issue, put more bluntly, is whether there is evidence of a tradeoff between helping and breeding, or whether breeding is the controlling factor evolutionarily and helping mainly the result of a lack of opportunity to breed.

We are looking for an aspect of social behavior that can serve as an indicator for the possible importance of one potential evolutionary route to cooperative breeding: the spread of cooperative traits through their effects on non-descendant kin, especially siblings, likely to share genes encoding those traits. Helping behavior itself has not proved very useful as such a diagnostic tool, for several reasons. When helping is universal among non-breeding philopatric individuals, it is very difficult to assess its cost. We have evidence that the cost of helping in survival is not negligible, but it is still difficult to say whether counterselection against helping, if it were of no fitness value in itself, would be strong enough to eliminate it. When helping behavior is nearly automatic and apparently indiscriminate, as in the wrens, it can mean either that it is of little selective importance or that its contribution to inclusive fitness is so reliable that it has been strictly programmed with only simple, easily satisfied, rules of thumb for its expression. Although discrimination and preferential helping of close kin, as in some species described in this book, argue that helping non-descendant kin has been evolutionarily important, lack of discrimination leading to scattered cases of helping non-kin (so long as that is not the norm) does not falsify the indirect selection hypothesis (any more than feeding brood parasites falsifies the hypothesis that natural selection favors parental behavior).

Increase in helping behavior with age among males suggests that the 'selfish' advantage of increasing aid in a group in which one is likely to inherit breeding status is probably important. On the other hand, females help as much as males in spite of their lack of selfish interest in the breeding potential of the natal group; this suggests involvement of 'indirect' selection. We turned next to another more powerful and incisive test: the dispersal 'decision' and its timing with regard to opportunity and circumstance in the natal group. It cannot be argued plausibly that this critical life history transition is of little adaptive significance, unlike providing food for another's young.

The great variation in lifetime reproductive success in the Samán population establishes that attempting breeding is a high-stakes risk in which timing and placement of effort are critical. The majority of adults spend their lives participating in breeding activities only in the role of non-reproductive helper, and nearly three-quarters fail to rear any young of their own. The social organization, demography, and breeding biology of this species combine to set up especially strong tradeoffs between philopatric helping and dispersing to breed. Dominance by principals apparently enforces a strict division between helping and breeding, and tight group membership constrains helping to the natal group. A strict seniority system further limits the options of young birds and the absence of complete habitat saturation opens up options for older birds. Poor colonization of available breeding habitat underscores the critical role that helpers play in facilitating breeding, and the sharp step–function relationship between reproductive success and group size creates clear categories of helping and breeding potential. Demographic calculations show that, given the low probability of successful breeding by young wrens, indirect delayed benefits of helping by themselves, produced by contributions of young to future breeding efforts of relatives, can be greater than those accruing to early attempted breeding (Rabenold 1985).

Testing the idea that helping is a viable alternative form of fitness enhancement is aided by the possibility of close behavioral observations to quantify motivational levels in helpers and would-be dispersers by measuring levels of effort in helping and contests for breeding status. The constancy of the breeding 'critical mass' makes it a plausible proposition that helping and dispersal tendencies could have been selected to optimize transitions among status/group-size categories. The density, viscosity and stability of the populations under study make experimentation possible on the key issues of mechanisms of reproductive enhancement and dispersal decisions, in addition to allowing the accumulation of natural dispersal data. Finally, given these facilitating factors, it should be possible to ask whether tradeoffs suggested by inclusive fitness theory are really important – whether dispersal might be a matter of trading fitness enhancement by sibling production for offspring production – and whether selection for optimization of this tradeoff has led to abilities to discriminate favorable life history transitions.

So far the answer is that Stripe-backed Wrens do discriminate breeding positions that are most likely to produce successful reproduction; considerable observational and experimental evidence, summarized in this paper, substantiates this. Furthermore, 'decisions' by dispersers, especially females, seem to be sensitive to losses of sibling production. Non-breeding females are statistically 'reluctant' to leave groups in which productivity would be jeopardized by the loss of one helper. We are continuing to test this phenomenon in our field experiments (Zack and Rabenold 1989). Female helping and choosiness in helping/breeding transitions suggest that kin selection in this population of Stripe-backed Wrens has contributed to molding tendencies to help and disperse. Indirect components of fitness contributed by helping behavior appear to be necessary to an evolutionary explanation for cooperative breeding in this species.

Of course, indirect fitness gains of the helping strategy are not sufficient to account for cooperative breeding in these wrens. Reciprocity is obviously important, especially for older male auxiliaries, but mainly takes the form of reliable helping by young birds for the older age classes that had earlier reared the younger. Reciprocity to specific donors (Ligon 1983) that aided in rearing a helper is uncommon. Similarly, habitat saturation (Stacey 1979; Koenig and Mumme 1987), while not complete, is undoubtedly a factor; much more widely available habitat would probably hasten age of first breeding, at least to some degree. Habitat saturation is probably most important in the more dense (palm savanna) Bicolored Wren population. However, the main requirement for young would-be breeders is a set of helpers, not just turf; this is argued by the continuing presence of uncolonized prime habitat. This is really just another kind of environmental constraint, imposed by predators. In fact, approximately one-third of the area in the Samán population that had been occupied by wrens in earlier years was unused in 1987; some of these areas had supported the most prolific groups, like the TM and WF areas (Fig. 6.3). Since the vegetation has not changed, we must conclude that young helpers are remaining in their natal territories for reasons other than the absence of suitable habitat for establishment of a new group.

It is difficult to say that cooperative breeding in Stripe-backed Wrens would have evolved without any of these selective forces, as Woolfenden and Fitzpatrick (1984) have been able to do for Florida Scrub Jays (see Chapter 8). We know at least that it is maintained without habitat saturation in both Stripe-backed and Bicolored wrens. Delayed dispersal undoubtedly results in part from the high level of competition for productive breeding positions with the 'critical mass' of helpers. However, the great effect of helpers that makes this competition so acutely important for lifetime reproductive success also means that helpers can contribute considerably to lifetime inclusive fitness through sibling production. Helping and dispersal tendencies among females appear to reflect such a tradeoff, suggesting at least that fitness of young birds is not determined solely by breeding; if it were, young females would be selected to take breeding opportunities as they arise regardless of the likely effects on their natal groups.

This volume, and Brown's (1987) review of cooperative breeding, establish that cooperative breeding is a heterogeneous phenomenon – critical selective forces vary greatly among species. Our study of *Campylorhynchus* has also demonstrated striking variability in potential selective forces at different times in the same populations (the current bottleneck in the Samán Stripe-backed Wren population), between congeneric populations in the same habitat (the great social differences between the two wren species in the Samán area) and between conspecific populations in different habitats (especially Bicolored Wrens in palm savanna compared to woodland). Such variability should inspire caution in interpreting short studies of single populations.

Because of the importance of the issue of the circumstances of dispersal, we are continuing the principal-removal experiments designed to test the responses of auxiliaries to breeding vacancies of varying quality. Field experiments like these are very costly in terms of time and labor, but in three years they have produced more information about the competitive process resulting in dispersal than was obtained in the previous seven years of observation (Zack and Rabenold, 1989). Further information on the sampling tactics of young females and the development of dominance status in young birds will also be instructive. Because of the great importance of the effect of helpers on reproductive success, the mechanism of this effect should be better understood. While some long-term effect of 'territory quality' (vegetation) on reproductive success is apparent, this must be better separated from the group size effect, in this study and in others. These improvements would clarify the importance of ecological and behavioral constraints on natal dispersal.

Of overriding importance is the development of greater insight into the genetic makeup of groups and populations. We need to understand better the nature of the pair bond, and the consequences of dispersal patterns and local environmental variation for population genetics. If reproduction were not strictly limited to one behavioral dominant of each sex in each group, then the advantages of philopatry would be quite different from those currently proposed. Our preliminary application of DNA fingerprinting techniques suggests that this could be the case. The production of non-descendant kin would remain as a potentially powerful impetus for cooperation, albeit redistributed, and the benefits of dispersing instead of philopatry could be diminished. If auxiliaries could produce offspring on the natal territory, the intense competition for principal status that we observe among females would be more difficult to understand. Of equal importance is greater observational sophistication in quantifying ecological constraints and motivational states of auxiliaries – their readiness to make life history transitions and the mechanisms of their assessment of alternatives.

Collaborators who have made the scope of this project possible are, in chronological order, R. Haven Wiley (who started it all), Carla Christensen, Patricia Parker Rabenold, Steven Austad, Steve Zack and Joseph Haydock. We are all (along with many other biologists) greatly indebted to Tomás Blohm for the biological station he has created at Hato Masaguaral. I thank Peter Stacey, Walt Koenig, Peter Waser, Steve Zack, Haven Wiley, Joey Haydock and Patty Rabenold for their helpful criticism of this paper. Rick Howard, Yvette Halpin, Indy Bakshi, Lee Elliot, Scott Wissinger and Hans Landel have helped in the field. The National Science Foundation supported this research with grants DEB-8007717 and BSR-8415926.

Bibliography

Anderson, A. H. and Anderson, A. (1973). *The Cactus Wren*. University of Arizona Press: Tucson, AZ.

Austad, S. N. and Rabenold, K. N. (1985). Reproductive enhancement by helpers and an experimental inquiry into its mechanism in the bicolored wren. *Behav. Ecol. Sociobiol.* **17**: 18–27.

Austad, S. N. and Rabenold, K. N. (1986). Demography and the evolution of cooperative breeding in the Bicolored Wren, *Campylorhynchus griseus. Behaviour* **97**: 308–24.

Brown, J. L. (1983). Cooperation – a biologist's dilemma. In *Advances in Behavior*. ed. J. S. Rosenblatt, Vol. 13, pp. 1–37. Academic Press: New York.

Brown, J. L. (1987). *Helping and Communal Breeding in Birds*. Princeton University Press: Princeton, NJ.

Koenig, W. D. and Mumme, R. L. (1987). *Population Ecology of the Cooperatively Breeding Acorn Woodpecker*, Monographs in Population Biology 24, Princeton University Press; Princeton, NJ.

Ligon, J. D. (1983). Cooperation and reciprocity in avian social systems. *Amer. Nat.* **121**: 366–84.

Rabenold, K. N. (1984). Cooperative enhancement of reproductive success in tropical wren societies. *Ecology* **65**: 871–85.

Rabenold, K. N. (1985). Cooperation in breeding by nonreproductive wrens: kinship, reciprocity and demography. *Behav. Ecol. Sociobiol.* **17**: 1–17.

Rabenold, K. N. and Christensen, C. R. (1979). Effects of aggregation on feeding and survival in a communal wren. *Behav. Ecol. Sociobiol.* **6**: 39–44.

Selander, R. K. (1964). Speciation in wrens of the genus *Campylorhynchus. Univ. Calif. Publ. Zool.* **74**: 1–224.

Stacey, P. B. (1979). Habitat saturation a communal breeding in the Acorn Woodpecker. *Anim. Behav.* **27**: 1153–66.

Thomas, B. T. (1979). The birds of a ranch in the Venezuelan llanos. In *Vertebrate Ecology in the Northern Neotropics*, ed. by J. F. Eisenberg, pp. 213–32, Smithsonian Institution Press: Washington, DC.

Wiley, R. H. (1981). Social structure and individual ontogenesis: problems of description, mechanism and evolution. In *Perspectives in Ethology*. ed. P. P. G. Bateson and P. H. Klopfer, vol. 4, pp. 261–93. Plenum Press: New York.

Wiley, R. H. and Wiley, M. S. (1977). Recognition of neighbors' duets by stripe-backed wrens *Campylorhynchus nuchalis. Behaviour* **62**: 10–34.

Wiley, R. H. and Rabenold, K. N. (1984). The evolution of cooperative breeding by delayed reciprocity and queuing for favorable social positions. *Evolution* **38**: 609–21.

Woolfenden, G. E. and Fitzpatrick, J. W. (1984). *The Florida Scrub Jay: Demography of a Cooperative-breeding Bird*, Monographs in Population Biology 20, Princeton University Press: Princeton, NJ.

Zack, S. and Rabenold, K. N. (1989). Assessment, age and proximity in dispersal contests among cooperative wrens: field experiments. *Anim. Behav.* (in press).

7 PINYON JAYS: MAKING THE BEST OF A BAD SITUATION BY HELPING

7

Pinyon Jays: making the best of a bad situation by helping

J. M. MARZLUFF* AND R. P. BALDA

During a casual visit to the pinyon–juniper woodland of the southwestern USA, a visitor is likely to miss seeing one of the most interesting and characteristic birds of this habitat type, the Pinyon Jay (*Gymnorhinus cyanocephalus*). This omission is simply explained as Pinyon Jays are not dispersed uniformly over the landscape, but are clumped into large flocks that maintain very large home-ranges measured in square kilometers. Thus finding them is often like locating the proverbial 'needle in a haystack'.

When a flock is located, however, an observer must be impressed by the noisy, raucous, gregarious, nature of these birds. Flocks of between 50 and 500 birds forage, roost, nest, and raise young together. Flocking is year round, and single birds are seldom seen. These tight units are particularly impressive when they become airborne and fly out of sight.

The Pinyon Jay, a robin-sized, blue bird ranges from Oregon and Montana to Baja California and Texas. As its name implies, this bird has a strong association with, and possible reliance on the pinyon pines (*Pinus edulis*, *P. monophylla*), which form an important component of the woodlands in the western USA. The relationship of the Pinyon Jay and the pinyon pines may be one of the best examples of plant–vertebrate co-evolution in North America. The bird receives nutrients, energy, nest and roost sites, and stimuli to breed from the pines, and in turn the tree relies on the bird for safe seed dispersal (Ligon 1978; Vander Wall and Balda 1981; Balda 1987).

To study such a wide-ranging mobile species, we introduced artificial feeding stations where a flock could be observed with ease, individuals color-banded, and the population censused. This flock, called the Town

* Author to whom correspondence should be addressed. Present address: Star Route, Box 2905 Dryden, ME 04255, USA.

Flock, inhabits the immediate environs of Flagstaff, AZ (Fig. 7.1). Sex, age, size, mates and offspring production of banded birds has been documented since 1972. Flock size, flock composition, dispersal, and mortality were determined by censusing all jays in the Town Flock every day that they visited feeding stations. Another flock, the Doney Park flock, which lives outside the city, was studied for five years and described in detail elsewhere (Balda and Bateman 1971, 1972; Bateman and Balda 1973).

Non-reproductive period
Flock composition and seed harvest
By early August most reproduction has ceased and young are independent of their parents. At this time the flock numbers are at a peak due to the infusion of young produced during the preceding spring. The Town Flock averages 191 jays (S.D. = ±91) during this period, but has

Fig. 7.1. Typical vegetation inhabited by Pinyon Jays in northern Arizona. Ponderosa pine dominates this habitat on the flanks of the San Francisco Mountains. Ponderosa pine habitat is not typical of all Pinyon Jay habitat. Pinyon pine–juniper woodland is more typical of lower elevation flocks.

varied from 90 in 1973 to 427 individuals in 1978 (Fig. 7.2). This variation is
primarily a result of the number of juveniles and the annual variation in
survivorship (survival rate) of yearlings. Juveniles are easily identified
because of their dull, grayish plumage, which contrasts with the deeper blue
hues of adults.

The annual infusion of young into the flock produces a seasonal cycle in
the size (Fig. 7.2) and age composition (Fig. 7.3) of jay social groups. In early
spring, before most juveniles have integrated into the flock, the flock is at its

Fig. 7.2. Seasonal changes in the size of a Pinyon Jay flock. Average
flock size from 1973 to 1982 is represented by each point. Vertical
lines show one standard deviation above and below the mean. Spring
includes March, April and May. Summer includes June and July. Fall
includes August, September and October. Winter includes November,
December, January and February.

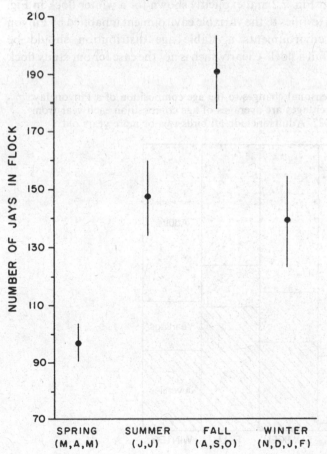

smallest size ($\bar{x} = 97.6$, S.D. $= \pm 24.6$) and composed of 75% adults and 25% yearlings (birds born the previous spring). Young produced by flock members make up one-third of the flock during the summer (Fig. 7.3) and their addition results in a 52% increase in summer flock size ($\bar{x} = 147.9$, S.D. $= \pm 31.2$) over spring flock size (Fig. 7.2). many of these juveniles disappear from the flock before the fall; however, flock size continues to increase through the fall as dispersers briefly join our study flock. At this time over 40% of the flock's members are juveniles. Flock size declines through the winter ($\bar{x} = 140.5$, S.D. $= \pm 49.7$) as many dispersers leave and some flock members die. Age composition throughout the winter is similar to that throughout the summer; approximately 30% of the flock members are juveniles, 20% are yearlings and half are adults (Fig. 7.3).

Size and composition of social units also varies from year to year. Yearly variation in flock size within each season is indicated by the large standard deviations given in Fig. 7.2 and explicitly shown for a winter flock in Fig. 7.4. This variation testifies to the variable environment inhabited by Pinyon Jays. In stable environments a stable age distribution should be approximated within a flock. Clearly such is not the case for our study flock

Fig. 7.3. Seasonal changes in the age composition of a Pinyon Jay flock. Percentages are averages of age composition each year from 1973 to 1982. Adults include all birds two or more years old.

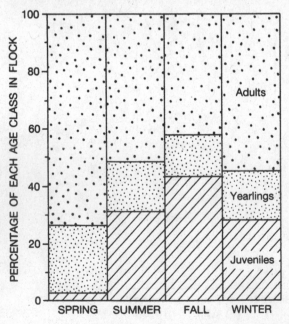

(Fig. 7.4). In particular, the numbers of juveniles vary from year to year, reaching peak abundance after exceptional productivity within the flock (1974, 1977) or after widespread immigration into the flock (1978). In contrast to this variability, there is a stable core of relatively old jays in the flock. Roughly 20% ($\bar{x}$ from 1975 to 1982 = 22%, s.D. = ± 8.2%) of the jays in a flock are five or more years old and 5% ($\bar{x}$ from 1980 to 1982 = 5%, s.D. = ± 0.25%) are 10 or more years old.

Flocks are very mobile during the autumn and travel many kilometers a day in search of ripening pinyon pine cones. Flocks leave their home ranges and often join with other flocks at areas of concentrations of ripe cones. Combined flocks may number thousands of birds (Bent 1946). Flock members recognize each other, so mixing is unlikely to result in significant dispersal between flocks (Balda and Balda 1978; McArthur 1982; Marzluff 1987). These long-distance movements occur because of the spatial distribution and temporal opening pattern of the pine cones.

Fig. 7.4. Detailed age composition of a winter flock of Pinyon Jays. The number of jays per age class is indicated by the vertical distance between the line above and the line below the age class. Ages followed by a 'plus' sign indicate jays that were at least two years old at the start of our study. This unknown-age cohort comprised 37.5% of the flock in 1973 and declined to 4.0% of the flock in 1982. Total flock size each winter equals the sum of all age classes and is shown by the uppermost point each year.

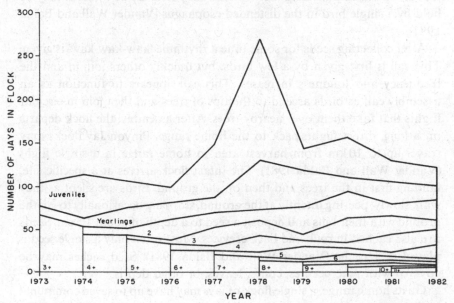

Pinyon pines produce seed crops with 'boom' or 'bust' frequency. Every five to seven years a heavy crop of cones is produced over a wide geographical area. In these years nearly every tree bears some cones. In marked contrast, in other years virtually no cones are produced over the same area, or few trees bear only a few cones. In these years, birds must search every hillside, canyon and wash for clumps of cone-bearing trees.

Flocks gather at these 'hot spots' and may spend 4–6 h foraging there. When located, a mast crop is first consumed. Pinyon pine seeds are relatively large, high in fats and carbohydrates, and contain adequate amounts of protein (Botkin and Shires 1948; Clark 1977). When cone density is low, flocks move slowly from tree to tree in search of full cones. Usually the direction of movement is the same for all flock members and flocks may criss-cross one another. Vocalizations at this time act primarily as contact calls to keep flock members informed of the whereabouts of other members. If cones are ripe but still closed, birds will twist the cones from the branch and carry them to sites conducive to holding the cone with their feet and hammering them with their strong, pointed bills. Open cones are visually inspected and edible seeds extracted with the bill. Pinyon Jays lack feathers covering the nostrils, which may aid the bird in extracting seeds from pitch-laden cones (Ligon 1978).

If cones are abundant the jays are soon satiated. Extracted seeds are now held in an expandable esophagus, which produces a noticeable bulge in the throat region, emphasizing the white throat patch. Up to 57 seeds may be held by a single bird in the distended esophagus (Vander Wall and Balda 1981).

After collecting seeds for some time a rhythmic 'kaw-kaw-kaw' is given. This call is first given by a few birds, but quickly others join in and the frequency and loudness increases. This call appears to function as an assembly call as birds ascend to the tips of trees and then join in circular flights that take them over nearby trees. After assembly, the flock departs on a long, direct flight, back to the home range. Pinyon Jay flocks may travel up to 10 km from harvest area to home range in a single flight (Vander Wall and Balda 1981). The intact flock arrives at a specific site, landing first in the trees and then on the ground. Birds are silent as they walk slowly, peering intently at the ground, stopping occasionally to jab the ground with their bills and deposit a seed to a depth of about 1.5 cm. Seeds are also cached in rock and bark crevices. Most often only a single seed is placed in a cache (Vander Wall and Balda 1981). Seed caches may be covered with pine needles, cones, leaves, or other debris.

On its home range, a single flock of jays may have up to seven communal caching areas, which are used periodically throughout the fall. Caching areas are often sparsely covered with vegetation, have patches of exposed soil and are exposed to winter sunlight. When trees are present in such areas, birds often cache seeds under the canopy, near the truck, and on the south side, where snow melt is greatest. Caching also occurs in areas where trees have been removed or where wildfires have occurred.

Birds continue to cache as long as seeds are available and caching areas remain snow free. Ligon (1978) calculated that a single flock of about 300 birds in central New Mexico cached up to 4.5 million seeds in a mast year. These seeds act as a larder of stored food for future use when other foods are scarce or absent.

Late summer and fall (July, August, September) is also the time of maximum social interaction within the flock, especially among the numerous juvenile birds. At feeding stations, juveniles engaged in aggressive interactions among themselves at frequencies far above those expected by chance, and at frequencies greater than among other cohorts except for older males (Balda and Balda 1978). It was obvious that older birds in the flock avoided conflicts with juveniles who were bold and testy in approaching the feeding platform. Mature birds often stopped feeding or moved to a new feeding location when juveniles approached. In terms of social status, juveniles appeared to be of equal stature with adult males, adult females and yearling males. Juveniles won 63% of their encounters with yearling females, a cohort that consistently deferred to them (Balda and Balda 1978). This picture changed dramatically in late autumn, however, as now adult males dominated young 72% of the time ($n = 254$ encounters), and juvenile–juvenile interactions declined dramatically to chance levels.

Waiting as a social consequence

One conspicuous behavior observed during intense foraging bouts was the presence of up to 15 birds sitting silently high in trees above the flock. These birds were sometimes deep in the foliage, sometimes exposed. Individuals were usually positioned around the periphery of the flock and appeared to be sentries as reported by Cary (1901) and Balda and Bateman (1971). At the approach of intruders these birds were first to give a loud rhythmic repeated 'krawk'. The flock ceased feeding immediately and took shelter in the trees. Flock members either mobbed the intruder vigorously, or started a 'din', which was then followed by a long flight. The flock appeared to be coordinated and integrated during these activities. In the field it was difficult to identify individual sentinels, except that most were adult birds.

An interesting variation of sentinel behavior occurred at feeding stations visited by the Town Flock. Visits to these concentrations of food were often short (5–10 min) and frantic, as birds scurried about to feed and collect seeds. Most often the flock arrived together. Dominant males normally fed unimpeded and consequently they were first to fill their esophagi. Subordinates had to wait before feeding, as also reported for Gray-breasted (Mexican) Jays (Craig *et al.* 1982). Dominants then left the feeder and sat high in the foliage as had the sentinels described above. They gave warning calls when approached by potential predators. When the flock left the feeder, these sentinels flew off with them. Thus dominants also must wait; they wait for subordinates to feed.

The apparent function of the sentinels is clouded by the observation that these birds may simply be waiting for the remainder of the flock to finish feeding. They may give danger calls more often than other birds simply because they are in a vantage point to see approaching intruders. Waiting for other birds may be a common behavior in animals living in social units as tightly integrated as a Pinyon Jay flock. At most times of the year it is highly unusual to see a single bird some distance from a flock. Predator protection and social facilitation may be two important reasons for waiting, as single birds may be in jeopardy.

Dispersal

Dispersal from the flock occurs during the fall. With levels of aggression relatively high and long-distance flights prevalent, it is not surprising that many juveniles (of both sexes) leave the flock at this time. Some of these birds (primarily females) become permanent members of flocks that they were not born in, while others (primarily males) return to their natal flock after several months to years of exploratory wandering. Most birds that disappear are never seen again.

Despite long-distance travel during the fall, most dispersal is over short distances, primarily between neighboring flocks. Long-distance dispersal is rare (Fig. 7.5).

Annual rates of dispersal are closely correlated with mate availability (Marzluff 1987). Young females were more likely to leave their natal flock when the operational sex ratio of the flock was only slightly male biased or when it was female biased (Fig. 7.6). In contrast, when the sex ratio was strongly male biased, many females did not emigrate but young males did. The fall exodus of juveniles and yearlings from their natal flock thus appears to follow an assessment of potential mating opportunities. As male-biased sex ratios prevail in most flocks (Ligon and White 1974; Marzluff and Balda 1988*a*), most young females find mates in new flocks, but most young males do not Males then return to their natal flock where they may (*a*) find a mate, (*b*) become a non-breeder, non-helper for a year or (*c*) help their parents to breed.

Survivorship

Fall is the time of heaviest mortality, perhaps because of increased activity in relatively unfamiliar areas associated with the fall pine seed harvest. Annual survivorship is relatively high in this species. We have followed the survivorship of 708 birds from 1972 through 1986. These represent 14 cohorts of juveniles born from 1972 to 1984. Averaging over

Fig. 7.5. Distribution of distances jays were known to disperse between birth and breeding. A distance of zero indicates breeding in the natal flock; a distance of 0.1–15 km indicates breeding in the flock neighboring the natal flock. Starred distances indicate that one jay of unknown sex was observed in a flock that distance from its place of birth (breeding uncertain). Data include jays banded in the Town Flock (*n* = 134) and all reported recoveries of jays banded throughout the western United States (*n* = 170).

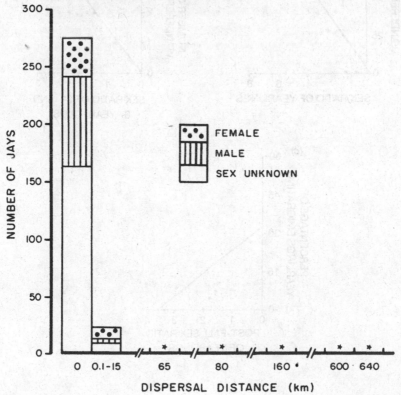

Fig. 7.6. Correlations between sex ratio (number of males per female jay) and various indices of dispersal. Dots represent sex ratio and dispersal per year from 1975 to 1982. Lines were fitted by least-squares estimation. (*a*) Percentage of wanderers banded in the fall that remained with the Town Flock through December plotted as a function of the sex ratio of yearlings during the fall ($r = 0.87$, $P = 0.01$). (*b*) Number of wanderers banded in the spring that remained with the Town Flock through December plotted as a function of the sex ratio of two- and three-year-olds in the Town Flock during the spring ($r = 0.83$, $P = 0.01$). (*c*) Percentage of yearlings in the Town Flock that emigrated during the spring as a function of the sex ratio of yearlings in the flock after fall immigration ($r = -0.52$, n.s., but notice that emigration was greatest during the two years when yearling sex ratios were female-biased). (Reprinted with permission from Marzluff, 1987.)

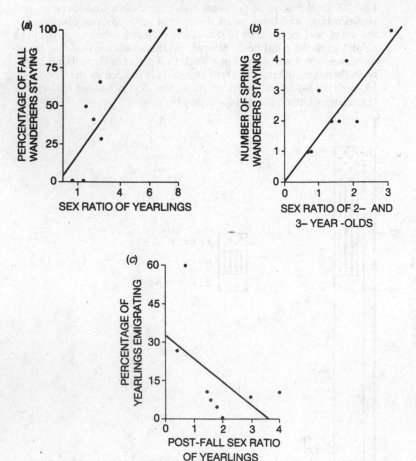

the years, 74% of adults, 62% of yearlings, and 41% of juveniles survived each year. Juvenile 'mortality' includes death and disappearance from the flock. We searched neighboring flocks and investigated banding returns nationwide to assess dispersal. Few dispersers have been located and we assume that the vast majority die.

Survivorship was significantly greater for males than for females (Fig. 7.7). Typically we do not determine sex until birds breed, thus Fig. 7.7

Fig. 7.7. Age-specific survivorship of Pinyon Jay males (solid lines, $n = 102$) and females (dashed lines, $n = 77$). Survivorship of males and females born from 1972 to 1982 was followed from 1972 to 1986. Survivorship curves were significantly different (Breslow's Wilcoxon = 5.56, d.f. = 1, $P = 0.02$).

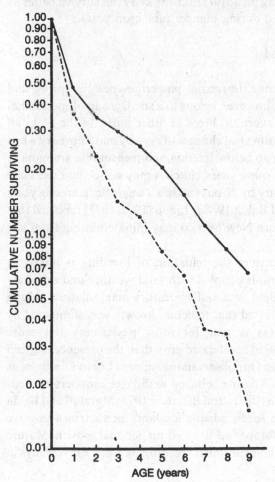

presents the mortality of breeders. Females may have lower survivorship because they perform all the incubation and brooding. Great-horned Owls (*Bubo virginianus*) kill some incubating females. In addition, increased energetic demands on incubating females, especially during the early spring, may weaken them prior to the annual fall molt and seed harvest.

All age classes have substantial year-to-year variability in survivorship. This variability is more closely correlated with weather than with abundance of pine seeds (Marzluff and Balda 1988a). Cold, snowy weather is not associated with low survivorship of Town Flock jays. On the contrary, years with high levels of precipitation, either in the form of snow or rain, are associated with high survivorship. Juveniles, but not adults, have higher survivorship following bumper pine crops. Juveniles survive best during winters following plentiful fall crops and even survive better as yearlings if they were raised during bumper pine crop years.

Reproductive period
Time of breeding
Most temperate-zone, terrestrial passerines nest in spring and early summer. Pinyon Jays, however, belong to a small group of species that can, when conditions are favorable, breed at other times. In the Flagstaff area, females may be incubating full clutches of eggs by mid February when night-time temperatures drop below freezing and measurable amounts of snow often accumulate. In other years, clutches may not be laid until late April. Thus, nesting may vary by 50 days within a single flock among years (see Fig. 13 of Marzluff and Balda 1988a). Ligon (1971, 1978) reported that Pinyon Jays in south-western New Mexico may initiate nesting from late January to mid August.

In most temperate passerines the initiation of breeding is linked to increasing vernal photoperiods coupled with mild weather and adequate food (Baker 1938). Using field data and laboratory manipulations, Ligon (1971, 1974, 1978) demonstrated that testicular growth was stimulated by increasing photoperiods (as is true for most passerines), but more importantly that testes showed accelerated growth in the presence of green cones and pinyon pine seeds. Our observations support Ligon's findings, as early breeding in northern Arizona is linked with large numbers of seeds cached the previous autumn (Balda and Bateman 1972; Marzluff and Balda 1988a). Thus, pinyon pine seeds, when abundant, interact in a positive manner with increasing photoperiod to speed up gonadal growth. Mature green cones actually stimulate autumnal breeding by some Pinyon Jays in New Mexico (Ligon 1974).

In the Flagstaff area, as in New Mexico, a selective premium seems to be placed on nesting early in the year. The timing of breeding in passerines is generally believed to occur when maximum reproductive success is possible (Lack 1968). Yet, incubating eggs and caring for helpless young in a snow storm hardly seems likely to increase this potential for success.

Early fledging may ensure that physical, social, and mental maturation occurs by late summer, when flock integration is paramount. Long-distance flights to harvest pine seeds presumably require well-developed and efficient flight muscles and coordination. The transport of vast numbers of pine seeds to specific caching areas requires highly coordinated and integrated flock behavior. These physical and mental characteristics may develop gradually over the course of the summer.

Pair bonds

Pinyon Jays mate for life. We observed only three 'divorces' out of 180 pair bonds (2%) observed over the past 15 years. Because pair bonds are permanent one would expect to see family units clearly defined within the flock during the breeding and non-breeding seasons. Pairs, however, were not apparent during the non-reproductive period nor did they appear to coordinate their movements or behaviors in any way.

One result of these permanent pair bonds is a high degree of synchrony in the onset of breeding. When courtship is initiated it is simultaneously performed by most adult birds. Thus, flock cohesion is maintained even during this critical period, and the time, energy and conflict normally associated with mate attraction and bonding are avoided. This synchrony also indicates that most pairs are energetically and physiologically ready to breed at about the same time.

Nesting

The Doney Park flock maintained a traditional nesting grounds of about 100 ha, but used no more than one-third of this in any year. We can find no physical or vegetative reason for continual use of this area, save the fact that it borders a productive pinyon pine woodland from which pine seeds were harvested. Nest sites were abundant throughout the total home range. In each year, nests were spaced so that, on average, there was one nest per hectare. On three occasions when nests were within 100 m of each other, one pair continually robbed nest material from the other, causing nest desertion in two cases and delayed breeding in the other. Thus, inter-nest distance may be related to the patchy distribution of nesting material. Nests spaced farther apart than the distance to naturally occurring materials may be uneconomical to pilfer.

The spatial distribution of nests in a colony may be related to nesting success in species with small inter-neighbor nest distances (Bongiorno 1970; Tenaza 1971; Patterson 1965; Coulson 1968; Parsons 1977) but has not been clearly demonstrated in Pinyon Jays (Balda and Bateman 1972). In some years nests on the periphery of the colony are more successful, whereas in other years nests in the center of the colony are more successful (Balda and Bateman 1973; Gabaldon 1978). These differences may be confounded by time of nesting and intensity of predation by different predators. Also, the age of breeders was not related to nest placement in the colony, as there was no tendency for either older or younger birds to nest in specific sectors of the colony (Gabaldon 1978).

In the Town Flock, where nesting areas changed yearly, there was no consistent nest placement by known pairs within the colony among years. Pairs that nested in the colony center one year were just as likely to nest at the periphery the next year, thus there appears to be no strong selective pressure operating for, or against, birds nesting in different sectors of the colony (Gabaldon 1978).

An interesting investigation conducted by Gabaldon (1978) on nest position with respect to relatives in the flock revealed a striking relationship. She divided all nests for a year into two categories. A nest was designated 'near' if the distance between relatives' nests was less than the mean of all inter-neighbor nest distances. Nesting 'far' from a relative was defined as nesting at a distance greater than the mean inter-neighbor distance. All relatives of both members of the pair were used in this survey. Over a three year period, a total of 19 pairs of nests of related birds were located. In seven cases where relatives' nests were located within the same colony and thus initiated at about the same time, the distances between these nests was close to the maximum inter-neighbor distance known for that year. Out of all 19 pairs only one pair nested 'near' a relative. She concludes that relatives avoid each other's nests. In the course of her study she noted that predation on nests frequently occurred in clumps within the colony over a time span of a day or two. Further observations of predation on 'dummy nests' containing quail eggs confirm this finding and suggest that nest predators, especially crows and ravens, forage in an area-restricted pattern. It would therefore be advantageous to nest as far as possible from all relatives to avoid contributing to a predator's search image or search intensity.

In the Flagstaff area, most early nesting attempts fail due to harsh weather conditions and severe predation (Marzluff 1988a). Nesting must not be a short-term energy drain, as by mid April some pairs have already lost three clutches of eggs. Pairs that fail in their first attempts in the colony do not re-nest therein but go off some distance and form satellite colonies (Balda and Bateman 1971). As the season progresses, fewer satellite colonies are formed and they diminish in size as pairs succeed. By mid to late June unsuccessful pairs may be nesting in isolation. This is the only time of year when lone birds are regularly (but rarely) seen.

Most females in a colony started incubating within 5 days of each other. Incubation normally began with the laying of the third egg, even in clutches of four or five eggs. During incubation females rarely left the nest, and were fed by their mates at a rate of about once every 73 min. This feeding rate is highly variable, however, and is determined in part by the males' ability to procure food. During cold weather with snow covering the ground, four incubating females were fed only 3.8 times per day. When conditions were exceptionally harsh, females deserted nests, sometimes eating young before departure (Balda and Bateman 1976).

During this period males of incubating females often foraged together as an all-male foraging flock. Foraging forays often took this subunit of the flock up to 2 km from the nests. After intense foraging both on the ground and in foliage, the males assembled high in one or two trees after a series of rhythmic contact calls were given. Males now flew silently as a group back to the nesting area. Upon approaching the area a characteristic 'near-er' call was given. Incubating females became alert and prepared for feeding. Males split up and flew to the vicinity of their individual nests. Feeding of females lasted no more than 45 s, often much less. During food transfer the silence of the nesting area was broken by the loud, begging calls. After feeding, the area again was serene and quiet. The progression from silence to loud begging to silence was pronounced and dramatic.

The integration and coordination of the males during this time is striking and may have two important functions. Group foraging may be beneficial in finding food and increasing vigilance for predators, thus benefiting the males directly and the females indirectly. The synchronous return to the colony means all females are fed at the same time. Predators thus have little time to use vocalizations to locate nests and females. Females also recognize the 'near-ers' of the mates and are therefore prepared for an efficient and rapid transfer of food (McArthur 1982; Marzluff 1988b). Only mates were seen feeding females during the incubation period, and the males were not accompanied by any other birds.

Incubation was 17 days, with eggs hatching over two or three days. Asynchronous hatching normally resulted in a brood of three young of equal size and one or two smaller young (Bateman and Balda 1973).

Pinyon Jay nestlings are fed a variety of arthropods as well as pine seeds that have been recovered by parent birds from caches made the previous fall. In 838 h of observations at 53 nests, feeding visits averaged 1.3 per hour (Marzluff 1983). Infrequent feeding visits were possible because parents brought large amounts of food to the nest each visit. Parents were seen approaching nests with their esophagi extended. Strong lateral shaking motions were used to dislodge food items into the bill. Food samples removed immediately from collared young were covered with copious amounts of a clear mucilagenous secretion (Bateman and Balda 1973).

During the latter stages of nestling life, an interesting behavior occurred during the warm mid-day periods. Groups of adults would fly slowly and quietly through the colony, stopping and peering into active nests. Sometimes young were preened or even fed (Balda and Bateman 1971). During these visits young often crouched in the bottom of the nest but occasionally they erupted into loud begging. These 'nursery visits' could serve to assess density and age of nestlings in the colony as well as the reproductive competence of the parents.

Most Pinyon Jays did not breed until two years of age (Fig. 7.8); however, relative to other cooperative breeders, breeding by yearling Pinyon Jays is substantial (Brown 1987). While most of the flock was nesting, a flock of first-year birds roamed the home range, often visiting the initial colony as

Fig. 7.8. Age of first breeding in Pinyon Jays. Percentages are based on $n = 63$ males (M) and $n = 66$ females (F). (Reprinted with permission from Marzluff and Balda, 1988b.)

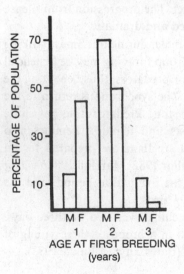

AGE AT FIRST BREEDING
(years)

well as subsequent satellite colonies. This non-breeding yearling flock was most often quiet and inconspicuous. These birds, however, were very quick to give the rhythmic danger calls and thus served as an alarm system for the nesting birds. They also promptly mobbed potential predators. Some of the birds in this group appeared to be establishing pair bonds as *Courtship feeding* and *Silent food transfer* were not uncommon.

Helping at the nest

Description of help. Shortly after the young hatch, a portion of the yearlings join the adult male foraging flock. All but one of 22 helpers were males. Data from 1973 to 1986 for the color-banded Town Flock revealed that all of these males ($n = 21$) were sons of the pair from the previous year. We never observed helpers feeding their mothers; however, after the female was emancipated from brooding, helping sons contributed substantially to the feeding of nestlings and to nest sanitation (Table 7.1). Helping sons typically visited the nest in the company of one or both parents (74% of 72 visits; Fig. 7.9). Helpers fed young more than they cleaned the nest, as is typical for males (Table 7.1). In fact, nestlings were fed more by helpers than by their mothers. Observations of helpers picking up food at feeding

Fig. 7.9. Frequency distribution of composition of visits to nests with Pinyon Jay helpers. A total of 124 visits at four nests is presented. Visits by males only (M), females only (F), helper only (H), male plus female (M F), male plus helper (M H), female plus helper (F H), and male, female and helper (M F H) are shown.

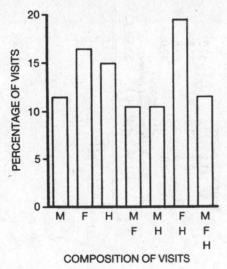

COMPOSITION OF VISITS

Table 7.1. *Nesting behavior of pairs of jays with helpers and pairs without helpers. All observations were conducted on nests with four young, from 1982 to 1984*

	Number of hours observed	Number of feedings per hour	Time(s) feeding per hour				Time(s) cleaning per hour		
			Total	Male	Female	Helper	Male	Female	Helper
Nests with helpers									
A	43.5	2.6	57.2	26.5	10.7	20.0	16.6	148.8	21.3
B	10.8	1.8	72.8	44.6	15.2	13.1	38.0	68.8	5.5
C	6.0	2.5	83.8	44.0	19.2	20.7	15.7	15.7	3.2
D	7.0	1.6	49.0	19.7	7.7	20.9	29.3	39.6	30.0
Average at nests without helpers									
	321.2	1.4[a] ± 0.3	53.3 ± 12.1	39.2 ± 13.3	14.0 ± 6.6	—	35.4 ± 23.4	76.4 ± 66.1	—

[a] $\bar{x} \pm$ S.D. of 17 nests.

stations suggest that helpers were selecting the same foods for young as were parents.

The primary effect of helpers was an increase in per-nest rate of feeding (Table 7.1). All four nests with helpers had a mean rate of feeding that was significantly greater than the average rate at nests without helpers (Table 7.1; Mann–Whitney $U = 160$, $P = 0.02$). Total time spent feeding per hour at three of four helper nests (Table 7.1; nests A, B, C) was higher than time spent feeding at the average nest without helpers. Average feeding time per hour was not significantly higher at nests with helpers. It appears, therefore, that pairs with helpers are provisioning their young more often, but not for significantly longer periods.

A reduction in time spent feeding by females with helpers may be the reason nestlings with helpers receive only slightly more food than unaided nestlings. Females at nests with helpers fed for a median of $12.9 \, s \, h^{-1}$ $(n = 4)$ but unaided females fed for a median of $20.3 \, s \, h^{-1}$ $(n = 15)$. These differences approach significance (Mann–Whitney $U = 169$, $P = 0.06$), suggesting that aided females reduce their participation in feeding nestlings. No difference in male feeding rates was evident (median: with helper $= 35.3 \, s \, h^{-1}$, without helpers $= 34.9 \, s \, h^{-1}$; Mann–Whitney $U = 171$, $P = 0.81$).

Lineage-specific helping. After six years of following the Town Flock, we began constructing 'family lineages' for the various extended families in the flock (Fig. 7.10). One fact became obvious immediately. Some family lineages were extremely successful in producing young, while others were not. Those lineages that were very successful were also the ones that occasionally had helpers. We call all members of lineages that have had helpers H-lineage birds (see case histories 1 and 4, pp. 233–4). It is important to realize that many H-lineage birds have not been helpers nor have they been helped. These individuals belong to an H-lineage by birthright because their relatives have participated in these behaviors (Table 7.2). Descendants of families that have never shown helping behavior are called NH-lineage birds.

H-lineage individuals had greater annual fecundity and lower mortality than did NH-lineage birds. We calculated the average annual fecundity

Fig. 7.10 (see pp. 218–19). Genealogical graph of Pinyon Jay Town Flock families from 1971 to 1986. Letters across the top of the figure are for reference only. See the text for explanation of symbols and for comments on specific family histories.

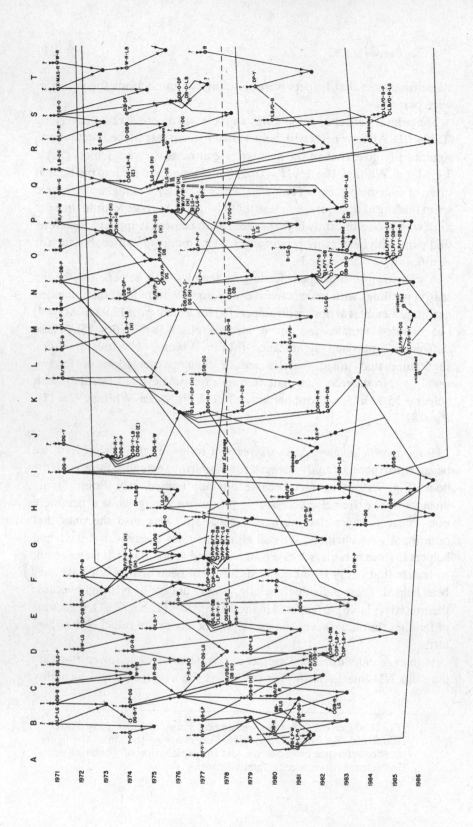

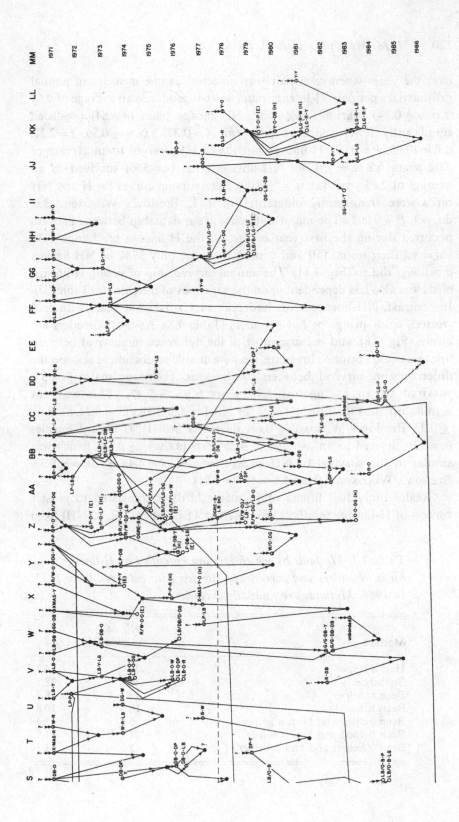

over the years when each pair was observed as one measure of annual productivity per pair. H-lineage pairs ($n = 64$) produced an average of 0.64 (s.d. = ± 0.84) yearlings per year. NH-lineage pairs ($n = 84$) produced significantly fewer yearlings per year ($\bar{x} = 0.37$, s.d. = ± 0.56, $t = 2.21$, d.f. = 103.5, $P = 0.03$). H-lineage birds ($n = 128$) survived to an average of 3.02 years (s.e.m. = ± 0.26). NH-lineage birds ($n = 506$) survived to an average of 2.13 years (s.e.m. = ± 0.09). Survivorship curves for H and NH birds were significantly different (Fig. 7.11; Breslow's Wilcoxen = 8.6, d.f. = 1, $P = 0.003$). The major difference in survivorship between lineages occurred during the first year of life. Of the H-lineage crechlings 47% survived their initial fall and winter; however, only 37% of NH-lineage crechlings did so (Fig. 7.11). The annual survivorship of young H-lineage birds was also less dependent upon the vagaries of the climate (Table 7.3). In contrast, NH-lineage birds had poor survivorship in years with cold winters, cold springs or hot summers (Table 7.3). As most breeders are adults (Fig. 7.8), and because most of the difference in survival between lineages occurs prior to breeding, we were unable to document sex-specific differences in survival between the lineages. H-lineage males ($n = 37$) survived to an average of 5.73 years (s.e.m. = 0.47). This was not significantly less than the average for NH-lineage males ($\bar{x} = 5.24$, s.e.m. = ± 0.33, Breslow's Wilcoxon = 0.68, d.f. = 1, $P = 0.41$). H-lineage females ($n = 12$) survived to an average of 4.25 years (s.e.m. = ± 0.68), which was similar to the survival of NH-lineage females ($\bar{x} = 4.48$, s.e.m. = ± 0.32, Breslow's Wilcoxon = 0.06, d.f. = 1, $P = 0.81$).

Greater individual fitness of H-lineage birds is translated to greater success of H-lineage families. We observed 41 families (9 H, 32 NH) from

Table 7.2. *Methods by which jays are classified as H-lineage birds. Numbers and percentages are for data collected from 1972 to 1986. Methods are mutually exclusive*

Method	No. of individuals	%
Had a helper	23	24.7
Birthright	36	38.7
Been a helper	17	18.3
Been helped	10	10.8
Been helped and been a helper	5	5.4
Been helped and had a helper	1	1.1
Been a helper and had a helper	1	1.1

1972 to 1986. H-lineage families produced significantly more emigrants and more breeders in the flock than NH-lineage families (median number of emigrants: H = 1.0, NH = 0.0, Mann–Whitney $U = 258.5$, $P = 0.03$; median number of breeders: H = 3.0, NH = 1.0, $U = 277$, $P = 0.006$). As of 1986 all nine H-lineage families were extant, but 18 of 32 NH-lineage families had gone extinct ($\chi^2 = 9.02$, d.f. = 1, $P < 0.005$).

Greater relative success of H-lineage families resulted in an increase in the abundance of H-lineage breeders in the Town Flock (Fig. 7.12). In 1973,

Fig. 7.11. Age-specific survivorship of H-lineage (solid line, $n = 128$) and NH-lineage (dashed line, $n = 506$) Pinyon Jays. Survivorship of males, females and birds of unknown sex born from 1972 through 1982 was followed from 1973 to 1986.

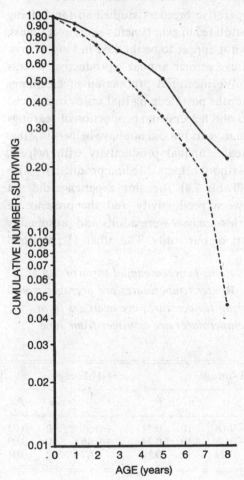

25% of 28 breeding pairs were H-lineage pairs. The proportion of H-lineage breeders reached a peak (16 of 33 breeders, 48%) in 1979 and 1980. Recently the proportion of H-lineage breeders has declined as many young NH-lineage pairs began to breed. Few recent H-lineage pairs have been as successful as their ancestors were in the mid 1970s. Over 14 years of study there has been a significant increase in the proportion of H-lineage breeders in the flock ($r = 0.57$, d.f. $= 12$, $P < 0.05$).

Pinyon Jays differ from other species detailed in this volume because helping at the nest is relatively rare and is restricted to certain extended families. Perhaps through understanding why some families have helpers, but others do not, we can gain insight into why helping is more common in other species.

Is helping helpful? In most cooperative breeders studied so far, helping and/or helped individuals are postulated to gain benefits in indirect fitness, direct fitness, or both. This does not appear to be the case in Pinyon Jays. Breeders with helpers did not have greater annual reproductive success than breeders without helpers. We measured production of crèchlings (young surviving two to three months post-fledging that reside in a crèche and are cared for by their parents and helpers) and production of yearlings by 16 pairs that had helpers in some years but did not have helpers in other years. Only four pairs had greater annual productivity with helpers, compared to their productivity without helpers. Median productivity with and without help was similar (Table 7.4). Breeding experience did not confound the relationship between productivity and the presence of helpers. Five of the 16 pairs in this analysis were adults and presumably experienced breeders at the start of our study. The other 11 pairs had

Table 7.3. *Lineage-specific correlations between annual survivorship of jays and annual weather factors. Winter temperatures are averages from December through February. Spring temperatures are averages from March through May. Summer temperatures are averages from June through August*

Weather	Age	H-lineage			NH-lineage		
		r	n	P	r	n	P
Winter temp.	Juvenile	−0.18	10	0.31	+0.61	10	0.03
Spring temp.	Yearling	−0.26	10	0.24	+0.46	10	0.09
Summer temp.	Yearling	−0.09	12	0.39	−0.67	12	0.01

similar years of breeding experience prior to having help ($\bar{x} = 2.5$ years, S.D. $= \pm 1.14$) and prior to not having help ($\bar{x} = 1.8$ years, S.D. $= \pm 1.35$; paired $t = -1.85$, d.f. $= 10$, $P = 0.09$).

Helping might be advantageous to helpers in several ways: (*a*) by providing breeding experience that would raise their reproductive success; (*b*) by removing them from the conspicuous, inexperienced, non-breeding flock of yearlings, which may suffer more predation and food shortages than the experienced flock of breeders; and (*c*) by enabling them to gain status in the eyes of potential mates and competitors, thus allowing helpers to secure higher-quality mates than birds not helping (Zahavi, personal communication). Our data did not support any of these hypotheses (Table 7.4). Breeding males that helped and males that did not help produced equal numbers of crèchlings and yearlings per breeding attempt. Helping, in fact was often costly. Nearly one-quarter of helpers (5 of 22, 24%) did not live long enough to obtain a mate and breed. In contrast, only one of 10 yearling males that did not help died before breeding. Although suggestive of a trend, the proportion of helpers versus non-helpers that eventually bred was not significantly different (Fisher's exact test, $P = 0.64$). Helping did not further the lifespan of yearling males. Lastly, helping did not allow males to obtain high-quality mates. Heavy female Pinyon Jays are preferred by males (Johnson 1988) and males mated with heavy females are very successful (Marzluff and Balda 1988c). Males that helped their parents did not obtain females larger than males who did not help.

Fig. 7.12. Annual variation in percentage of the Pinyon Jay Town Flock breeders that belonged to the H-lineage. Number of breeders censused per year from 1973 to 1986 are: 28, 36, 46, 42, 34, 34, 31, 33, 45, 41, 42, 39, 38, 38.

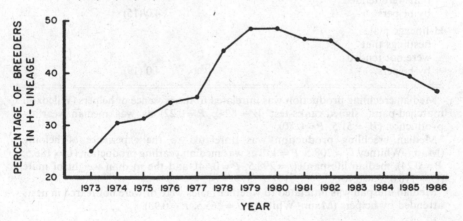

Helpers appear to increase the rate at which nestlings are fed (Table 7.1). Perhaps these nestlings survive better than nestlings that are not helped. We showed that helpers do not increase production of crèchlings or yearlings, therefore survival of nestlings over the first few months or first year of post-nestling life was not enhanced by the actions of helpers. Lifespans of yearling males raised with helpers also did not differ from lifespans of yearling males raised only by their parents (Table 7.4).

When does helping occur? We provide details on many of the variations in helping behavior in our last section, which deals with the genealogical

Table 7.4. *Impact of helpers on annual productivity and longevity of Pinyon Jay breeders, helpers, and helped offspring*

	Median crèchlings per breeding attempt (n)	Median yearlings per breeding attempt (n)	Median lifespan (years) of yearlings (n)	Median body weight (g) of mate (n)
Pairs with helpers[a]	1.0 (16)	0.25 (16)		
Pairs without helpers[a]	1.0 (16)	0.63 (16)		
Males that helped[b]	0.80 (16)	0 (16)	4.0 (21)	101.0 (10)
H-lineage males that lived 1 year and did not help[b]	0.67 (9)	0.20 (9)	6.0 (8)	100.0 (11)
Male nestlings that were tended by helpers[c]			4.0 (15)	
H-lineage male nestlings that were not tended by helpers[c]			4.0 (19)	

[a] Median crèchling production was unrelated to the presence of helpers (Wilcoxon matched-pairs, signed-ranks-test $W = 82.0$, $P = 0.22$) as was median yearling production ($W = 51.5$, $P = 0.70$).
[b] Median crèchling production was unrelated to the experience of helping (Mann–Whitney $U = 208.5$, $P = 1.0$) as was median yearling production ($U = 186.5$, $P = 0.24$), median lifespan ($U = 298.5$, $P = 0.44$), and the median weight of mate ($U = 116.0$, $P = 0.70$).
[c] Median lifespan was unrelated to whether males were or were not reared in nests attended by helpers (Mann–Whitney $U = 265.5$, $P = 0.93$).

structure of the Town Flock. As a summary of that information we provide the following description of the occurrence of helping. At a proximate level, helping by sons is obviously contingent upon production of sons. H- and NH-lineages differ dramatically in this respect; H-lineage pairs produced over five times as many male yearlings as female yearlings (33 males: 6 females) whereas NH-lineage pairs produced only twice as many males as females (25 males: 13 females) ($\chi^2 = 4.1$, d.f. = 1, $P < 0.05$). Overproduction of males by H-lineage pairs may indicate that NH-lineage pairs simply produce many females, most of which die or disperse before breeding (when identification of sex is made). This does not appear to be the case. As we showed above, H-lineage families produced more emigrants than NH-lineage families. Some yearling sons born to H-lineage parents did not help (8% of 34 sons (36%) did not; Table 7.5). Sons that survived long enough to help and that were reared without the aid of helpers were no less likely to help than were sons reared with the aid of helpers (41% of 32 unhelped sons became helpers and 38% of 13 helped sons became helpers). Most importantly, although NH-lineage pairs produced several yearling sons potentially able to help ($n = 25$), none did (Table 7.5). Instead, many of these NH-lineage sons bred as yearlings (8 of 25 sons; 32%). Fewer H-lineage sons bred as yearlings (4 of 34 sons; 12%). Lineage differences in the occurrence of breeding by yearling sons approached significance ($\chi^2 = 3.65$, d.f. = 1, $P < 0.10$).

The frequency of helping at all nests varied greatly between years (Fig. 7.13). During the course of this study helping occurred at as few as 2% of the nests (1982) and at as many as 47% of the nests (1977). J. D. Ligon (personal communication) also reports a low frequency of helping in New Mexico. A strong correlation existed between frequency of helping and the fall sex ratio of yearlings ($r = 0.86$, d.f. = 6, $P = 0.05$). Helping was most prevalent in years when the sex ratio was strongly biased in favor of males.

Table 7.5. *Family lineage differences in behaviors of Pinyon Jay Town Flock male yearlings*

Lineage	Number of sons	Helped parents		Bred as a yearling		Non-breeder, non-helper	
		n	%	n	%	n	%
H	34	22	65	4	12	8	24
NH	25	0	0	8	32	17	68

Helping may be an avenue for helpers to escape the potential rigors of life in the small, conspicuous, non-breeding yearling flock. If this hypothesis were correct, we would expect helping to be especially prevalent in years when this flock is very small and benefits of flocking presumably low. To the contrary, there was no significant correlation between the percentage of nests with helpers each year and the number of non-breeding, non-helping yearlings in the flock ($r = 0.08$, d.f. $= 6$, not significant (n.s.)).

Why does helping occur? Helping in Pinyon Jays is correlated with the abundant production of yearling males by certain families (H-lineages) and with periodic scarcity of unmated females. The interesting point is that, unlike most other avian cooperative breeders that have consistently high levels of helping, Pinyon Jays have extremely variable levels of helping between years and between families in a given year. Mate availability

Fig. 7.13. Annual variation in the percentage of Pinyon Jay pairs that had helpers in the Town Flock. No data are presented for 1979, 1980, 1986 and 1987 because few nests were found in 1979 and 1980 and nests were not frequently observed in 1986 and 1987. Sample sizes of nests censused from 1974 to 1985 are: 31, 23, 40, 15, 22, 26, 52, 50, 43. 39.

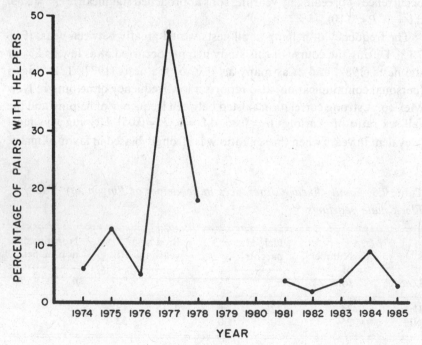

fluctuates annually, thus providing an impetus for annual variation in helping, but why do only a few non-breeding yearling males help?

Usual explanations for helping do not appear to be relevant to Pinyon Jays. Reciprocity of helpers' deeds (Ligon and Ligon 1978) does not appear to be a viable explanation for helping because helpees rarely help their helpers (Table 7.2) nor do they noticeably form any type of coalitions with siblings that helped them. Indirect genetic benefits to helpers or direct genetic benefits to parents or helped nestlings also appear to be non-existent because pairs do not have higher reproductive success in years when they have helpers compared to when they breed unassisted. Helpers also do not appear to be enhancing their lifespan or their future mating opportunities by helping.

We suggest that helpers are making the best of a bad situation. When adult sex ratios are male-biased many two and three year old males mate with yearling females. Male yearlings, who are typically subordinate to adult males have few mating opportunities. Dominant male yearlings should obtain the few available females. Less dominant males would then have two options: join the non-breeding yearling flock or help their parents and flock with the breeding adults, which may enhance access to cached food stores. An advantage of remaining in the yearling flock is that some unpaired females may be there. All yearling females do not breed, even in years when unmated adult females are scarce. Yearling males may compete for unmated females throughout the breeding season while in the non-breeding flock. Such a flock would be a perfect arena for the honest advertisement of a potential mate's abilities.

If mate shortage and subsequent mate competition drives helping, then it would seem obvious that helpers are males unable to obtain mates during the winter or unable to compete successfully for mates in the non-breeding flock. We propose that these helpers may have low dominance. We do not have consistent dominance data for the tenure of our study, but we do have substantial yearly samples of body weights. Larger male Pinyon Jays tend to dominate small ones, at least in captivity (Johnson 1988). Helpers should be small, relative to the other young produced in a given year. We tested this hypothesis by calculating the relative weight (percentage of crèchlings each year that were heavier than the focal bird) of helping versus non-helping, surviving male descendants of H-lineage pairs. We did not know the sex of all crèchlings used in this comparison. All weights were taken during the late summer or fall, when jays were two to five months post-fledging. These weights are representative of size when young jays are beginning to establish themselves in the flock (Balda and Balda 1978) and should be indicative of their ability to compete for mates with adults and yearlings. We found that helpers were smaller on average than 30.4% ($n = 19$, S.D. = ± 19.7) of their potential competitors, but that non-helping, H-lineage sons were only smaller than 17.8% ($n = 13$, S.D. = ± 14.9) of their potential competitors. Helpers were smaller, relative to competitors, than were non-helpers ($t = 2.07$, d.f. = 29.6, $P < 0.05$). Absolute weight of helpers was not significantly lighter than the absolute weight of non-helpers ($\bar{x}_H = 103.1$ g, $N = 19$, S.D. = ± 5.4; $\bar{x}_{NH} = 106.0$ g, $n = 13$, S.D. = ± 6.4; $t = -1.34$, d.f. = 22.9, $P = 0.19$). It appears that weight relative to potential competitors in a given year may influence a yearling's 'decision' to leave the non-breeding flock and help.

Helping by males and breeding by yearling females should be more frequent in years of mate shortage. Thus, we predicted that the percentage of nests with female yearling breeders and the percentage with helpers should be positively correlated. These measures are positively, but not significantly, correlated ($r = 0.27$, $n = 9$ years, n.s.). The lack of a stronger relationship may be because the incidence of breeding by yearling females was related to the sex ratio of two to three-year-olds (more breed when males are abundant $r = 0.8$), but the incidence of helping was related to the sex ratio of yearlings (see above). In the future we need to determine the sex of all yearlings for several years before we can determine exactly what percentage of yearling males help and what percentage of female yearlings breed. Without these data our predictions about mate shortage and helping or breeding by yearlings cannot be directly addressed.

Helping is not solely a decision made by the helper. Parents must permit or tolerate help. Given that a pair's productivity is not enhanced by helping, why should they allow it? Allowance of helping may be a form of parental facilitation (Brown and Brown 1984) whereby relatively small, possibly subordinate yearlings are nurtured in the relatively safe, breeding flock. Helpers thus attain longevity and fecundity equal to that obtained by their (perhaps) more dominant yearling competitors. The lightest crèchlings, especially H-lineage ones, are typically lost from the flock each year (Table 7.6). This loss certainly represents some dispersal by light-weight females, but also presumably indicates that many of the lightest young jays die. Perhaps some helpers would have died if they did not help.

Viewing helping from the parents' point raises interesting interpretations of H- and NH-lineage families. H-lineage parents invest in long-term parental care of some of their many sons. Perhaps this is a bet-hedging strategy that allows some sons (non-helpers) to try for a quick reproduction payoff (or to obtain a high-quality female) as yearlings in the mating lottery,

Table 7.6. *Crèchling weights of surviving (for one year) and dying Pinyon Jays*

Year	Surviving H-lineage crèchlings Mean weight (g)	n	±S.D.	P^a	Dying H-lineage crèchlings Mean weight (g)	n	±S.D.	Surviving NH-lineage crèchlings Mean weight (g)	n	±S.D.	P^b	Dying NH-lineage crèchlings Mean weight (g)	n	±S.D.
1972	105.0	2	7.1	—	None died			97.4	30	7.2	n.s.	95.5	131	8.5
1973	107.0	4	7.3	—	94.0^c	1	—	104.3	7	10.6	n.s.	100.0	29	7.5
1974	101.3	11	3.8	*	95.5	4	5.1	101.9	12	9.5	**	94.8	76	8.7
1975	99.4	8	9.9	n.s.	97.6	5	5.3	98.4	8	8.3	n.s.	99.5	39	7.9
1976	103.0	5	6.8	*	96.6	10	7.7	102.0	14	6.0	n.s.	96.6	33	9.0
1977	98.8	10	6.8	**	92.7	8	7.1	99.4	46	8.2	**	94.1	47	6.4
1978	100.7	6	6.3	**	92.1	4	3.6	99.3	21	6.6	n.s.	97.7	159	7.2
1979	109.8	5	5.4	***	92.0	4	3.6	99.2	15	9.9	n.s.	95.5	27	6.9
1980	104.8	2	7.9	—	94.2^c	1	—	96.8	31	7.7	n.s.	101.0	75	8.5
1981	97.4	8	7.2	n.s.	97.9	13	10.3	95.9	12	6.8	n.s.	96.9	26	8.1

[a] One-tailed *t*-tests comparing H to dying H. * = $P < 0.10$, ** = $P < 0.05$, *** = $P < 0.01$.
[b] One-tailed *t*-tests comparing surviving NH to dying NH. * = $P < 0.10$, ** = $P < 0.05$.
[c] Crèchling that died was the smallest of the year.

and simultaneously allows other sons (helpers) to improve their breeding skills while garnering protection and resources necessary for survival and successful reproduction in the future. When parents rarely reproduce successfully they do not hedge their bets but only practice the rapid payoff strategy. This appears contrary to life history theory (see e.g. Stearns 1976), which predicts extensive care to be especially evident when few young are reared.

We do not understand why only certain pairs tolerate helpers. NH-lineage male crèchlings were not significantly larger than H-lineage male crèchlings (comparing weights of males surviving one year: $\bar{x}_H = 104\,g$, $n = 35$, S.D. $= \pm 6.5$; $\bar{x}_{NH} = 104.7$, $n = 41$, S.D. $= \pm 7.1$; $t = -0.43$, d.f. $= 73.5$, $P = 0.67$), nor were they larger than males that helped ($t = -0.83$, d.f. $= 46.2$, $P = 0.41$). Our data suggest that helping is cultural, not genetic (see case history 2), yet despite the rapidity with which cultural traits can spread, it is still relatively rare in the Town Flock. Perhaps the apparently small or non-existent benefits of helping are only profitable when many sons are produced. Only very successful breeders may be able to withstand the potential detriments of having more birds at the nest (increased conspicuousness) or allowing some potential breeders to withdraw from the yearly mating lottery. As long as a few pairs are extremely successful, helping will remain at low levels.

Phylogenetic, demographic, and social influences on helping. Helping at the nest is common among New World jays (Brown 1987; Skutch 1987), yet it is infrequent in Pinyon Jays. Pinyon Jays are permanent residents with high annual adult and juvenile survivorship, important precursors to helping (Brown 1987). Apparent lack of benefits from helping to breeders and helpers is at the root of infrequent helping behavior. One could easily argue that the occurrence of helping is simply due to phylogenetic inertia (Wilson 1975). We find this explanation unsatisfactory for three reasons. First, helping is restricted to certain families of Pinyon Jays and within these families helping is fairly common. Second, the incidence of helping fluctuates inversely with the availability of breeding options in the flock. Lastly, within the closely related *Aphelocoma* (Scrub) Jays, the incidence of helping differs among species and subspecies and is closely correlated with costs of dispersal and the probability that yearlings may obtain breeding status (Woolfenden and Fitzpatrick 1984). We propose that the unique social structure of Pinyon Jay populations rarely favors helping. Pinyon Jays spend their entire lives in large flocks, nest colonially where nest sites are not limited, and do not defend breeding territories. This social structure affords many mating opportunities in the natal home range and does not necessitate large-scale dispersal. As a consequence, a relatively large number of yearlings breed (Fig. 7.8) and many do not disperse (Fig. 7.5). This is in stark contrast to jays that defend small territories which saturate the available habitat. Consequently, few individuals can breed (Brown 1970; Woolfenden and Fitzpatrick 1984). Yearlings and young adults of these territorial species rarely have breeding options on the natal territory and must either undergo costly dispersal or remain in the territory. The latter is often a more economical option. The options for a young male Pinyon Jay are very different. They can remain in the natal flock and breed as yearlings, join a non-breeding, non-helping subset of the flock, or help their parents. We believe natural selection favors the first two options for most yearlings. Yearling breeding directly enhances individual fitness because these breeders suffer neither significantly lower reproductive success nor lower longevity than adult breeders (Marzluff and Balda, 1988*b*). Remaining in the non-breeding, non-helping yearling flock also may directly enhance fitness. Yearlings in this flock may be establishing their social status and their individual identity within their age cohort. This age cohort is an important source of initial and subsequent mates as most pair bonds, even those formed between older jays, are assortative for age (Marzluff and Balda, 1988*c*). An individual's status among similar-age flock members may thus be important throughout its life and may be established initially as a yearling. Helping may disrupt the socialization of yearlings because helpers flock with breeders not with non-breeding yearlings.

We have portrayed helping as a costly option for yearling males, but this cost is not equal for all males. Males at the bottom of the yearling social structure have little to gain by staying there. Their potential gain fluctuates in direct proportion to the availability of mates. When mates are rare, presumably only the dominant yearlings will obtain them. Mateless, low-ranking males may actually have a lower survival by remaining in the non-breeding flock, because flock members are inexperienced and conspicuous. Joining the breeding flock as a helper may be the best option for low-ranking males. Parental facilitation may be favored because the quality of low-ranking males (helpers) may be enhanced even though the quantity of young produced may not increase (Brown and Brown 1984; Brown 1987). Extended care should especially be favored in very successful breeders who already have high fitness relative to their flock members. Helpers may feed and defend nestlings rather than just passively accompany their parents because of their hormonal conditions and the stimuli presented by begging nestlings (Woolfenden and Fitzpatrick 1984).

The pattern of helping in Pinyon Jays appears to be evolutionarily intermediate between uncooperative- and cooperative-breeding New World jays. This may be misleading. Helping occurs at intermediate levels because Pinyon Jays live in an extremely unique social setting that rarely favors abstention from breeding or removal from the mate pool by yearlings. As in other cooperative breeders, limited breeding opportunities favor helping; however, delaying breeding but remaining with a group of non-breeding peers may provide long-term benefits to relatively dominant yearlings thus lowering the overall level of cooperative breeding in Pinyon Jays.

Genealogical structure of a Pinyon Jay flock

The Town Flock is composed of monogamous pairs and members of their extended families. Our observations began in 1971 and 1972 when flock members were uniquely color-banded. At that time their relationship to one another was unknown; however, many pair bonds were identified in 1973 and offspring of these pairs were followed throughout their lives. Here we show pairs, the production of yearlings (young surviving for one year), and the formation of pair bonds in a genealogical graph (Fig. 7.10). The majority of individuals (73% of 246 jays) and pair bonds (72% of 180 bonds) that we have studied are illustrated in this figure. Beginning with any individual in Fig. 7.10, the biological and affinal (marital) relationships of that individual to any other jay in the figure can be traced. All jays in Fig. 7.10 are at least affinally joined; thus it is evident that the flock we have studied is composed of extended families that are linked to each other through pair bonds. We can best summarize this social organization using terms usually reserved for primate societies: Pinyon Jays live in troops consisting of numerous clans.

We follow the notation of Cannings and Thompson (1981) in the genealogical graph. Two types of nodes are shown: open circles represent individuals and filled circles pair bonds. These nodes are linked by two types of arcs: ─➤─ joins mated individuals, and is called a marriage arc; ─➤➤─ joins a pair bond to an individual produced by the pair and is called a reproductive arc. Two types of relationship are shown, biological and affinal (through pair bonding). Individuals are biologically related if they are connected by a series of alternating marriage and reproductive arcs. Individuals are affinally related if they are linked only by a marriage arc. Inbreeding results in a closed loop of arcs. The coefficient of relatedness, r, equals the reciprocal of the number of marriage and reproductive arcs that link two biologically related individuals. Adjacent to each open circle node we list the color-band combination of that individual, its sex ($\ominus$ = female or $\oplus$ = male or $\bigcirc$ if unknown), whether it helped (H = yes, none = no), and whether it emigrated (E = yes, none = remained in flock). A reproductive arc with a question mark preceding it (? $\rightarrow\!\!\!\rightarrow\!\!\!\rightarrow$) indicates that the parents of this individual are not known.

In the following, we describe case histories of some of the extended families shown in Fig. 7.10. Our objectives are: (1) to illustrate our H- versus NH-lineage classification scheme, emphasizing that when sons are produced by NH-lineage pairs, they do not help; (2) to illustrate the sporadic occurrence of helping through several generations of a family; and (3) to illustrate possible cultural inheritance of helping behavior.

Case history 1 (Fig. 7.10, A–D). DP-DG was born in 1974 and nested with R-DB-O in 1976. Parents of DP-DG were not known, but R-DB-O had an unusual family background. She was born in 1975 to R-DB and DG-B. DG-B had been mated with LP-LG in 1973 and 1974 and had produced a surviving yearling (DG-B-Y) with him. However, in late 1974 they divorced and LP-LG paired with a 1974 female, Y-O, while DG-B paired with a recent widower, R-DB. This is the first documented case of divorce in this flock of jays. Only two others have been observed since this time.

DP-DG and R-DB-O produced three fledglings from their second nesting attempt in 1976; however, none survived its first winter. They successfully produced yearlings in 1977, 1978, 1979 and 1981. All of their surviving yearlings were males; two helped to rear their siblings the following year, while the other two did not. DP-DG and R-DB-O are thus a H-lineage pair, and all their descendants belong to the H-lineage. R-DB-O died in 1981 and DP-DG remated with DP-LB, a female born in 1979. This pair produced three yearlings in 1982, all of which survived for only one year. These yearlings were H-lineage birds because they were descendants of DP-DG; however, their mother, DP-LB, was a NH-lineage female because her parents were not known and she was never helped while mated with DP-DG or her previous mate, R-W-Y.

Case history 2 (Fig. 7.10, D–F). A nestling exchange experiment was performed by Pat McArthur in 1978, in which OGRE-B-LB was moved from his natal nest to the nest of a H-lineage male, R/W-DG-LB (Fig. 7.10, Z). OGRE-B-LB helped his foster parents to rear his foster siblings, thus suggesting that helping behavior may be culturally, not genetically, controlled.

Case history 3 (Fig. 7.10, F–H). The mother of this extended family (P/P-B) was born in 1972 and first bred as a yearling in 1973. She exhibits a pattern of mating common in this flock – pairing between individuals of similar age. All of her three mates were also born in 1972, even though she mated with them when they were one, four and six years old, respectively. Her first two sons (P/P-B-Y and P/P-B-LB) show a behavior rare in this flock; both simultaneously helped their parents to rear their siblings in 1975. Her last mate, LP-P (Fig. 7.10, 1972-U), remated after her death with a female born in 1975, LP-LP-DP. This female had previously been mated with LP-P's stepchild, P/P-B-LB.

Case history 4 (Fig. 7.10, X–Z). The longest-lived female, P-P, was born in 1971 and died 14 years later in 1985. She was a H-lineage bird because in 1977 and 1978 she and her second mate, a subordinate, XMAS-Y, were helped at their nest by their yearling sons, P-P-R and XMAS-Y-O. Both of these sons are alive. It should be noted that an emigrating offspring (R/W-O-O) of XMAS-Y, produced with R/W-O in 1974, is also considered to be a H-lineage bird, because her father was helped when mated with P-P. Her mother, R/W-O, was a NH-lineage female.

Case history 5 (Fig. 7.10, Y–AA). The tendency for H-lineage sons to help and H-lineage daughters to emigrate was exemplified in the family started by P-O and R/W-DG in 1972. P-O was the beta male of the flock for many years. Both birds were adults in 1972 and both survived until 1976, during which time they produced four yearlings; two sons helped, another son bred at age two years without helping, and one daughter emigrated. P-O-LP was a helper son that bred very successfully with LB-LP-LG. Their sons did not help, and their daughter emigrated. R/W-DG-DB, also a helper son, produced two yearlings in three years of marriage to LP-DB. One son, R/W-DG-LB, bred with a female of unknown parentage, OGRE-B, in 1976. They produced two yearlings of their own and successfully reared a foster child, OGRE-B-LB, who, as discussed above, helped. One of their biological sons, R/W-DG/LB-O is alive and breeding successfully with O-LG. Their surviving offspring from 1983, O-O-DG, helped at a late 1984 nest. In this family lineage we trace helping behavior through three generations, and importantly, virtually every son produced helped before breeding.

Future research dealing with helping behavior

Further elucidation of the causes and consequences of helping will be challenging for future students of Pinyon Jays. The challenge comes from the relative infrequency of helping. This should not dissuade potential researchers. We have raised many more questions about helping than we have answered. At least some of these questions are testable. Our ideas suggest that future research will need to focus on the non-breeding, non-helping yearling flock. How is this flock linked to the main flock? Is survival in the yearling flock low, and more importantly is survival related to weight and dominance? Are pair bonds formed in this flock? Are males that leave this flock to help subordinate to those that remain and are those that leave to breed dominant to those that remain? Are hormone levels of non-breeders, helpers and breeders different? Detailed observations of several cohorts of nestlings as they form crèches, join the entire flock and subdivide to form the non-breeding flock could provide answers to many of these questions. Helping could be experimentally investigated by manipulating the sex ratio of the flock. Removing adult females is feasible. Many or all yearling females should respond to this manipulation by breeding. In contrast, only dominant male yearlings should breed. Helping should accordingly increase in frequency.

Many of our hypotheses were untestable or only weakly tested because we have not determined the sex of nestlings. Such determination would allow future researchers to determine the sex composition of the non-breeding flock and assess its potential as an arena for mate choice. The temporal formation and possible breaking of pair bonds by juveniles and yearlings could also be investigated. When is it evident that all yearling males will not obtain mates? Are helpers mateless for months prior to helping and does this contrast with yearling breeders or yearlings that do not breed but remain in the non-breeding flock?

We thank all the previous researchers who added to our understanding of the Town Flock: Gene Foster, Jane Balda, Katherine Bartlett, Dan Cannon, Larry Clark, Diana Gabaldon, Pat McArthur, Steve Vander Wall, Mike Morrison, Larry Pyc, Scott Winterstein, Mike Munoz, Mike Horn, John Lugenbuhl and Gary Bateman. David Ligon broadened our views by constantly reminding us that all Pinyon Jays do not reside in Flagstaff. George Merrick Jr, L. M. Baylor, Ronald Ryder, Nathaniel R. Whitney and Kathy Klimkiewicz supplied us with banding return records. Emilee Mead slaved over the genealogical graph. Gil Pogany provided the photograph of our study area. Bill and Judy Burding, Gene Foster, and Katherine Bartlett allowed us to monitor the Town Flock from their homes. Unlike the jays, Colleen Marzluff helped constantly throughout this project making this research possible and enjoyable. Financial support was provided by Northern Arizona University, The Frank M. Chapman Memorial Fund, The Wilson Ornithological Society, A.R.C.S., Sigma Xi and Jospeh and Elizabeth Marzluff.

References

Baker, J. R. (1938). The evolution of breeding seasons. In *Evolution: Essays on Aspects of Evolutionary Biology*, ed. G. R. Beer, pp. 161–77. Oxford University Press: London.

Balda, R. P. (1980). Are seed caching systems co-evolved? *Proc. 17th Intern. Ornith. Congr.* 1185–91.

Balda, R. P. (1981). Sociobiology as learned from a drab, blue bird. *Auk* **98**: 415–17.

Balda, R. P. (1987). Avian impacts on pinyon-juniper woodlands. In *Proceedings–Pinyon–Juniper Conference*, compiled by R. L. Everett, pp. 525–33. General technical report INT-215. Intermountain Research Station: Ogden, UT.

Balda, R. P. and Balda, J. (1978). The care of young Piñon Jays (*Gymnorhinus cyanocephalus*) and their integration into the flock. *J. Ornith.* **119**: 146–71.

Balda, R. P. and Bateman, G. C. (1971). Flocking and annual cycle of the Piñon Jay. *Condor* **73**: 287–302.

Balda, R. P. and Bateman, G. C. (1972). The breeding biology of the Piñon Jay. *Living Bird* **11**: 5–42.

Balda, R. P. and Bateman, G. C. (1973). Unusual mobbing behavior by incubating Piñon Jays. *Condor* **75**: 251–2.

Balda, R. P. and Bateman, G. C. (1976). Cannibalism in the Piñon Jay. *Condor* **78**: 562–4.

Balda, R. P., Bateman, G. C. and Foster, G. (1972). Flocking associates of the Piñon Jay. *Wilson Bull.* **84**: 60–76.

Balda, R. P., Morrison, M. L. and Bement, T. R. (1977). Roosting behavior of the Piñon Jay in autumn and winter. *Auk* **94**: 494–504.

Bateman, G. C. and Balda, R. P. (1973). Growth, development, and food habits of young Piñon Jays. *Auk* **90**: 39–61.

Bent, A. C. (1946). Life histories of North American jays, crows, and titmice. *U.S. Mus. Bull.* **191**.

Bongiorno, S. F. (1970). Nest-site selection by adult laughing gulls (*Larus atricilla*). *Anim. Behav.* **18**: 434–44.

Botkin, C. W. and Shires, L. B. (1948). The composition and values of piñon nuts. *New Mexico Exp. Stn Bull.* **344**: 3–14.

Brown, J. L. (1970). Cooperative breeding and altruistic behavior in the Mexican Jay, *Aphelocoma ultramarina*. *Anim. Behav.* **18**: 366–78.

Brown, J. L. (1987). *Helping and communal breeding in birds*. Princeton University Press: Princeton, NJ.

Brown, J. L. and Brown, E. R. (1984). Parental facilitation: parent–offspring relations in communally breeding birds. *Behav. Ecol. Sociobiol.* **14**: 203–9.

Cannings, C. and Thompson, E. A. (1981). *Genealogical and Genetic Structure*. Cambridge University Press: Cambridge.

Cannon, F. D., Jr (1973). Nesting energetics of the Piñon Jay. M.Sc. thesis, Northern Arizona University.

Cary, M. (1901). Birds of the Black Hills. *Auk* **18**: 231–8.

Clark, L. (1977). Bioenergetics of nestling Piñon Jays. M.Sc. thesis, Northern Arizona University.

Clark, L. and Balda, R. P. (1981). The developing of effective endothermy and homeothermy by nestling Piñon Jays. *Auk* **98**: 615–19.

Clark, L. and Gabaldon, D. J. (1979). Nest desertion by the Piñon Jay. *Auk* **96**: 796–8.

Coulson, J. C. (1968). Differences in the quality of birds nesting in the centre and in the edges of a colony. *Nature* (*Lond.*) **217**: 478–9.

Craig, J. L., Stewart, A. M. and Brown, J. L. (1982). Subordinates must wait. *Z. Tierpsychol.* **60**: 275–80.

Gabaldon, D. J. (1978). Factors involved in nest site selection by Piñon Jays. Ph.D. thesis, Northern Arizona University.

Gabaldon, D. J. and Balda, R. P. (1980). Effects of age and experience on breeding success in Piñon Jays. *Amer. Zool.* **20**: 787.

Johnson, K. (1988). Sexual selection in piñyon jays. I. Female choice and male–male competition. *Anim. Behav.* **36**: 1039–47.

Lack, D. (1968). *Ecological adaptations for breeding in birds.* Methuen and Company Ltd: London.

Ligon, J. D. (1971). Late summer-autumnal breeding of the Piñon Jay in New Mexico. *Condor* **73**: 141–53.

Ligon, J. D. (1974). Green cones of the piñon pine stimulate late summer breeding in the piñon jay. *Nature (Lond.)* **250**: 80–2.

Ligon, J. D. (1978). Reproductive interdependence of piñon jays and piñon pines. *Ecol. Monogr.* **48**: 111–26.

Ligon, J. D. and Ligon, S. H. (1978). Communal breeding in green woodhoopoes as a case for reciprocity. *Nature (Lond.)* **276**: 496–8.

Ligon, J. D. and White, J. L. (1974). Molt and its timing in the Piñon Jay. *Condor* **76**: 274–87.

McArthur, P. D. (1979). Parent–young recognition in the piñon jay: mechanisms, ontogeny, and survival value. Ph.D. thesis, Northern Arizona University.

McArthur, P. D. (1982). Mechanisms and development of parent–young vocal recognition in the piñon jay (*Gymnorhinus cyanocephalus*). *Anim. Behav.* **30**: 62–74.

Marzluff, J. M. (1983). Factors influencing reproductive success and behavior at the nest in Pinyon Jays (*Gymnorhinus cyanocephalus*). M.Sc. thesis. Northern Arizona University.

Marzluff, J. M. (1985). Behavior at a Pinyon Jay nest in response to predation. *Condor* **87**: 559–61.

Marzluff, J. M. (1987). Individual recognition and the organization of Pinyon Jay societies. Ph.D. thesis, Northern Arizona University.

Marzluff, J. M. (1988*a*). Do pinyon jays use prior experience in their choice of a nest site? *Anim. Behav.* **36**: 1–10.

Marzluff, J. M. (1988*b*). Vocal recognition of mates by breeding pinyon jays. *Anim. Behav.* **34**: 296–8.

Marzluff, J. M. and Balda, R. P. (1988*a*). Resource and climatic variability: influences on sociality of two southwestern covids. In *The Ecology of Social Behavior*, ed. C. N. Slobodchikoff, pp. 255–83. Academic Press: New York.

Marzluff, J. M. and Balda, R. P. (1988*b*). Advantages of, and constraints forcing, mate fidelity in Pinyon Jays. *Auk* **105**: 286–95.

Marzluff, J. M. and Balda, R. P. (1988*c*). Pairing patterns and fitness in a free-ranging population of Pinyon Jays: what do they reveal about mate choice? *Condor* **90**: 201–13.

Marzluff, J. M. and Balda, R. P. (1989). Causes and consequences of female-biased dispersal in a flock-living bird, the Pinyon Jay. *Ecology* **70**: 316–28.

Parsons, J. (1977). Nesting density and breeding success in the herring gull (*Larus argentatus*). *Ibis* **118**: 537–46.

Patterson, I. J. (1965). Timing and spacing of broods in the black-headed gull (*Larus ridibundus* L.). *Ibis* **107**: 433–60.

Skutch, A. F. (1987). *Helpers at Birds' Nests.* University of Iowa Press, Iowa City, IA.

Stearns, S. C. (1976). Life-history tactics: a review of the ideas. *Q. Rev. Biol.* **51**: 3–47.

Tenaza, R. (1971). Behavior and nesting success relative to nest location in Adelie Penguins. *Condor* **73**: 81–92.

Vander Wall, S. B. and Balda, R. P. (1981). Ecology and evolution of food-storage behavior in conifer-seed-caching corvids. *Z. Tierpsychol.* **56**: 217–42.

Wilson, E. O. (1975). *Sociobiology.* Belknap Press: Cambridge, MA.

Woolfenden, G. E. and Fitzpatrick, J. W. (1984). *The Florida Scrub Jay.* Princeton University Press: Princeton, NJ.

8 FLORIDA SCRUB JAYS: A SYNOPSIS AFTER 18 YEARS OF STUDY

8

Florida Scrub Jays: a synopsis after 18 years of study

G. E. WOOLFENDEN AND J. W. FITZPATRICK

The Florida Scrub Jay (*Aphelocoma coerulescens coerulescens*) is a disjunct population of a species that otherwise inhabits western North America from Oregon and Colorado south to southern Mexico. Two other, largely Mexican species complete the genus *Aphelocoma*. The genus is unusually revealing for studying the evolution of cooperative breeding, because it contains closely related populations that exhibit varying degrees of sociality both between species and within species. The hypothesis that these differences are associated with different environmental and demographic conditions is central to most current work on social evolution in jays (Brown 1974, 1978; Woolfenden and Fitzpatrick 1984; Fitzpatrick and Woolfenden 1986).

The Scrub Jay is known to exhibit cooperative breeding only in the Florida population, where it occurs in scattered, often small populations, restricted to relict patches of oak scrub. This population has almost certainly been isolated from the vast western populations since at least the Pleistocene (Pitelka 1951). In contrast to the Florida population, Scrub Jays in the west disperse from the natal territory before reaching the age of one year. Here, we describe the social system of the Florida Scrub Jay, placing it in its environmental context as a means of emphasizing our interpretation of how the system evolved.

Study area and methods

Several hundred Florida Scrub Jays reside permanently on the property of the Archbold Biological Station, located in Highlands County near the southern terminus of the central Florida hill country (known as the Lake Wales Ridge). The station is a research institution with excellent facilities dedicated to long-term studies of the native habitats within its

2000 ha reserve (Fig. 8.1). In 1969 we began studying the jays by marking them individually with metal and colored plastic bands. Having quickly established that the jays were permanently territorial, we adopted a field protocol that has not changed appreciably in nearly 20 years. In a study tract of about 500 ha we band every jay for individual recognition. We attempt to record all events pertinent to survival, movement and reproduction within this sample of jays. Annually we find all nests, band all offspring, and map all territories ($n = 28$ to 35). Monthly we census all jays in the study tract. All immigrant jays are captured and marked soon after their appearance. Periodically we census scrub habitats peripheral to the study tract in search of dispersers. Many of the jays are extremely tame, allowing us to perform a variety of behavioral experiments. However, we do not manipulate the population in any way that affects natality or mortality. Recently, a colleague, Ronald L. Mumme, has established nearby a separate, marked population in which experimental perturbations including removals are being conducted.

Habitat

Aptly named, the Scrub Jay inhabits open scrub lands. It shuns forests with closed canopies, which in North America are occupied by *Cyanocitta* and *Perisoreus* jays. No North American jays inhabit pure grasslands. Scrub Jays and Steller's Jays (*Cyanocitta stelleri*) or Blue Jays (*C. cristata*) are truly sympatric only in woodland or parkland habitats, which are intermediate in structure between scrub and forest. Over most of their range, Scrub Jays occupy habitats that are dominated by oaks.

Florida Scrub Jays live in one of North America's rarest habitats, a relict formation of stunted oak scrub endemic to the peninsula of Florida. Sometimes known as sand pine scrub or rosemary scrub (after *Pinus clausa* and *Ceratiola ericoides*, two indicator species in the scrub), the habitat typically includes abundant oak shrubs of several species (*Quercus inopina, Q. geminata, Q. chapmanii, Q. minima, Q. myrtifolia*). Extensive patches of scrub occur in the central peninsula, especially in Marion and Highlands counties, and on a thin strip of sandy hills along the Atlantic coast. The oak scrub of Florida grows on old sand dunes, and its distribution is explained by the persistence of old Recent and Pleistocene dunes, reflecting modern and ancient seashores. The substrate is well drained, often pure-white sand. The natural vegetation of most of the Florida peninsula is pine flatwoods, with smaller areas of mixed hardwoods, wetlands and prairies.

The topography of the Florida scrub reflects its origin as sand dunes – gradual slopes from sandy ridgetops to interdunal depressions. Despite the abundant rainfall of present-day Florida, the porous soil drains rapidly, producing a xerophytic habitat. Natural fires are frequent, usually ignited by lightning. Fire is necessary to maintain the scrub in its natural condition, which, except for widely scattered sand pines and slash pines (*Pinus elliottii*), rarely exceeds 1 to 2 m in height. The stunted oaks that characterize Florida scrub grow only on the dune tops and slopes, where they are mixed with low-growing palmettos, rosemary, several other sclerophyllous woody shrubs, and scattered prickly-pear cactus. Patches of bare white sand are frequent. The interdunal depressions, sometimes filled with standing water during the summer wet season, support a variety of rank grasses mostly shunned by the jays.

The sharp boundaries of the scrub habitat, defined by the distribution of sandy ridgetops, are of critical importance to the demography and natural history of the Florida Scrub Jay. In the space of a few meters one can walk downslope from open oak scrub into wet grassland, or into stands of pines or bay trees. Compared to other oak-dominated habits in North America, the oak scrub is relatively sharply bounded, and of generally uniform quality. Large expanses of marginally suitable habitat do not exist.

Fig. 8.1. Mature oak scrub at Archbold Biological Station, showing a single 8 m sand pine (right) and scattered, taller slash pines in the distance. Note bare white sand (lower right) and an adult Florida Scrub Jay perched atop an oak shrub (left).

Habitats of Scrub Jays in south-western North America are widespread and far more varied in character and composition (Bent 1946).

Where fire has been suppressed, late successional scrub becomes tall and dense, often with an overstory of scattered pines. Florida Scrub Jays tend to avoid these unburned habits. During the 1970s we witnessed the abandonment of a large tract of formerly suitable scrub that had been prevented from burning since the 1930s. Reproductive success and survival of both juveniles and adults were significantly lower in this area, and some remaining jays actually fought their way out into the open, burned habitats of the main study tract (Fitzpatrick and Woolfenden 1986). These observations reinforce our conclusions that the habitat tolerance of Florida Scrub Jays is exceedingly narrow, and that this constraint is fundamental to the entire demographic and social regime of the species in Florida.

Food

Florida Scrub Jays eat a wide variety of animal food. Insects and other arthropods living in the shrubby vegetation form the bulk of their diet. Orthoptera and lepidopteran larvae are especially frequent food items. In addition, we have seen the jays capture and consume about 20 species of vertebrates, mostly frogs, lizards, and snakes. The jays also feed on bird eggs, nestlings, and carrion of larger vertebrates when encountered.

A variety of berries and other small fruits are taken occasionally, but one plant food stands out as crucial in the diet. Every autumn each individual Florida Scrub Jay caches several thousand acorns plucked from the oaks within its territory. Acorns are eaten throughout the year, constituting nearly half the food intake in certain seasons. Acorns are plucked directly from the shrubs while they last and unearthed from cache sites in the sand during the winter, spring and summer (DeGange *et al.* 1989).

Predators

Although a careful study of predation on Florida Scrub Jays has not been made, we have accumulated evidence that implicates certain species of snakes, mammals and birds. Various of these feed on jay eggs, small young, adults, or combinations of all three. All large snakes are mobbed by the jays, and the eastern coachwhip (*Masticophis flagellum*) and indigo snake (*Drymarchon corais*) are known to eat fledglings.

Mammalian predators include bobcats (*Lynx rufus*) and raccoons (*Procyon lotor*). Avian nest predators include two owls (*Bubo virginianus, Otus asio*), two raptors (*Buteo jamaicensis, Circus cyaneus*), and possibly three corvids (*Corvus ossifragus, C. brachyrhynchos, Cyanocitta cristata*). Three other diurnal raptors seem to be important predators on adult jays: Merlin (*Falco columbarius*), Sharp-shinned Hawk (*Accipiter striatus*), and Cooper's Hawk (*A. cooperii*). These three hawks and the harrier are encountered regularly during migration and winter, and a family of jays may experience as many as one pursuit per day by these raptors when they are present.

Spacing system
Territories

With rare and brief exceptions, all Florida Scrub Jays live in permanently defended territories, which contain all the resources necessary for living. Territory size is relatively large for the jays' body size (79.2 g), averaging 9.0 ha (range 1.2–20.6 ha) within our study tract (excluding the larger grassy depressions, which often are incorporated into territorial boundaries). Some territories are consistently larger than others, regardless of family size, probably because of differences in habitat quality. This variation is the subject of current study.

Annual mean territory size fluctuates very little. The one year when territories were significantly smaller (1972) occurred two years after the most successful breeding season on record (1970). This reflects the frequent establishment of new territories by two year old males (see below). New territories within the main study area develop only at the expense of other families' holdings. In our study tract, all open scrub habitat is occupied and defended by jays every year. Therefore, competition for space appears to be ongoing and severe.

A territory is 'owned' by a monogamous pair of jays, and ownership is passed on to a sequence of replacement breeders or offspring heirs indefinitely. Besides the breeding pair, territories also may contain up to six adult non-breeders. About 98% of these are jays that have not yet bred. About 95% have not yet dispersed from their natal ground. A dominance system exists among these 'helpers' (defined below) in which males dominate females and older jays dominate younger ones. In the rare cases when both breeders disappear at once, the dominant non-breeding male, if one resides in the territory, acquires a mate from outside, inherits the territory, and becomes a breeder.

All jays occupying a territory help to defend it. The breeders are only slightly more active defenders than the helpers. During certain seasons, especially autumn and early spring, territorial displays and even fights occur many times per day. Most jays spend many of their daylight hours on exposed perches overlooking the open scrub. Besides its function in predator surveillance, this sentinel behavior allows rapid location of intruders and also displays territorial ownership to neighboring families.

Density

The density of territories and breeding pairs within the study tract remains remarkably stable from year to year (Fig. 8.2). The 17 year average since 1971 has been 3.2 territories per 40 ha. Variance about this average is typically low, although densities following an epidemic which occurred in 1979 have not recovered to pre-epidemic levels. An extensive fire in 1984 burned about half the study tract, and may be helping to keep destinies lower. Total population density, including juveniles and older non-breeding helpers, is more variable. Relative constancy in breeding numbers, the existence of a large surplus population of non-breeders, and confinement

Fig. 8.2. Densities of Florida Scrub Jays in April, at the onset of breeding, in the main study tract at Archbold Biological Station, 1971–87. Densities are calculated within the outer boundaries of the perimeter territories mapped each year (variable areas mapped are shown above *x*-axis). Breeding densities were stable before (CV = 6%) and after (CV = 5%) the 1979 epidemic, but remain significantly lower since that event ($\bar{x}$ ($\pm$s.d.) = 7·57) ($\pm$0.49) before versus 5.91 ($\pm$0.31) after the epidemic). A large fire in September 1984 affected two-thirds of the study tract, and post-fire succession may be keeping densities lower.

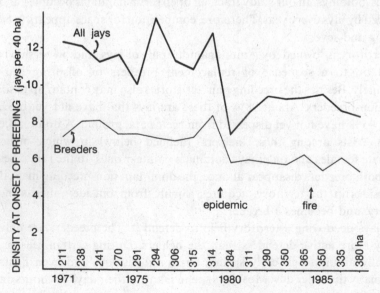

to a sharply bounded, uniform habitat lead us to conclude that competition for breeding space is almost always severe within the Florida Scrub Jay population.

Territory growth

When first established, territories are relatively small (see below). After one or two years, however, their sizes are determined by a combination of habitat factors (currently under study) and family size. As family size grows through successful reproduction, territory size grows at the expense of neighboring, less successful territories. Territory size grows asymptotically among families containing more than four jays. Average territory size is about equally high for families of six, seven and eight jays. Because no new space is available to defend, territory establishment and growth can occur only as other territories shrink or disappear. Territories disappear when both members of a lone pair die. Territories sometimes merge when adjacent, widowed jays pair with each other.

Territorial budding

Of new territories, 90% are established through a process we have termed 'territorial budding' (Woolfenden and Fitzpatrick 1978), a method of dispersal essentially restricted to males. The mean age of males budding a new territory is 2.6 years. With assistance from parents and other family members, a male helper actively expands his natal territory. This is accomplished by fighting with neighbors and gaining new ground that contains good ridgetop scrub. The young male then pairs with a female from outside the family, and the two begin to defend the newly acquired ground, sometimes including a portion of the original parental territory. The parents and other family members retreat to the now smaller original territory. Although the new pair vigorously defends its budded territory against outside neighbors, defense against the male's family develops more slowly. Displays at this boundary are perpetrated mostly by the immigrant female. The female breeders rarely if ever move across this boundary, but father and son often traverse into each other's territory for several years. Foraging and territory defense is typically separate between the two pairs, except in the infrequent cases of 'double families' (see below). Death of one or the other causes the boundary to become permanent.

Newly budded territories encompass typically about 35% of the original, expanded territory. They average only 4.5 ha, or about half of the size of well-established territories, but their growth is rapid. Second-year territories do not differ significantly from the mean for the population.

Multi-pair territories

About one in 50 territories consists of two breeding pairs and their offspring, jointly defending a large territory. The typical example consists of father and son, with separate mates, in which the son has budded a territory, but a boundary is especially slow to develop and the offspring use both territories indiscriminantly. One example involved two brothers, mated to a mother and daughter. We have never seen a breeder attend the nest of the other pair in such cases, but they do feed the fledglings of both nests equally. The offspring, in turn, attend both nests the following year. Although these examples bear a striking resemblance to 'plural nesting' in the related Mexican Jays of Arizona (*Aphelocoma ultramarina arizonae*) (see Chapter 9), they are temporary exceptions rather than the rule. These 'double families' typically break up upon the death of any one of the breeders involved, usually within two to three years.

Social system

The basic social unit of Florida Scrub Jays is the monogamous breeding pair. Each year almost half of the territories are occupied by just the two breeders (Fig. 8.3). The remainder are augmented by helpers, usually the resident offspring from previous years. Discounting the young of the year, the maximum family size has been eight jays.

Fig. 8.3. Relative frequencies of Florida Scrub Jay group sizes at onset of breeding, 1969–87. Of the 'groups', 45% consist of a simple, mated pair with no helpers. Actual counts are shown above each bar.

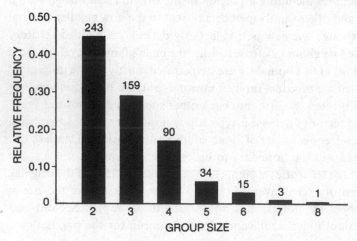

Evidence for monogamy

Our evidence for monogamy is behavioral and circumstantial, and is based upon 18 breeding seasons in which we have scrutinized the jays' behavior during the nesting season. We have never witnessed copulation between jays not paired with one another. Until an aberrant case of two breeding females laying and incubating in the same nest was discovered by R. L. Mumme in 1987, no jay other than the single resident breeder had been found to lay eggs in a nest. Our sample consists of over 800 nests, usually monitored during laying, when the breeding female adds one egg per day. Only one female incubates at each nest. We eagerly await a safe biochemical technique for confirming that parentage is not complicated by cuckoldry, but we have no reason to suspect that it is.

Mate fidelity

Except in rare cases of 'divorce' (see below), breeders remain paired with the same jay until one of them dies. Mortality of breeders averages 21% per year. Therefore, mated pairs often remain together for many successive years, the longest surviving pair having bred for eight years together.

Reproductive success increases with duration of the pair bond, from 1.7 fledglings per pair initially to 3.0 after six years together (Woolfenden and Fitzpatrick 1984). We also suspect that established pairs show slightly higher annual survivorship (survival rate) than newly formed ones, although this has not been tested. As non-breeding offspring may remain in the natal territory for up to several years, the probability that a pair will have helpers increases with duration of the pair bond, further enhancing the expected breeding success as the years go by. These factors contribute to enhancing the lifetime reproductive output of breeders, thereby favoring mate fidelity.

'Divorce'

Divorce accounts for about 4% of the broken pair bonds, death accounting for the remaining 96% ($n = 285$). Although their causes are often not apparent, many divorces appear to follow the inability of one member of the pair to perform its normal tasks as a breeder. When injury or illness impairs normal functioning of the breeder the healthy mate may pair with a new jay, and the impaired jay may regain its health but not its mate.

Fates of widowed jays

Jays whose mates have died almost always remain in the territory and pair with an immigrant. Simultaneous deaths of opposite sexed breeders in adjacent territories can result in a merger of the two territories as the widowed jays pair with one another. A few widowed females have lost their territory outright, as they are somewhat less able than males to defend their borders alone. Of 93 female breeders that lost their mates and survived to the following breeding season, 83 (89%) gained a replacement mate at home or in an adjacent territory. Of the remainder, 9 (10%) lost their territory to neighboring groups and 1 (1%) was apparently expelled by her son, a three year old male helper. Evicted widows usually disappear while searching the surrounding neighborhood for breeding vacancies.

Incest avoidance

Florida Scrub Jays virtually never pair with close relatives. We suspect that the mechanism is an aversion to pairing with a member of the same functional group. With one exception in 18 years, when a breeder dies a resident helper never pairs with the widow if the latter is its parent. In the single anomaly a male helper paired with his mother long enough to produce a clutch of fertile eggs, then the female was killed by a predator on the nest. Several pairs have formed between more distant relatives such as first cousins or aunt–nephews, but these occurred between jays of neighboring groups where neither jay had previously experienced its relative as a member of its own group.

Rarely, certain circumstances do lead to a jay pairing with a breeder it has helped. If a jay attempting to join a family group as a replacement breeder fails, sometimes it remains as a helper. Always in such cases it is subordinate to the resident breeders and helpers. Then, if it still is a helper in the family when the appropriately sexed breeder dies, it may become the replacement. Immediately it becomes dominant to all helpers in the territory. We suspect that these immigrants sometimes actually disrupt the breeding activities of the pair with which they associate. Such apparently premature dispersals may represent a strategy for queuing to become a breeder in a favorable territory (Wiley and Rabenold 1985).

Age at first breeding

First breeding is most common at age two years (58% for females, 47% for males). By age three years, 98% of the females and 77% of the males that eventually breed have paired. The remaining males in the sample first bred at ages four through seven years, the frequency diminishing with age (Woolfenden and Fitzpatrick 1986).

During the 18 years of our study only one male and three females have bred at age one year. An additional few of each sex began pairing before reaching age one year, but did not build nests or attempt to raise their own young. Yearlings comprise about 62% of the helpers, and two-year-olds comprise another 24%. We suspect that most or all yearlings are physiologically capable of breeding.

Division of labor

Division of labor between paired jays is well developed. Only breeding females incubate, brood and shade the young in the nest. Prior to and during this time, the breeding male provides her with much of her food intake. He also delivers much of the food to nestlings, especially during the first few days post-hatching.

The presence of helpers reduces somewhat the foraging load on the male breeder, who in turn spends proportionately more time as a sentinel (see below). This behavioral shift changes the nature of parental investment by the male, but does not necessarily reduce it.

Away from the nest, and during the non-breeding months, division of labor also is evident between the sexes. Males are more active defenders of the territory, engaging more frequently than females in physical battles with neighboring jays. Both sexes tend to be more aggressive to rival jays of their own sex. Males also perform more mobbing behavior, and spend more time than females on sentinel duty. Despite these asymmetries, male and female breeders suffer exactly equal annual mortality, which has averaged about 21% over our 18 years of study.

Helpers

Essentially all Florida Scrub Jays fall into one of three social classes: breeder, helper or dependent young. Because we can easily observe the social relationships of every jay in the study tract, we have confirmed that no long-term floaters exist. All jays work to establish themselves into a group. Dispersing or displaced jays may move about from territory to territory for several months before either vanishing or becoming associated with a pair or group. With rare exceptions we categorize any adult, non-breeding jay as a helper. We emphasize that helpers participate alongside the breeders in virtually all activities within the territory. Sometimes, non-kin individuals associated with a group may be repeatedly chased by group members, which inhibits their participation in group activities, especially feeding dependent young. Rare examples actually appear to have hindered group activities. These 'adjunct' individuals (about 8 in 18 years) always have been excluded from our analyses of helper activities or effects.

Most helpers assist both parents (64%) or one parent and a step-parent

(24%). When both parents die, helpers sometimes join neighboring families as immigrant helpers. As a result, a small percentage of helpers assists distant relatives (8%) or even unrelated jays (4%).

The data that follow regarding numbers of helpers are based on counts early in the nesting season, when group membership is most stable (see Fig. 8.3). Slightly over half (55%) of the breeding pairs have helpers. This percentage varies annually from 41 to 80%, depending mainly upon average reproductive success during the previous one or two years. The average number of helpers per pair ranges from 0.65 to 1.52 annually, with a mean of 1.00. Excluding breeders without helpers, the annual mean number of helpers per pair ranges from 1.55 to 2.16. Only 10% of pairs have more than two helpers, and the maximum number we have observed is six ($n = 1$).

Sex of helpers

We determine the sex of Scrub Jays primarily through behavior. By age one year we know the sexes of most jays, and sex ratio at that age appears to be even (131 males, 153 females, 27 unknown sex, probably mostly males).

Females disperse earlier and further than males (see below). Dispersal attempts apparently cause high mortality, so that Florida Scrub Jay populations overall are biased slightly toward males. This bias appears only among the class of helpers older than age one year. Fifty-four per cent of helpers two years old, and 77% of helpers three years old or older, are males.

Dominance

A dominance system exists within families that has far-reaching effects on the social system. Males dominate females, and within each sex breeders dominate non-breeders. Same-sex helpers have linear hierarchies, usually with the older jay dominant. Males are more aggressive than females, with clearer dominance relationships, but we suspect that female helpers have hierarchies as well (Woolfenden and Fitzpatrick 1977).

We hypothesize that males dominate females ultimately because of the behavioral asymmetries that characterize breeding pairs. Males are, on average, slightly larger (Pitelka 1951), and they clearly take a more aggressive role in territory defense. Females select mates on the basis of their holding a territory, hence they may actually select for aggressiveness. In contrast, successfully dispersing females must move from territory to territory in a subordinate role while searching for an opening, tolerating and avoiding aggression from the local territory holders. Male dominance may be a by-product of these asymmetries (Woolfenden and Fitzpatrick 1986).

Dominant helpers tend to disperse or inherit breeding space earlier than their subordinates (Woolfenden and Fitzpatrick 1977, 1984). Moreover, older male helpers that live in groups containing subordinate helpers show nearly three times the probability of breeding in the following year compared to those without subordinates (Fitzpatrick and Woolfenden 1986). This important advantage in having siblings results mostly from the increased probability of budding by males that have siblings.

Sentinel and mobbing behavior

We suspect that virtually all Florida Scrub Jays die through predation except during rare epidemics (see below). The daily risk of being detected by a snake or hawk, combined with the low and open structure of the habitat, presumably selects for some kind of surveillance system. Indeed, a well-developed coordinated sentinel system is employed during daylight hours throughout the year (McGowan and Woolfenden 1989).

Although jay territories are large, family members tend to remain close to one another during the daytime (except when the female is on the nest). A jay on sentinel sits on an exposed perch, a few meters above the ground, from which it scans the horizons. Upon sighting a dangerous hawk, the sentinel gives a distinctive warning call, causing all jays within earshot to dash for dense cover. Sentinels see more hawks, and give more warning calls, than non-sentinels. Sentinels are more frequently on post during migration and winter seasons, when the dangerous hawks are most prevalent. During winter, the most hazardous season, breeding males perform more sentinel behavior than do other family members.

Florida Scrub Jays have an elaborate mobbing system which includes loud scolding, advances toward the predator by hopping or flying, and sometimes outright attacks by pecking. Experiments using tethered snakes reveal that breeding males are usually more aggressive than other family members during mobbing. The jays mob terrestrial predators and sometimes perched raptors, especially owls. We suspect that mobbing may dissuade some predators near the nest and may alter the searching behavior of others. Snakes, for example, often escape down holes while being mobbed. In addition, mobbing certainly must teach naive jays which animals in their environment should be feared.

Dispersal

Dispersal consists of shifting permanently from one territory to another. Breeders and young of the year rarely move to another territory, even after the death of a mate or parent. Most dispersals are performed by helpers older than twelve months. First-year jays, especially brown-headed juveniles before their post-juvenile molt, occasionally wander from their territory, but only rarely do they stay away for more than a few hours.

Florida Scrub Jays are extremely sedentary. Virtually all dispersal occurs within a few kilometers of the natal territory. Females disperse significantly farther than males (means: 1180 m versus 456 m), and they tend to do so at an earlier age (means = 1.6 versus 2.4 years; Woolfenden and Fitzpatrick 1986). Many males become breeders by inheriting all or part of their natal territory; females rarely do so. Male dominance may be the proximate factor that causes these asymmetries.

The most common route to becoming a breeder is by replacing a lost breeder: 48% of the first breeders gain a mate in this fashion. However, 4% become breeders by establishing a territory *de novo*, away from home and between existing territories. The two remaining methods for becoming a breeder involve males directly inheriting the entire natal territory (10%) or budding a segment of it (38%). Males that inherit space pair with females from outside their own group. Because females rarely inherit breeding space, they become breeders by pairing with widows or inheritors. These they locate through dispersal forays. When on forays, females leave the natal territory, sometimes for one or more days, and wander through the neighborhood. They may travel to several kilometers away, and they appear to pay special attention to areas where active boundary displaying indicates a territorial shift or a mate loss. These forays presumably account for increased mortality among non-breeding females. Only three females in 18 years have acquired breeding space through budding.

Dispersal, both attempted and successful, shows two peaks annually. The most active period is just prior to the spring breeding season, when most territorial budding occurs. The other peak is in autumn, after molt and during the peak of the acorn season. Most long-distance dispersals by females occur in the autumn of their second year.

Dispersal by sibling groups of like sex is essentially non-existent. Relatively few families contain more than one dispersal-aged helper of the same sex. Sometimes these will compete for the same breeding vacancy nearby, and the loser either disappears or returns home. In three instances a male has become a helper to his brother, but all three involved complicated social circumstances. It is clear that, compared to certain other cooperative breeders that practice group dispersal (e.g. Acorn Woodpeckers (see Chapter 14) and Green Woodhoopoes (see Chapter 2), Florida Scrub Jays practice dispersal as a solitary and competitive event.

Reproduction

The nesting season of Florida Scrub Jays is confined to a three month period from mid March to mid June. By the first week in April most first-clutches have been laid. Re-nesting occurs after failed attempts until about mid May. True second-brood attempts (i.e. a clutch laid while young are still alive from an earlier attempt in the same season) are rare ($n = 23$ in 524 pair-years).

Mean clutch size is 3.33 eggs, with the following distribution: 1 egg ($n = 1$), 2 ($n = 65$), 3 ($n = 349$), 4 ($n = 259$) and 5 ($n = 19$). Factors that influence clutch size include age and experience of the breeders, time in the nesting season, heredity (probably) and annual food supply (presently under study). Presence of helpers does not affect clutch size.

Nest failures almost always occur through predation. Inclement weather, desertion, fire and starvation account for fewer than 2% of the losses. Most losses of young from continuing broods also reflect predation. Of eggs that 'survive' to hatching date, 11% fail to hatch. Intriguingly, the probability of hatching by eggs that reach hatching age is higher in nests of pairs with helpers than in the nests of unassisted pairs.

Probability of nest success decreases rapidly through the breeding season. Nests begun in March have about a 70% chance, those in May have less than 40%. We suspect that this rapid decline in success rate, combined with an elevated risk of mortality while breeding (see below), may contribute to the relatively brief nesting season of the Florida Scrub Jay. With high life expectancies, jays are better off investing in breeding over many years rather than laying all their eggs in one basket.

Annual reproductive success fluctuates greatly. Over 17 years, average fledgling production has varied from 0.9 to 2.9 fledglings per pair. Yearly means average 1.97 (s.d. $= \pm 0.55$) fledglings per pair. Yearly variation in predation pressure appears to be the most important proximate cause of this variation in average reproductive success.

Survival of fledglings

Survival of fledglings to age one year (when they become yearlings) has fluctuated from 21% to 44% during all but one year ($\bar{x} = 34\%$). In the fall of 1979, following a highly successful nesting season, a presumed epidemic swept through the study population and killed all but one of the 93 juveniles. This extraordinary event also killed 45% of the breeders, more than double their average yearly mortality.

Mortality of young is always high immediately after fledging. Juvenile survivorship increases steadily for about eight months, then levels off near the adult rate. Survival over the first two months post-fledging correlates slightly with weight as a nestling (McGowan 1987). Annual survival of fledglings in the overall population correlates strongly with survival of breeders and with mean fledgling production. Again, these patterns implicate predation as the principal cause of fledgling deaths.

Helping

As young jays mature they integrate into the activities of the family. Fledglings appear to be nutritionally independent by age three months (July). Social integration of these helpers continues to develop for many months thereafter. Helpers assist with the care of young, sentinel duty, mobbing and territory defense (McGowan 1987). They do not participate in building nests, incubating eggs, or brooding nestlings. As mentioned, we have no evidence that helpers ever contribute gametes to the young that they help to raise.

Effects of helpers

Breeders with helpers raise significantly more young than those without ($\bar{x} \pm$ s.d. = 2.4 $\pm$ 1.5 versus 1.6 $\pm$ 1.3) fledglings per pair, respectively; Fig. 8.4). The advantages of having helpers appear at several independent stages: the hatching of eggs, the fledging of nestlings and the survival of fledglings to independence (McGowan 1987). The survival of breeders also correlates with presence of helpers. While virtually all our comparisons suggest that helpers contribute actively to the production of additional offspring, we have not performed controlled experiments to demonstrate this unequivocally. Colleague R. L. Mumme is currently conducting the relevant experiments in a study tract near the historical one.

Measured as the proportion of nests producing at least one fledgling, nest success was higher for pairs with helpers in all but one of the 18 breeding seasons in our study. Presence of helpers correlates with higher daily survivorship of nest contents, thereby increasing the probability of a given nest succeeding. This effect is more pronounced during years in which average nest success is low; helpers apparently dampen the negative effects of poor breeding years.

McGowan (1987) demonstrated that helpers enhance the survival of fledglings during a 30 day period of dependence about 10 days after leaving the nest. Mortality of fledglings during their first 10 days out of the nest is extremely high (36%), and at this time no effect of helpers is evident. During the next 30 days, fledglings from families with helpers survive at a higher average rate than those from families without helpers. No effect is evident thereafter.

Neither the number of helpers beyond one nor the sex of helpers correlates with fledging success. Families with two, three, or four helpers of either sex are relatively common, and their average fledgling production does not differ from those with only one helper (Fig. 8.4). Older helpers appear to be more attentive to the nestlings than one-year-old helpers, and a non-significant effect of helper age upon fledgling production does exist.

The effects of helpers upon reproductive success diminish as the nesting season progresses. As mentioned above, the probability of fledging young from a nest falls steadily from over 70% to less than 40% through the season. Late in the season, the contributions made by helpers appear to be swamped by the seasonal increase in nest predation, which is by far the most important cause of nest failure.

Fig. 8.4. Reproductive success as a function of group size in the Florida Scrub Jay (means ±S.E.M.). Sample sizes as in Fig. 8.3. Independent (Ind) young are those alive in mid July, about age two months. Reproductive success is significantly lower between group size 2 and group sizes 3, 4, 5, or combined. No significant differences exist among any group sizes greater than 2.

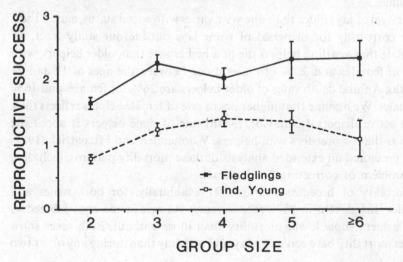

How helpers help

Several lines of evidence suggest that the food provided to nestlings and fledglings by helpers is only a minor cause of increased nest success and fledgling survival. Starvation is rare, and fledgling survival correlates only weakly with weight as a nestling. Virtually all nest failures occur through predation, including losses of eggs (which need no food). The presence of helpers correlates with decreased rate of nest predation. The effects are most pronounced during years of low average fledging success, corresponding to years when predation pressure is highest. We suspect that helpers increase reproductive success principally by reducing the probability that eggs, nestlings and fledglings are detected and consumed by predators. This contribution may come about through participation in sentinel activities, predator mobbing and passive nest attendance.

Foraging and provisioning by helpers at the nest significantly reduces the time and energy expended in these activities by the breeding pair (Stallcup and Woolfenden 1978). By affecting the breeders' time budgets, as well as by helping to spot predators, helpers also increase the survival of breeders. Breeders without helpers experience significantly higher annual mortality than those with helpers (23% versus 15%).

Survivorship

Average survival rate of juvenile jays through the first 12 months post-fledging is 34%. Annually, this figure has varied from 21% to 44%, excluding 1979, when an apparent epidemic killed all but one of 93 fledglings.

Survival of jays older than one year varies with social status, age and sex. After correcting for dispersal of some jays outside our study tract, we calculate that yearling helpers die at a higher rate than older helpers, with 35% of females and 20% of males dying between the ages of 12 and 24 months. Annual death rates of older helpers are 26% for females and 16% for males. We assume that higher death rate of female helpers reflects their more active dispersal behavior. Death rate of male helpers is about the same as that of breeders with helpers. Woolfenden and Fitzpatrick (1984) have presented an extended analysis of these mortality patterns, including the problem of correcting for dispersal.

Mortality of breeders averages 21% annually, for both males and females (186 deaths in 883 breeder-years). As mentioned above, breeders with helpers show lower mortality than those without. Both sexes show higher mortality between mid May and mid July than during any other two month period of the year. In this period jays are engaged in feeding dependent fledglings and in molting (Bancroft and Woolfenden 1982). Both activities are energetically stressful, and may also place the jays at greater risk of being taken by predators.

A surprising result emerged from our comparison of breeders that helped for one year versus two or more years before becoming breeders. In both survival and average reproductive success, the earlier dispersing breeders were superior. Furthermore, these earlier dispersing males show higher average reproductive success. These patterns are mostly evident among males. They strongly suggest that any advantage of experience gained by helping does not accumulate after the first year of life. Moreover, they raise the possibility that individuals that remain longer as helpers may be in some way inferior to those that disperse early. We are examining this possibility carefully.

Death rate of breeders does not vary with overall population density. However, a direct correlation exists between the density of older helpers during June and the death rate of breeders during the ensuing 12 months. We have not yet fully explained this correlation, but it may result from increased active competition for breeding space during years when the density of potential breeders is highest.

Senescence

Survivorship of breeders is age independent for the first nine years of breeding. The few breeders that have survived for nine breeding seasons show decreased survivorship thereafter. No breeder has survived for more than 11 breeding seasons. The oldest jay known to us died at age 14 years 5 months. This male helped for three years and bred for 11. Senescence appears to characterize reproductive capacity as well as survivorship. The oldest breeders show significant decreases in clutch size, fledging success and fledgling survival (Fitzpatrick and Woolfenden 1988).

Age structure

Because of annual variation in both reproduction and survival, the population's age structure varies considerably from year to year. On average, about 30% of the population consists of pre-breeders, of which two-thirds are in their first year of life. Among breeders, modal age is three years (15% of breeders) and median age is five years. Between 15% and 20% of the breeders are age 10 years or older. Mean life expectancy for a jay that becomes a breeder is 4.3 breeding seasons.

Lifetime reproductive success

We have accumulated data on reproductive success of 135 jays (68 males, 67 females), now dead, whose entire breeding histories were known, and which began breeding before 1983. This sample provides a representative set of cohorts not biased towards shorter-lived breeders. Of this sample, hearly half failed to produce even a single surviving offspring (Fig. 8.5). About 20% produced about 65% of the new recruits into the breeding population. Males and females show identical patterns of variance in lifetime reproduction, as would be expected in a monogamous population with equal mortality between the sexes.

Variance among individuals in lifetime reproduction is increased by several features of Florida Scrub Jay natural history (Fitzpatrick *et al.* 1989). (1) Breeding success increases with breeding experience for several years, but most breeders die before reaching peak productivity. Therefore,

Fig. 8.5. Lifetime production of independent (age two months) and breeding offspring among 67 Florida Scrub Jays that became breeders between 1971 and 1977 (34 males, 33 females). (From Fitzpatrick *et al.*, 1989.)

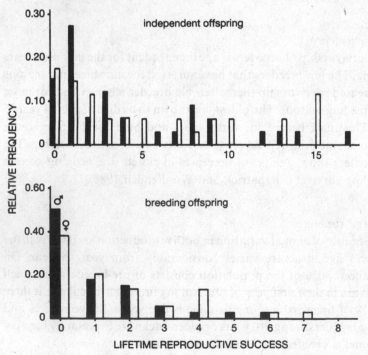

the longer-lived breeders are proportionately more productive than shorter-lived ones. (2) Success during a given breeding year produces helpers during subsequent years. Helpers increase both reproductive success and survival of breeders. Therefore, reproductive success is compounded as a breeding lifetime proceeds. (3) Certain years are unusually good ones for overall reproduction, while other years are poor for both reproduction and adult survival. Jays lucky enough to breed during peak years receive the randomly acquired bonus of increased survival and fledgling production, not just during the good year but also during subsequent years (owing to the effects of helpers). (4) Certain family lineages show comparatively high average rates of breeder survival, or greater success at producing offspring that win competitions for breeding vacancies, or both. These important traits may be partly heritable, and could represent genetic variation within the population for parameters important in lifetime fitness. (5) Average reproductive success appears to be uneven across the microhabitat patches within our study area. Certain local areas may be of superior habitat quality, causing lineages that hold them to demonstrate higher average success than jays of other lineages. Separating this environmental component from the just-mentioned genetic component will be difficult. As discussed below, this challenge is an important direction for future attention.

One result of the strong skew in reproductive success among breeders is that, over many years, local neighborhoods develop in which numerous breeders in contiguous territories represent direct descendants of the same few highly successful breeders. Our study area of about 35 territories, studied for 18 years, is currently dominated by the descendants of four extremely successful male breeders and their mates, banded at the onset of the study.

Speculations and unanswered questions

The preceding synopsis of Florida Scrub Jay social and population biology leaves numerous theoretical issues unanswered. Many of them continue to loom before us as challenges for further work and thought. Here, we summarize briefly our ideas on the evolution and maintenance of cooperative breeding in the Florida Scrub Jay, highlighting some of the unsolved problems still facing us.

Essentially all Florida Scrub Jays belong to a territorial pair or group. Most remain in their natal group until they disperse to a new group to breed. Social events sometimes leave non-breeders without a natal group, while others may leave the natal group inexplicably. Even in these exceptional cases, the jays soon struggle to work themselves into some other group, often as subordinate, non-breeding helpers. Living in a group (even if the group is a simple pair) appears to be an essential ingredient to survival for Florida Scrub Jays.

Whenever the opportunity arises, all group members will deliver food to dependent young. This behavior is independent of kinship, as unrelated helpers usually feed nestlings and fledglings of their group just as actively as do close relatives. These observations suggest to us that the *act of feeding young* during the nesting season may be a simple behavioral response to the presence of begging nestlings or juveniles. Feeding dependent offspring within the territory may be a by-product of living in a group, and not necessarily an adaptive consequence of its effects on the fitness of the helper or its relatives. We offer this null hypothesis as a model to be tested and refuted. In its defense, evidence is extremely weak that the food provided to dependent young affects the number or survival of those young.

Brown (1987) considers that understanding cooperative breeding requires answering two separate questions. First, why do certain individuals delay breeding and remain at home? Second, while at home, why assist the resident breeders in the care of their young? This second activity represents Brown's (1975) operational definition of 'helping behavior'. He points out that, on the first question, little disagreement remains: delayed breeding provides direct benefits in demographic environments where opportunities for successful dispersal are limited. He contends that the second question is different, and argues that important indirect benefits are involved in selecting for 'helping behavior'. We consider Brown's dichotomy to be valid in concept, but misleading in practice. The substance of 'helping behavior' cannot be measured simply by counting visits to nests. Virtually all activities practiced by non-breeders in their natal territory, including merely being present, affect the fitness of every member of the group. Within the group, helpers assist adults and juveniles, breeders and non-breeders, relatives and non-relatives, all year round. All their activities have purely selfish components as well. The fact that direct and indirect benefits exist to both delayed breeding and helping makes it difficult to separate these two independent 'stages' in the evolution of cooperative breeding. In many respects, delayed breeding *is* 'helping'.

Our analyses suggest that helpers raise the reproductive output of breeders, but our tests are only correlational because to date we have conducted no removal experiments within the study tract. Therefore, we have difficulty controlling for the confounding influences of variation in habitat quality and genetic endowment among lineages. R. L. Mumme's experimental studies are designed specifically to address this unanswered question by controlling for group size among random samples of families and territories. In the absence of differences in territory quality, do helpers genuinely improve the reproductive success of the breeders they assist? The definitive answer, in this and most other studies of cooperative breeding, still is not available.

If Mumme's experiments indicate limited or no influence of helpers on fledgling production, this would constitute evidence that habitat variation has confounded many of our previous analyses of the effects of helpers. In our historical study tract, patches of high-quality habitat may have permitted certain jay families consistently to produce more young, and therefore more helpers. We are commencing an intensive investigation into this and related questions: how does habitat quality influence reproductive success, both within a breeding season and over the lifetimes of individuals?

We have emphasized elsewhere (Woolfenden and Fitzpatrick 1984) that limited expected success at early dispersal, combined with the opportunity to obtain breeding space by remaining home, can select for delayed dispersal. Emlen (1978) and others have shown that the maximum expected gains in inclusive fitness accrued through helping while remaining home are small compared to those gained from breeding. Consequently, non-breeders can be expected to follow a course that maximizes their probability of breeding. For Florida Scrub Jays, intense competition for breeding space, limited opportunity to survive and breed in marginal habitat, relatively high survivorship when living in a group, and the potential to improve the chances of becoming a breeder without having to leave home, select for delayed dispersal. Is living and interacting for extended periods in a natal group tantamount to helping? From the point of view of maximizing expected lifetime reproductive success, all the other options appear to be worse. Helping ultimately is a selfish strategy.

From time to time we witness events that tax our own belief in our explanation for cooperation in the Florida Scrub Jay. The most nagging observations involve non-breeders that seemingly 'decline' apparently available breeding space nearby, and instead remain as non-breeding helpers. The epidemic in 1979 provided several such cases, as numerous breeder deaths made large areas of scrub available for non-breeders to colonize and defend. Many did so, but others did not. Breeder density has remained below the ten-year average for several years afterward (Fig. 8.2). Even in typical years we sometimes see older helpers who do not avail themselves of apparent opportunities to pair and defend a territory. If our basic interpretation of the system is not flawed, two unanswered questions arise from these anomalous non-breeders: do jays sometimes forgo breeding opportunities because they detect features of the available habitat, or available mates, that they judge to be substandard? Conversely, are such individuals themselves substandard? Do they represent biologically inferior individuals, chronic losers that remain visible to us simply because of the sedentary and relatively tame nature of this jay? Our preliminary evidence suggests that the latter may be true. If so, is this acquired or innate variation? Is the presence of chronic losers typical of other bird populations?

We speculate that establishing and maintaining a territory is unusually difficult for Florida Scrub Jays, for reasons already outlined. Survival as a non-breeder, on the other hand, is relatively easy in the comparative safety of the home territory. Moreover, chances for successful dispersal can improve through time, especially if group size can be made to grow. A relatively long lifespan is expected once breeding status is attained. Because most breeders fail to replace themselves genetically before dying, producing even a few offspring that become breeders represents genetic success for a Florida Scrub Jay. Therefore, investing heavily and continuously in a small number of offspring is favored. Garnering sufficient space in the territory to help to provide one or more older offspring with the chance to breed on a 'starter territory' may therefore be one of the most important adaptations to have arisen within the population.

Most of the implicit assumptions just made about territory size and resource supply necessary for breeding are still untested. Is size of territory correlated with resource base? Do Florida Scrub Jays often defend more land than their own survival and nesting efforts require? Must territories hold a certain minimum set of resources before they are acceptable to potential dispersers? Indeed, what ultimately determines territory size in this *singular-nesting* species?

The last question leads directly to a broader one of more theoretical interest, which should be asked of all cooperative breeders, both singular and plural nesters: what ultimate and proximate factors keep non-breeding family members from breeding at home? Is some minimum and dependable resource base necessary to allow plural nesting to evolve? In effect, do offspring of plural nesters such as Mexican Jays halve the resources available to their parents' nest by pairing and breeding within the natal group? Why is this fundamental demographic event so rare in the closely related Florida Scrub Jay? We suspect that answers to these key questions involve interactions between resource characteristics (especially annual variation, patchiness and defendability) and demography. These and numerous related questions are current focal points in our continuing study.

We remain indebted to the Archbold Biological Station and its staff for long-term cooperation in making this study possible. Our students and coworkers over the past 18 years have made significant contributions to our understanding of Scrub Jay biology. Bobbie Kittleson and Jan Woolfenden deserve special thanks for their continued assistance with field work. Fred Lohrer is of constant help in the library, and aided in the preparation of this manuscript. Robert Curry, Walt Koenig and Peter Stacey made valuable comments on the manuscript. We thank the H. B. Conover Fund of the Field Museum of Natural History for partial support of both field work and data analysis. Current field work is supported under NSF grant BSR-8705443.

Bibliography

Bancroft, G. T. and Woolfenden, G. E. (1982). The molt of Scrub Jays and Blue Jays in Florida. *Ornith. Monogr.* **29**.

Barbour, D. B. (1977). Vocal communication in the Florida Scrub Jay. M. A. thesis, University of South Florida.

Bent, A. C. (1946). Life histories of North American jays, crows, and titmice. *U.S. Nat. Mus. Bull.* **191**.

Brown, J. L. (1974). Alternate routes to sociality in jays – with a theory for the evolution of altruism and communal breeding. *Amer. Zool.* **14**: 63–80.

Brown, J. L. (1975). *The Evolution of Behavior*. W. W. Norton and Co. Inc.: New York.

Brown, J. L. (1978). Avian communal breeding systems. *Ann. Rev. Ecol. Syst.* **9**: 123–56.

Brown, J. L. (1987). Review: *the Florida Scrub Jay: Demography of a Cooperative Breeding Bird. Auk* **104**: 350–2.

DeGange, A. R. (1976). The daily and annual time budget of the Florida Scrub Jay. M.A. thesis, University of South Florida.

DeGange, A. R., Fitzpatrick, J. W., Layne, J. N. and Woolfenden, G. E. (1989). Acorn harvesting by Florida Scrub Jays. *Ecology* **70**: 348–56.

Emlen, S. T. (1978). The evolution of cooperative breeding in birds. In *Behavioural Ecology: An Evolutionary Approach,* ed. J. R. Krebs and N. B. Davies, pp. 245–81. Blackwell: Oxford.

Fitzpatrick, J. W. and Woolfenden, G. E. (1981). Demography is a cornerstone of sociobiology. *Auk* **98**: 406–7.

Fitzpatrick, J. W. and Woolfenden, G. E. (1984a). Staying around the nest. In *Science Yearbook*, pp. 13–25. Worldbook/Childcraft Inc.: Chicago.

Fitzpatrick, J. W. and Woolfenden, G. E. (1984b). The helpful shall inherit the scrub. *Nat. Hist.* **93**: 55–63.

Fitzpatrick, J. W. and Woolfenden, G. E. (1986). Demographic routes to cooperative breeding in some New World jays. In *Evolution of Animal Behavior,* ed. M. Nitecki and J. Kitchell, pp. 137–60. University of Chicago Press: Chicago.

Fitzpatrick, J. W. and Woolfenden, G. E. (1988). Components of lifetime reproductive success in the Florida Scrub Jay. In *Reproductive Success,* ed. T. Clutton Brock, pp. 305–20, University of Chicago Press, Chicago.

Fitzpatrick, J. W., Woolfenden, G. E. and McGowan, K. J. (1989). Sources of variance in lifetime fitness of Florida Scrub Jays. *Proc. 19th Intern. Ornith. Congr.* 876–91.

McGowan, K. J. (1987). Social development in young Florida Scrub Jays (*Aphelocoma c. coerulescens*). Ph.D. dissertation, University of South Florida.

McGowan, K. J. and Woolfenden, G. E. (1989). A sentinel system in the Florida scrub jay. *Anim. Behav.* **37**: 1000–6.

Pitelka, F. A. (1951). Speciation and ecological distribution in American jays of the genus *Aphelocoma. Univ. Calif. Publ. Zool.* **50**: 195–464.

Stallcup, J. A. and Woolfenden, G. E. (1978). Family status and contribution to breeding by Florida Scrub Jays. *Anim. Behav.* **26**: 1144–56.

Wiley, R. H. and Rabenold, K. N. (1985). The evolution of cooperative breeding by delayed reciprocity and queuing for favorable social positions. *Evolution* **38**: 609–21.

Woolfenden, G. E. (1974). Nesting and survival in a population of Florida Scrub Jays. *Living Bird*, **12**: 25–49.

Woolfenden, G. E. (1975). Florida Scrub Jay helpers at the nest. *Auk* **92**: 1–15.

Woolfenden, G. E. (1976a). Co-operative breeding in American birds. *Proc. 16th Intern. Ornith. Congr.*: 674–84.

Woolfenden, G. E. (1976b). A case of bigamy in the Florida Scrub Jay. *Auk* **93**: 443–50.

Woolfenden, G. E. (1978). Growth and survival of young Florida Scrub Jays. *Wilson Bull.* **90**: 1–18.

Woolfenden, G. E. (1981). Selfish behavior by Florida Scrub Jay helpers. In *Natural selection and social behavior*, ed. R. D. Alexander and D. W. Tinkle, pp. 257–60. Chiron Press: New York.

Woolfenden, G. E. and Fitzpatrick, J. W. (1977). Dominance in the Florida Scrub Jay. *Condor* **79**: 1–12.

Woolfenden, G. E. and Fitzpatrick, J. W. (1978a). The inheritance of territory in group-breeding birds. *BioScience* **28**: 104–8.

Woolfenden, G. E. and Fitzpatrick, J. W. (1978b). Authors' reply. *BioScience* **28**: 752.

Woolfenden, G. E. and Fitzpatrick, J. W. (1980). The selfish behavior of avian altruists. *Proc. 17th Intern. Ornith. Congr.*: 886–9.

Woolfenden, G. E. and Fitzpatrick, J. W. (1984). *The Florida Scrub Jay: Demography of a Cooperative-breeding Bird*, Monogr. Pop. Biol. no. 20. Princeton University Press: Princeton, NJ.

Woolfenden, G. E. and Fitzpatrick, J. W. (1986). Sexual asymmetries in the life history of the Florida Scrub Jay. In *Ecological Aspects of Social Evolution: Birds and Mammals*, ed. D. Rubenstein and R. W. Wrangham, pp. 87–107. Princeton University Press, Princeton, NJ.

9 MEXICAN JAYS: UNCOOPERATIVE BREEDING

9

Mexican Jays: uncooperative breeding

J. L. BROWN AND E. R. BROWN

Studies of unbanded Mexican Jays* (*Aphelocoma ultramarina*), which began with Gross (1949), Wagner (1955) and Hardy (1961), reported the presence of more than two birds around nests and hinted at cooperation. The first intensive study based on color-banding of entire flocks, however, revealed mainly uncooperative rivalry (Brown 1963a). Here, we summarize published knowledge of the behavioral ecology of this species and discuss the apparent contradiction of the presence in the same social units and even the same individuals of both cooperative and uncooperative behavior. A more detailed report covering 20 years of data is currently in preparation.

Habitats and study areas

We have studied Mexican Jays at three principal localities in the United States and a variety of places in Mexico. We began our work in the Santa Rita Mountains of Arizona in 1958. For our long-term study beginning in 1969, however, we chose a base at the Southwestern Research

---

* The Mexican Jay (*Aphelocoma ultramarina*) has two common names, both given to it by committees of taxonomists. A few words of explanation seem desirable. The Mexican Jay was given its English name as a species in 1957 when the American Ornithologists' Union switched from a policy of providing common names for each subspecies to one of having a common name only for the species (American Ornithologists' Union 1957). Mexican Jay became the established name in the scientific literature and the field guides. Recently, however, the taxonomists on the committee for nomenclature of the American Ornithologists' Union (AOU) voted, for unstated reasons, to change the long-established name to Gray-breasted Jay (American Ornithologists' Union 1983). The result is that the AOU has burdened us with two officially sanctioned names. Because Mexican Jay is used in virtually all the literature on the behavior and ecology of the species we shall continue to use this name in order to provide nomenclatural stability and continuity in the scientific literature. In addition to being established it has the virtues of priority and of being usefully descriptive.

Station of the American Museum of Natural History in the Chiricahua Mountains of Arizona, working with *A. u. arizonae*. To gain information about other subspecies we have made brief studies in the Chisos Mountains of Texas and in the Sierra Madre of Mexico. Thus, most of our work has been done on *A. u. arizonae*, but we have some field data on nearly all the other subspecies.

Mexican Jays throughout their range are typically associated with temperate-zone oaks (*Quercus* spp.) and mountains. In our study area at 1640 m elevation in the Chiricahua Mountains, three species of oak tree are common (*Quercus emoryi*, *Q. arizonica*, *Q. hypoleucoides*) and other tree species of oaks occur rarely (*Q. gambelli*, *Q. robusta*) along with a few scrub oak species. Acorns from these oaks are harvested and stored during August and September. A juniper (*Juniperus deppeana*) is the most abundant tree and three pines are common (*Pinus leiophylla chihuahuaensis*, *P. engelmanni*, *P. cembroides*). A typical view of the vegetation and topography is shown in Fig. 9.1.

The jays on our study area experience a climate that is dominated by strong late-summer or monsoon rains, hot dry springs, and cold, snowy winters, with night temperatures routinely well below freezing. Annual reproductive success varies drastically as a function of rainfall (Brown 1986).

Foods and foraging

In Arizona, the Mexican Jay varies its diet with the seasons. In August the species competes effectively with Acorn Woodpeckers for ripening acorns (see Chapter 14), which the jays commonly store in the ground in scattered locations. In October the nuts of *Pinus cembroides* are similarly harvested and stored. Throughout winter and even into the following breeding season, stored nuts are retrieved and eaten. When eating an acorn, a jay typically pierces it with the lower mandible, grasps a piece of the husk using the slightly hooked upper mandible, and tears it off. After opening a sufficient hole it then breaks off pieces of the nut meat in a similar fashion.

In April when the weather warms, hairy caterpillars appear in numbers. These are avoided by small birds generally because of their toxic hairs (E. Greene, personal communication), but Mexican Jays have a detoxifying behavior. Grasping the caterpillar in the bill they fly to the ground, then rub the caterpillar in the sandy soil again and again, leaving an arc of about 13 cm in the soil. This procedure gradually removes the hairs.

When hot weather arrives in late May and nestlings and fledglings are begging, cicadas emerge. Quantitative studies of the foods given to nestlings have shown that cicadas form roughly half of their diet at this time (J. L. Brown and H. Alvarez, unpublished data). Most cicadas are caught shortly after emergence while their wings are drying and before they can fly.

Throughout the year jays forage mainly on the ground, where they catch grasshoppers, cicadas, flying ants (Westcott 1969; Brown 1983) and a great variety of other insects, as well as recovering stored foods. They also forage in trees for nuts and large insects and occasionally engage in aerial flycatching. They commonly catch lizards (J. L. Brown, personal

Fig. 9.1. Study area for the Mexican Jay in the Chiricahua Mountains, Arizona. The fence separates the Southwestern Research Station on the left from the national forest land on the right. Grazing by cattle is permitted on national forest land but not on the research station. Trees shown in the figure include oaks and junipers. (Photo by Dana Slaymaker.)

observations) and occasionally take eggs, nestlings or even adult passerines (Roth 1971).

Flock size and group territoriality

In North America, social groups of birds tend to be seasonal and transient. For example, social groups of Blue Jays (*Cyanocitta cristata*; Hardy 1961) and Steller's Jays (*C. stelleri*; Brown 1963*b*) tend to be temporary aggregations around a predator, a female, or a food source. To see whether or not this was true also for the Mexican Jay, we color-banded and observed two flocks of Mexican Jays in the Santa Rita Mountains of Arizona in 1958. Just as we had for the Steller's Jay, we plotted the winter home range of each bird individually. The difference between the species was striking. Steller's Jays lived as pairs with home ranges that overlapped extensively. Non-breeders roamed through the entire study area. In contrast, all the members of a flock of Mexican Jays, breeders and non-breeders, had the same home range; and the breeding pairs shared all of it

Fig. 9.2. Sizes of social units of the Mexican Jay in Chiricahua Mountains, Arizona, and Chisos Mountains, Texas. Counts were made in late spring or early summer. They include all regular group members except the young of the year. *n*, number of groups; $\bar{x}$, mean group size. Data for Arizona from Brown and Brown (1984); for Texas, from Strahl and Brown (1987).

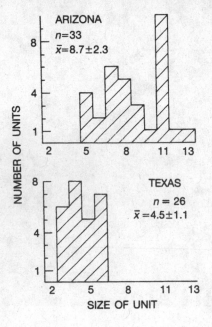

with no hint of separate core areas for each pair. These shared home ranges proved to be defended areas. When we returned in the spring, the same individuals were still together in the same flocks on the same territories. This, therefore, was a case of group territoriality, one of the first avian examples to be verified through study of marked individuals.

Having verified that each flock was actually a social unit that remained nearly the same size and composition from day to day and month to month, we wanted to determine the range and variability of unit size. We examined variability in space over the elevational range of the Mexican Jay in the Chiricahua Mountains, and variability over time in a subset of this population over a 14-year period (Brown and Brown 1984). Both seasonal and yearly variation were studied. In our study area in Arizona, 33 social units ranged in size from five to 13 birds in one spring (Fig. 9.2), averaging 8.7 ($\pm$2.3) birds then and 11.2 ($\pm$3.3) birds after the breeding season in August of the same year.

Yearly variation was impressive. From 1972 through 1982 average unit size ranged from 6.7 to 17.5 individuals in spring (Fig. 9.3). Compared to most other communally breeding species, average unit size in this population is relatively large.

Plural breeding

The nest of a Mexican Jay is an open cup of sticks lined with plant and animal fibers. A clutch of four or five eggs is normal. Joint-nesting of

Fig. 9.3. Yearly variation in unit size of Mexican Jays in the Chiricahua Mountains paralleled population size, while the number of social units remained relatively stable. (After Brown 1986.)

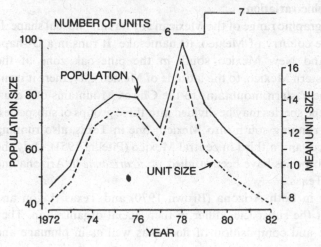

two females in one nest has been recorded, but it is extremely rare. The normal situation is for only one female to participate in building a nest with a single male, for that female to do all the incubation and brooding, and for this pair to be the most active and reliable feeders of the nestlings. Operationally, we identify the brooding female as the mother and the male builder and consort as the primary male. Sexually receptive females are sometimes followed by more than one male, especially if the pairs are asynchronous in nesting or if there is an excess of males in the group. These female-following males include only the older males (over three years old) and not yearlings. At some nests paternity-exclusion analyses have revealed that the primary father can be excluded as the genetic father for some of the young (B. S. Bowen, R. R. Koford and J. L. Brown, unpublished data). At some nests more than one mature male feeds the young (e.g. Fig. 9.5), but in no case does more than one male build the nest (unless usurpation occurs).

In some populations of the Mexican Jay two or more females commonly attempt to breed in the same unit at the same time, a condition known as plural breeding (Brown 1978). In our study area in the Chiricahuas, plural breeding has occurred in every flock (six or seven) nearly every year over the 18 years of our study. Plural breeding, therefore, is the typical condition for this population. Not only may two to five females begin to nest at the same time, but separate nests in the same unit are often *successful*. Many fail, but in 30 of 69 unit-years from 1971 to 1982 two to four nests fledged young in the same unit in the same year (Brown 1986). Plural breeding appears to be the common pattern also in the Santa Rita Mountains of Arizona (Brown 1963a), and in most of Mexico (Brown and Horvath 1989).

Geographic variation

The geographic range of the Mexican Jay has an unusual shape. It lies mainly in the country of Mexico, its namesake. It runs in a U-shape from Arizona and New Mexico south in the pine–oak zone of the mountains of western Mexico, to the latitude of Mexico City; there it turns back north in the eastern mountains to the Chisos Mountains of Texas. Taxonomically, the species may be divided into three groups of subspecies; one in Arizona, running south into Mexico; one in Texas, also running south into Mexico; and a third in central Mexico (Pitelka 1951). Only the two northern subspecies have been studied, *A. u. arizonae* in Arizona and *A. u. couchii* in Texas.

Helpers occur in both Arizona (Brown 1970) and Texas (Ligon and Husar 1974), but the subspecies differ in their social organization. They differ in the size and composition of flocks, as well as in plumage and behavior. Strahl and Brown (1987) found that flocks in Texas were distinctly smaller than those in Arizona, ranging in size from three to six with a mean of 4.5 (S.D. = 1.14), as shown in Fig. 9.2. Even more interesting, current evidence strongly suggests that *couchii* is singular breeding. Therefore, *A. u. couchii* in its social structure resembles the Florida Scrub Jay (see Chapter 8) more than the Arizona population of its own species.

The ontogeny of reproductive success in Mexican Jays

Like most other species of cooperatively breeding birds, Mexican Jays normally delay breeding. In many species only a fraction of the population delays, often only males or a fraction of the males; but in Mexican Jays both sexes delay, and the length of the delay is extreme (Brown 1985). In Fig. 9.4, which shows the pattern of reproduction as a function of age, both sexes are combined; females, however, tend to breed a little earlier than males. Virtually no birds in this 14 year sample bred in their first spring; even at age two years, few individuals attempted breeding, and hardly any were successful. Not until age four years did the percentage of birds attempting to breed level off; and asymptotic levels of fledging success were not achieved until age 5 years. Only about half of the individuals had fledged any young in their entire lifetime by that age.

Fig. 9.4. Age-specific reproductive success in the Mexican Jay, 1969–82. The percentage of banded, known-age jays that attempted to breed at least by beginning to build a nest is shown by the upper curve (Attempted). The percentage that fledged one or more young is shown by the lower curve. (After Brown 1986.)

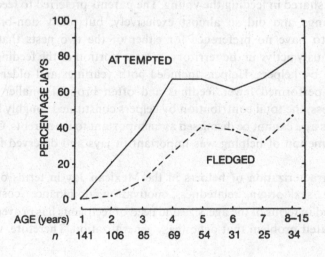

Beyond age five years some adults were still not breeding. These were usually males for whom no females were available, but females also failed to breed if males were unavailable in their unit.

As in other species of cooperatively breeding birds that show conspicuous delays in breeding, *A. u. arizonae* shows conspicuous 'signals of immaturity' (Brown 1985). In Arizona, but not in Texas, the bills of juveniles retain much of the light coloration of the nestlings. This is still conspicuous at age one year, and a remnant of the nestling color is usually present until age two years. It is tempting to speculate that these visual correlates of a low probability of successful breeding are adaptive, but such hypotheses are untested. Because the bills of *couchii* attain adult color at fledging we predict that delayed breeding will be found to be much less marked in *couchii* than in *arizonae*.

Helping

When Hamilton's (1963) theory of inclusive fitness appeared, the phenomenon of helping, which was long known in birds (Skutch 1935), appeared to provide material suitable for field-testing certain predictions of the theory. The first step was to confirm the occurrence of helping in color-banded flocks and to estimate its importance in terms of effort for the various classes of individuals. This had not been done before for any species of bird except by Rowley (1965). Therefore, we resumed our study of Mexican Jays in 1969, and began by describing the phenomenon of helping as it occurred in jays of the Chiricahua Mountains.

We found both breeding helpers and non-breeding helpers in *arizonae* (Brown 1970, 1972). As shown in Fig. 9.5. all individuals in the first units to be studied shared in feeding the young. The parents preferred to feed their own nestlings and did so almost exclusively, but many non-breeders appeared to have no preference for either of the two nests that were simultaneously active in the territory. A large portion of the feedings was performed by helpers. Helpers included both yearlings and older birds. Yearlings performed fewer feedings and often brought smaller items. Nevertheless, the total contribution by helpers constituted roughly half of all feedings and cannot be dismissed as unimportant to the parents. Clearly the phenomenon of helping was important in jays and deserved further study.

The characterization of helpers in the Mexican Jay in terms of their effort, age, sex, origin, relatedness, motivation, dominance, costs and benefits and in terms of the age, size and need of recipients has proven to be a complicated problem that defies easy generalization. Therefore, we feel that such statements should be deferred until elaborate analyses – now in progress – are completed. Nevertheless, a few preliminary statements can be made. Our data on dispersal (Fig. 9.6) and genealogies (Fig. 9.8) show that many individuals stay for many years in their natal units. Thus, many potential helpers are related to their nestling-recipients. On the other hand, these same figures also reveal that immigration is common. It can be said, at least, that helpers may be of any age, sex, origin, relatedness and dominance rank. Furthermore, no simple hypothesis explains all cases of helping.

This is not surprising because the inclusive-fitness benefits to individuals probably vary widely depending on most of these factors. We illustrate with three examples, 'reciprocal' helping by successful breeders, helping by polyandrous males at a single nest, and helping by non-breeders. In all three cases relatedness is probabilistic. In the first two examples, direct-fitness benefits are probably predominant. In the case of non-breeders, however, indirect fitness is likely to be important. Clearly helpers in these three cases have acquired their uncertainty of relatedness in very different ways.

Fig. 9.5. Relative contributions of each member of the Station flock of Mexican Jays to the feeding of young in each of two nests in two successive years. An asterisk designates a significant ($P < 0.01$) preference for delivering food to one of the nests. Year of hatching is indicated by CLASS. *n*, number of feedings. (From Brown 1972.)

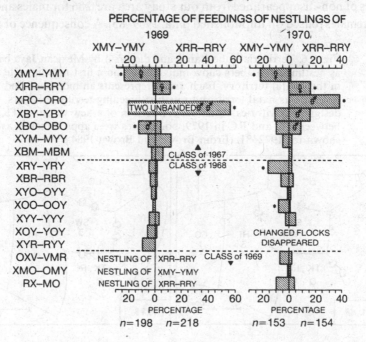

Demography and dispersal

By 1970 it became obvious to us that the technique of saturation color-banding in the Mexican Jay would provide high-quality data on survival, dispersal, age structure and demography in general. After a preliminary look at results on these topics in the Mexican Jay we stressed the importance of demography for understanding the ecology of communally breeding birds and predicted that they would be found to have delayed breeding, increased survival rate, lowered reproductive rate and diminished dispersal relative to non-cooperative species (Brown 1974). We also invoked demographic factors in the strategies of non-breeders for acquiring breeding status (Brown 1969, 1974).

These predictions have virtually all received strong support, especially for species with many non-breeding helpers (Brown 1987a). For example, dispersal in the Mexican Jay proved to be as conservative as in any passerine bird known. Figure 9.6 shows for both sexes that many individuals did not leave their natal flock to breed; instead they stayed and bred. Many spent their entire lives there. With only one known exception, those individuals that did leave home and were recorded breeding went no further than a neighboring territory.

Survival proved to be quite high. Our current estimates for adult annual rates of non-disappearance from our study area are 0.86 for males and 0.81 for females (based on unpublished data, 1969–85). A consequence of a high

Fig. 9.6. Territorial inheritance and dispersal by Mexican Jays banded as nestlings. Numbers show individuals whose first breeding site was in their natal territory. Each arrow represents an individual and connects its natal territory to its first breeding territory. Letters designate territories. Aside from insertion of a new territory, TK, between SW and RC in 1979, boundaries were approximately as shown for 1972–81. (From Brown and Brown 1984.)

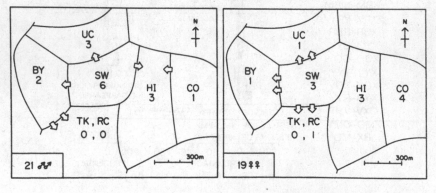

rate of survival is that relatively many individuals live a long time. Our oldest jay was 18 years old in 1987, and roughly 5% of the population is 10 years or older.

A summary of our population data for the period 1971–82 in Fig. 9.7 shows the history of the age structure of the population in a way that brings out the importance of certain year classes (Brown 1986). Birds hatched in 1973 and 1977 were numerous and survived their first winter well, causing each of these year classes to predominate numerically in the population for several years. This year-class phenomenon shows that the population is not typified by a stable age distribution and serves as a warning not to rely heavily on population data from only one or two years. During the period 1969–82 the population fluctuated by a factor of two. Nevertheless, the number of territorial social units changed relatively little (Fig. 9.3).

Genetic structure

One of the consequences of delayed dispersal and absence of dispersal in some individuals is that siblings live together in their natal territory with one or two parents for a longer period than if they were to

Fig. 9.7. Year classes in the Mexican Jay. The number of jays in a given year class each year on 1 May may be read as the vertical difference between the upper and lower bounds. Each year class enters the diagram in the year following its hatching. Hatching-year jays are not shown. Data are for the entire study population including a small number of unbanded jays. (From Brown 1987a.)

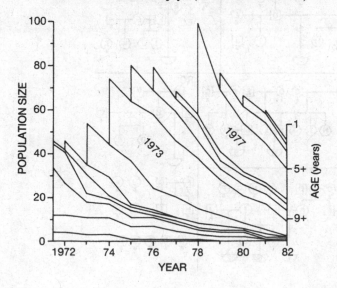

leave at an early age. Genetically this leads to elaborate pedigrees and extended families in some social units (Brown and Brown 1981). In this respect the Mexican Jay differs from singular-breeding species, since in the latter the social unit typically contains only a single nuclear family or perhaps a parent and step-parent. These pedigrees for individuals that were present in 1979 are shown in Fig. 9.8. In the CO unit a grandfather, C, coexisted with his grandoffspring, I, J and K, who probably helped to feed their uncles and aunts in the nest of C and B. Cousins may be found in some units; in HI L, M, and N are cousins of O, all coexisting also with an aunt or uncle.

Fig. 9.8. Genealogies of six social units of Mexican Jays. All individuals present in May 1979 are indicated by symbols with solid lines. Birds with dashed lines were either dead or in another unit. Squares, males; circles, females; hexagons, unsexed; semicircles, nests; +, present at start of study; ×, in neighboring unit; ●, immigrant. (From Brown and Brown 1981.)

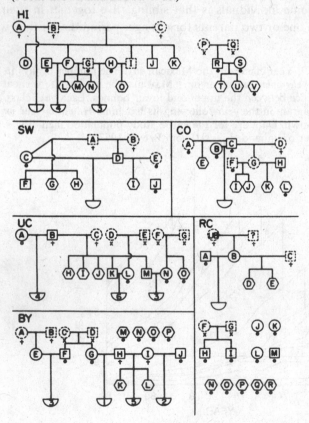

In the pedigrees, immigrants can be identified by the black dots. It is probable that two immigrants in the same flock are unrelated because siblings seldom disperse together in this species. Thus, if two immigrants are breeding at separate nests in the same flock, such as E at nest 4 in HI and G and H at nest 3, none of the four parents at these nests is likely to be related to each other.

Uncooperative breeding

If breeding is interpreted as including courtship, nest-building, incubation, brooding and feeding of nestlings, then the relationship between most separate pairs in the Mexican Jay is certainly not cooperative. These activities are generally not shared between pairs (e.g. see Fig. 9.5), nor are non-breeders allowed to participate, except in feeding young. Worse yet, one pair may actively interfere in the nest of another in the same unit. Pairs often rush to their nests with the apparent intention of protecting them from robbery and destruction. And with good reason, for robbery of nest-lining is common (Brown 1963*a*), and nests in earlier stages have been totally destroyed by rivals. Eggs too have a tendency to disappear, and intraspecific destruction of eggs is not uncommon (Trail *et al.* 1981).

Dominance hierarchies are important in the lives of Mexican Jays (Barkan *et al.* 1986), and competition for food is clearly influenced by rank (Craig *et al.* 1982). It should then not be very surprising that males compete aggressively for females and that females are involved in destruction of nests of rivals. In our opinion, because of the prominence of uncooperative behavior, cooperative breeding is not the right phrase to describe this highly competitive situation.

'Reciprocity'?

In spite of their rivalries, breeders actually do under some circumstances feed each other's young. In general, parents in the same unit do not feed each other's *nestlings*; but when the young are large and beg loudly, the breeders may feed young not their own. If two pairs in the same unit have young about the same age, mutualistic feeding of each other's young begins shortly before fledging (Brown and Brown 1980). After fledging, no evidence of a preference for feeding their own *fledglings* remains in such parents. The interesting point is not that the breeders do or do not help or show a preference; it is that a *transition* occurs between a strong preference for feeding one's own young, a selfish and uncooperative strategy, to helping with no manifest preference for one's own offspring, a cooperative strategy. Should this be called 'reciprocity'? The question is discussed below under Kin-group selection.

The evolution of helping and communal breeding in the Mexican Jay

The Mexican Jay is unique, so far as is known, in having some populations that are plural breeding and some that are very likely singular breeding (Strahl and Brown 1987). Therefore, we must consider the transition between singular and plural breeding, as well as the transition between pair breeding with helpers (*couchii*) and pair breeding without helpers. These transitions can go from the less social to more social or the reverse; indeed, since most tropical jays have helpers, it is possible that those populations of *Aphelocoma* that lack helpers have lost them. Nevertheless, for simplicity we discuss only the direction toward greater sociality.

Phase one: K-selection

It is useful to distinguish three phases in the evolution of sociality in *Aphelocoma* jays (Brown 1974). The first phase is a period of intense competition for space, resulting in territoriality and a surplus of potentially breeding birds (Brown 1969), a condition now known as habitat saturation. These conditions are typical of K-selection, and they were presumed to be responsible not only for delayed breeding, but also for the higher survival rates of Mexican Jays, their reduced dispersal, and comparatively old age structure. A key to this effect is that families must be able to survive on or near their territories during the non-breeding season so that they can stay and retain control of them.

These processes require a stable vegetation form, as postulated in 1974. They do not, however, require predictable and stable yearly rainfall. It is entirely feasible for these processes to occur in semi-arid areas with rainfall that is unpredictable in amount (e.g. Arizona), as was pointed out by Brown (1978). Consequently, the theory is quite applicable to the Mexican Jay and to other populations living in such climates.

Phase two: indirect selection

Delayed breeding, to use our 1974 phrase, 'sets the stage' for helping. The processes that are made possible by delayed breeding, allowing the nutritionally independent offspring to remain with their parents, occur in the second phase. These are primarily related to indirect selection. Efforts by the non-breeder to obtain a territory, a mate and breeding status are included in phase 1. Phase 2 tends to follow phase 1, but they can overlap, According to our 1974 theory, the main reason why helping is conspicuous in non-breeders of singular-breeding populations, such as *A. u. couchii*, is that helpers are rearing full siblings, half-siblings, or at least relatives of some sort most of the time. Because helpers would be rearing kin, selection against helping would not be expected.

Do non-breeders 'automatically' feed any begging baby birds? Or do they feed any nestlings within their own group (situation dependent)? Perhaps indirect selection is unnecessary. The evidence suggests that helping is not 'automatic'. Non-breeders in most species do not help, by which we mean, they do not normally feed any nestlings or fledglings. Many passerine species have significant numbers of non-breeders; and even when they are known to have restricted home ranges, as in the Cape Sparrow (*Zonotrichia capensis*; Smith 1978), helping is unknown, or perhaps rare. Similarly, interspecific helping has been reported only sporadically, even though individual birds are frequently exposed to begging young of other species (Shy 1982). Are there also species in which situation dependence is adequately controlled and where a strong case can be made for the necessity of indirect selection? There certainly are. Studies on the Pied Kingfisher (see Chapter 17), European Bee-eater, and White-fronted Bee-eater (see Chapter 16) have demonstrated distinct preferences for singling out kin over non-kin as recipients of help; and in some cases non-breeders refuse to help non-kin (see references and discussion in Brown 1987*a*, *b*).

At times unusual circumstances arise, leading some individuals to help; but for helping to become a regular and conspicuous part of the behavior of the species seems to require more than just being around begging young as a member of a group. This idea, 'the membership hypothesis', is discussed further by Brown and Brown (1980) for the Mexican Jay.

Phase three: kin-group selection

In the third phase, 'because of the retention of offspring... the most productive territories of an area are handed down from generation to generation within the ownership of the same genetic lineages' (Brown 1974). This is territorial 'inheritance', an especially conspicuous phenomenon in the Mexican Jay (Fig. 9.6). Although the genotypes of flock members change due to immigration more than originally envisaged, it may be seen in the pedigree shown in Fig. 9.8 that up to three generations of the same lineage have been recorded in the same territory.

The retention of non-breeding offspring requires an increase in tolerance by the parents, relative to species in which all offspring leave before breeding. We have interpreted this as a case of parents facilitating later dispersal by their offspring, on the condition that the fitness of the parents not be reduced (Brown 1969). Anthropomorphically, this amounts to 'paying for staying' (for a model of this concept, see Brown 1982); however, we have seen no evidence that non-helpers are evicted.

We see no increase in parental intolerance here, and we see no evidence that parents can effectively prevent their offspring from breeding off the parental territory. The restraint on breeding is provided by competition outside the home territory. Therefore, the term parental manipulation does not properly describe the situation. Furthermore, a simple mathematical model shows that parental manipulation, a form of *variance-enhancement*, is a more expensive way to evolve helping than *variance utilization* (Brown and Pimm, 1985; Brown 1987a), which simply means that parents utilize aid offered by offspring that are prevented from breeding outside their natal territory by factors beyond control of the parents.

The presence of *breeding* offspring, as in *arizonae*, requires a further increase in the tolerance of offspring by their parents relative to singular breeding. As in the previous case, this may be interpreted as *parental facilitation*, namely aid to nutritionally independent offspring, beyond normal parental care, that facilitates reproduction by the offspring (Brown and Brown 1984).

Why plural breeding occurs in some species of jay and not others is unclear. Environmental factors alone might favor plural breeding when territories are severely limited, but few data exist to back up this hypothesis. One hypothesis that might repay further study is as follows. In *arizonae* a high rate of predation on begging juveniles may be important. The transition in parents from feeding almost exclusively their own offspring to indiscriminate feeding of seemingly any juvenile member of their social unit occurs at the age when loud begging of the young makes them easy prey for Cooper's Hawks (*Accipiter cooperi*), which abound in the Chiricahua Mountains. The sharing of feedings at this time may benefit all young and even parents by reducing the frequency of attraction of hunting hawks. This situation has been modeled mathematically as a two-player game by Caraco and Brown (1986). A simplified presentation is given by Brown (1987a, chapter 14) and will not be reviewed here. This blend of self-interest and mutual feeding of offspring may under certain circumstances be termed *by-product mutualism*. This is not the same as reciprocity in the usual sense (a type of score-keeping mutualism), since the payoff matrix of the game differs from that in a prisoner's dilemma.

At present we do not have a well-defined theory for the origin of plural breeding in jays. Among the factors possibly involved are shortage of suitable territories, parental facilitation, predation leading to by-product mutualism, and simply a high population density.

Developmental constraints: the membership hypothesis

In recent decades the main theme of behavioral ecology in general and of the study of helping behavior in particular has been the application of economic, or cost–benefit, selectionist thinking. Most of the theorizing about cooperative breeding has been of this sort. This approach has been fruitful, but other approaches may now also be useful.

Both genetic and developmental approaches are needed, but the latter may be more insightful at present, given our ignorance of the genetics of parental and alloparental behavior in birds. A useful way to begin may be to consider a null hypothesis, one that explains variation solely in terms of proximate conditions and cues and not in terms of natural selection.*

Instead of considering how helping was selected for, let us consider how it might be maintained or selected against. This approach, which we have used earlier (Brown and Brown 1980), is not unreasonable, since helping is probably of ancient origin in New World jays and occurs in the Unicolor Jay (*Aphelocoma unicolor*; J. Brown and T. Webber, unpublished observations), which could be interpreted as an ancestral species (Pitelka 1951).

The feeding of nestlings is a conservative trait in passerines, having been lost only in brood-parasitic species. The feeding of nestlings may be negatively influenced by testosterone. This hormone normally peaks during the sexual phase of breeding and is low during the period of parental care (Wingfield *et al.* 1987), but in the Brown-headed Cowbird (*Molothrus ater*), a brood parasite, testosterone remains at high levels while its young are being fed by foster parents (Dufty and Wingfield 1986). Do helpers have lower testosterone levels than non-helpers? Is the seasonal decline in competitive behaviors and increase in sharing of feedings effected by testosterone levels? These questions are currently under investigation in the Mexican Jay (L. McKean, personal communication) and other species (R. Hegner, personal communication). Perhaps young males by staying in their natal territories do not have elevated testosterone levels because they do not face the 'challenge' thought to be responsible developmentally for elevated testosterone levels (the challenge hypothesis of Wingfield *et al.* 1987), namely competition for territories and females.

The above scenario may be described as a 'non-adaptationist story' based upon events in development. In it, a mere change in social context resulting from the decision *Stay home* leads to lower testosterone levels and in turn to a more favorable hormonal condition for feeding young.

---

* See Craig and Jamieson, Chapter 13 – Eds.

A second developmental constraint may also influence this scenario. According to the skill hypothesis, which has been described as an alternative to the prevailing habitat saturation hypothesis (Brown 1987a), yearling Mexican Jays may not be sufficiently skillful at foraging and other behavior to justify an attempt at dispersal and breeding. This hypothesis is currently being investigated by G. Keys and J. Brown.

Are such contextual changes together with their physiological consequences sufficient to explain helping in jays? Perhaps any jay in a similar context would respond by feeding young, at least those in its own social group – without selection for helping. This was the rationale for our 'membership hypothesis' (Brown and Brown 1980). There is a major problem, however, with null hypotheses of this sort. They fail to explain the *selectivity* shown by feeders.

In Mexican Jays some feeders are highly selective; they feed almost exclusively at certain nests and not others. These include the mother and father(s) at a given nest. Non-breeding jays may also have preferences, although the reasons remain unclear. In some other species, non-breeders are more likely to help if kin are recipients (for a review, see Brown 1987a). The developmental basis of such preferences in the Mexican Jay and other communal breeders is an important problem, that of kin recognition.

Indiscriminate feeding of young may suffice for parents in groups with no other breeders, but, in a large group, parents that failed to discriminate would be selected against. It is precisely in large social groups where refined kin recognition is needed and where it is found. This fact is not easily explained without invoking natural selection.

Conclusions

We have reviewed briefly a wide variety of processes that might influence the social systems of two subspecies of Mexican Jay, *couchii* in the Chisos Mountains of Texas and *arizonae* in the various mountain ranges of south-eastern Arizona. Not only have we learned much since 1958, when we first began our studies of Mexican Jays, but the kinds of theoretical constructs against which we compare our data have changed with time. Thus the approaches we have employed cover a wide spectrum, including correlations with ecology (Brown 1969, 1974), applications of inclusive-fitness theory (Brown 1974, 1978), optimality theory (Brown 1982), parent–offspring relations (Brown and Brown 1984; Brown and Pimm 1985), and game theory (Caraco and Brown 1986). We see these various approaches as complementary, and, in our opinion, all are valuable. Even after 10 years of our long-term study, 1969–79, important insights continued to emerge, and this process still continues. The level of complexity that we now comprehend after 19 field seasons far exceeds our early expectations, and we still have 'miles to go before we sleep'.

We thank profusely the many persons who have helped us in the field over the years. In recent years these have included C. Barkan, J. Craig, H. Douglas, L. Elliott, A. Heise, E. Horvath, K. Johnson, R. Russell, S. Russell, D. Siemens, A. Stewart, S. Stoleson, S. Strahl, P. Trail and many others. For permission to work at the Southwestern Research Station and on surrounding properties we are grateful to the American Museum of Natural History, M. Cazier and E. Bagwell. For comments on this manuscript we thank D. Lemmon and G. Keys. Our work has been made possible by a series of grants from the US Public Health Service and the National Science Foundation (currently BNS 8410123).

Bibliography

American Ornithologists' Union (1957). *Check-list of North American Birds.* Port City Press: Baltimore, MD.

American Ornithologists' Union (1983). *Check-list of North American Birds.* Allen Press: Lawrence, KS.

Barkan, C. P. L., Craig, J. L., Strahl, S. D., Stewart, A. M. and Brown, J. L. (1986). Social dominance in communal Mexican Jays, *Aphelocoma ultramarina. Anim. Behav.* **34**: 175–87.

Brown, J. L. (1963a). Social organization and behavior of the Mexican Jay. *Condor* **65**: 126–53.

Brown, J. L. (1963b). Aggressiveness, dominance and social organization in the Steller Jay. *Condor* **65**: 460–84.

Brown, J. L. (1969). Territorial behavior and population regulation in birds. *Wilson Bull.* **81**: 293–329.

Brown, J. L. (1970). Cooperative breeding and altruistic behavior in the Mexican jay, *Aphelocoma ultramarina. Anim. Behav.* **18**: 366–78.

Brown, J. L. (1972). Communal feeding of nestlings in the Mexican Jay. (*Aphelocoma ultramarina*): interflock comparisons. *Anim. Behav.* **20**: 395–402.

Brown, J. L. (1974). Alternate routes to sociality in jays – with a theory for the evolution of altruism and communal breeding. *Amer. Zool.* **14**: 63–80.

Brown, J. L. (1978). Avian communal breeding systems. *Ann. Rev. Ecol. Syst.* **9**: 123–55.

Brown, J. L. (1982). Optimal group size in territorial animals. *J. Theoret. Biol.* **95**: 793–810.

Brown, J. L. (1983). Communal harvesting of a transient food resource in the Mexican Jay. *Wilson Bull.* **95**: 286–7.

Brown, J. L. (1985). The evolution of helping behavior – an ontogenetic and comparative perspective. In *The Evolution of Adaptive Skills: Comparative and Ontogenetic Approaches*, ed. E. S. Gollin, pp. 137–71. L. Erlbaum Assoc. Inc., Hillsdale, NJ.

Brown, J. L. (1986). Cooperative breeding and the regulation of numbers. *Proc. 18th Intern. Ornith. Congr.*: 774–82.

Brown, J. L. (1987a). *Helping and Communal Breeding in Birds: Ecology and Evolution.* Princeton University Press: Princeton, NJ.

Brown, J. L. (1987b). Testing inclusive fitness theory with social birds. In *Animal Societies: Theories and Facts*, ed. Y. Ito, J. L. Brown and J. Kikkawa, pp. 103–14. Japan Sci. Soc. Press: Tokyo.

Brown, J. L. and Brown, E. R. (1980). Reciprocal aid-giving in a communal bird. *Z. Tierpsychol.* **53**: 313–24.

Brown, J. L. and Brown, E. R. (1981). Extended family system in a communal bird. *Science* **211**: 959–60.

Brown, J. L. and Brown, E. R. (1984). Parental facilitation: parent–offspring relations in communally breeding birds. *Behav. Ecol. Sociobiol.* **14**: 203–9.

Brown, J. L. and Horvath, E. G. (1989). Geographic variation of group size, ontogeny, rattle calls, and body size in *Aphelocoma ultramarina. Auk* **106**: 124–8.

Brown, J. L. and Pimm, S. L. (1985). The origin of helping: the role of variability in reproductive potential. *J. Theoret. Biol.* **112**: 465–77.

Caraco, T. and Brown, J. L. (1986). A game between communal breeders: when is food-sharing stable? *J. Theoret. Biol.* **118**: 379–93.

Craig, J. L., Stewart, A. M. and Brown, J. L. (1982). Subordinates must wait. *Z. Tierpsychol.* **60**: 275–80.

Duffy, A. M. and Wingfield, J. C. (1986). Temporal patterns of circulating LH and steroid hormones in a brood parasite, the brown-headed cowbird, *Molothrus ater.* I. Males. *J. Zool. Lond.* **208**: 191–203.

Gross, A. O. (1949). Nesting of the Mexican Jay in the Santa Rita Mountains. *Condor* **51**: 241–9.

Hamilton, W. D. (1963). The evolution of altruistic behaviour. *Amer. Nat.* **97**: 354–6.

Hardy, J. W. (1961). Studies in behavior and phylogeny of certain New World jays (Garrulinae). *Univ. Kansas Sci. Bull.* **42**: 13–149.

Ligon, J. D. and Husar, S. L. (1974). Notes on the behavioral ecology of Couch's Mexican Jay. *Auk* **91**: 841–3.

Pitelka, F. A. (1951). Speciation and ecologistic distribution in American jays of the genus *Aphelocoma. Univ. Calif. Publ. Zool.* **50**: 195–464.

Roth, V. D. (1971). Unusual predatory activities of Mexican Jays and brown-headed cowbirds under conditions of deep snow in Southeastern Arizona. *Condor* **73**: 113.

Rowley, I. (1965). The life history of the superb blue wren, *Malurus cyaneus. Emu* **64**: 251–97.

Shy, M. M. (1982). Interspecific feeding among birds: a review. *J. Field Ornith.* **53**: 370–93.

Skutch, A. F. (1935). Helpers at the nest. *Auk* **52**: 257–73.

Smith, S. M. (1978). The underworld in a territorial sparrow: adaptive strategy for floaters. *Amer. Nat.* **112**: 571–82.

Strahl, S. D. and Brown, J. L. (1987). Geographic variation in social structure and behavior of *Aphelocoma ultramarina. Condor* **89**: 422–4.

Trail, P. W., Strahl, S. D. and Brown, J. L. (1981). Infanticide and selfish behavior in a communally breeding bird, the Mexican Jay (*Alphelocoma ultramarina*). *Amer. Nat.* **118**: 72–82.

Wagner, H. O. (1955). Bruthelfer unter den Vögeln. *Veröff. Überseemus. Bremen Reihe A* **2**: 327–30.

Westcott, P. W. (1969). Relationships among three species of jays wintering in southeastern Arizona. *Condor* **71**: 353–9.

Wingfield, J. C., Ball, G. F., Dufty, A. M. Jr, Hegner, R. E. and Ramenofsky, M. (1987). Testosterone and aggression in birds. *Amer. Sci.* **75**: 602–8.

10 GALAPAGOS MOCKINGBIRDS: TERRITORIAL COOPERATIVE BREEDING IN A CLIMATICALLY VARIABLE ENVIRONMENT

10

Galápagos mockingbirds: territorial cooperative breeding in a climatically variable environment

R. L. CURRY* AND P. R. GRANT

During the voyage of the *Beagle*, Darwin collected mockingbirds on four different islands in the Galápagos. He determined that his specimens represented three different varieties, all unique to the archipelago, and that no two of these occurred together. Finding different mockingbirds on islands within sight of each other contributed to Darwin's realization that similar species could replace each other geographically, an idea that sparked his thinking about evolutionary processes. Four forms are now recognized: *Nesomimus trifasciatus*, the Floreana (Charles Island) Mockingbird, *N. melanotis*, the San Cristóbal (Chatham Island) Mockingbird, *N. macdonaldi*, the Española (Hood Island) Mockingbird, and *N. parvulus*, the Galápagos Mockingbird.

For the past 11 years, we have investigated population ecology and social organization in this endemic genus. Most Galápagos mockingbirds, like many other cooperatively breeding species, live in groups holding collective territories. Mockingbirds in the Galápagos also experience a climate that varies widely and unpredictably. Conditions range from severe droughts to extraordinarily wet years associated with El Niño-Southern Oscillation (ENSO) events; both extremes occur on the Galápagos on average once every four years (Grant 1985). A major focus of our study has been to investigate how territorial behavior and climatic variation interact to produce a complex form of cooperative social organization. This complexity has provided challenges for deciphering and explaining patterns of social behavior amid large environmental and demographic

---

* Present address: Archbold Biological Station, PO Box 2057, Lake Placid, FL 33852-2057, USA.

variation, as well as excellent opportunities for testing hypotheses about the evolution of cooperative breeding. In this chapter, we summarize our understanding of ecological and evolutionary influences on cooperative breeding in the genus, and the unanswered questions that remain.

Phylogeny, distribution and general behavior

Galápagos mockingbirds are closely related to mockingbirds (*Mimus* spp.) of North, Central and South America, but the identity of their closest mainland relative is unknown. Two species, the Long-tailed Mockingbird (*Mimus longicaudatus*), which inhabits the arid coastal zone of Ecuador and Peru, and the Bahama Mockingbird (*M. gundlachii*), a resident of the Bahama Archipelago and southern Jamaica, have the most characteristics in common with the Galápagos populations (Gulledge 1975) – and interbreeding between *M. longicaudatus* and *N. parvulus* has occurred in captivity (Bowman and Carter 1971). The four species of Galápagos mockingbirds are nevertheless more similar to each other than to any mainland relatives, although the separation of *Nesominus* from *Mimus* has sometimes been questioned; an alternative treatment considers the four Galápagos species as a superspecies with *Mimus* (see e.g. Steadman 1986).

The four species of *Nesomimus* are distributed allopatrically (Fig. 10.1). *N. trifasciatus* survives today on just two small islets adjacent to the larger island of Floreana. This species was extirpated from Floreana during the last decades of the nineteenth century, probably as a result of predation by introduced black rats, *Rattus rattus*, or habitat destruction by goats, or both (Curry 1986; Steadman 1986). *N. melanotis* is endemic to San Cristóbal, and *N. macdonaldi* inhabits only Española and a satellite islet. *Nesomimus parvulus* occupies almost all other islands in the archipelago. Davis and Miller (1960) treated all the Galápagos populations as members of a single species, but they differ considerably in size and they vary as much in shape, plumage and behavioral displays, including song, as do many different mimid species elsewhere (Table 10.1; Swarth 1931; Bowman and Carter 1971; Abbott and Abbott 1978). In all populations, the sexes are indistinguishable on the basis of plumage, but females are smaller on average than males; measurements of adult wing length can be used to determine the sex of most individuals (Table 10.1). A distinctive juvenal plumage, characterized by spots on the breast, is retained in each of the four species until birds are six to nine months of age. There are no reliable morphological indicators of relative age among older birds.

Galápagos mockingbirds are abundant residents of the arid lowlands on most islands. The semi-desert scrub habitats of these dry zones are dominated by a few deciduous tree species, arborescent cacti (*Opuntia* and *Jasminocereus*), and various shrubs and vines in the understory. This environment is highly seasonal. In most years, rain falls only during the warmest months, usually between January and April, and green vegetation is present on most trees and many shrubs only after these rains. Mockingbirds also inhabit less seasonal transition forest at middle elevations, and on some islands (e.g. Fernandina, Isabela, Santiago, Pinta and San Cristóbal), they can be found in moist forest on the highest volcanoes, though they seem to be absent from the highest zones on Santa Cruz.

Fig. 10.1. Distribution of the four allopatric species of *Nesomimus* mockingbirds in the Galápagos Archipelago. Mockingbirds inhabit nearly all islands (stippled), but they have been extirpated (E) on two islands and they are absent from a third (?) where their original status is unknown. Names of islands where we have studied mockingbird social organization are shown in bold face.

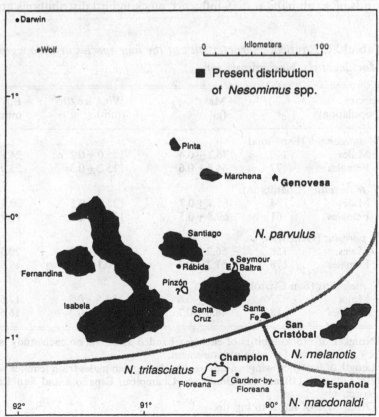

In all habitats, Galápagos mockingbirds are essentially omnivorous. They are predominantly terrestrial insectivores, but they also feed regularly on arboreal insects, fruit, nectar, pollen, centipedes, crabs, flying fish (regurgitated by boobies, *Sula* spp.), lizards, and, on some islands, on ticks gleaned from iguanas (Christian 1980) and on blood taken from living iguanas or seabird nestlings (Curry and Anderson 1987). The only major type of food that mockingbirds seldom eat are seeds, which in the dry seasons make up the bulk of the diet of many Darwin's finches occupying the same habitats.

The mockingbirds are subject to no obvious ecological limitations other than the physical extent of scrub habitat on the smaller islands. Their open-cup nests, which can be built in just a few days, are placed in a wide variety of shrubs, trees, and cacti and are not used for roosting. They do not make 'improvements' to their territories by storing or caching food, unlike the cooperatively breeding Acorn Woodpeckers (see Chapter 14) or *Aphelocoma* jays (see Chapter 9). As explained below, the availability of different kinds of scrub habitat may influence mockingbird distributions and their

Table 10.1. *Physical characteristics of the four species of Galápagos Mockingbirds*, Nesomimus *spp.*

Species (Population)	n^a	Mass[b] (g)	Wing length[c] (mm)	Bill length[d] (mm)
N. macdonaldi (Española)				
Males	140	76.1 ± 0.4	125.0 ± 0.2	24.1 ± 0.1
Females	77	64.8 ± 0.6	115.2 ± 0.3	23.0 ± 0.1
N. trifasciatus (Champion)				
Males	34	65.7 ± 0.7	125.9 ± 0.5	20.4 ± 0.2
Females	61	59.8 ± 0.5	116.8 ± 0.3	19.9 ± 0.1
N. parvulus (Genovesa)				
Males	150	56.2 ± 0.4	118.0 ± 0.2	20.3 ± 0.1
Females	178	51.2 ± 0.3	110.1 ± 0.1	19.6 ± 0.1
N. melanotis (San Cristóbal)				
Males	27	53.2 ± 0.6	114.5 ± 0.5	17.6 ± 0.1
Females	18	48.0 ± 0.7	106.6 ± 0.3	16.6 ± 0.1

[a] Number of mockingbirds of each sex banded as adults on each study island.
[b] $\bar{x} \pm$ S.E.M. shown for each measurement.
[c] Length of flattened wing; values used to distinguish males from females were 114, 120, 120, and 110 mm for Genovesa, Champion, Española and San Cristóbal, respectively.
[d] Distal edge of nares to bill tip.

social organization. Predators may influence mockingbird abundance on some islands. Short-eared Owls (*Asio flammeus*) and Yellow-crowned Night-Herons (*Nycticorax violacea*) are known predators on mockingbird nests, and Lava Herons (*Butorides* [*striatus*] *sundevalli*) may also take eggs or nestlings. Galápagos Hawks (*Buteo galapagoensis*) occasionally kill mockingbirds on the few islands where the hawks survive (see Chapter 12). Introduced black rats are known or suspected nest predators on islands having human settlements (e.g. San Cristóbal and Santa Cruz), where feral cats may also kill some mockingbirds.

Study sites

Mockingbird helpers in the Galápagos were first reported in *N. parvulus* on Santa Cruz by Hatch (1966). We began studying cooperative breeding in the same species on Isla Genovesa in 1978 (Grant and Grant 1979). Genovesa is a low island (maximum elevation 76 m) about 17 km² in area. Our study area has covered up to 50 ha in the vicinity of a small tourist landing beach on the inner shore of Darwin Bay. The habitat consists of remarkably uniform scrub woodland (Fig. 10.2; see also Plage and Plage 1988, pp. 131–2): an open canopy of palo santo trees (*Bursera graveolens*) stands above an open shrub layer dominated by *Croton scouleri*, *Cordia lutea* and *Lantana peduncularis*. Cacti (*Opuntia helleri*) are scattered throughout the island, becoming more dense near the coasts.

We marked 1675 mockingbirds in all on Genovesa and monitored them, up to 200 color-banded birds at a time, from 1978 through 1988. We attempted over this interval to collect detailed information on the behavior of every individual in the study area during each breeding season, which lasted usually from January through May. Breeding extended into July or June in two El Niño years, 1983 and 1987, respectively, and no breeding took place in two drought years, 1985 and 1988. Along with observations on dominance interactions, mating, conflict within groups, and territorial defense, we monitored nearly every nest during 1 h watches at two to three day intervals during the nestling period to quantify feeding of nestlings by breeders and helpers. We define helpers in the narrow sense of allofeeders, birds that feed dependent young (nestlings or fledglings) that are not their own offspring (Brown 1987). Mockingbird helpers (and parents) also attempt to defend young from predators. We use this definition of helping because not all residents care for nestlings within their territory: some individuals neither breed nor help, and others do both, either simultaneously or sequentially, during a single breeding season (see Curry 1988*a*, and below). The remarkable tameness of Galápagos mockingbirds made all of our data collection relatively easy. We made additional observations in the dry seasons of most years to document behavior during non-breeding periods.

To gain comparative perspective, we also studied mockingbirds on three other islands. We monitored the entire population of *N. trifasciatus* on Champion from 1980 through 1988. Our studies on this arid islet (maximum elevation 46 m; area 10 ha) consisted of visits of up to 10 days roughly twice each year except in 1984, when our study lasted for five months. In 1984 we also studied *N. macdonaldi* at three sites on Española, another arid island (maximum elevation 198 m; area 58 km^2), and *N. melanotis* at one highland and two lowland sites on San Cristóbal, a larger and ecologically more diverse island (maximum elevation 715 m; area 552 km^2).

Throughout this chapter, we present averages as $\bar{x} \pm$ s.e.m. Statistical tests are two-tailed unless otherwise noted, and χ^2-tests with 1 d.f. include correction for continuity.

Fig. 10.2. View of arid scrub habitat in the Darwin Bay study site on Isla Genovesa at the end of a typical wet season. Predominant vegetation consists of an open canopy of palo santo trees (*Bursera graveolens*), which are leafless through the dry months of the year (May–December), and an understory of scattered *Opuntia* cacti and semi-deciduous shrubs. Small patches of mangroves and salt-tolerant shrubs occur along the shore.

Spacing system

Galápagos mockingbirds are strictly and permanently territorial. We define each group territory as the area occupied by a dominant (alpha) male and defended by him against other alpha males. We classify all additional birds resident in each alpha male's territory, and subordinate to him, as members of his group. Territorial residents chase solitary intruders and defend borders of the collective territory through frequent displays we call flick fights. During flick fights, members of adjacent groups line up, usually on the ground or in low vegetation, along a common border (Fig. 10.3). Females participate less frequently and less actively than males (Kinnaird and Grant 1982), and each flick fight seldom involves more than six birds from a particular group. While giving distinctive calls, which may rally other group members to join the display, opposing birds face each other, flick their wings out from the body, and bob their tails and bodies. These displays sometimes escalate to actual fights in which birds grapple and peck each other violently. More commonly a flick fight gradually lessens in intensity until the two groups separate. Flick fights occur almost daily at

Fig. 10.3. Interaction of mockingbirds at territorial border on Genovesa. The three birds at left exhibit the stereotyped postures used in flick-fight displays, while the fourth member of their group grapples with a member of the opposing group of four birds.

some territorial borders, particularly where a boundary extends across open areas (Fig. 10.3).

Nearly every bird is a member of a territorial group. Counting each adult in the Genovesa study area once per breeding season, only 1% of males ($n = 512$ bird-years, 1980–7; range among years, 0–2%) and 4% of females ($n = 509$; range 0–7%) wandered without attachment to a particular group. Most wanderers are young females apparently searching for breeding vacancies. All other birds remain in their territories for breeding, roosting, resting and nearly all feeding. In territories containing several nesting pairs (see Social system, below), members of each pair tend to forage near their respective nests, but non-breeding birds move throughout the entire territory, and both they and the breeders continue to participate in flick fights on all sides of the inclusive group territory. Flick fights rarely occur between birds sharing a territory; we have seen this happen only twice.

The spacing system changes only slightly during the driest months. Territories are maintained during these non-breeding periods, but residents occasionally leave for a few hours at a time to feed in 'hot spots' with locally plentiful food (e.g. the intertidal zone or areas where *Opuntia* plants bear numerous fruits). These visitors are attacked by residents of the territories where they intrude, and they spend most of their time feeding in their regular territories. Some females also return to their natal groups for extended periods before moving back to their breeding territories at the start of the next breeding season. Because such movements involve birds moving between permanently maintained territories, they have little influence on the spacing system as whole.

Territories fill all habitat on Genovesa almost constantly. Normally, space that becomes vacant when its owners die is annexed by adjacent groups within hours. An exception arose in only one year of our study (1983), when a few territories were left vacant for up to two months before being filled. This occurred at a time when population density had been reduced by an epidemic (Curry 1985), and nearly every surviving bird was already breeding (see Demography, below). Territories ranged in size (measured during the breeding season) from 0.1 to 3.5 ha and averaged 0.94 (± 0.05) ha ($n = 142$, 1980–5). Territory size increased with the number of adults in each group ($F = 69.9, P < 0.005, r = 0.62, n = 116$, 1980–4). Because all habitat was continuously occupied, the relationship between group and territory size changed with variation among years in population density, but the former two variables were always positively correlated. The number of territories defended, their size, and the locations of their boundaries fluctuated as groups changed in size through mortality and recruitment.

Mockingbirds on other islands in the Galápagos are territorial, but the spacing of territories in relation to habitat differs among populations. As on Genovesa, territories on both Champion and Española fill all terrestrial habitats, which on these low, arid islands are limited to relatively uniform interior scrub and small areas of littoral vegetation. In contrast, habitats on San Cristóbal are more varied, and on this island, mockingbirds defend territories in only a few of the available habitats; other nearby habitats differing in vegetation structure and composition are not occupied by territorial birds. Territories we measured in 1984 were significantly larger on San Cristóbal than elsewhere (Fig. 10.4; Mann–Whitney U-tests, $P < 0.05$). Territory size increased with group size on each of the four islands, but among the islands group size varied inversely with mean territory size (Fig. 10.4). As first reported by Hatch (1966), groups on Española were exceptionally large, particularly within seabird colonies, and groups on Champion were similar in size to those on Genovesa. In contrast, three birds at most occupied the large territories on San Cristóbal, and average group size there was significantly smaller than on all other islands (Mann–Whitney U-tests, $P < 0.05$). The inverse relationship between average group and territory size among the islands gave rise to large inter-island variation in population density: breeding habitat on San

Fig. 10.4. Geographic variation in mockingbird group and territory size. Population averages ($\pm$S.E.M.) were measured during simultaneous studies on the four islands during the breeding season (January–April) in 1984. Territory size increases with group size on each island, but groups on San Cristóbal maintain exceptionally large territories.

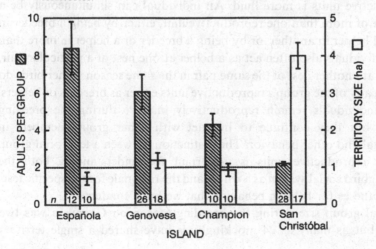

Cristóbal held just 0.6 birds ha^{-1}, compared with 4.4, 4.9, and 9.2 birds ha^{-1} on Genovesa, Champion, and Española, respectively.

In the short time (six weeks) that we worked at one lowland site on San Cristóbal, we saw three males establish new territories, and succesfully attract mates, in vacant areas at the edges of a cluster of already-established territories. At least one of these males had previously been a subordinate bird in a group of three. Two of the females joining these males had been seen previously wandering through surrounding undefended areas; the third came from a nearby territory after being replaced as breeder by an immigrant female. These observations and the correlational evidence described above support the idea that group territoriality is promoted by the lack of space to which young birds could disperse: where excess habitat exists, as on San Cristóbal, groups are smaller. We do not yet know whether lower density on San Cristóbal results from birds dispersing out of preferred breeding habitat into surrounding areas or from predation on nests by introduced black rats or both (Curry 1989).

Social system

The basic social unit on Genovesa is the territorial group, which includes all birds regularly resident within each territory as defined above. Within these groups, the basic reproductive unit consists of the male and female parents at each nest and their associated nest helpers. Each territorial group can contain more than one breeding female and thus more than one reproductive unit (plural breeding; Brown 1978). Membership in territorial groups is comparatively closed and stable: birds remain members of the same group for months or years at time. Membership in reproductive units is more fluid. An individual can simultaneously be a member of more than one reproductive unit, either by being a breeder in one and helper in another, or by being a breeder or a helper in more than one. Individuals also often act as a helper at one nest of a particular pair, but not at another nest of the same pair in the same season. Other birds do not join any of the group's reproductive units either as breeders or helpers. These individuals remain reproductively inactive during the breeding season, but they continue to interact with other group members in territorial and other behavior. The distinction between a territorial group and its reproductive units is important for understanding both the mockingbird social system as a whole and the rationale for the specific tests of hypotheses for helping behavior that we have conducted.

Modal group size during the breeding season on Genovesa was two adults, but as many as 24 mockingbirds have shared a single territory

(Fig. 10.5; the exceptionally large group of 24 birds arose through the rare fusion of two groups, and this group occupied the largest territory, 3.5 ha, measured during the study). Mean group size varied with changes in density among years ($r_s = 0.76$, $n = 8$ years, 1980–7, $P < 0.05$); fission of groups occurred infrequently, so groups were largest when density was high. Birds in a group engage in few collective behaviors except for communal defense of territorial boundaries. Each bird forages individually at all times except when a large food item (such as a flying fish) may be shared by several birds. On Española, a dozen or so mockingbirds can feed on blood from a single booby nestling. We have seen no evidence of a coordinated anti-predator sentinel system either on Genovesa or on any of the other islands.

Mockingbird groups are maintained primarily through recruitment of young born in the group territory. Dispersal, especially among males, is

Fig. 10.5. Distribution of sizes of all mockingbird groups on Genovesa (top) and of groups containing at least one breeding female (bottom). Group size was measured during the middle of the usual breeding season (March) in each year. Shading indicates the proportion of breeding groups of each size that contained one to four nesting females. Because polygyny occurs, breeding females can make up more than half of a group's members.

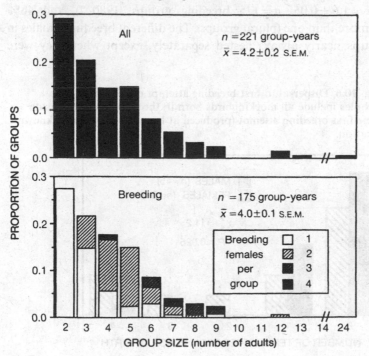

limited in distance and delayed. Of those birds alive on Genovesa in the first breeding season after their birth, most males (74%, $n = 88$) and fewer females (56%, $n = 66$) remained in their natal groups ($\chi_1^2 = 4.6$, $P < 0.05$). The proportion of yearlings dispersing varied inversely with population density ($r_s = -0.74$, $P < 0.05$, $n = 8$ cohorts, sexes combined); when density was low, more yearlings left their natal territories. Older adults dispersed less often. Most adult males (96%, $n = 308$ bird-years, 1981–7) remained in the same territory from one year to the next (range of annual values 83–100%), as did fewer adult females (77%, $n = 263$, range 58–91%; $\chi_1^2 = 41.1$, $P < 0.005$). Dispersing females moved relatively freely among groups because they were courted by unpaired males, but adult males were usually prevented by territoriality from joining neighboring groups, except when all of its males had died. Mockingbirds nearly always dispersed individually; we have recorded only two instances where two birds moved simultaneously from one group to another. Most of the males and nearly half of the females on Genovesa bred at least once in their natal territory (Fig. 10.6). Most females eventually moved one to three territories from their place of birth (a few moved much farther), whereas males usually remained in their natal territory throughout their lives.

Each territorial group on Genovesa contained up to four breeding females ($\bar{x} = 1.6 \pm 0.05$, $n = 175$ breeding groups, 1980–7), and 46% contained more than one (plural groups). The different breeding females in plural groups nearly always nested separately, except when they were

Fig. 10.6. Dispersal to first breeding attempt on Genovesa. The samples include all mockingbirds born in the study area that made their first breeding attempt (produced at least one clutch) at a known location.

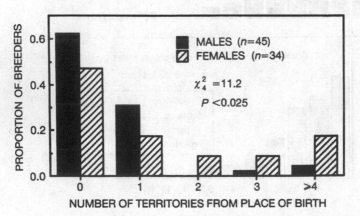

paired polygynously with the same male (see *Mating system*, below). The relative frequency of plural groups in each year (range 14–76%) was determined more by the age structure of the population than by its size: relative frequency of plural groups increased with increasing proportions of yearlings in the population ($r_s = 0.89$, $n = 8$ years, $P < 0.01$, excluding 1984, when few paired birds nested). Neither the proportion of plural groups ($r_s = 0.31$) nor the proportion of yearlings in the population ($r_s = 0.26$) varied significantly with density ($n = 8$ years, $P > 0.1$). Not surprisingly, plural groups were larger on average than groups with only one breeding female (singular groups; Fig. 10.5).

Fig. 10.7. Structure of a typical mockingbird territorial group on Genovesa over an eight year interval. Birds are identified by their band numbers. All birds shown are adults, except for juveniles present in the second half of the prolonged 1983 breeding season. Immigrants are shown in bold text. Superscripts denote birds that either moved up in rank within the group (r), emigrated (e), or died (†) between breeding periods. Boxes enclose breeding pairs and trios; heavy lines denote breeders that successfully produced at least one fledgling in a season. Arrows indicate birds that acted as helpers at one or more nests of a particular breeding pair or trio, and descending curves connect breeders and offspring. This group exhibited many common features of the Genovesa social system, including both plural and singular breeding, occasional polygyny, predominantly female dispersal, and simultaneous helping and breeding.

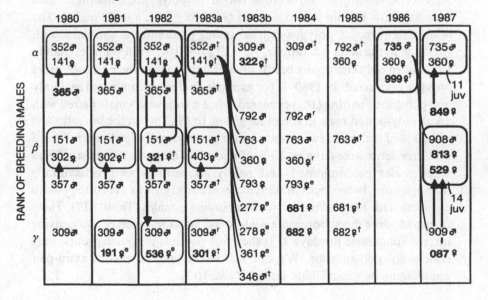

Linear dominance hierarchies exist among birds of each sex within each mockingbird group. Dominant birds frequently displace, peck at, or chase subordinate birds in their groups, and subordinates back away and give ritualized displays, similar to the begging of juveniles, whenever they are approached by a more dominant bird. The alpha male in each group is invariably its oldest male, and rankings among subordinate males correspond to their relative ages (Curry 1988b). Females typically rank lower than adult males in their group, except for their sons, until the latter are one or two years old. Most breeders pair with birds of similar relative rank, but exceptions sometimes occur when an alpha male pairs with a young mate in a group that also contains an older female. Such situations can confuse dominance relationships among females, but in most situations we consider a breeding female's rank to be the same as that of her mate.

An example of the changes in structure of one group on Genovesa over eight years is illustrated in Fig. 10.7. During this interval, the birds in this group exhibited nearly the entire range of social interactions that we have observed on Genovesa.

Breeding system
Mating system

Galápagos mockingbirds have a variable mating system. Most pairs on Genovesa were monogamous, but polygyny also occurred: during seven breeding seasons in the interval 1980–8, 7% of males ($n = 283$) and 33% of females ($n = 254$) were involved in polygynous matings. Each breeding male had up to three mates ($\bar{x} = 1.08 \pm 0.02$, $n = 183$ males). The relative frequency of polygyny varied among years and was correlated with variation in the adult sex ratio ($r_s = -0.90$, $P < 0.01$, $n = 6$ years, excluding 1984, when only eight pairs bred): when females outnumbered males, more polygyny occurred. In 1980–2, for example, males predominated and only one polygnous pairing (1%) occurred, when a widowed female paired with an already-paired male in the same group. In 1983, when the sex ratio was even, 8% of males had two mates simultaneously, and in 1986, when 58% of adult residents were females, 29% of males had more than one mate. Even when males predominated and nearly all pairings were behaviorally monogamous, however, dominant males in plural groups were often able to copulate with females paired with subordinate males (Table 10.2). These occurred while the subordinate male stood just a few meters away, giving intense submissive displays. It is therefore possible that such groups were genetically polyandrous. We did not record any successful extra-pair copulations by subordinate males (Table 10.2).

Most females on Genovesa laid eggs in their own separate nests, but joint nests, those where more than one female lays eggs and cares for the young, also occurred (4% of all nests, $n = 510$). The proportion of joint nests varied among years ($\chi_6^2 = 25.0$, $P < 0.001$, range 0–14%) and was positively correlated with the annual incidence of polygyny ($r_s = 0.78$, $P < 0.025$, $n = 7$ years). A direct association between pairing and mode of nesting gave rise to this correlation: joint nests occurred in 71% of groups where two or more breeding females shared a mate ($n = 17$), but females monogamously paired to different males in the same plural group rarely nested together (2%, $n = 59$; $\chi_1^2 = 39.5$, $P < 0.001$).

Mean age of first breeding on Genovesa was 1.9 (± 0.1) years for males ($n = 54$) and 1.6 (± 0.1) years for females ($n = 42$; Mann–Whitney U-test, $0.05 < P < 0.1$). Differences in the climatic and demographic conditions experienced by each cohort complicated the calculation of age of first breeding. Most birds born in 1983, for example, first bred in 1986, after being prevented from breeding during the previous two years by drought conditions, whereas El Niño conditions in 1987 and a low density of older adults enabled many yearlings to breed. Consequently, ages of first breeding were unevenly distributed: 50% and 64% of males and females, respectively, began breeding as yearlings; 9% and 7% did so at 2 years of age; 39% and 29% at 3 years of age; and 2% and 0% at 4 years of age.

Because dispersal on Genovesa is reduced and delayed in both sexes, and because birds can begin breeding in the first year, related males and females often breed in the same group territory. As a result, at least 7% of pairs on Genovesa ($n = 156$, with each unique pairing counted once, regardless of

Table 10.2. *Outcome of within- and extra-pair copulation attempts of Galápagos mockingbirds in relation to dominance*[a]

Type of copulation	Attempt successful[b]	Attempt unsuccessful	P[c]
Within-pair			
Dominant pair	13	5	
Subordinate pair	6	13	<0.05
Extra-pair, birds from same group			
With mate of lower-ranking male	6	7	
With mate of higher-ranking male	0	19	<0.005

[a] After Curry (1988b).
[b] Cloacal contact observed or male mounted acquiescent female for at least 10 s.
[c] χ_1^2 test with correction for continuity.

duration) have been between siblings or between parent and offspring. This is a relatively high frequency of incest for a wild vertebrate (Ralls *et al.* 1986). The 11 incestuous pairings included six between siblings, three between father and daughter, and two between mother and son. Preliminary evidence indicates that incestuous pairs can reproduce successfully: one brother–sister pair produced two young that survived to independence, and a father–daughter pair fledged one. Most incestuous pairings lasted for only one season because the females involved usually left their natal group to pair with an unrelated male elsewhere.

Our data from eight years of study on Champion indicate that the breeding system there is generally similar to that on Genovesa. The most significant differences between these populations concern dispersal, age at first breeding, and the breeding structure of groups. As described above, birds on Genovesa frequently breed in their natal territory in their first year, while their dominant parents remain alive, which contributes to a high frequency of plural groups (46%). In contrast, plural groups are comparatively rare on Champion (5% of all group-years, $n = 78$; $\chi_1^2 = 38.3$, $P < 0.001$) because birds there rarely breed early in life as subordinates. Instead, males on Champion wait longer to begin breeding ($\bar{x} = 3.2$ (± 0.3) years, $n = 13$), and they usually do so (58%) after inheriting their natal territory from their father (or an older brother). Champion females do not delay breeding as long as males ($\bar{x} = 2.3$ (± 0.2) years, $n = 20$), and most (79%) begin breeding by filling a vacancy away from their natal territory. Higher rates of survival of adult males on Champion ($\bar{x} = 81.2$ (± 5.6)%, $n = 8$ years) than on Genovesa (see Demography, below; Wilcoxon test, $P < 0.05$) may contribute to the differences in age of first breeding and dispersal. The degree to which the difference in island size is also involved is not yet clear.

Helpers

With helpers defined as birds that provision nestlings other than their own offspring, helpers attended 34% of all nests on Genovesa ($n = 450$ nests, 1978–87). Of those nests with helpers, 78% had only one, 20% had two, and 2% had three; we did not observe more than three helpers at any nest. Counting each helper once for each pair it assisted during a season on Genovesa, non-breeding adult males made up nearly half (48%) of all helpers ($n = 151$). Other helpers consisted of 25% breeding males, 9% breeding females, 9% juvenile males, 5% non-breeding females, and 3% juvenile females (juveniles, birds less than seven months of age, did not help during their first 30 days after fledging). Helpers contribute little to nest construction and they do not share incubation. Helpers feed and protect nestlings and fledged young. Unlike plurally breeding Gray-breasted (Mexican) Jays (Brown and Brown 1980), mockingbirds that do not care for the young in the nest virtually never begin acting as helpers after the young fledge. Even when more than one set of fledglings is produced within a plural group, each set is fed only by the same birds (parents and helpers) that cared for it in the nest.

We reiterate that, by our definition, some birds neither breed nor help, and some do both. Variation of this kind is unusual in a territorial cooperative breeder, and it provides an informative way to investigate factors influencing helping behavior: by assuming that each bird is a potential helper at all nests (other than its own) within its group, we can examine factors influencing the proportion of birds that either do or do not help when they have an opportunity to do so. This analysis indicates that sex, age and breeding status influence the tendency of birds to act as helpers (Table 10.3). Juvenile and adult females became helpers equally rarely, but males became helpers more often, and adult males did so more often than juveniles. The proportion of birds helping otherwise varied little with age (Curry 1988a). As few female breeders as non-breeders acted as helpers, but non-breeding males become helpers disproportionately more often than did breeding males (Table 10.3). More male breeders (44%, $n = 45$) became helpers when they had a chance to do so after their own breeding had failed

Table 10.3. *Influence of age, sex, and breeding status on the percentage of mockingbrids helping. Each bird of known age is counted once for each opportunuty it had (n = helper-seasons, 1980–4) to help at one or more of a particular pair's nests during a breeding season*[a]

| | Juveniles | | | Adults | | | | | | |
| | | | | Non-breeders | | | Breeders | | Total[b] | |
	% helped	n		% helped	n		% helped	n	% helped	n
Males	16	55	*	39	102	*	25	129	32	303
Females	13	39	n.s.	7	98	n.s.	9	27	7	169
P^c		n.s.			**			**		**

[a] After Curry (1988a). n.s., not significant.
[b] includes some adults of unknown breeding status.
[c] χ^2_2 tests of independence; * $P < 0.05$, ** $P < 0.01$.

than when their own breeding was successful (14%, $n = 36$; $\chi_1^2 = 7.4$, $P < 0.01$).

Different associations between helping and breeding in each sex, combined with variation in the proportions of males and females in the Genovesa population, produced a complex pattern of helping as a function of the adult sex ratio (Fig. 10.8). Sex ratio appears to have a causal effect on the occurrence of helping – relatively more birds of both sexes helped when birds of the opposite sex were in short supply – but the sex ratio influences males and females through different mechanisms. The proportion of males that acted as helpers was significantly correlated with increasing sex ratio because non-breeding males, which most often become helpers (above), were relatively more abundant when males predominated in the population. This correlation arose because, in the absence of polyandry, the proportion of males prevented from breeding also increased with increasing sex ratio. In a population where non-breeding and helping are inseparable, this result

Fig. 10.8. Influence of adult sex ratio on the proportion of mockingbirds of each sex breeding or helping in each year. Correlation coefficient (r) and significance of regression (*, $P < 0.05$; **, $P < 0.005$) are indicated beside each least-squares line. Helping and breeding by males are inversely affected by shifts in adult sex ratio because non-breeding males most often help; the relationships differ for females because help is most often performed by females that also breed (see the text).

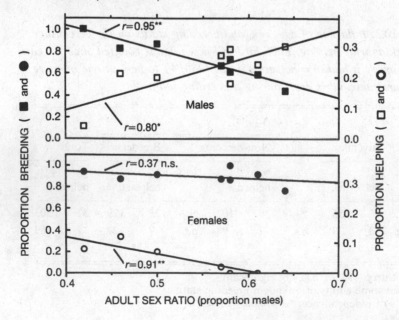

would be trivial, but in the mockingbird social system the correlations are meaningful because some breeders also help and some non-breeders do not. Sex ratio variation had a different influence on helping by females. Because of their ability to pair polygynously, breeding by females was not dependent, over the range of sex ratios we observed, on the relative abundance of males (Fig. 10.7). Furthermore, females became helpers most often when they were breeders in polygynous mating associations: 32% of females with a chance to give help at the nest of the other female in a polygnous trio did so ($n = 37$), but only 1% of females helped to raise the young of females paired with a different male in the same group ($n = 145$; $\chi^2_1 = 35.8$, $P < 0.001$). Therefore, although the proportion of females that help in each year did increase when females outnumbered males, the correlation did not arise because females were then prevented from breeding. It was associated instead with increased frequency of polygyny.

The observation that breeding females most often become helpers for the female with which they share a mate suggests that their help is related in some way to the maintainance of polygnous social bonds. It also seems possible that the helping by females in polygnous associations may facilitate their joint-nesting: in 80% of cases ($n = 10$) where one of the females first acted as a helper at an early nest of the other female, they subsequently nested jointly, but polygnously paired females nested jointly less often (27%, $n = 11$) when neither first acted as a helper for the other (Fisher's Exact test, $P = 0.03$, two-tailed).

Kinship and helping behavior

Helpers on Genovesa were usually close relatives of the nestlings that they fed: most helpers (60%) were older siblings, half-siblings, uncles, or aunts (Table 10.4). Those adults that helped to raise young of no known relationship ($r \leq 0.25$; $n = 24$ adult males, 11 adult females, or 51% of adult helpers) fall into several categories. Some dominant males helped at nests of subordinate pairs in their groups ($n = 3$); these males may have fathered young in these nests through extra-pair copulations. Other males ($n = 16$) fed the young of other males in their natal group that were probably relatives (i.e. older brothers or uncles, or sons born before the study started). If this is correct, the proportion of males that help to feed related young ($r \geq 0.12$) could have been as high as 84%. Few adult males ($n = 7$) helped to feed broods in which circumstances suggest they could have no direct or indirect genetic contribution, and this sample was attributable largely to one male that helped the same pair of unrelated breeders over four years. Because females more often emigrated from their natal groups, a higher proportion of female helpers fed young likely to be completely unrelated. Most such females ($n = 8$ of 11) helped to raise young of their polygynous mate and one of his other females.

Flexible helping behavior, along with natural variation in kin relationships among and within groups on Genovesa, provided an excellent opportunity for investigating the influence of kinship on helping behavior. As would be expected if indirect fitness benefits were important in the evolution of helping, kin discrimination appears to exist: both males and females more often became helpers when the nestlings available to be fed are close relatives than when they are more distant relatives or unrelated birds (Table 10.5). This effect was independent of the influences discussed above (sex, age, and breeding status; Curry 1988a). Another way to test for an effect of kinship on helping is to ask whether help is always directed to

Table 10.4. *Occurrence of helping in relation to the mockingbird helper's age, sex, and relationship to the recipients. Each helper is counted once for each breeding season in which it fed young in one or more nests of a particular pair of breeders (all years, 1978–87, are included)*

Relationships of breeders to helper	r^a	Age and sex of helper				Percent-age of helpers[b]
		Juveniles		Adults		
		Males	Females	Males	Females	
Parent × full sibling	>0.50	0	0	4	0	4.5
Both parents	0.50	13	4	14	5	42.0
Parent × half-sibling	0.38	0	1	0	0	1.1
Parent × unknown	0.25–0.5	0	0	1	0	1.1
Parent × non-relative	0.25	0	0	4	0	4.6
Full sibling × non-relative	0.25	0	0	2	1	3.4
Offspring × non-relative	0.25	0	0	2	1	3.4
Non-relative × unknown	0–0.25	0	0	6	2	9.0
Both non-relatives	≈0	0	0	18	9	30.7
Relationships unknown	?	0	0	61	2	—

[a] Coefficient of relatedness between potential helper and nestlings, as estimated from relationship between the potential helper and the putative parents of the nestlings fed by the helper. Immigrants are assumed to be unrelated to all other group members, barring contradictory evidence.
[b] Percentage of sample ($n = 88$) for which at least partial information about relationship between helper and recipients was available.

the more closely related of two potential sets of recipients (see Chapter 16). In our study, all juvenile mockingbirds ($n = 8$) with chances to help at two different nests helped only at the one containing more closely related nestlings (binomial test, $p < 0.01$); each aided its parents (or, in one case, its mother paired with a half-brother) in preference to another pair. Conducting the same test for adults is difficult because few had opportunities to help to feed more than one set of nestlings of known kinship, and the results were equivocal: two yearlings males helped to raise full siblings in preference to another brood of more distantly related young, but another adult male fed non-relatives when it could have helped to raise the young of its brother and an immigrant female. Four other birds fed only one brood when they could also have helped to raise another equivalently related brood.

The mechanism by which kin discrimination in helping is achieved appears to be associative learning. Birds more often became helpers at nests belonging to breeders who fed them (as a parent or a helper) when they were nestlings (Table 10.6). The observed pattern of helping better matches that expected from a mechanism based on association than from kinship alone: 56% of potential helpers fed unrelated nestlings whose parents had fed the potential helper ($n = 9$) but none fed related nestlings ($r \geq 0.12$) whose parents had not fed the potential helper in the nest ($n = 6$ potential helpers; Fisher's Exact test, $P = 0.042$, one-tailed). These results suggest that

Table 10.5. *Influence of relatedness on percentage of mockingbirds helping. Each bird is counted once for each pair it could have helped in a breeding season* (n = *helper-seasons, 1980–4*)[a]

	Sex of potential helper					
	Males		Females		Total[b]	
Relatedness category[c]	% helping	n	% helping	n	% helping	n
$r \geq 0.5$	41	46	18	34	28	94
$0.38 \geq r \geq 0.12$	11	37	12	16	11	53
$r \approx 0$	25	83	3	62	16	145
P[d]	**		*		*	

[a] After Curry (1988a).
[b] Includes birds of unknown sex.
[c] Estimated minimum relatedness between potential helper and nestlings.
[d] χ_2^2 tests of independence; * $P < 0.05$, ** $P < 0.01$, *** $P < 0.005$.

mockingbirds follow a rule of thumb for helping that is based on the identity of the breeders, rather than the identity of the nestlings. Such a rule would require more discriminatory ability than other rules helpers might follow, such as to feed at any nest within their territory (Brown and Brown 1980; Rabenold 1985). Theoretically, a mechanism for discriminatory helping based on social learning is vulnerable, once established, to deception: unrelated birds could act as helpers in order to deceive nestlings into recognizing them as relatives, to which the nestlings would then direct their help when they became older (R. C. Connor, unpublished results; Curry 1988a). At least four of the unrelated helpers mentioned above did receive help from the birds they helped to raise.

Demography
Climatic conditions

As described above, several aspects of the social organization of Galápagos mockingbirds, including mating and helping behavior, are influenced by changes in the size and structure of the population. Climatic variation directly or indirectly produces these changes. Here, we summarize the sequence of climatic changes that occurred over the study period and the resulting changes in mockingbird reproduction, survival and population structure.

Timing, duration, and amount of rainfall all vary unpredictably at lowland sites throughout the Galápagos. During the 11 years of our study,

Table10.6. *Influence of past association between mockingbird breeders and potential helpers on the percentage of birds helping. Potential helpers are included if the identities of the birds that raised them (as a parent or helper) was known* (n = *helper-seasons)[a]*

	Age of potential helper					
	Juveniles		Adults		All	
Number of breeders that fed potential helper in the nest	% helped	n	% helped	n	% helped	n
Both	21	72	58	24	30	96
One	0	11	23	22	15	33
Neither	0	22	9	82	7	104
P^b	*		***		***	

[a] After Curry (1988a).
[b] χ_2^2 tests of independence; * $P < 0.05$, *** $P < 0.005$.

Genovesa experienced seven years with moderate rainfall, two drought years (1985 and 1988), and two ENSO years (1983 and 1987; Fig. 10.9). The 1983 El Niño event was extraordinary: it came an unusually long time (seven years) after the previous El Niño, and it began earlier, lasted longer (eight months), and brought more rain to the islands than any other event on record. The 1987 El Niño event was more typical, lasting five months.

Reproduction

The timing, amount and duration of rainfall influence mockingbird reproductive success. The number of clutches produced per female varied among years (Kruskal–Wallis test, $P < 0.005$); females produced more clutches (up to six) in years with heavy rainfall (Fig. 10.9), when rain and nesting both extended over as much as eight months. We did not

Fig. 10.9. Temporal variation (*a*) annual rainfall on Genovesa and (*b*) by the number of mockingbird clutches and fledglings produced per female ($\bar{x} \pm$ S.E.M.; $n =$ number of breeding females). Rainfall totals for 1978 and 1979 are estimates based on incomplete data, and the total for 1983 includes a minimum estimate of rain falling in December 1982. Dashed lines and question marks denote estimated reproduction in 1979, when the population was studied during only part of the breeding season.

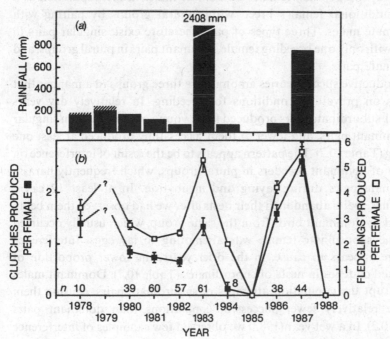

measure food supplies systematically, but our casual observations suggest that prolonged availability of arthropods accounts for extended breeding in wet years. On average, each breeding female produced 1.8 (± 0.06) clutches ($n = 314$) during each breeding season. Variation in the number of clutches produced is primarily responsible for variation in seasonal production of fledglings per female (Fig. 10.9(b); Kruskal–Wallis test, $n = 8$ years, $P < 0.005$). The average number of young produced by each female in a season varied from none in drought years, when breeding was not attempted, to 5.2 in years with prolonged rains (total $n = 316$, $\bar{x} = 2,3 \pm 0.1$). Average clutch size (mode $= 4$ eggs) increased in wet years (Kruskal–Wallis test, $n = 8$ years, $P < 0.005$), but was offset by lower hatching success, caused in part by storms, in El Niño years. As a result of variation in clutch size and hatching success, the average number of fledglings produced per nest also varied significantly among years (Kruskal–Wallis test, $n = 8$ years, $P < 0.005$).

Within-group conflict and climatic conditions

The presence in plural groups on Genovesa of multiple breeders differing in dominance status gives rise to competitive as well as cooperative interactions within groups. Dominant males nearly always obtain mates and breed; except when females outnumber males in the population as a whole, additional females breed within plural groups by pairing with subordinate males. Three types of pairs therefore exist: singular pairs in groups with only one breeding female, dominant pairs in plural groups, and subordinate pairs.

Reproductive success varies among these three groups in a manner that depends on prevailing conditions for breeding. In relatively dry years (1980–2), subordinate pairs produced fewer young on average than singular and dominant pairs, and fewer of them successfully produced at least one fledgling (Table 10.7). The pattern appears to be the result of interference on the part of dominant breeders in plural groups, which frequently harasss subordinate pairs during laying and incubation. In at least 12 cases, subordinate pairs abandoned their nests after we had observed them being attacked by dominant birds from the same group, which usually occurred while the subordinate female was attempting to lay eggs. Interference therefore appears to cause, in the drier years, the lower proportion of completed clutches in nests of subordinates (Table 10.7). Dominant males also disrupt the copulation attempts of subordinate pairs, causing them to have relatively fewer successful copulations than dominant pairs (Table 10.2). In a wet year (1983) we observed few examples of interference by dominants, and in the same year reproductive success did not vary among the three dominance categories (Table 10.7).

One hypothesis for the occurrence of conflict proposes that dominant birds suffer costs by allowing subordinate pairs to breed in their territory, and that they can reduce these costs by interfering at subordinate nests. Measurements of territory size are consistent with this hypothesis. Territory size increased with the number of breeding females present, but, assuming that the territory was shared equally, each breeding female in a plural group occupied significantly less space than females in singular groups (Fig. 10.10). Dominant breeders may therefore disrupt the nests of subordinates so that a higher proportion of the territory's food resources will be available for the dominant pair's nestlings. Under such a hypothesis, more conflict would be expected in dry years: the costs of sharing a plural territory would then be relatively more costly, and the benefits of disrupting subordinate nesting greater, than in wet years when resources are probably more abundant (Curry 1988b).

Table 10.7. *Reproductive success in relation to dominance status of mockingbird breeders in comparatively dry (1980–2) and wet (1983) years[a]*

Component of reproductive success	Status of pair	Year	
		1980–2	1983
Percentage of clutches completed[b]	Singular	96% (73)	93% (73)
	Plural dominant	100% (43)	89% (28)
	Plural subordinate	78% (50)	95% (20)
	χ^2 test	$P < 0.001$	n.s.
Fledglings per nest[c]	Singular	1.4 ± 0.15 (73)	0.8 ± 0.13 (73)
	Plural dominant	1.3 ± 0.20 (43)	1.1 ± 0.27 (28)
	Plural subordinate	1.1 ± 0.15 (50)	1.0 ± 0.27 (22)
	Kruskal–Wallis test	$P < 0.005$	n.s.
Percentage of pairs successful[d]	Singular	76% (63)	78% (27)
	Plural dominant	64% (42)	73% (16)
	Plural subordinate	40% (48)	62% (13)
	χ^2 test	$P < 0.005$	n.s.

[a] After Curry (1988b). n.s., not significant.
[b] (n, clutches); clutches classified as completed if breeders initiated incubation.
[c] $\bar{x} = \pm$ S.E.M. (n, nests); all nests included.
[d] Pairs producing one or more fledglings during a season classified as successful (n, pairs).

Helping and reproductive success

Helpers on Genovesa affect the reproductive success of the breeders they assist primarily by increasing nestling survival. The number of hatched young did not differ between nests with and without helpers, but nests with helpers produced an average of 1.9 (± 0.1) fledglings ($n = 153$ nests), an increase of 20% over nests without helpers ($\bar{x} = 1.6$ (± 0.1) fledglings, $n = 297$; Mann–Whitney U-test, $P < 0.05$; Fig. 10.11). Relatively more nests with helpers (66%, $n = 101$) successfully fledged at least one young compared with nests not attended by helpers (46%, $n = 101$; $\chi_1^2 = 10.8$, $P < 0.005$; Fig. 10.11). This did not occur because of any influence of helpers on total brood loss: predators (mainly Short-eared Owls) destroyed 13% of broods attended by helpers ($n = 97$) and 15% of broods of unassisted pairs ($n = 134$; $\chi_1^2 = 0.1$, n.s.). Higher nestling survival apparently resulted instead from an increase in the amount of food delivered to the nestlings. Male and female breeders both fed nestlings at lower rates when aided by helpers, but helpers more than compensated, resulting in a significantly higher total feeding rate at helped nests (Kinnaird and Grant 1982; Curry, unpublished results). We did not detect significant increases in average growth rate of nestlings fed by helpers, but

Fig. 10.10. Territory size in relation the numbers of breeding females in each mockingbird group on Genovesa (after Curry 1988*b*). Histograms denote means ($\pm$S.E.M.) for total territory size and for area per breeding female; figures inside bars indicate sample sizes. All means differ significantly (Mann–Whitney U-tests, $P < 0.05$) except those connected by horizontal bracket ($P > 0.1$).

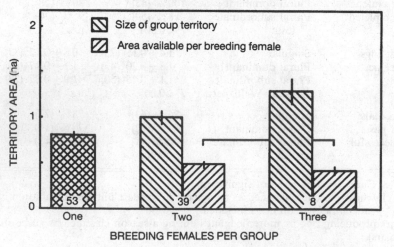

the increased feeding rates attributable to helpers could directly influence nestling survival by reducing starvation; partial losses from broods occurred commonly on Genovesa, and in many cases the young that disappeared were apparently starving.

The participation of additional helpers beyond one did not result in significant increases in feeding rates (Curry, unpublished results), nor did it confer further improvement in reproductive success in terms of either the average number of fledglings produced (nests with one helper compared to those with more than one, Mann–Whitney U-tests, n.s.; Fig. 10.11) or the proportion of nests that were successful (65% for 77 nests with one helper, 71% for 24 nests with more than one; $\chi_1^2 = 0.1$, n.s.). Helpers also did not influence the number of broods that recipient pairs produced in a season. In 1982, for example, 17% of pairs re-nested after receiving help in raising their first brood ($n = 24$), as did 14% of pairs that did not receive help with their first brood ($n = 21$; $\chi_1^2 = 0.2$, n.s.).

The dominance status of the breeding pair and prevailing climatic conditions further influence the effect helpers have on recipient reproductive success. In relatively dry years (1980–2), helpers significantly increased the proportion of nestlings that fledged, especially in nests of subordinate pairs (Table 10.8). In wet years, help had less influence on fledging success, regardless of dominance status of the breeders. This pattern, together with

Fig. 10.11. Influence of helpers on average production of mockingbird fledglings and success of nests on Genovesa, 1978–84. The number of helpers include all juveniles and adults other than parents that fed nestlings (n = number of nests; joint nests excluded). Nests were classified as successful if at least one fledgling was produced.

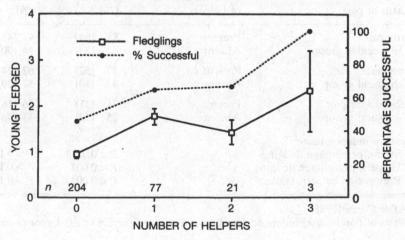

data indicating that subordinate males whose nests fail later often become helpers for the dominant pair, leads to another possible explanation for the timing of reproductive conflict: dominant breeders may disrupt subordinate nests in dry years in order to increase their own chances of receiving help at times when such help is most likely to improve the dominant pair's reproductive success (Curry 1988b; see also Emlen 1982b). This hypothesis differs from the alternative based on competition for resources discussed earlier, but the two hypotheses are not mutually exclusive. Both mechanisms could be responsible for the occurrence of conflict in plural groups in years with relatively poor climatic conditions.

Effect of help on donors

Even where helpers increase recipient reproductive success, determining whether the helping behavior is also advantageous to the helper, and is as a result favored by direct or indirect selection (Brown 1980) or both, is difficult. The flexible behavior of mockingbirds on Genovesa has enabled us to consider the effects of help on donors by comparing birds that did and did not help when they had opportunities to do so. This analysis suggests that providing help confers no detectable cost in terms of survival:

Table 10.8. *Percentage of mockingbird hatchlings that successfully fledged (* n, *hatchlings) in relation to status of pair and presence or absence of helpers at the nest*[a]

		Year	
Status of pair	Presence/absence of helpers	1980–2 (%)	1983 (%)
Only pair in singular group	Present	88 (64)	47 (45)
	Absent	47 (133)	46 (90)
Dominant pair in plural group	Present	77 (52)	62 (24)
	Absent	42 (60)	52 (29)
Subordinate pair in plural group	Present	71 (31)	56 (16)
	Absent	25 (52)	33 (45)
Tests of independence[b]			
Status, percentage fledging		$P < 0.05$	$P > 0.1$
Helpers, percentage fledging		$P < 0.001$	$P > 0.1$
Presence of helpers, status		$P < 0.001$	$P > 0.1$

[a] After Curry (1988b).
[b] Tests of conditional independence between pairs of factors in a 3-factor G-test.

78% of the adults that helped ($n = 81$) survived to the following year, as did 67% of adults that did not help ($n = 202$; $\chi_1^2 = 2.8$, n.s.). Helping also does not appear to function as a way for birds to improve their chances for becoming a breeder. Many helpers were already breeding (above), and at least 25% of males and 60% of females that became breeders in our study did so without ever acting as a helper. Of non-breeding males that helped, 6% ($n = 85$) later paired with a female whose breeding they assisted, but 3% ($n = 158$) also paired with females they did not help when they could have ($\chi_1^2 = 0.5$, n.s.). The percentage of non-breeding males that became breeders in the year after they had a chance to help was as high among those that did not help (45%, $n = 29$) as among those that did (32%, $n = 40$; $\chi_1^2 = 0.6$, n.s.). One reason why helping does not appear to improve the helper's chances of breeding is that the number of young produced within a group does not determine subsequent changes in group or territory size (Curry, unpublished results). This means that mockingbirds probably do not improve their chances of obtaining territory space by helping to produce additional young, as has been proposed for Florida Scrub Jays (Woolfenden and Fitzpatrick 1978). Benefits in the form of 'reciprocated' help also appear to be small: only 6% of all helpers ($n = 81$), and 15% of those that bred after they had helped ($n = 33$), received help from the birds they had helped to raise, and none received help from the breeders it had assisted.

For most mockingbirds, therefore, the direct fitness benefits of helping are small. Helpers that feed related nestlings augment their indirect fitness, but on average these benefits also appear to be small: helpers do not have a large effect on the number of young produced, and relatedness between donors and recipients is often reduced through mortality and dispersal of relatives.

Survival

Annual survival of adults, measured from one January to the next, averaged 60.5 (± 5.5)% ($n = 10$ years; total $n = 1969$ bird-years). Adult survival was lowest during both the wettest and driest conditions of the study. Survivorship (survival rate) dipped to 47% in 1983 and 41% in 1984 during an epizootic. We suspect that the disease was avian pox, and that mosquitoes acted as vectors to spread the disease; mosquitoes were unusually abundant during the 1982–3 El Niño. Adult survival was also low (37%) following the 1985 drought. In years with intermediate conditions, adult survival ranged from 63% to 93%. Survival of breeders did not differ significantly from that of adult non-breeders, and adult survival did not vary with age. Female breeders aided by helpers had significantly higher survival (83%) than unaided females (57%) in one year (1981), but breeder survival was otherwise unaffected by receipt of help: in total, 63% of females that received help survived to the next year ($n = 79$, 1980–4) compared to 59% of unaided females ($n = 126$; $\chi_1^2 = 0.4$, n.s.), and 66% of male breeders receiving help survived ($n = 82$), as did 67% of unassisted males ($n = 129$; $\chi_1^2 = $ n.s.). Survival of juveniles to January of the year after their birth averaged 34.7 ($\pm 5.1\%$ ($n = 7$ years; total $n = 796$ fledged young). The pattern of variation among years was similar to that of adult survival, except that juveniles born had very poor survival (9%) in one moderately dry year (1981) when adult survival was relatively high (67%). Juvenile survival was almost twice as variable ($CV = 47\%$) as annual adult survival ($CV = 29\%$).

Data on survival, as well as direct observations of longevity, suggest that males live longer on average than do females. The sex ratio is approximately even at fledging: 51% of the birds we banded as nestlings and remeasured later to determine their sex (e.g. Table 10.1) were males. Survival of juveniles to the January of their first year of life (after which we classified them as adults) averaged 39.2 (± 7.6)% for males and 30.1 (± 4.8)% for females ($n = 8$ years; Wilcoxon test, $0.05 < P < 0.1$). Annual survival of adult males ($\bar{x} = 62.4$ (± 5.2)%) averaged slightly higher than that of adult females ($\bar{x} = 58.9$ (± 6.0)%, $n = 10$ years; Wilcoxon test, $P < 0.05$). Survivorship curves of the two sexes consequently diverge such that estimated maximum longevity of males (calculated as the age by which 99.9% of fledged young will have died) is 12.1 years for males and 9.0 years for females. No birds of either sex survived through our entire study. One male attained at least eight years of age (and was still alive at the end of the study) and three others lived for at least seven years. Four females attained at least six years of age, and five others lived for five years.

Average survivorship values obscure differences in the temporal pattern of survival of the sexes. Adult males survived better than females prior to the 1982–3 El Niño and again between 1985 and 1988. During the intervening years, females survived better than males. Juvenile males survived better than juvenile females in seven of eight years.

Population size and structure

The size of the population resident in our original study area of Genovesa (Fig. 10.12(a)) varied as a result of changes in natality and survival. Because territories on this island constantly filled all available space, population density measured over the entire area studied in each year followed an almost identical pattern (Fig. 10.12(b)). By both measures, the size of the breeding population was more stable. Two years, 1984 and 1985, were exceptional in that many paired birds did not nest and were therefore not classified as breeders. Excluding these years, breeder density ($\bar{x} = 2.9\,(\pm 0.6)$ birds ha^{-1}, $n = 8$ years) was less than one-fourth as variable ($CV = 21\%$) as the density of non-breeding adults ($CV = 93\%$). This pattern, reinforced by the observation that nearly every bird breeds in years when density is low, is consistent with the hypothesis that a relatively fixed upper limit on breeding density exists, and that some birds are excluded from breeding when total density exceeds this level.

Fig. 10.12. Temporal variation in mockingbird population size and density. (*a*) The number of breeding and non-breeding adults residing each year in a core area (approximately 6 ha) occupied by nine territorial groups at the start of the study. Variation in the number of territories completely or partially occupying the original area is indicated along the horizontal axis. (*b*) Density of breeding and non-breeding adults over the entire study area. The total area (ha) included in each year is indicated along the horizontal axis. Numbers and density were both estimated in January of each year, at the start of the usual breeding season. Birds were classified as breeders if they produced at least one clutch; breeding numbers and density were exceptionally low in two years when few (1984) or none (1985) of the resident pairs produce eggs.

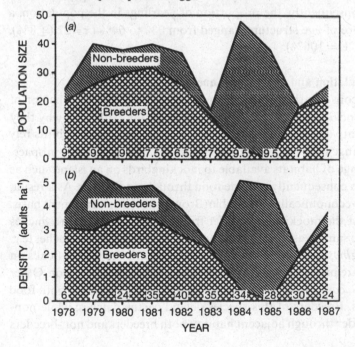

Because of independent changes in demographic parameters, the structure of the Genovesa population varies considerably through time. During our study, the adult sex ratio changed as a consequence of variation in the mortality rates of the two sexes. Between 1978 and 1982, when males survived better than females, the sex ratio ranged from 57% to 64% males. Increased mortality of males during the epizootic in the 1982–3 El Niño produced an even sex ratio in 1983 and then an excess of females in 1984 (42% males) and in 1985 (39% males). Because of higher male survival during and after the 1985 drought, the sex ratio swung back to 42% males in 1986, 46% in 1987, and 52% in 1988. As discussed earlier, this variation in the adult sex ratio influenced both the mating system and the occurrence of helping behavior.

The age structure of the population, determined by the outcome of breeding in the preceding year and by the subsequent survival of juveniles and adults, also varied. The age structure was relatively even from 1978 through 1983, as a result of high adult survival and intermediate levels of both fledgling production and juvenile survival. Yearlings predominated in 1984 following prolonged successful nesting and high adult mortality in 1983. Two subsequent years without successful reproduction caused a gap in the age structure and, after the next year with successful breeding (1986), the population was almost evenly divided between yearlings and three-year-olds. Consequently, the proportion of yearlings in the population, a simple measure of age structure, ranged from 0% to 64% ($\bar{x} = 22 \, (\pm 8\%)$, $n = 8$ years; $CV = 106\%$).

Speculation and unanswered questions

We consider the foraging requirements and spacing behavior of Galápagos mockingbirds to be the starting point for explaining why they breed cooperatively. Because of their omnivorous diet, mockingbirds rely on a broad range of foods, most of which are distributed broadly in space. The entire range of habitats available to mockingbirds on an island such as Genovesa can consequently provide food throughout the year. As a result, territories are economically defendable (Brown 1964) on a permanent basis. In this respect the mockingbirds differ from the various Darwin's finches inhabiting the same seasonal environments (Grant 1986). Some finches (e.g. *Geospiza fuliginosa*) abandon their breeding territories to feed in flocks on seeds, which are probably patchily distributed, during the dry season. Other species (e.g. *G. conirostris*) rely on clumped patches of cactus for both food and nest sites. Breeders of this species defend cactus patches, while non-breeders wander through adjacent habitat; both breeders and non-breeders concentrate in the cactus patches during the non-breeding season. Therefore omnivory in mockingbirds promotes both their occupancy of the full range of available terrestrial habitats on islands such as Genovesa, and their year-round defense of space within these habitats.

The next step toward cooperative breeding is formation of groups larger than two within permanently defended territories. Here demographic parameters are important. Specifically, adult survival rate that is usually high produces a population that tends to remain at high density. As a result, the population often surpasses the level at which all birds would be able to breed (Brown 1969). This argument assumes that an upper limit on density of breeders exists, which is supported on Genovesa by the presence of non-breeding females as well as males in years of high density.

We therefore consider Galápagos mockingbirds to be a prime example of birds in which the combined constraints of habitat and territorial behavior force individuals into group-living. Where all habitat is occupied and vacancies occur rarely because of high adult survival rates, birds may have higher inclusive fitness if they remain in their natal group than if they compete for space against already established territory holders. For Galápagos mockingbirds, the physical limits of habitat on small islands pose a barrier beyond which a bird would have a negligible chance of successfully dispersing. Their situation therefore represents an extreme example of the conditions faced by several other cooperative breeders, where the chances of successfully dispersing, breeding, and surviving are prohibitively low outside some preferred habitat (Koenig and Pitelka 1981; Emlen 1982a; Fitzpatrick and Woolfenden 1986; Zack and Ligon 1986; see also Chapter 8).

The preceding conclusion is strengthened by the comparative results described earlier: the majority of mockingbirds live in groups of just two birds only on one island (San Cristóbal) where the population does not fill all habitats. An important unanswered question is why *N. melanotis* is unable to use all the habitats available there. Apart from the possible role of predators mentioned above, there seems to be little difference in foraging behavior or ecological requirements that can explain why, for example, San Cristóbal mockingbirds do not hold breeding territories in scrub woodland dominated by the tree *Bursera graveolens* (which is occupied by mockingbirds on the other islands) while they do hold territories in adjacent scrub dominated by a different tree (*Piscida cathagensis*; Curry, 1989). Whether the extra birds, which are all males, in trios on San Cristóbal ever act as nest helpers is still not known; we did not have a chance to observe normal nesting during our studies there, or on Española.

Helpers have been recorded at nests on Española (S. Groves, unpublished results), and we observed helpers at seven of nine nests in our study on Champion.

Population density and habitat availability are clearly not the only ecological factors influencing on social organization. A mockingbird population identical in size with that on Genovesa, but with an even sex ratio, could be distributed in such a way that all birds lived in groups of two defending pair-only territories. The adult sex ratio on Genovesa, however, is seldom even: through differential mortality, it tends to become male-biased in drier years and female-biased after epizootics. Sex ratio biases in either direction promote group-living, through interactions with the mating system. When females predominate on Genovesa nearly all breed, many by pairing polygynously. Doing so necessarily entails maintaining a group of more than two birds. Therefore female-biased sex ratios also promote group maintenance by forcing females to share mates. More frequently, males on Genovesa outnumber females. As a consequence, many males are excluded from breeding, because they seem incapable of forming polyandrous associations. Males without mates could leave their group and establish new territories, but they would have to do so by competing for space already defended by other mockingbirds. Such a step is unlikely to be favored, because the costs and risks of obtaining a territory would not be balanced by fitness benefits from breeding until the bird obtained a mate. On San Cristóbal, in contrast, some unpaired males do leave their groups to establish territories in unoccupied habitat. It is probably advantageous for them to do so because of the low cost of defending such space compared to the more competitive situation on Genovesa.

Even though modal group size on Genovesa is two birds, some larger groups nevertheless occur even when density is low and males and females are present in equal numbers. In these situations, dominance relationships among birds of different ages appear to be responsible for the persistence of large groups. Relative dominance among males is so closely tied to age that young birds, especially yearlings, are usually unable to split away and to defend territories against the older males of their group and other nearby groups. In 1987, for example, only one yearling male established his own independent territory, even though almost 50% of the males alive were yearlings. The largest group we recorded during our study, 24 birds, is also illustrative: this group formed late in 1983 through the fusion of what had been three separate groups; almost every adult in these groups had died and, by 1984, the 23 yearlings living in the area were all subordinate to the single oldest survivor, a male at least five years of age. In 1986, after this older male died, the group split up and three of the surviving males, each then three years of age, established separate territories.

The persistent dominance of older males over younger birds on Genovesa is responsible for the occurrence of a system involving plural breeding with separate nests, a relatively unusual feature among cooperatively breeding birds (Brown 1987). Plural groups arise when subordinate birds in larger groups attempt to breed without waiting to replace older dominant breeders. As a consequence of age-dependent dominance, the relative frequency of plural groups increases in years when a high proportion of the population consists of yearlings, which attempt to breed as subordinates. Plural groups almost never arise in this way in most other territorial cooperative breeders with similar age-related dominance systems (e.g. Florida Scrub Jays; Fitzpatrick & Woolfenden 1986). Instead, birds become breeders only after they have developed the ability to defend an independent territory. To understand why plural breeding occurs on Genovesa, the question that must be addressed is: if young mockingbirds are unable to split off part of their former territory to breed independently, why do they attempt to breed as subordinates?

Plural breeding appears to occur because breeding by birds too young to defend independent territories is nevertheless favored by unpredictable climatic variation. The frequent but unpredictable occurrence of wet years, when disproportionate numbers of young can be produced, increases the value of being able to breed early in life: yearlings that did not breed on Genovesa in 1983, for example, had to survive two subsequent years of poor conditions before getting another chance. Furthermore, young birds breeding as subordinates have low reproductive success in dry years because of interference from dominant group members. Nevertheless, the advantages of breeding are large in those years when favorable conditions arise because the rains persist. There seem to be few cues, however, by which a bird could predict, at the start of the breeding season, whether conditions will be favorable (wet) through the rest of the season, and the survival costs of attempting to breed also appear to be small (Curry 1988*b*). Consequently, birds that can obtain mates will almost always benefit by attempting to breed, on the chance that it will be a wet year, even if many must do so as subordinates.

The preceding explanation for plural breeding assumes that yearlings cannot defend their own independent territories. Why should such an ability not have evolved? We hypothesize that the indirect effects of climatic variation on mockingbird population structure provide a partial answer (Curry and Grant, 1989). In Galápagos mockingbird populations, favorable climatic conditions often occur at times when, because of the large variation in the population's age structure, many yearlings are able to obtain mates. In many other years, however, yearlings are outnumbered by older, experienced birds, against which they would have little chance of competing for mates or territories. If climatic conditions were less variable and, because of the population's demographic characteristics, yearlings were usually able to obtain mates and breed, they would be expected to have evolved the ability to defend independent territories. On the other hand, if climatic and demographic conditions consistently made breeding by yearlings unprofitable, they would not be expected to have developed the ability to hold a territory; this appears to be the case for other cooperative breeders such as the Florida Scrub Jay. The changing demographic environment of Galápagos mockingbirds therefore appears to favor the ability to breed before the age when birds develop the ability to defend territories, which results in plural breeding.

This preceding hypothesis does not explain why dominant birds should tolerate the presence of subordinate breeders in their territories. Factors that may be involved include kinship and opportunities for extra-pair copulations. Because dispersal is reduced in males, subordinate male breeders in plural groups are often the sons of the dominant breeding pair. The dominant pair may therefore permit their son to breed as a form of extended parental care (Brown and Brown 1984). Dominant males may also benefit by tolerating or even facilitating breeding by additional pairs in their territory, if by doing so they can father young through extra-pair copulations with subordinate females. Conflict may arise between the dominant breeders as a result, because dominant females do not appear to gain similarly; they do not make any direct genetic contribution to subordinate nests, as they might through egg parasitism (e.g. see Chapter 16). Even when if a dominant female might benefit genetically by permitting her son to breed, her interest would be reduced or eliminated if the dominant male is the genetic father of the subordinate pair's nestlings (Curry 1988b). Another possibility is that tolerating the presence additional group members may in some way benefit the dominant pair (e.g. in territorial encounters), but plural groups arise simply because dominant pairs are often too busy with their own nesting to prevent subordinates from attracting mates and attempting to breed. The frequent disruption of subordinate nests by dominant birds is consistent with this view.

Other aspects of mockingbird social organization are flexible, including both mating and helping behavior. Flexibility in mating is demonstrated by the shift from monogamy to polygyny as females become relatively more abundant (flexible mating systems are common to several other cooperative breeders, e.g. Hedge Accentors (Dunnocks); see Chapter 15). Polygyny is an opportunistic way for a female that would otherwise be unable to obtain a mate to breed. Why, though, do these females so often nest jointly? Our analyses of joint nesting is still underway, but at first inspection it appears that females laying in a joint nest have lower reproductive success than separately nesting females: hatching success is low in joint nests (Curry 1988*b*), perhaps because incubating an enlarged clutch is difficult. Separate nesting entails other costs, however, at least for subordinate females. Their separate nests are vulnerable to disruption by other more dominant females in the same group. By laying in a joint nest, a subordinate female may in effect 'hide' her eggs among those of the other female, making disruption, in the absence of the ability to discriminate among each other's eggs, nearly impossible. Subordinate females might thereby produce more young than they would if they nested separately in a territory containing another, more dominant female.

Incestuous pairing further indicates that mating behavior is opportunistic. Nearly all matings between close relatives on Genovesa occur when a young bird remains in its natal territory, in the presence of relatives of the opposite sex. Climatic conditions seem to favor the ability to begin nesting very rapidly once the rains fall; in relatively dry years, the rains end in a matter of days, and any nests begun later are unlikely to be successful. As a consequence, mockingbirds appear to spend little time in choosing a mate, often pairing with the closest available potential mate in a local area. Because of reduced dispersal in both sexes, these mates are frequently relatives. Our conservative estimate of 7% for the frequency of close inbreeding on Genovesa is unusually high for a vertebrate (Ralls *et al.* 1986), but surpassed by the Splendid Fairy-wren, *Malurus splendens*, a species with a similar social organization in several respects (see Chapter 1). Presumably the advantages of breeding quickly outweigh any disadvantages of incestuous pairing.

Acting as a nest helper or alloparent is also flexible in Galápagos mockingbirds. This flexibility is indicated by the tendency for mockingbirds to help after their own nests fail, after they successfully produce young of their own, or even while they are feeding young in their own nest. Old, established breeders act as helpers after breeding successfully for several years. Clearly, alloparenting in this population is not simply a behavior performed by birds endeavoring to become breeders, although birds that have not yet bred do become helpers most often. The lack of measurable survival costs from helping, along with the variable circumstances in which it occurs, suggest that helping is an inexpensive behavior used by many individuals in the complex mockingbird society.

A question for which we do not yet have a complete answer is whether helping in Galápagos mockingbirds is an evolved adaptation or, in other words, whether natural selection has increased the frequency or efficacy of help. According to an alternative non-adaptive hypothesis (Craig and Jamieson 1987), provisioning of young by helpers is an incidental result of birds living and breeding together in social groups, and the behavior has no fitness consequences for donors or recipients. The improved reproductive success of breeders that receive help argues against this hypothesis. The positive effect of Galápagos mockingbird helpers on nestling survival is small compared with some other species in which helpers reduce predation rates significantly, but the effect is consistent among breeder dominance categories and among years. A plausible mechanism by which helpers improve recipient reproductive success exists: as a consequence of the variable climatic and feeding conditions, nestlings on Genovesa frequently starve, and the extra food brought by helpers appears to increase the proportion of nestlings that survive. Selection should favor tolerance of helpers by breeders, as long as the benefits of having helpers are not offset by other costs. However, we cannot rule out the possibility that variation in the quality of territories accounts in part for the higher success of pairs with helpers. It has also been difficult for us, because of analytical complexities and our relatively small samples, to assess breeder age and experience as determinants of reproductive success, but preliminary results suggest that the effects of these factors are minor relative to the effect of helping.

Even if helpers have a positive effect on recipient reproductive success, non-adaptive hypotheses for their behavior are difficult to reject without information on, for example, the heritability of helping. The increased probability of helping as a function of kinship we observed, however, would not be expected unless indirect selection had favored alloparenting. By improving the reproduction of relatives, many helpers increase their indirect fitness. This assumes that providing help does not confer costs that negate indirect fitness gains. The lack of options for mockingbird helpers is consistent with this assumption: most are young birds that have not bred or that have tried and failed, with little chance of re-nesting successfully because of the brevity of the typical Galápagos breeding season. These helpers do not appear to gain significant direct fitness benefits. In other situations, direct benefits to helpers could be important. Dominant males may augment their immediate direct fitness by feeding young in the nests of subordinate pairs, if in fact they father young in those nests through extra-pair copulations. How often such copulations result in fertilization, and whether subordinate males are ever able to father young through copulations with the mates of dominant males, remain to be determined. If joint nesting does provide a benefit to some females, and if their help is a prerequisite for joint nesting, then breeding females in polygynous associations may also receive direct fitness benefits by helping.

In conclusion, Galápagos mockingbirds in several ways experience ecological stability: they occupy relatively uniform, undisturbed habitat on small islands, they are omnivorous and they are permanently territorial. Their populations are also subject a large degree of climatic and, secondarily, demographic variation. By favoring flexible breeding and helping behavior, these sources of variation give rise to the unusual and complex features of Galápagos mockingbird social organization.

Our studies have been carried out with the permission of the Programa Nacional Forestal, Quito, Ecuador, and with the help of the Servicio Parque Nacional Galápagos and the Charles Darwin Research Station. Much of the research was conducted while we were both at the University of Michigan, which we thank for financial support. Margaret F. Kinnaird conducted the field study during the breeding seasons of 1979 and 1980; we thank her for access to unpublished data. We also thank: D. J. Anderson, C. B. Chappell, S. H. Curry, J. L. Gibbs, B. R. Grant, K. T. Grant, N. Grant, A. E. Heise, M. Iturralde, W. Johnson, K. Nelson, S. H. Stoleson, J. R. Waltman, and T. C. Will for field assistance; N. Deyrup of the Archbold Biological Station for photographic help; and the World Wildlife Fund (USA), the Frank M. Chapman Memorial Fund of the American Museum of Natural History, and Sigma Xi, the Scientific Research Society, for financial contributions. Additional financial support derived from a NSERC Canada postgraduate scholarship to R.L.C., and from NSF grants to P.R.G. This is contribution No. 429 of the Charles Darwin Foundation for the Galápagos Islands.

Bibliography

Abbott, I. and Abbott, L. K. (1978). Multivariate study of morphological variation in Galápagos and Ecuadorean mockingbirds. *Condor* 80: 302–8.

Bowman, R. I. and Carter, A. (1971). Egg-pecking behaviour in Galápagos mockingbirds. *Living Bird* 10: 243–70.

Brown, J. L. (1964). The evolution of diversity in avian territorial systems. *Wilson Bull.* 76: 160–9.

Brown, J. L. (1969). Territorial behavior and population regulation in birds: a review and re-evaluation. *Wilson Bull.* 81: 293–329.

Brown, J. L. (1978). Avian communal breeding systems. *Ann. Rev. Ecol. Syst.* 9: 123–55.

Brown, J. L. (1980). Fitness in complex avian social systems. In *Evolution of Social Behavior: Hypotheses and Empirical Tests*, ed. H. Markl, pp. 115–28. Verlag Chemie GmbH: Wienheim.

Brown, J. L. (1987). *Helping Behavior and Communal Breeding in Birds.* Princeton University Press: Princeton, NJ.

Brown, J. L. and Brown, G. R. (1980). Reciprocal aid-giving in a communal bird. *Z. Tierpsychol.* **53**: 313–24.

Brown, J. L. and Brown, E. R. (1984). Parental facilitation: parent–offspring relations in communally breeding birds. *Behav. Ecol. Sociobiol.* **14**: 203–9.

Christian, K. A. (1980). Cleaning/feeding symbiosis between birds and reptiles of the Galápagos islands: new observations of inter-island variability. *Auk* **97**: 887–9.

Craig, J. L. and Jamieson, I. (1987). Critique of helping behavior in birds: a departure from functional explanations. In *Perspectives in Ethology*, vol. 7 ed. P. P. G. Bateson and P. H. Klopfer, pp. 79–98. Plenum Press: New York.

Curry, R. L. (1985). Breeding and survival of Galápagos Mockingbirds during El Niño. In *El Niño in the Galápagos Islands: The Event of 1982–1983*, ed. G. Robinson and E. del Pino, pp. 449–71. Charles Darwin Foundation for the Galápagos Isles: Quito.

Curry, R. L. (1986). Whatever happened to the Floreana Mockingbird? *Not. Galapagos* **43**: 13–15.

Curry, R. L. (1988a). Influence of kinship on helping behavior of Galápagos Mockingbirds. *Behav. Ecol. Sociobiol.* **22**: 141–52.

Curry, R. L. (1988b). Group structure, within-group conflict, and reproductive tactics in cooperatively-breeding Galápagos Mockingbirds, *Nesomimus parvulus. Anim. Behav.* **36**: 1708–28.

Curry, R. L. (1989). Geographic variation in social organization of Galápagos Mockingbirds: ecological correlates of group territoriality and cooperative breeding. *Behav. Ecol. Sociobiol.* (in press).

Curry, R. L. and Anderson, D. J. (1987). Interisland variation in blood drinking by Galápagos Mockingbirds. *Auk* **104**: 517–21.

Curry, R. L. and Grant, P. R. (1989). Demography of the cooperatively breeding Galápagos Mockingbird, *Nesomimus parvulus*, in a climatically variable environment. *J. Anim. Ecol.* **58**: 441–64.

Davis, J. and Miller, A. H. (1960). Family Mimidae. In *Check-list of Birds of the World*, vol. 9, ed. E. Mayr & J. C. Greenway, pp. 440–58. Museum of Comparative Zoology, Harvard University: Cambridge, MA.

Emlen, S. T. (1982a). The evolution of helping. I. An ecological constraints model. *Amer. Nat.* **119**: 29–39.

Emlen, S. T. (1982b). The evolution of helping. II. The role of behavioral conflict. *Amer. Nat.* **119**: 40–53.

Fitzpatrick, J. W. & Woolfenden, G. E. (1986). Demographic routes to cooperative breeding in some New World jays. In *Evolution of Animal Behavior: Paleontological and Field Approaches* ed. by M. H. Nitecki and J. A. Kitchell, pp. 137–60. Oxford University Press: Oxford.

Grant, P. R. (1985). Climatic fluctuations on the Galápagos Islands and their influence on Darwin's Finches. *Ornith. Monogr.* **36**: 471–83.

Grant, P. R. (1986). *Ecology and Evolution of Darwin's Finches.* Princeton University Press: Princeton, N.J.

Grant, P. R. and Grant, N. (1979). Breeding and feeding of Galápagos mockingbirds, *Nesomimus parvulus. Auk* **96**: 723–36.

Gulledge, J. L. (1975). A study of the phenetic and phylogenetic relationships among the mockingbirds, thrashers and their allies. Ph.D. thesis, City University of New York.

Hatch, J. J. (1966). Collective territories in Galápagos Mockingbirds, with notes on other behavior. *Wilson Bull.* **78**: 198–207.

Kinnaird, M. F. and Grant, P. R. (1982). Cooperative breeding by the Galápagos Mockingbird, *Nesomimus parvulus. Behav. Ecol. Sociobiol.* **10**: 65–73.

Koenig, W. D. and Pitelka, F. A. (1981). Ecological factors and kin selection in the evolution of cooperative breeding in birds. In *Natural Selection and Social Behavior*, ed. R. D. Alexander and D. W. Tinkle, pp. 261–80. Chiron Press: New York.

Plage, D. and Plage, M. (1988). Galápagos wildlife under pressure. *Nat. Geog.* **173**: 122–45.

Rabenold, K. N. (1985). Cooperation in breeding by nonreproductive wrens: kinship, reciprocity, and demography. *Behav. Ecol. Sociobiol.* **17**: 1–17.

Ralls, K., Harvey, P. H. and Lyles, A. M. (1986). Inbreeding in natural populations of birds and mammals. In *Conservation Biology*, ed. M. E. Soulé, pp. 35–56. Sinauer: Sunderland, MA.

Steadman, D. W. (1986). Holocene vertebrate fossils from Isla Floreana, Galápagos. *Smithsonian Contrib. Zool.* **413**: 1–103.

Swarth, H. S. (1931). The avifauna of the Galapagos Islands. *Occasional Papers Calif. Acad. Sci.* **18**: 1–299.

Woolfenden, G. E. and Fitzpatrick, J. W. (1978). The inheritance of territory in group-breeding birds. *BioScience* **28**: 104–8.

Woolfenden, G. E. and Fitzpatrick, J. W. (1984). *The Florida Scrub Jay.* Princeton University Press: Princeton, N.J.

Zack, S. and Ligon, J. D. (1986). Cooperative breeding in Lanius Shrikes. II. Maintenance of group-living in a nonsaturated habitat. *Auk* **102**: 766–73.

11 GROOVE-BILLED ANIS: JOINT-NESTING IN A TROPICAL CUCKOO

11

Groove-billed Anis: joint-nesting in a tropical cuckoo

R. R. KOFORD,* B. S. BOWEN* AND S. L. VEHRENCAMP

Species and study area

The Groove-billed Ani (*Crotophaga sulcirostris*) has a rare type of breeding system that sets the species apart from most other cooperative breeders. Breeding groups consist of two or more females and their mates. Females in these groups lay their eggs in a single nest and the joint clutch is cared for by all members of the group. This type of breeding system is known as joint-nesting plural breeding (Brown 1978). Plural breeders are species in which social units may contain two or more breeding females. Joint-nesting occurs when these females lay their eggs in a single nest. Other examples of joint-nesting plural breeders are the Acorn Woodpecker (*Melanerpes formicivorus*) and the Pukeko (*Porphyrio porphyrio*), which are discussed in Chapters 14 and 13, respectively. Those chapters and this one illustrate the considerable differences that exist among the species that share this breeding system, particularly in their mating systems and the numbers of non-breeding helpers.

Joint-nesting in the Groove-billed Ani probably did not evolve recently. The Smooth-billed Ani (*C. ani*), the Greater Ani (*C. major*) and the Guira (*Guira guira*) are also joint-nesting plural breeders (Davis 1940a,b, 1941; R. Macedo, personal communication). All four of these species are members of the Crotophaginae, in the family Cuculidae. Cuckoos have a range of breeding systems, including monogamy, polyandry and interspecific brood parasitism.

The distribution of the Groove-billed Ani extends from southern Texas to northern Argentina. In Costa Rica, Panama and South America, its distribution overlaps that of the Smooth-billed Ani, which is apparently

* Present address: Northern Prairie Wildlife Research Center, Jamestown, ND 58402, USA.

expanding its range northward. We observed Smooth-billed Anis at our study site in Guanacaste province, Costa Rica, where they had not been observed earlier. In addition to part of Central America, the distribution of the Smooth-billed Ani extends to northern Argentina and to Florida via the Greater and Lesser Antilles. The Greater Ani occurs in Panama and South America, and the Guira is confined to South America.

The Groove-billed Ani inhabits primarily open country, although individuals living adjacent to closed-canopy forest may use forest trees as roost sites during the day. In its habitat requirements, it is similar to the other members of the Crotophaginae, all of which live in marshes, along the banks of rivers and lakes, or in grassy areas with scattered trees (Davis 1942). The Groove-billed Ani and Guira are found in drier conditions more often than are the others, and the Greater Ani seems to be tied most closely to water. The destruction of forests that has accompanied the expansion of agriculture and ranching in Central and South America has increased the amount of savannah-like habitat for these species, as has the settlement of grasslands (Davis 1940*b*). Drainage of marshes during the same period has decreased the amount of mesophytic habitat. In Costa Rica, Groove-billed Anis are abundant and nearly ubiquitous in open country at low elevations. The life history of the species was described first by Skutch (1959).

We studied the Groove-billed Ani in north-western Costa Rica, an area that experiences a severe dry season for almost half the year. Most of the original tropical dry-deciduous forest in this area has been cleared and converted to cattle ranches and farms. Away from the remaining stands of forest, trees occur in gallery forests along rivers, in 'living fences', and in pastureland, where they provide shade for cattle and other animals (Fig. 11.1). The pastures typically have a savannah-like appearance, with areas that have been invaded by thorny scrub (primarily *Mimosa*). A study in 1971–3 focused on a comparison between pastureland and marshy areas (Vehrencamp 1978) and laid the groundwork for another study, in 1978–82. Because the marshy areas were drained before the later study started, the focus was shifted to a closer examination of the pastureland.

Typical food items for anis are orthopterans, other insects, spiders and small vertebrates that occur on or under grass and bushes. Individuals forage by walking or hopping along looking for prey. Grasshoppers and other insects that take flight when disturbed by a foraging ani are chased and often caught in the air. Anis sometimes follow livestock, chasing prey that move or fly up. Hard-bodied prey are held in the bill and macerated by repeated rapid bites before being swallowed or fed to the young. Unless they are in very short grass, foraging individuals probably cannot see other individuals much of the time. The difficulty of maintaining visual contact would make direct cooperative foraging difficult. However, during foraging bouts individuals periodically fly to perches on fences or trees. There, they can locate other individuals and perhaps find profitable foraging areas. Members of a social unit usually forage near each other. Although not always within sight of each other, they communicate through contact calls and other vocalizations. They frequently call while flying, thus revealing their location to others. If threatened, alarm calls either bring the social unit together as they fly to high perches or cause them to stay low in the vegetation.

Social organization

Social units consisted typically of a breeding unit, either a pair or a communal group comprising two or three (rarely four) pairs, and their immature offspring. About 20% of the units observed had an odd number of adults, some of which did not breed. The distribution of breeding unit

Fig. 11.1. The study area, showing the open habitat favored by anis.

sizes in our study is shown in Fig. 11.2. During the breeding season most anis formed monogamous pair bonds. These bonds were most obvious during the egg-laying period, when males defended their mates from the advances of other males and when members of different pairs tended not to associate. Offspring remained in their natal units for the remainder of the breeding season in which they hatched. The breeding season corresponded to the wet season, June through December, when prey items were most abundant. During the breeding season, there could be multiple nesting attempts, with some breeding units raising two or three broods. Each year, we could expect to see a couple of groups with more than 20 members by the end of the breeding season. The immature offspring from early broods helped to feed nestlings and fledglings from later broods. During the non-breeding season, which was dry and windy, the size of social units decreased as most of the young hatched the previous year disappeared from their natal units.

During the five breeding seasons of our study, we monitored the dispersal fates of 196 young anis that had survived until the end of the breeding season in which they hatched. We refer to these potential dispersers as 'independent young'. At the beginning of the subsequent breeding season, 48 (25%) of the independent young were still in the study area. Of those, 25 (13% of all independent young) dispersed and bred within the study area and 15 (8%) bred in their natal units. All but one of those that bred in their natal units were males. The remaining eight independent young in the study

Fig. 11.2. Distribution of ani breeding unit sizes. The number of adults was determined in 171 unit-years over five years.

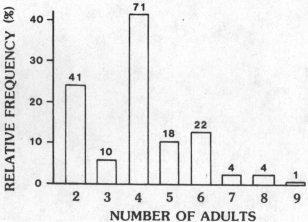

area (4%) remained in their natal units as non-breeders. Of the 196 independent young, 75% disappeared from the study area before the start of the subsequent breeding season. These must have died on the study area or dispersed. In Relatedness and dispersal (see below) we estimate the percentage that we think bred successfully outside of our study area.

Most of the adult anis in our study area bred in communal groups. We knew nothing about the formation of groups that existed when we began our study. Many individuals and pairs joined communal groups during the study, often following the disappearance of a breeder, but these changes in group composition usually did not change the size of the breeding unit. We occasionally observed the formation of a communal group on a territory that had previously been occupied by a pair. In three cases, the original pair produced offspring in one year and, in the next year, bred as a communal group with one of their sons and his mate, which had been attracted from outside the unit. In other cases, communal groups formed when two pairs coalesced into a group of four; the adults were not known to be related.

Territorial spacing was apparent among social units during the breeding season. All members of a social unit used the entire territory. Territorial aggression was rarely observed. When we observed it, it accompanied a shift in a territorial boundary and involved a breeding male from each group displaying close to the boundary. Social units generally stayed within their territories during the breeding season. The major exception to this pattern involved units that successfully fledged many young. The large social units that resulted tended to wander more widely than they had earlier in the season. During the dry season, home range overlap among social units generally was greater than it was during the breeding season. In some years, groups in the drier parts of the study area temporarily abandoned their territories during the dry season, going to the nearby gallery forest to forage.

Spatial variation in social organization occurs along with this temporal variation. Within the space of a few hundred meters, breeding units can vary in size from two to eight and territory size can vary from 1 to 10 ha ($\bar{x} \pm$ s.e.m. $= 3.3$ (± 0.33), $n = 49$). Marshy areas with low brushy vegetation tend to have large breeding units and small territory sizes relative to pastureland (Vehrencamp 1978). The population density is thus much higher in the marshy areas, suggesting that these areas are preferred. The somewhat higher prey abundance in these areas does not appear to be sufficient to account for the difference in density. Our later study, which concentrated on variation within the pastureland, also revealed relationships between breeding unit size, territory size, and ecological features such as the amount of tree cover on the territory (Koford *et al.* 1986), but these relationships were not as strong as those between habitat types.

Unit size and fitness

We used the variation in size of the breeding units to examine how unit size was related to survival rates and reproductive success. The relationships between unit size and fitness components were used as an indication of how a change in the size of a breeding unit would affect the fitness of the members of the unit, i.e. whether a given change in breeding unit size would be favored by natural selection. Because fitness components were correlated with characteristics of the territories (e.g. territory size was related directly to reproductive success), we performed multivariate analyses to determine the independent effect of unit size on fitness components.

The first fitness component we considered was reproductive success. Our measure of reproductive success was determined by dividing the number of young birds surviving to the end of the breeding season (i.e. 'independent young') by the number of females in the breeding unit at the beginning of the breeding season. We used this measure, the average annual reproductive success per female (rather than per adult), because the size of a communal clutch is closely tied to the number of breeding females. For the same reason, the statistical analyses use the number of females as an index of breeding unit size. Figure 11.3 shows the relationship between reproductive success and the number of females in a breeding unit. In multivariate analyses using reproductive success as the dependent variable, we used the values of several independent variables that were determined

Fig. 11.3. Annual reproductive success (mean and estimated s.e.m.) in relation to number of female anis in the breeding unit. The number of young is the number, at least one month old, that were alive at the end of the breeding season.

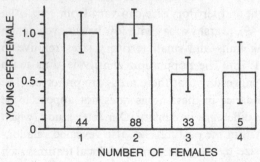

each year for each breeding unit. These variables included the number of females in the breeding unit, territory size, the number of trees in the territory, the areal extent of the tree canopy in the territory, percentage tree cover in the area, percentage brush cover, the number of sites known to have been used for nesting and the number of sites judged to be suitable for nesting. In a stepwise multivariate analysis that treated number of females as a categorical variable (thus allowing for a non-linear effect by this variable), we found no significant relationship with reproductive success. In a stepwise multivariate linear regression analysis (no allowance for possible non-linear effects), unit size had a significant negative effect on reproductive success (Vehrencamp *et al.* 1988). Another analysis suggested that the quality of the territory determined whether or not reproductive success decreased with increasing breeding unit size. There was no decrease on territories with large amounts of tree canopy area, whereas there was a decrease on territories with small amounts of tree canopy area (Koford *et al.* 1986).

The second fitness component we investigated, survival rate, was analysed only for the breeding season. During the dry season, meaningful analysis was precluded by the variable sizes of the social groups and by movements of groups. For both sexes, our samples exhibited trends of increases in survival rate during the breeding season with increases in breeding unit size, as indexed by the number of females in each breeding unit (Figs. 11.4 and 11.5). Suspecting that survival was probably related to incubation effort, which generally decreases as unit size increases, we analyzed breeding season survival with an ordered χ^2 contingency test that

Fig. 11.4. Survival rates of adult female anis during the breeding season in relation to number of females in the breeding unit.

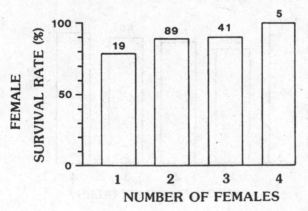

takes into account possible linear effects (Everitt 1977). We found no significant relationship between survival rate and unit size, either with the sexes pooled ($\chi^2 = 3.04$, 1 d.f. $n = 411$, $0.05 < P < 0.1$) or tested separately. In a previous analysis with the sexes tested separately, females were found to have higher survival rates in larger units (Vehrencamp *et al.* 1988). The previous analysis used unit size classes that were based on the number of adults in a unit, rather than the number of females. Larger sample sizes are needed to resolve the inconsistency between these analyses.

These analyses suggest that there are complex relationships between unit size, reproductive success, survival rate and territory quality. Breeders may increase their likelihood of surviving by allowing the size of their breeding unit to increase. However, an increase in unit size may lead to a decrease in reproductive success unless the birds are on one of the uncommon territories with a high amount of tree cover.

Breeding roles and fitness

Further complexity in the relationships discussed above is revealed when we account for individual differences in reproductive success and survival among the members of communal groups. In the analyses discussed above, our measure of annual reproductive success per female for each breeding unit was an average value. This average is not a true reflection of individual reproductive success, as not all females in communal groups contribute equally to the clutch (Vehrencamp 1977). Individual males and females in communal groups tend to be consistent over time in certain behavioral characteristics related to nesting. We refer

Fig. 11.5. Survival rates of adult male anis during the breeding season in relation to number of females in the breeding unit.

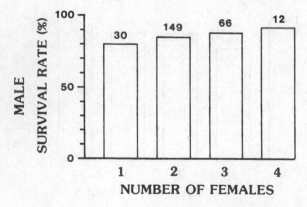

to these behavioral characteristics as breeding roles. A complete understanding of how individual fitness is affected by joint-nesting requires knowledge of the relative egg contributions of females with different roles, the relationship between breeding role and survival, and the pairing pattern between females and males with different breeding roles.

We distinguished breeding roles for females on the basis of the order in which laying was initiated. From nest to nest within a season, this order was consistent except when rapid re-nesting occurred following the loss of a nest during the egg-laying stage, when the order appeared to be random. In a three-pair group, there would thus be a first, second and last layer. A two-pair group would have a first and last layer. Females removed (tossed) the eggs of earlier layers from the nest before they began laying (Vehrencamp 1977). Once all females were laying, the clutch accumulated rapidly, with each female contributing three to five eggs at the rate of one every other day. The number of eggs laid, tossed and incubated for females of different laying orders is shown in Table 11.1. Early layers had the disadvantage of having to lay more eggs in all, in addition to being late in starting to contribute to the incubated clutch. As a result, last layers contributed significantly more eggs to the clutches that we monitored. The relatively small differences among joint-nesting females in their contributions to communal clutches indicate that the main effect of tossing is to equalize the contributions of the females. If a last layer did not toss, her contribution to a communal clutch by the time of initiation of incubation would be much smaller than that of the first layer.

Table 11.1. *Number of eggs laid, tossed, and incubated* $\pm$ s.e.m. *for two-female and three-female groups of anis as a function of laying order*

Laying order	Laid	Tossed	Incubated
Two-female groups ($n = 51$ nests)			
First	5.98 ± 0.19	1.94 ± 0.17	4.04 ± 0.15
Second	4.75 ± 0.12	0.08 ± 0.05	4.67 ± 0.13
Paired t-test	$t = 5.81$	$t = 11.24$	$t = 3.64$
analysis	$P < 0.001$	$P < 0.001$	$P < 0.001$
Three-female groups ($n = 9$ nests)			
First	6.22 ± 0.32	2.78 ± 0.55	3.44 ± 0.38
Second	5.56 ± 0.47	1.78 ± 0.43	3.78 ± 0.40
Third	4.56 ± 0.18	0	4.56 ± 0.18
Regression	$F = 78.7$	$F = 11.9$	$F = 5.72$
analysis	$P < 0.001$	$P < 0.005$	$P < 0.025$

The relationship between breeding role and survival through the breeding season differed for the two sexes. We found no significant effect of breeding role on survival for females, but male survival through the breeding season was significantly related to breeding role. We have distinguished two male breeding roles: nocturnal incubators and those that are not nocturnal incubators. Males perform all of the nocturnal incubation, whether in pairs or communal groups, and only one male in a communal group has the role of nocturnal incubator. Both types of males, as well as females, perform diurnal incubation. Nocturnal incubators (including those in pairs) survived to the end of the breeding season at a rate lower than that of males that were not nocturnal incubators (Fig. 11.6, $\chi^2 = 4.35$, $n = 194$, 1 d.f., $P < 0.05$). This difference is apparently due to nocturnal nest predation, when the predator takes both the clutch and the incubator (Vehrencamp *et al.* 1988).

Nocturnal incubators were more likely to be paired with last layers (68% of 50 groups) than with other females (one-tailed binomial test, $P = 0.008$). In some cases, nocturnal incubators were paired with early-laying females that contributed more eggs than the last layer, resulting in an even greater average reproductive advantage for the nocturnal incubator. In 32 groups in which the females had unequal contributions to the incubated clutches, the nocturnal incubator's mate had the greater contribution in 25 cases (76%; $\chi^2 = 7.74$, 1 d.f., $P < 0.01$). This reproductive advantage for nocturnal

Fig. 11.6. Survival rates of adult male anis during the breeding season in relation to breeding role. Nocturnal incubators bred in pairs or communal groups; males that were not nocturnal incubators necessarily bred in groups.

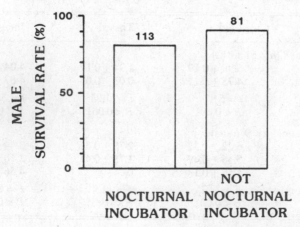

incubators in communal groups may offset the relatively high mortality risk faced by them, compared to other males in the group.

Relatedness and dispersal

Another subject to be considered in evaluating how breeding in communal groups affects fitness is kin selection. If the members of a communal group are related, behaviors such as tossing eggs and performing the nocturnal incubation can affect the indirect component of inclusive fitness (Brown and Brown 1981). We obtained information on levels of relatedness by marking adults and nestlings and following these individuals through time. As reported above, about 25% of the independent young bred in the study area in their first breeding season. About one half of them dispersed within the study area and the remainder stayed in their natal groups. We found no evidence of joint dispersal of nestmates, such as occurs in the Acorn Woodpecker (Koenig and Pitelka 1979). The young birds that bred in their natal units either replaced a breeder that had disappeared or caused an increase in the size of the breeding unit. These birds, mostly males, usually mated with immigrant females, but two cases of possible inbreeding were detected (Bowen *et al.* 1989). In one case a young male remained in his natal group following the disappearance of an adult male during the non-breeding season. The young male mated with one of the two adult females that had bred the previous year. Because this was a communal group, there is a probability of approximately 0.5 that he mated with his mother. This group remained stable for the next four breeding seasons, producing several successful clutches. In the other case, two brothers and a sister inherited their parental territory following the disappearance of their parents. Together with two immigrant females, these birds bred communally and produced at least two clutches of eggs. Observations of the sister indicated that she was gravid during one of the nesting attempts (the lower abdomen of an ani female about to lay an egg has a characteristic bulge). We conclude that sibling–sibling mating occurred in this group.

Our dispersal data indicate that relatedness is not extremely high in ani groups and suggest that indirect selection does not play a central role in affecting reproductive competition or cooperation among group members. The only possible effect of kinship that we observed concerned male breeding roles in groups with presumed fathers and sons. Males that bred in their natal units were always non-nocturnal incubators (breeding roles were known for six such groups); males that dispersed could have either role. Because nocturnal incubators in groups do not have lower survival rates than nocturnally incubating males in pairs, parental males that perform the nocturnal incubation presumably do not have a decreased survival rate after accepting a son into the unit. Non-dispersing sons avoid any possible cost of dispersal. The overall effect on the fitness of fathers and sons depends on the reproductive success of the larger unit, which varies with territory quality. If reproductive success does not decline with increasing breeding unit size, fathers and sons may both benefit. Relatedness due to males remaining in their natal units may also affect selection on a mother's behavior if she removes eggs from the nest. In a two-female breeding unit, the eggs she tosses belong to her son's mate. The importance of this kind of relatedness obviously depends on the percentage of young males that breed in their natal units. We do not have a direct measure of this percentage, as we do not know how many young males bred outside of the study area.

We have, however, developed a population model that provides indirectly an estimate of the percentage of young birds that bred in their natal units, and that dispersed and bred away from their natal territories. Dispersers unable to acquire breeding status either die or become floaters. In our study area, we only detected a floater occasionally, and these were adults that had bred previously. The model treats a breeding population of constant size that is in equilibrium with surrounding populations. Dispersing individuals can leave the population (emigrate) or enter it (immigrate). The number of emigrants that are successful in establishing themselves as breeders equals the number of immigrants that breed. Breeding opportunities are made available only by mortality among the breeders. Non-breeders occur only in social units, not as floaters. Dispersal occurs during a distinct period between breeding seasons. During this dispersal period, young birds have the same daily survival rate as adults as long as they are on their natal territories and once they are established on a new territory. We believe that this model is essentially representative of the real population we studied.

Empirical data from 1978 to 1982 provided the values we used in this model. Annual reproduction was summarized by the geometric mean; for every breeder, 0.45 independent young were produced. We assume a 1:1 sex ratio among these young. The estimated annual mortality rate averaged 0.28 deaths per adult. The actual rate of disappearance of adults was somewhat higher than this, but we attributed some disappearances to dispersal. Allowing for 87% survival of these young on their natal territories during the non-breeding season, the number alive at the start of the subsequent breeding season would have averaged 1.40 times the number of breeding opportunities if there were no mortality related to dispersal.

Our estimates of the fates of all independent young are presented in Table 11.2. We estimate that a young female was about twice as likely to breed outside of our study area as was a young male. If we consider only those birds that we estimate were alive (breeding or not) at the beginning of the breeding season after they hatched, we estimate that 21% of males and 2% of females bred in their natal groups, 32% of males and 6% of females dispersed within the study area, 41% of males and 86% of females dispersed out of this study area. For both sexes, 6% remained in their natal groups as non-breeders.

Habitat saturation

An important factor selecting for cooperative breeding in many singular breeders is habitat saturation or, more generally, ecological constraints (Brown 1974; Emlen 1982). A population is said to exist under conditions of habitat saturation if the appropriate breeding habitat is filled with breeders, i.e. if territorial behavior prevents some young from breeding. We examined the possibility that our study population existed under conditions of habitat saturation.

Some evidence suggests that the population that we studied existed under conditions of habitat saturation (Koford *et al.* 1986). First, more young were produced each year than the number of adult deaths, creating an excess number of young. With 1.40 potential dispersers per breeding vacancy created by adult mortality, the overall probability of obtaining a breeding opportunity was 0.72. Second, the density of breeders remained quite stable over a period of five years (Fig. 11.7; $CV = 6.7\%$) despite considerable variation in reproductive success among years. The average numbers of independent young produced per adult in 1978–82 were 1.19, 0.35, 0.29, 0.58 and 0.25, respectively ($CV = 132\%$). This stability in density of breeders suggests that there was density-dependent mortality of

Table 11.2. *Estimated distribution of dispersal fates of independent young anis the year after they hatch, based on a demographic model*

	Males	Females
Breeder in the natal unit	0.14	0.01
Successful dispersal within the study area	0.21	0.04
Successful dispersal out of the study area	0.27	0.57
Non-breeder in the natal unit	0.04	0.04
Death	0.34	0.34

dispersers. Third, we often observed aggression and repulsion of dispersers in the months preceding the breeding season. These three lines of evidence suggest that there is reproductive competition for breeding opportunities and that some young are prevented from breeding.

Other evidence suggests that the level of reproductive competition was lower in anis than in 'typical' habitat-saturated cooperative breeders such as the singular-breeding Florida Scrub Jay (*Aphelocoma c. coerulescens*; Woolfenden and Fitzpatrick 1984, and see Chapter 8). Anis apparently do not occupy suitable habitat as fully as do the jays. Parts of our study area with more than 5% tree cover were essentially fully occupied, whereas those with less than 5% tree cover were not (Koford *et al.* 1986). Communal groups had good reproductive success on territories with high and medium amounts of tree canopy area and poor reproductive success on territories with low amounts of tree canopy area. Pairs had good reproductive success on territories with high, medium and low amounts of tree canopy area. Some of the unoccupied parts of the study area with less than 5% tree cover therefore may have been suitable for breeding by pairs, but probably not by communal groups. The amount of suitable habitat available for colonization would be inversely related to the amount of reproductive competition for breeding opportunities among young birds. We did not determine how much of the unoccupied habitat was suitable for breeding by pairs. The possibility therefore exists that a substantial amount was suitable and that the habitat was not saturated.

The probability of dispersing successfully is only one component of the inequality that expresses the conditions under which dispersal or non-

Fig. 11.7. The number of adult anis in parts of the study area that were studied in all five years.

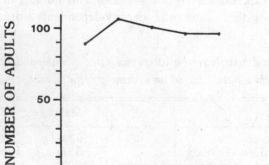

dispersal will be favored for newly mature individuals. Non-dispersal and helping as a non-breeder will be favored over dispersing and breeding as a solitary pair when

$$dw_s r_s < \Delta w r_n,$$

where d is the probability of dispersing successfully, w_s is the fitness of a disperser, r_s is the relatedness of the disperser to its young, Δw is the increment in fitness caused by the helper, and r_n is the relatedness of the helper to the nestlings it assists in raising. If d is very small (e.g. 0.5), non-dispersal can be favored even if there is little increment in fitness caused by the helper. This is the condition envisioned by the habitat saturation model as it has been applied to singular breeders; this condition may occur in the Florida Scrub Jay. If d is relatively high (e.g. 0.75) and the helper provides a slightly larger increment, non-dispersal also will be favored. This may be the condition for some other species with non-breeding helpers. Considering a hypothetical ani population in which all breeding units are pairs and in which rates of reproduction and survival are the same as we observed in our study population, d would be relatively high (0.72). There was no evidence of any positive increment in fitness with an additional non-reproductive individual. Annual reproductive success was 1.0 ± 0.11 ($\bar{x} \pm$ S.E.M., $n = 154$) young/female for all breeding units with even sex ratios and 0.89 ± 0.28 ($n = 18$) young/female for all groups with an extra male or apparent non-breeder of unknown sex. Under these conditions, dispersal is favored over remaining in the natal unit as a non-breeder. We found that most young dispersed. Many of those that did remain as non-breeders were exceptionally young, having been hatched in the last month of the breeding season. For them, d may have been lower than average, making it advantageous to postpone dispersal until a time when d was higher.

Young anis have another option: they can attempt to remain in the natal unit and attract a mate from outside the group. Whether they attempt to remain will be dictated by the same fitness inequality we applied to non-breeding helpers above. For this inequality, the increment in group fitness (Δw) is via the direct (individual) fitness of the young bird. If the young bird is a male retained by a pair on a territory with large tree canopy area, on which average reproductive success is the same for pairs and two-pair groups, his fitness (Δw) is equal to that of a pair male (w_s). Because r_s equals r_n in this case, the inequality would favor remaining in the natal group. On a territory with tree canopy area small enough that his expected reproductive output is below the level needed to balance the probability of dispersing successfully (0.72), the inequality would favor leaving. He probably would never get to make that choice, because at a level of territory quality where he breaks even, his parents will benefit by excluding him from the breeding unit.

Behaviorally dominant older adults might not allow a young male to remain even if he attempted to do so. Adults would be expected to allow a male to remain if their territory had a large tree canopy area and if he was closely related to them. On a territory with large tree canopy area, on which average reproductive success is the same for pairs and communal groups, the fitness of a pair would be increased by allowing one of their offspring to remain and breed on the territory. Their survival rates might be increased and they would benefit by ensuring that at least one of their offspring would not face a dispersal risk. Communal groups, regardless of the amount of tree canopy area on the territory, would be less inclined than pairs to allow a given male to remain, as all but two of the adults would not be his parents. Under the conditions of our study, in which relatively few pairs occupied territories with medium or large tree canopy area (Koford *et al.* 1986), one would expect to observe few cases of young males remaining in their natal units and thereby increasing the size of the breeding unit. Most young males and females would benefit by dispersing. They could find either unoccupied habitat that would support a breeding unit or occupied habitat inhabited by communal groups that had reproductive vacancies.

Together, our analyses of reproduction, mortality, habitat relations and social behavior reveal multiple advantages of breeding in communal groups. Shared diurnal incubation may lower the mortality rate of breeders in communal groups. Males that do not perform the nocturnal incubation are spared the risk that accompanies that behavior. Nocturnal incubators are compensated for their increased risk by fathering a larger proportion of the group's offspring than the other males. Finally, parents that allow one of their offspring to remain and breed on their natal territory save the offspring from facing the risk of dispersal. The advantages associated with incubation would favor the evolution of joint-nesting plural breeding. The advantage associated with the risk of dispersal would favor the evolution of plural breeding, although not necessarily joint-nesting. If habitat saturation played a role in the evolution of joint-nesting, it did so by facilitating the operation of some other advantage(s).

Comparison and discussion

Why does the Groove-billed Ani have this breeding system? The system presumably evolved long ago, probably in some common ancestor of the anis. The system has been maintained under rapidly changing ecological conditions in many parts of the ranges of the species, suggesting that the operation of strong selective factors favoring joint-nesting. Our analyses indicate that birds breeding in groups may benefit by having higher rates of survival, that groups and pairs have approximately the same per capita reproductive success on good territories, and that habitat saturation may restrict breeding opportunities.

Joint-nesting must have evolved independently several times, as it occurs in several avian families. This independent evolution raises the possibility that this type of breeding system has resulted from evolutionary convergence. Convergence could be inferred if all joint-nesters were known to share some unusual ecological condition or behavioral characteristic that had been important in shaping the breeding system. Is there any such condition or characteristic shared by joint-nesters that makes these species distinct from other species, particularly other cooperative breeders?

Three ecological conditions were found to be associated with advantages for joint-nesting anis. First, the risk of predation while incubating, during the day and night, is important because it creates advantages associated with joint-nesting and shared incubation. However, as the rate of nest predation is relatively high in most tropical avian species, with the possible exception of cavity nesters, this condition is probably not very distinctive. Female survival is enhanced somewhat in communal groups of the Acorn Woodpecker (Koenig and Mumme 1987, and see Chapter 14), but this survival difference is unlikely to be related to nest predation, which is relatively rare in this species. Second, habitat saturation may have facilitated the evolution of joint-nesting in anis. The level of reproductive competition in our study population was apparently less than in the Florida Scrub Jay (Woolfenden and Fitzpatrick 1984), but a low level of reproductive competition does not characterize all joint-nesters. The Acorn Woodpecker (Stacey 1979; Koenig and Pitelka 1981) and, perhaps, the Pukeko (Craig 1979) exhibit habitat saturation with a higher level of reproductive competition. Non-breeding helpers are common in the Acorn Woodpecker and floaters are common in the Pukeko.

The third ecological condition we found to be important is the existence of ani territories that will support more than one breeding pair with little or no decrease in reproductive success compared to one pair. This ability to support multiple pairs is not likely to be due solely to territory size increasing proportionately as unit size increases; larger units still have only one primary defender, the male that performs the nocturnal incubation. If a territory of a given size were much larger than the size necessary to contain sufficient food resources for one breeding pair, it might be able to support additional breeders with no increase in size. Vehrencamp (1978) found that, in the densely populated marshy areas, the rate of nest predation was directly related to nest density. Anis might benefit by defending large territories, thereby reducing nest density. The observation that ani territories may be larger than needed to supply food requirements is not unique among cooperative breeders. Woolfenden and Fitzpatrick (1984) similarly report that territories of the Florida Scrub Jay seem to be larger than needed. We suspect that this condition may characterize most cooperative breeders, especially those with year-round territoriality. In species that experience annual variation in food abundance, the size of food-based territories is likely to fluctuate less in species with year-round territoriality than in species with seasonal or feeding territories. The size of feeding territories would be influenced mainly by intruder pressure (e.g. Myers *et al.* 1979), whereas the size of year-round territories would have the additional influence of traditional, long-established boundaries. In many Acorn Woodpecker populations, territories are not only held year-round, but they have fixed granary sites. Both of these factors would limit the ease with which new territories could be squeezed between existing territories in years of good acorn crops. Acorn Woodpeckers therefore might have unusually stable territorial boundaries that would facilitate the evolution of joint-nesting. In the case of the Groove-billed Ani, there was no indication that territorial boundaries were unusually stable relative to singular-breeding cooperative breeders.

Although no ecological condition clearly distinguishes joint-nesters from other cooperative breeders, one behavioral characteristic does: male incubation. Male Rheas (*Rhea americana*) perform all of the parental care (Bruning 1974). Male Ostriches (*Struthio camelus*) perform the nocturnal incubation duties and a major hen incubates during the day (Bertram 1980). Males perform much of the diurnal incubation and all of the nocturnal incubation in the Magpie Goose (*Anseranas semipalmata*; S. Davies, personal communication concerning captive birds), Moorhen (*Gallinula chloropus*; Gibbons 1986), Pukeko (Craig 1980), Acorn Woodpecker (MacRoberts and MacRoberts 1976), Groove-billed Ani (Skutch 1959), and presumably the other anis. Some cooperative breeders without nocturnally-incubating males do exhibit joint-nesting: the Brown Jay (*Cyanocorax morio*; Lawton and Lawton 1985), Gray-crowned Babbler (*Pomatostomus halli*: Counsilman 1977; Brown and Brown 1981), and White-winged Chough (*Corcorax melanorhamphus*; Rowley 1978). In these species, however, joint-nesting is confined to some populations and is uncommon where it does occur. Cooperative breeders that do not nest jointly do not exhibit a strong male commitment to incubation.

How might male incubation lead to joint nesting? The evolutionary route to joint-nesting could have started either with intraspecific brood parasitism of nesting pairs or with the admission of additional individuals to social units prior to nesting. Parasitism is known to occur in several joint-nesters: the Groove-billed Ani (Vehrencamp 1978), the Ostrich (Bertram 1980) and the Moorhen (Gibbons 1986). Male incubation might make it easier for females to parasitize nests. A female whose mate was incubating a newly completed clutch would have more time and energy to invest in parasitizing nearby nests than a female that had to perform more of the incubation. For a laying female that was going to perform little of the incubation, the cost of having her nest parasitized might be less than it would be otherwise. Brood parasitism could be used to gain entry into a breeding unit in high-quality habitat that was filled with territories. Once a pair had been parasitized, their best option might be to let the parasite (and perhaps a mate) into the social unit, so that they could get assistance in caring for the young. The other possible route, involving pairs joining an existing breeding unit, might occur if physiological changes in males, related to incubation, decrease their aggressiveness in territorial defense. Especially under conditions of habitat saturation and a shortage of good nest sites, a pair might seek entry by providing assistance in caring for the offspring of the breeding unit that they were attempting to enter. Ultimately, a breeding pair would only be expected to allow such entry if their fitness were not thereby decreased.

Speculation about the factors contributing to the evolution of joint nesting is valuable if it leads to testable hypotheses. Our tests have resolved some issues and suggested further tests. We found some support for the hypothesis that joint-nesting females have higher rates of survival, presumably because of the shared incubation in communal groups. It is very likely that joint-nesting originated in anis under conditions of high rates of nest predation. We found inconclusive evidence concerning the hypothesis that habitat saturation limits breeding opportunities for young anis. Observations should be made on the suitability of unoccupied habitat for breeding and on the survival rate of breeders on territories with low amounts of tree canopy cover. The hypothesis that certain ani territories held by pairs can support additional breeders with little change in size should be tested using careful observations on territories that appear to be of high quality. The idea that male incubation facilitates joint-nesting requires additional development and might be tested by comparing normal and doubled clutches in species with male incubation and species with only female incubation. Such tests and field manipulations should help to resolve hypotheses developed from observed correlations between joint-nesting and possible causal factors.

Bibliography

Bertram, B. C. R. (1980). Breeding system and strategies of Ostriches. *Proc. 17th Intern. Ornith. Congr.* 890–4.

Bowen, B. S., Koford, R. R. and Vehrencamp, S. L. (1989). Dispersal in the communally breeding Groove-billed Ani (*Crotophaga sulcirostris*). *Condor* 91: 52–64.

Brown, J. L. (1974). Alternate routes to sociality in jays – with a theory for the evolution of altruism and communal breeding. *Amer. Zool.* 14: 63–80.

Brown, J. L. (1978). Avian communal breeding systems. *Ann. Rev. Ecol. Syst.* 9: 123–155.

Brown J. L. and Brown, E. R. (1981). Kin selection and individual selection in babblers. In *Natural Selection and Social Behavior: Recent Research and New Theory*, ed. R. D. Alexander and D. W. Tinkle, pp. 244–56. Chiron Press: New York.

Bruning, D. F. (1974). Social structure and reproductive behavior in the Greater Rhea. *Living Bird* 13: 251–94.

Counsilman, J. J. (1977). Groups of the Grey-crowned Babbler in central southern Queensland. *Babbler* 1: 14–22.

Craig, J. L. (1979). Habitat variation in the social organization of a communal gallinule, the Pukeko, *Porphyrio porphyrio melanotus. Behav. Ecol. Sociobiol.* 5: 331–58.

Craig, J. L. (1980). Pair and group breeding behaviour of a communal gallinule, the Pukeko, *Porphyrio p. melanotus. Anim. Behav.* 28: 593–603.

Davis, D. E. (1940a). Social nesting habits of the Smooth-billed Ani. *Auk* 57: 179–218.

Davis, D. E. (1940b). Social nesting habits of *Guira guira. Auk* 57: 472–84.

Davis, D. E. (1941). Social nesting habits of *Crotophaga major. Auk* 58: 179–83.

Davis, D. E. (1942). The phylogeny of social nesting habits in the Crotophaginae. *Quart. Rev. Biol.* 17: 115–34.

Emlen, S. T. (1982). The evolution of helping. I. An ecological constraints model. *Amer. Nat.* 119: 29–39.

Everitt, B. S. (1977). *The Analysis of Contingency Tables.* John Wiley & Sons: New York.

Gibbons, D. W. (1986). Brood parasitism and cooperative nesting in the Moorhen, *Gallinula chloropus. Behav. Ecol. Sociobiol.* 19: 221–32.

Koenig, W. D. and Mumme, R. L. (1987). *Population Ecology of the Cooperatively Breeding Acorn Woodpecker.* Princeton University Press, Princeton, NJ.

Koenig, W. D. and Pitelka, F. A. (1979). Relatedness and inbreeding avoidance: counterploys in the communally nesting Acorn Woodpecker. *Science* 206: 1103–5.

Koenig, W. D. and Pitelka, F. A. (1981). Ecological factors and kin selection in the evolution of cooperative breeding in birds. In *Natural Selection and Social Behavior*, ed. R. D. Alexander and D. W. Tinkle, pp. 261–80. Chiron Press: New York.

Koford, R. R., Bowen, B. S. and Vehrencamp, S. L. (1986). Habitat saturation in Groove-billed Anis (*Crotophaga sulcirostris*). *Amer. Nat.* 127: 317–37.

Lawton, M. F. and Lawton, R. O. (1985). The breeding biology of the Brown Jay in Monteverde, Costa Rica. *Condor* 87: 192–204.

MacRoberts, M. H. and MacRoberts, B. R. (1976). Social organization and behavior of the Acorn Woodpecker in central coastal California. *Ornith. Monogr.* 21: 1–115.

Myers, J. P., Connors, P. G. and Pitelka, F. A. (1979). Territory size in wintering Sanderlings: the effects of prey abundance and intruder density. *Auk* **96**: 551–61.

Rowley, I. (1978). Communal activities among White-winged Choughs, *Corcorax melanorhamphus*. *Ibis* **120**: 178–97.

Skutch, A. F. (1959). Life history of the Groove-billed Ani. *Auk* **76**: 281–317.

Stacey, P. B. (1979). Habitat saturation and communal breeding in the Acorn Woodpecker. *Anim. Behav.* **27**: 1153–66.

Vehrencamp, S. L. (1977). Relative fecundity and parental effort in communally nesting anis, *Crotophaga sulcirostris*. *Science* **197**: 403–5.

Vehrencamp, S. L. (1978). The adaptive significance of communal nesting in Groove-billed Anis (*Crotophaga sulcirostris*). *Behav. Ecol. Sociobiol.* **4**: 1–33.

Vehrencamp, S. L., Bowen, B. S. and Koford, R. R. (1986). Breeding roles and pairing patterns within communal groups of groove-billed anis. *Anim. Behav.* **34**: 347–66.

Vehrencamp, S. L., Koford, R. R. and Bowen, B. S. (1988). The effect of breeding-unit size on fitness components in Groove-billed Anis. In *Reproductive Success*, ed. T. H. Clutton-Brock, pp. 291–304. University of Chicago Press: Chicago.

Woolfenden, G. E. and Fitzpatrick, J. W. (1984). *The Florida Scrub Jay: Demography of a Cooperative-breeding bird*. Princeton University Press: Princeton, NJ.

12 GALAPAGOS AND HARRIS' HAWKS: DIVERGENT CAUSES OF SOCIALITY IN TWO RAPTORS

Galápagos Hawks

Harris' Hawks

12

Galápagos and Harris' Hawks: divergent causes of sociality in two raptors

J. FAABORG AND J. C. BEDNARZ*

Hawks (Order Falconiformes) and owls (Order Strigiformes) are long-lived, territorial, and often show delayed maturation, traits common in cooperative breeders. Yet, of over 250 species of hawks and nearly 140 species of owls, fewer than a dozen species possess some form of group breeding. This low level of sociality may reflect characteristics of food limitation or aggressiveness distinctive to raptors (see below), or it may indicate our limited knowledge of reproductive behavior in this group. In particular, little is known about the nesting behavior of tropical raptors, yet it is there that one might expect cooperative breeding to occur. The observation that a third of the known group-breeding hawks are tropical or subtropical in occurrence suggests strongly that generalizations about raptor sociality should be made carefully.

Those hawks where more than two breeders have been recorded at a nest generally fit into three categories:

(1) Unusual, probably aberrant threesomes that occur in typically monogamous species, as occurs rarely in the Red-tailed Hawk (*Buteo jamaicensis*: Santana *et al.* 1986).

(2) Helper-at-the-nest systems typified by the regular occurrence of juvenile or young-adult helpers, as occurs in the Mississippi Kite (*Ictinia mississippiensis*; Parker and Ports 1982) and Merlin (*Falco columbarius*; James and Oliphant 1986).

(3) Multiple-adult breeding groups that are maintained between breeding attempts.

In this paper, we focus our attention on two members of the latter categories, the Galápagos Hawk (*Buteo galapagoensis*), which apparently

* Present address: Hawk Mountain Sanctuary Association, Route 2, Kempton, PA 19529, USA.

mates polyandrously, and Harris' Hawk (*Parabuteo unicinctus*), which may have helpers at the nest and multiple-adult breeding groups. Early work suggested that these species each exhibited cooperatively polyandrous breeding systems (Faaborg and Patterson 1981); however, recent observations indicate that more differences than similarities may actually exist between the species. We first describe the breeding behavior of each hawk and then compare the ecological factors affecting the development of group breeding. We end with some thoughts on how raptors contribute to our overall understanding of cooperative breeding in birds.

Cooperative breeding in the Galápagos Hawk

The Galápagos Hawk is the only diurnal raptor that breeds on the Galápagos Islands. Endemic to the archipelago, it is common to abundant on islands where settlement or military activity did not lead to its extirpation earlier this century. Although a typical *Buteo*, several mainland forms have been suggested as its ancestor. This confusion results in part because the Galápagos Hawk is different from mainland forms in the absence of a light phase in adults and by its more pronounced sexual dimorphism in size.

The Galápagos Hawk is very much a generalist, occupying all terrestrial habitats and feeding on virtually all available foods. Major prey items include terrestrial birds such as Darwin's finches and the ubiquitous lava lizards, which are captured with the talons in typical *Buteo* fashion. Scavenging is also common, especially in territories that include sea lion breeding colonies, where placentae are available. On islands with introduced herbivores such as goats and donkeys, the hawks scavenge on corpses of these animals when available.

Most of the natural history of the Galápagos Hawk was studied by the Dutch ecologist (now residing in Ecuador) Tjitte de Vries between 1965 and 1971 (see de Vries 1973). When my students and I initiated further studies on the evolution of group breeding in this species in 1977, we chose the islands of Santa Fe and Santiago as study sites partly because they already possessed many birds banded by de Vries, and partly because his work had shown that they offered contrasting situations in terms of hawk densities, non-breeding populations, and frequency of breeding groups (see Faaborg *et al.* 1980).

Santiago (also called James or San Salvador) is a large (58 464 ha), high (914 m elevation) volcanic island that originally possessed the full Galápagos range of vegetation zones, from cactus-dominated thorn scrub in the lowlands, through moist forest, to fern-bracken scrub in the highlands (Fig. 12.1). Unfortunately, this vegetation has been highly affected by introduced herbivores, such that the forests are more open in structure than in the past and the highland scrub has been replaced by grasslands. In contrast, Santa Fe is small (2413 ha) and low (239 m maximum elevation) and is rather uniformly covered with thorn scrub typical of arid regions in the Galápagos.

Territorial behavior and group composition

Galápagos Hawks defend breeding territories in dry, lowland vegetation up to about 300 m elevation. Breeding above this elevation may not occur naturally, because the thick vegetation of the upper slopes makes prey less available to hawks. In recent years, though, the clearing of these forests by introduced goats has extended the altitudinal breeding range of this hawk on a few islands. Breeding territories are defended through the year, although the intensity of defense is highest just before breeding is

Fig. 12.1. Typical breeding habitat of the Galápagos Hawk in the lowlands of Santiago.

initiated. The territories are classical all-purpose territories; birds associated with them do not seem to leave the territory for any reason. Territory size appears to be fairly uniform (Fig. 12.2), with boundaries often coinciding with natural features such as ridges. Because territories are defended through the year and the hawks are long-lived, territorial boundaries do not change much from year to year.

In addition to territorial birds, the population of each island includes a number of non-breeding birds in higher elevation habitats or in lowland areas that are not defended (Fig. 12.2). The number of non-breeders varies from island to island (see below) and year to year, with peak numbers regularly exceeding the breeding population. These non-breeding floaters

Fig. 12.2. Extent of breeding and non-breeding area (shaded) of Galápagos Hawks on Islas Santa Fe (top) and Santiago. Approximate territorial boundaries are shown for Isla Santa Fe.

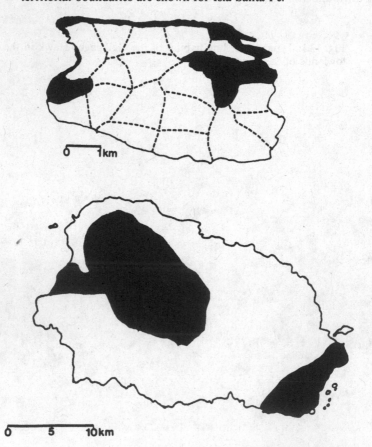

make regular intrusions into territories, where they are repulsed by the territorial birds. With its extensive areas of higher elevation habitat, Santiago supports large numbers of non-breeding birds, primarily juveniles and adult females. Estimates of over 200 total birds and 75 adult females have been made during August; these numbers undoubtedly increased as juveniles entered this population shortly thereafter. On Santa Fe, non-territorial birds live in those few areas that are not parts of territories, or by literally sneaking around the margins of territories and the coastline. Both because of this limited non-breeding habit and the smaller number of breeding territories on Santa Fe, the non-breeding population there has been estimated to vary at between five and 15 birds, mostly juveniles.

The Galápagos Hawk is distinctive because many territories are defended by a group of adults consisting of one female and up to five males (Fig. 12.3). Although we (Faaborg and coworkers) have not seen a group of males take possession of a territory, observations on marked individuals suggest that once a group forms, no new males are added as group members either die or, possibly, leave the territory (which we have never observed). Rather, a group of four males becomes a group of three when a male dies, then two, then one. This means that many of the currently monogamous pairs we observe on Galápagos Islands may include old male survivors of groups. Once this male dies or is usurped by a group of males, the new group takes over until death or replacement by a still younger group. Once

Fig. 12.3. Social unit size of Galápagos Hawks for Islas Santa Fe and Santiago, measured as the number of males in a breeding group.

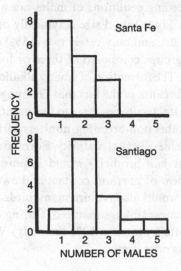

a female becomes established on a territory, she also stays until she dies. We have recorded three cases where the territorial female apparently died and was replaced but the male group was unchanged, and two examples where a female with one mate one year was mated with a new group the next. We have never seen a bird of either sex change territories or allow new members of the same sex into the group once established.

These dynamics of group formation result in long-term cycles in the number of breeding birds on a territory that may be unrelated to aspects of territory quality or size. Territory size seems more or less fixed, while territory quality varies primarily with rainfall. Any differences in average territory quality that do exist affect group size, but only over long periods of time by influencing mortality rates of the resident birds.

The frequency and size of breeding groups varies greatly from island to island (Fig. 12.3). On Santiago, we have recorded groups (two or more males and a female) on 87% of the territories examined, with an average of 2.6 ± 0.9 (S.D.) males per group. Santa Fe, in contrast, has had pairs on 50% to 67% of its territories in different years, with 2.4 ± 0.5 (S.D.) males per territory. These differences are primarily the result of the mechanism of group formation and disintegration, probably coupled with the demographics of the different islands. When a territory is open for a group of males to occupy on Santiago, there are always many males available to join the social unit. New groups on Santiago are probably always large because of the number of available floaters. In contrast, few extra males exist on Santa Fe at any time. Newly formed groups may be smaller and it is possible that even single males may gain access to a territory on occasion.

Several factors suggest that these breeding coalitions of males are not composed of closely related individuals. Since brood size is typically one nestling (81% of all nests monitored by us), and only rarely two (15%) or three (4%), it is nearly impossible that groups composed of three or four brothers from the same nest could occur. The formation of sibling alliances away from the natal territory is difficult because of the fact that Galápagos Hawk offspring are expelled from their natal territories three to five months after fledging, after which they live with other non-breeding, non-territorial birds for the three years that it takes to achieve sexual maturity. Formation of coalitions composed of brothers (or half-brothers) would require a mechanism of kin recognition independent of previous contact and away from the natal territory. Such a system would also require many birds to wait, perhaps for several years, while younger siblings matured. During all of this waiting, these birds have high mortality rates (> 50% annually). We have no evidence that any of these behaviors occur.

Breeding behavior of the Galápagos Hawk

The basic breeding biology of the Galápagos Hawk is typical of *Buteo*s, particularly with regard to the general roles of the sexes. Clutch size varies from one to three ($\bar{x} = 1.96 \pm 0.61$; $n = 49$ nests; de Vries 1973), depending on rainfall, and the female does most of the incubation, although the male(s) may assist in this activity. Once the eggs hatch, the female is primarily responsible for brooding and feeding the young while the male(s) do most of the hunting. As the nestlings develop, the female spends progressively less time brooding and more time foraging. The young fledge at about seven weeks of age, then remain on the natal territory for another three to five months before leaving.

When the territory contains a group of males, they participate approximately equally in all behaviors. de Vries (1973) has shown that all males share in territorial defense, copulations, incubation and feeding the young. While he found some differences in the types of food brought to nestlings by different males, these differences seemed to reflect individual variation in foraging behavior rather than differences in the amount of effort expended.

Unlike most cooperative breeders, no sign of a dominance hierarchy within a group of males has been observed by either de Vries or Faaborg; aggression between group males is rare. This apparent lack of a dominance hierarchy is surprising, considering that in most social systems individuals attempt to secure positions of advantage which would allow them access to greater fitness benefits. This should be especially crucial in a polyandrous system, where kinship is not a factor and direct genetic benefits are at stake. If a dominance order exists, it must be quite subtle or, perhaps, is expressed only at critical times of the breeding cycle.

Confirmation that this mating system is truly cooperative polyandry will require evidence that each male has some probability of fertilizing each egg produced by the female. We have watched copulations by groups over periods of several days, during which we saw no consistent order of copulation by males within the group, and no variation in the behavior of the female with different males. Group copulations generally are noisy affairs, with much calling by the males as they either wait or take their turn with the female. Nothing we have seen suggests that one male may be actually copulating with the female while the others are participating in some form of pseudo-copulation (as in Harris' Hawks (see below)). One male could be dominating successful fertilizations through sperm competition, or, perhaps, by some undescribed mechanism employed by the female. If such is the case, the system should be considered monogamy with non-breeding helpers; the helpers would presumably then work their way up the dominance ladder with the mortality of the breeding male. Only actual tests of paternity (presently underway) can determine whether multiple males sire offspring within breeding groups such that the mating system is truly cooperative polyandry.

Breeding success in the Galápagos Hawk

Measuring potential breeding success in groups of Galápagos Hawks can be difficult. Males may have equal chances of fathering each offspring, but the number of young in each clutch is often fewer than the number of potential fathers, such that some males are effectively helpers rather than fathers during a particular breeding cycle. Additionally, while knowledge of the success of groups relative to pairs is critical to understanding the evolution of this behavior, these pairs may, in fact, include old males that once were members of larger groups.

Reproductive success of Galápagos Hawks varies greatly with climatic conditions. In wet years, all groups or pairs may breed successfully, while in dry years few breed at all. We have censused breeding success in one very dry year (1981), one wet year (1979), and one intermediate year (1977), although in all years conditions were drier on Santa Fe than Santiago. Because of limitations on time, we often could not ascertain if young fledged. Rather, we computed two reproductive success measures, one that included only old nestlings or fledged young, and one that included all reproductive activity from eggs to fledged young. The general results from both techniques are the same.

Counting all nests on the two study islands and all stages of

Table 12.1. *Reproductive success in the Galápagos Hawk*

Location	Mating system	No. of nests	No. of males	No. of young	No. young per nest	No. young per male
Santa Fe	Monogamy	21	21	4	0.2	0.2
	Polyandry	10	23	1	0.1	0.04
Santiago	Monogamy	8	8	9	1.1	1.1
	Polyandry	17	48	23	1.4	0.5
Combined	Monogamy	29	29	13	0.45	0.45
	Polyandry	27	71	24	0.89	0.34

reproduction, polyandrous groups produce more young per year than monogamous pairs by a factor of almost 2 (Table 12.1). Deducting eggs or young nestlings from this summary reduces the difference between the two categories, but groups retain an overall advantage in production. This difference, though, is due to the success of groups in one year (the wet year of 1979); in the other two years pairs did as well or better than polyandrous groups. These data suggest that it is generally advantageous for females to breed with a group of males rather than a monogamous male, although in dry years (such as 1981) this may not be the case (Faaborg 1986).

To understand the potential reproductive success of males when parentage is equally shared, we need to partition the nest success by the number of potential fathers. When we do this, we find that in all years monogamous males outproduce polyandrous males. Even in the wet year of 1979 when groups were quite successful, success of group-breeding males did not match that of monogamous males on a per capita basis.

Survivorship

The above data suggest that cooperative polyandry does not present short-term reproductive rewards to male Galápagos Hawks; its adaptive value must involve some sort of long-term payoff. To gain insight into how this might occur, we must look at our information on survivorship (survival rate) of various categories of hawk. In 1977 and 1979 we banded 21 birds on territories on Santiago and 56 non-territorial birds. From these banded birds, plus some banded by de Vries in 1974, we have been able to estimate survivorship among both territorial and floating birds on Santiago (Faaborg *et al.* 1980; Faaborg 1986).

Territorial birds show extremely high survivorship; over 90% of our banded territorial birds survived each year on Santiago. In contrast, maximum annual survival rate estimates for non-breeding, non-territorial birds are about 50%. Smaller numbers of non-breeding birds have been marked on Santa Fe and we have seen survivorship over two year periods to vary from 0 to 33%. Although these estimates are provisional, the take-home points from our information are obvious. There are more Galápagos Hawks than breeding territories can support; floaters face a high probability of mortality; breeding, territorial birds experience relatively low mortality.

The evolution of polyandry in the Galápagos Hawk

With the above information, we can speculate that cooperative polyandry is an adaptive strategy for males where they trade the annual reproductive success they might attain as monogamous males for increased survivorship probabilities that are attained by joining a group and gaining access to a territory earlier in life. In a purely monogamous system where territory acquisition was determined by some form of age-related dominance, most young males probably faced small probabilities of acquiring a breeding territory, even though those that did may have been very successful. Males joining a group may have been able to breed over a longer period of time (through lack of a long waiting period to acquire dominance and a monogamous territory), such that the annual losses were offset in terms of lifetime reproductive success.

Although some sort of trade-off like that outlined above must be at work in the evolution of group breeding in the Galápagos Hawk, several possibilities exist as the key factor leading to this system. For example, territory acquisition may be critical. Once groups evolved that were willing to cooperate in gaining space, single males probably had no chance at reproduction. In its most extreme form, this results in a 'prisoner's dilemma' situation (Craig 1984), where a male hawk is more or less forced to cooperate and join a group, even if this is not the optimal reproductive strategy. A somewhat less severe alternative may be that the monogamous situation is the only way a male can attain optimal reproduction, but the only pathway to monogamy under these circumstances is by joining a group to acquire a territory initially. In the above cases, we might expect males in groups to adopt strategies that minimize their own risks relative to fellow group members in an attempt to outlive them. The least extreme scenario suggests that all males (on the average) benefit from polyandry compared to monogamy because there are lifetime rewards from group breeding. In this case, cooperative behavior might be adaptive.

All of the above scenarios can be described as 'making the best of a bad situation', which is a common theme among cooperative breeders. Certain similarities exist between the situation in the Galápagos Hawk and that found in the acquisition of prides by male lions (see Bygott *et al.* 1979). While polyandry must be adaptive for males on a lifetime basis, the fact that new males are not added as groups decay suggests that there is nothing inherently advantageous about groups. Otherwise new males should be added. Rather, the group may be a way for the young male to avoid long periods with potential high mortality away from breeding territories. Without kinship between cooperating males, group formation in Galápagos Hawks becomes a case of an obligate reciprocity (cf. Ligon 1983) or mutualism.

The apparent lack of kinship, the long pre-reproductive period, and, perhaps, a low cost/benefit ratio for keeping young raptors as helpers may be the chief reasons why such an unusual system with equal-status males has evolved, rather than the breeder-helper systems seen most often in group-breeding birds. If there were a dominant male breeder but with similar conditions of group dynamics and demography, the most subordinate male helper would face a severe situation of many years of helping before acquiring breeding opportunities with no helpers of his own. This cost, perhaps accompanied by an inability of single males to attain territories when competing with male groups, may have forced the evolution of an equal status, cooperatively polyandrous breeding system.

Cooperative breeding in the Harris' Hawk

Another hawk social system that has attracted considerable attention is that of the rather enigmatic Harris' Hawk. At least seven different research efforts have revealed incomplete and conflicting insights into the social behavior of this species. Reasons for this slow progress include the inherent difficulties of studying long-lived, often wary, predators that are found in relatively low densities, and the fact that Harris' Hawks may facultatively vary behavior, adjusting their social system to exploit current environmental situations (Bednarz 1987*b*).

The Harris' Hawk is taxonomically included in the 'sub-buteos', which are distributed chiefly in tropical America (Brown and Amadon 1968). Based on phenotypic similarities, Amadon (1982) suggested that the Harris' Hawk is most closely allied to the Savannah Hawk (*Buteogallus meridionalis*) of South America.

Harris' Hawks occupy a variety of sparse woodland and thornscrub semi-desert environments from southern Chile and central Argentina (where it is known as the Bay-winged Hawk) north to the south-western United States. Essentially all pertinent field research on the Harris' Hawk has been carried out in the northern periphery of the species' range. Studies of varying length have been carried out by Mader (1979), Whaley (1986), and Dawson (Bednarz *et al.* 1988) in Arizona, Griffin (1976), Hamerstrom and Hamerstrom (1977), and Brannon (1980) in Texas, and Bednarz (1986) in New Mexico. Most of the information presented here will be based on data collected in the last study.

Mader (1975) first brought attention to the cooperative breeding system of the Harris' Hawk when he reported observing extra adults at 46% of the nests ($n = 50$) that he monitored. He inferred that these auxiliaries were polyandrous, on the basis of intensive study at one nest and opportunistic observations at four others. Since that time, the notion that Harris' Hawks engage in cooperative polyandry has been widely accepted. When Griffin and Brannon found extra attendants at Texas Harris' Hawk nests (5% and 14%, respectively), they assumed that these represented polyandrous males.

Bednarz's long-term study of the Harris' Hawk was initiated in 1980 in the Los Medaños Raptor Area near Carlsbad, New Mexico. As of this writing five years (1981–3 and 1985–6) of intensive field data have been collected, but only three years of data have been analyzed completely. The information reported here is based primarily on the first three years of study, supplemented by preliminary analysis of data obtained during the latter two years of work. The study population varied from 24 to 29 breeding units per year, which mostly reflects variation in work effort rather than population fluctuations. The study site is typified by open shrubland vegetation situated on rolling sandy hills. Scattered throughout the study area are dwarfish mesquite (*Prosopis glandulosa*) and soapberry (*Sapindus drummondii*) 'trees' (4–6 m in height) that provide nesting locations and perches for the hawks (Fig. 12.4). Weather conditions can be described only as extreme; summer temperatures regularly top 40°C, while winter readings below −12°C are not unusual. Violent storms accompanied by

Fig. 12.4. A soapberry tree (*Sapindus drummondii*) supporting a nest used by three Harris' Hawks in the Los Medaños study area, New Mexico.

high winds occur unpredictably throughout the summer. Snow and ice storms periodically transform the study site into a spectacular snow-covered desert at any time between October and April. Spring winds exceeding $64\,\mathrm{km\,h^{-1}}$ occur regularly in March and April.

Reproductive behavior of the Harris' Hawk

The Harris' Hawk inhabits and successfully breeds in this rigorous environment in relatively high numbers. During the first five years of research, 306 Harris' Hawks were color-marked on the study area. The hawks breed in simple pairs and in social groups of up to five individuals (Fig. 12.5). The average composition of 30 groups (more than two individuals) between 1981 and 1983 included 2.4 adults and 0.9 immatures. Auxiliary hawks tend to be males, particularly among older age classes (Fig. 12.6), but female auxiliaries occur. Sometimes females disperse during their first three to 12 months of life, whereas males tend to stay with their natal group through their first year..

The Harris' Hawk is unusual in that it is known to breed almost throughout the year, at least in the northern periphery of its range. I (Bednarz 1987a) have documented Harris' Hawks in New Mexico to lay eggs from 17 February until late October. The latest successful fledging occurred during the first weeks of December.

Fig. 12.5. Social unit size of Harris' Hawks, as determined by the number of hawks in the vicinity of the first nesting attempt of the year. UNK, unknown.

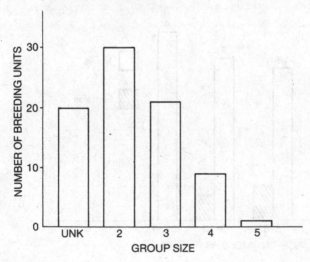

The actual incidence and length of breeding varies between years (Fig. 12.7). Statistical analyses have revealed that the number of second nests was significantly correlated with the number of cottontails (*Sylvilagus auduboni*) recorded on prey censuses, suggesting that the availability of the food is an important factor influencing the length of breeding season.

After successful breeding attempts, many social units swell to five or six members with the addition of fledglings. All these birds maintain close social ties, often occurring within a small area and commonly perching on the same tree. When not breeding, the entire group participates in group hunts for large or aggregated prey (e.g. flocks of birds). Average social unit size declines throughout the fall and winter as a result of mortality and the dispersal of some group members. In the spring, the remaining young females and males make exploratory forays from their natal range into surrounding areas. After such expeditions, immature-plumaged hawks often return to the natal nesting range and associate with other group members. However, bonds to the group during spring nesting seem to be weak, presumably because primary breeders are occupied with current broods. As yearlings molt into a more adult-like plumage, they again associate more closely with current-brood fledglings and breeders. Thus, most of the adult-plumaged auxiliaries are hawks hatched within the nesting range the previous year. The result is a typical Harris' Hawk group in late summer consisting of one adult female breeder, one adult male

Fig. 12.6. Sex composition of Harris' Hawks marked as nestlings and later observed on their natal breeding range.

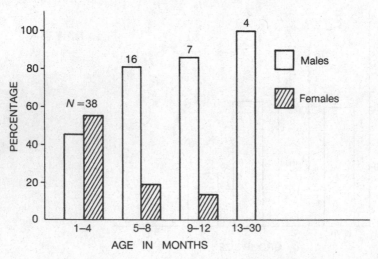

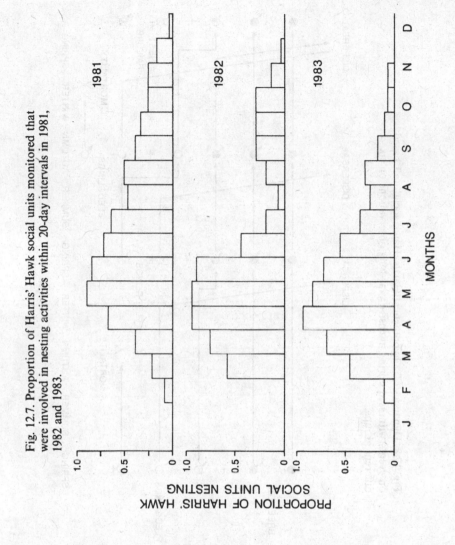

Fig. 12.7. Proportion of Harris' Hawk social units monitored that were involved in nesting activities within 20-day intervals in 1981, 1982 and 1983.

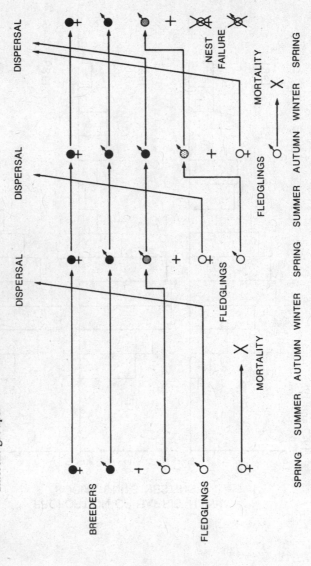

Fig. 12.8. Dynamics of a hypothetical Harris' Hawk social unit based on events witnessed during intensive year-long observations on six different groups.

breeder, one adult-plumaged male auxiliary (usually a previous young), and one to four hawks born that year (Fig. 12.8).

Both breeders invest heavily in the current brood of nestlings. The female assumes the typical *Buteo* role of guardian of the brood and remains close to the nest. The breeding male regularly relieves the female of these duties for short periods, but his primary function is that of provider of food. Adult auxiliaries also regularly bring prey to the female or deposit prey in the nest. Although I (J.B.) have been unable to quantify adequately the differences in delivery or attendance rates by adult males, it is my impression that the breeding male spends more time at the nest and makes more prey deliveries than adult auxiliaries (Bednarz 1987*b*). Since Harris' Hawks regularly share food among group members, the delivery of a prey item does not necessarily indicate that the kill was made by that hawk. Immature-plumaged auxiliaries are only loosely tied to the breeding area when active nests are present. These hawks may provide indirect 'help' as sentinals in the nest area and undoubtedly some prey that they have killed is brought to the nest by adult group members. Only once has an immature been observed actually bringing food to the nest.

Is the Harris' Hawk polyandrous?

My information from New Mexico suggests that polyandry is probably rare and, in fact, may be rather exceptional (Bednarz 1987*b*). First, as described above, the adult auxiliary male(s) is most likely the offspring of the breeding female (four of four cases involving banded hawks). Thus, if polyandry is common, the Harris' Hawk would have a highly incestuous breeding system which has never been observed in birds. In addition, paternity analyses completed on two New Mexican groups have shown that only one male sired the broods of two young. This result, however, could have occurred by chance ($P = 0.25$), assuming that both males had an equal probability of siring an individual offspring. Finally, copulations only occurred between the two adult breeders during intensive observations made on four groups prior to and during egg-laying. Auxiliaries showed no interest in copulating with either of the breeding adults. These limited data suggest that polyandry is not the prevalent mating strategy of Harris' Hawks in New Mexico.

Since Mader (1975) published his information on the mating system of the Harris' Hawk, considerable research effort has uncovered no collaborative evidence for polyandry (Bednarz *et al.* 1988). The Harris' Hawk does exhibit *Backstanding* behavior (one hawk alights and quietly stands on the back of another hawk), which when observed from a distance could be misinterpreted as copulatory behavior. Both the breeding and auxiliary males will *Backstand* on the breeding female. While Mader's (1979) intensive observations of one pair suggest polyandry in that group, all other evidence suggests that auxiliary hawks pursue a more standard cooperative breeding strategy of remaining on their natal nesting range and assuming the role of non-breeding helpers to the primary breeders.

Electrophoretic analysis of muscle tissue from one group of New Mexico Harris' Hawks indicated that the two adult-plumaged males were homozygous for different alleles at two different presumptive loci and thus could not have been father and son, and were probably not brothers ($P < 0.008$; Bednarz 1987*b*). This trio could have resulted from the retention of a male hawk hatched on that range, coupled with the replacement of the auxiliary's father. Alternatively, unrelated males may form coalitions and at times successfully form a social bond with one adult female. In the case described above, though, paternity analysis showed that one of the males did not contribute genetically to the offspring, but the probability of finding this pattern was 0.5, assuming that both males had an equal chance of fathering each nestling. Should polyandry occur, it would probably be among groups that are composed of unrelated adult members, but the evidence available at this time suggests that such groups are rare.

Territoriality

Unlike many cooperatively breeding birds, Harris' Hawks are apparently non-territorial. In five years of study only occasional intraspecific aggressive interactions have been observed; most were of mild intensity and were probably between group members. Generally, Harris' Hawks seem to move freely through their neighbor's nesting ranges without being challenged, although all hawk species, including Harris' Hawks, or potential predators are evicted from the immediate vicinity of the nest.

Reproductive success of groups and pairs

The first step involved in understanding any cooperative system is the examination of relationships of reproductive success and group size. In the Harris' Hawk, both clutch size and fledging success is similar between pairs and larger groups (Table 12.2). Some differences, however, do exist in the reproductive output of pairs and groups. Nests tended by groups failed significantly ($P < 0.05$; see Bednarz 1987*b*) more often (34%) than the nests of pairs (14%). Failures occurred primarily in the incubation phase of breeding; it is possible that the social competition within the larger groups might lead to increased probabilities of interference or neglect of the clutch.

Groups were able to compensate for losses due to nest failure by initiating more second broods (40%) than pairs (12%; $\chi^2 = 4.4$, $P = 0.03$, d.f. = 1). In fact, groups initiated second nests after fledging young in each year between 1981 and 1983, whereas pairs attempted second broods only in 1981, when the highest numbers of prey were observed. This suggests that auxiliary Harris' Hawks may decrease the stress on the primary breeders associated with rearing the first brood, thus enhancing their ability to undertake a second breeding attempt.

Initial evidence suggests that groups produce slightly larger fledglings than pairs. Estimated asymptotic sizes for three measurements in females (one was significant) and two of three in males were larger in young from groups (Bednarz 1987b). Whether this slight difference in estimated size confers any fitness benefits to the fledged young is unknown, but this pattern may indicate that conditions may be slightly better at group nests than those of pairs.

The overall differences in productivity measures of pairs and groups are extremely slight. On a per capita basis, however, individuals in groups rear significantly fewer young than do members of pairs (Wilcoxon test, $z = 2.62$, $P < 0.01$; Table 12.2). It seems unlikely that Harris' Hawks could justify the forfeiture of any potential of independent breeding by auxiliaries in order to gain kinship benefits.

Explanations for cooperative breeding in the Harris' Hawk

In an effort to understand the evolution and maintenance of cooperative breeding in Harris' Hawks, I (J.B.) examined two mutually exclusive hypotheses. The first hypothesis is that ecological saturation may have favored the evolution of cooperative breeding (Koenig and Pitelka

Table 12.2. *Mean number of fledglings reared by pairs and groups of Harris' Hawks per year*

	Pairs		Groups	
	Mean ± S.E.M.	n	Mean ± S.E.M.	n
1981	2.29 ± 0.36	7	2.00 ± 0.46	8
1982	1.83 ± 0.31	6	2.17 ± 0.52	12
1983	1.69 ± 0.30	16	1.70 ± 0.33	10
Combined	1.86 ± 0.20	29	1.97 ± 0.26	30
Per capita	0.93 ± 0.10	29	0.60 ± 0.09	30

S.E.M., standard error of the mean.

1981). This might occur if space or suitable habitat for breeding is unavailable or extremely limited when animals are attempting to disperse. In this case, it would be adaptive for the young to stay on the parental range until a breeding vacancy became available. Larger groups could be maintained on better-quality territories, or might even be needed to defend such territories. Both patterns would result in positive correlations between group size and territory quality (Emlen and Vehrencamp 1983), possibly measured as the availability of food.

A competing hypothesis suggested by Gowaty (1981) extended the so-called polygyny threshold model by supposing that all 'good'-quality habitat is occupied and excess birds must use poor habitats. If it is difficult or risky for pairs to raise young in these habitats, cooperative groups may be adaptive to improve long-term survival and increase fitness. This proposal predicts that pairs should occupy better territories than groups.

These hypotheses were tested by the use of measures of available prey resources consumed by Harris' Hawks and a broad comparison of habitats used by groups and pairs. By counting rabbit fecal pellets (over 70 000), I found that group ranges contained non-significantly ($P > 0.1$) more pellets (mean $= 108 \text{ m}^{-2}$) than areas used by pairs (100 m^{-2}) in the summer. This pattern was reversed in the winter, and none of these differences was statistically significant (Bednarz 1986). Other analyses also revealed no important differences between ranges of groups or pairs for 21 habitat variables (including measures of perch and nest site resources). Additionally, monitoring of nesting ranges indicated that at least 12% of the potential nesting habitat was always vacant (Bednarz 1986). Previously-used ranges were often abandoned, while unoccupied areas were continually being colonized.

These observations fail to support either hypothesis for the evolution of cooperative breeding in the Harris' Hawk. The similarity in habitat quality for groups and pairs does not fit the Gowaty hypothesis, while the existence of empty space (in addition to the lack of territorial behavior) does not fit the saturation hypothesis.

Cooperative hunting

The apparent failure of the above hypotheses to explain cooperative breeding in the Harris' Hawk may involve the unusual cooperative hunting behavior displayed by this species. Although several previous studies have suggested that Harris' Hawks participate in cooperative 'wolf pack' hunts when subduing prey, no one has examined this behavior closely. To do this, we (J.B. and assistant Tim Hayden) attempted to radio-track and to monitor foraging of groups and individuals through the year. To date, most of our success has come during the non-breeding season, in part because breeding birds are extremely sensitive to disturbance.

Our first discovery was that individual Harris' Hawks, at least when not nesting, are highly social and most often in company of other group members. We found Harris' Hawks to be perched alone only 31.2% of the time ($n = 4560$ spot-observations made on four individuals in three different groups). Groups were generally subdivided into subgroups of one to three hawks that kept constant visual contact with one another and with other subgroups. When hunting, these subgroups 'leap frog' throughout the home range, searching for prey. When a prey item (most often a cottontail, but woodrats (*Neotoma* spp.), birds in loose flocks, and jackrabbits (*Lupis californicus*) are also taken) is discovered, several hawks might swoop on it simultaneously. If the potential prey item finds cover, several birds will surround it, while one or two hawks will walk or fly into the vegetation where the prey item hides. Once a bird(s) has killed a prey item, all group members congregate at it and feed. All six hawks of the social unit fed during two complete lagomorph feeding episodes and a mean of 3.2 hawks (range two to four hawks) consumed portions of individual avian prey during five feeding episodes witnessed.

The obvious benefit of taking relatively large prey (an adult jackrabbit may weigh more than three times a Harris' Hawk) is that one kill will provide adequate food for the social unit for an entire day or longer. A typical cottontail should be able to supply the food requirements of at least five Harris' Hawks for a day, while a jackrabbit will feed even more. Observations have been made of hawks capturing smaller prey items (especially quail and doves in loose flocks), in all cases while foraging in groups, but large items seem to dominate the diet of the Harris' Hawk (Bednarz 1987*a*).

A factor that accentuates the importance of group foraging may be the limited temporal availability of large prey items to foraging birds. Cottontails and woodrats are active during a brief period of the day (*c.* 30 min before to 30 min after sunrise and sunset); only smaller food items are available during the heat of the day. Thus, group foraging not only may be important as a means of capturing large prey items, but being a group member increases the probability of capturing at least one large prey item in any given morning or evening during this window of food availability. Groups may be able to procure prey on a more consistent basis than pairs, which perhaps leads to increased survival and fitness benefits. These benefits may be particularly important to young members of the group that are less proficient foragers than are the adults. If this explanation of the Harris' Hawk social system is correct, then the cooperative breeding observed may be a consequence of cooperative hunting behavior.

Comparisons of hawk social systems

The Galápagos and Harris' Hawks constitute two of the taxonomically and ecologically most similar species examined in this volume. Yet, our work suggests that their social systems are widely divergent in composition, stability and function. The apparent cooperative polyandry found in the Galápagos Hawk seems to be a result of pressures caused by saturated populations and limited space. The equal-status, all-male, no-juvenile composition of the groups is highly unusual. The monogamous breeding pairs with helpers in Harris' Hawks seem to be related to the group hunting behavior in this species. Although Harris' Hawk group composition resembles that commonly found in cooperative breeders, the absence of territorial limitations is unusual among non-colonial species.

What does the the existence of such different social systems among these highly social raptors tell us? First of all, it suggests that avian predators are basically asocial. Two highly social species plus a couple of species with occasional nest helpers among nearly 400 species of raptors falls well below the 3% cooperative breeding species suggested for birds overall by Emlen and Vehrencamp (1983). The differences in the ecological and evolutionary factors leading to these systems re-emphasizes (as do other chapters in this volume) the futility of searching for a single model for the evolution of cooperative breeding in birds.

These two hawks do share some similarities. Group size in both is most likely limited by food supply. Observations on Galápagos Hawks have shown that groups are less likely to breed during dry years, presumably because of food limitations. Packing more than four or five adult hawks on a territory is probably not feasible; should it occur, higher mortality rates may rapidly reduce group size. Although Harris' Hawks are not as constrained to the food available within an area, evidence suggests that the dynamics of group foraging may put an upper lid to group size. As we stated earlier, a cottontail can easily be food for five adult hawks for a day; groups much larger than this might result in individuals who would not get a meal from a kill. This would either lead to their mortality or force them to join a group elsewhere.

Our studies of these hawks are not as complete to date as many studies of other species in this volume. Work is underway that will, we hope, clarify many of the ideas suggested above. The Galápagos Hawk work has suggested that the lack of kinship among males is a prime mover in the evolution of equal-status breeders; if non-kin male groups of Harris' Hawks occur, we might expect them to show more equal subdivision of roles among males. Harris' Hawks have rather clearly defined divisions of labor, with breeding birds working harder during nesting while 'helpers' seem most intent on ensuring their own survival. Among large groups of Galápagos Hawks, a male might want to balance efforts directed towards young who may not be his with those directed at his own survival. In both species, we know little about long-term patterns of group formation. How do male Galápagos Hawks form alliances in the process of territory acquisition? At what point should a male Harris' Hawk leave a group to try to establish his own? Finally, in both species we still need to determine the actual mating system.

Many similarities exist between the behavioral ecology of prides of lions and the social systems of these hawks (Bygott *et al.* 1979). Galápagos Hawks are like lions in the way that lion prides regularly have multiple males usurping single males to gain control of the pride. Harris' Hawks resemble lions in the manner that their foraging allows them to exploit larger prey items then would be possible by foraging alone, while reducing daily variation in food supply through social foraging. The fact that these three rather unusual predators possess social systems that differ from all of their close relatives seems to suggest that unusual sets of conditions are necessary for social behavior among top level carnivores. Perhaps for the Galápagos Hawk the key is space limitation coupled with the prospect of vast disparities in survival expectations for territorial and non-territorial birds. With the Harris' Hawk, possibly the combination of the non-migratory habit affording the opportunity for social association and the limited temporal availability of relatively large desert prey may have favored the maximization of efficient foraging via cooperative hunting, and thus cooperative living.

Work on the Galápagos Hawk has been done with the support and cooperation of the Servicio Parque Nacional Galápagos and Charles Darwin Research Station. Tjitte de Vries was instrumental in helping with this work. Financial support was provided by the Research Council of the Graduate School, University of Missouri-Columbia. I thank Curt and Bridget Griffin, Robin Kennedy, Margaret Parker, Janice Winters, Leslie Burger, and Susana Struve for help in the field.

For the Harris' Hawk study, we especially thank J. David Ligon and Tim Hayden for their multiple contributions. Many students and others assisted in the field, especially Joey Haydock, Bill Howe, William Iko, and Dirk Freeman. Ken Fischer, Jess Juen, Stan Van Velsor, and Charles Reith provided encouragement and suggestions. Peter Stacey, Walt Koenig, and Bruce Woodward reviewed an earlier version of the manuscript. Harris' Hawk research was supported by the U.S. Department of Energy (Waste Isolation Pilot Plant Studies), the Bureau of Land Management, Frank M. Chapman Memorial Fund, and the University of New Mexico.

Bibliography

Amadon, D. (1982). A revision of the sub-Buteonine hawks (Accipitridae, Aves). *Amer. Mus. Novit.* **2741**: 1–20.

Bednarz, J. C. (1986). The behavioral ecology of the cooperatively breeding Harris' Hawk in Southeastern New Mexico. Ph.D. dissertation, University of New Mexico.

Bednarz, J. C. (1987*a*). Successive nesting and autumnal breeding in the Harris' Hawk. *Auk* **104**: 85–96.

Bednarz, J. C. (1987*b*). Pair and group reproductive success, polyandry, and cooperative breeding in Harris' Hawks. *Auk* **104**: 393–404.

Bednarz, J. C. (1988*a*). Cooperative hunting in Harris' hawks (*Parabuteo unicintus*). *Science* **239**: 1525–7.

Bednarz, J. C. (1988*b*). A comparative study of the ecology of Harris' and Swainson's hawks in southeastern New Mexico. *Condor* **90**: 311–23.

Bednarz, J. C. (1988*c*). Harris' hawk subspecies: is *superior* larger or different than *harrisi*? In *Proceedings of the Southwest Raptor Management Symposium and Workshop*, ed. R. L. Glinski, B. Giron Pendleton, M. B. Moss, M. N. LeFranc Jr, B. A. Millsap and S. W. Hoffman, pp. 294–300, National Wildlife Federation, Washington, DC.

Bednarz, J. C., Dawson, J. W. and Whaley, W. H. (1988). Harris' Hawk. In *Proceedings of the Southwest Raptor Management Symposium and Workshop*, ed. R. L. Glinski, B. Giron Pendleton, M. B. Moss, M. N. LeFranc Jr, B. A. Millsap and S. W. Hoffman, pp. 71–82. National Wildlife Federation, Washington, DC.

Bednarz, J. C. and Ligon, J. D. (1988). A study of the ecological bases of cooperative breeding in the Harris' Hawk. *Ecology* **69**: 1176–87.

Brannon, J. D. (1980). The reproductive ecology of a Texas Harris' Hawk (*Parabuteo unicinctus harrisi*) population. M.Sc. thesis, University of Texas.

Brown, L. and Amadon, D. (1968). *Eagles, Hawks and Falcons of the World*, vol. 2. McGraw-Hill: New York.

Bygott, J. D., Bertram, B. C. R. and Hanby, J. P. (1979). Male lions in large coalitions gain reproductive advantages. *Nature (Lond.)* **282**: 839–41.

Craig, J. L. (1984). Are communal Pukeko caught in the prisoner's dilemma? *Behav. Ecol. Sociobiol.* **6**: 289–95.

de Vries, Tj. (1973). *The Galapagos Hawk*. Free University Press: Amsterdam.

Emlen, S. T. and Vehrencamp, S. L. (1983). Cooperative breeding strategies among birds. In *Perspectives in Ornithology*, ed. A. H. Brush and G. A. Clark Jr, pp. 93–120. Cambridge University Press: Cambridge.

Faaborg, J. (1986). Reproductive success and survivorship of the Galapagos Hawk *Buteo galapagoensis*: potential costs and benefits of cooperative polyandry. *Ibis* **128**: 337–47.

Faaborg, J., de Vries, Tj., Patterson, C. B. and Griffin, C. R. (1980). Preliminary observations on the occurrence and evolution of polyandry in the Galapagos Hawk (*Buteo galapagoensis*). *Auk* **97**: 581–90.

Faaborg, J. and Patterson, C. B. (1981). The characteristics and occurrence of cooperative polyandry. *Ibis* **123**: 477–84.

Gowaty, P. A. (1981). An extension of the Orians–Verner–Willson model to account for mating systems besides polygygny. *Amer. Nat.* **118**: 851–9.

Griffin, C. R. (1976). A preliminary comparison of Texas and Arizona Harris' Hawk (*Parabuteo unicinctus*) populations. *Raptor Res.* **10**: 50–4.

Hammerstrom, F. and Hammerstrom, F. (1978). External sex characteristics of Harris' Hawks in winter. *Raptor Res.* **12**: 1–14.

James, P. C. and Oliphant, L. W. (1986). Extra birds and helpers at the nests of Richardson's Merlin. *Condor* **88**: 533–4.

Koenig, W. D. and Pitelka, F. A. (1981). Ecological factors and kin selection in the evolution of cooperative breeding in birds. In *Natural Selection and Social Behavior: Recent Results and New Theory*, ed. R. D. Alexander and D. Tinkle, pp. 261–80. Chiron Press: New York.

Ligon, J. D. (1983). Cooperation and reciprocity in avian social systems. *Amer. Nat.* **121**: 366–84.

Mader, W. J. (1975). Biology of the Harris' Hawk in southern Arizona. *Living Bird* **14**: 59–85.

Mader, W. J. (1979). Breeding behavior of a polyandrous trio of Harris' Hawks in southern Arizona. *Auk* **96**: 776–88.

Morizot, D. C., Bednarz, J. C. and Ferrell, R. E. (1987). Sex linkage of muscle creatine kinase in the Harris' Hawk. *Cytogenet. Cell Genet.* **44**: 89–91.

Parker, J. W. and Ports, M. (1982). Helping at the nest by yearling Mississippi Kites. *Raptor Res.* **16**: 14–17.

Santana, C. E., Knight, R. L. and Temple, S. A. (1986). Parental care at a Red-tailed Hawk nest tended by three adults. *Condor* **88**: 109–10.

Whaley, W. H. (1986). Population ecology of the Harris' Hawk in Arizona. *Raptor Res.* **20**: 1–15.

13 PUKEKO: DIFFERENT APPROACHES AND SOME DIFFERENT ANSWERS

13

Pukeko: different approaches and some different answers

J. L. CRAIG AND I. G. JAMIESON

New Zealand has at least eight species of birds that regularly have more than a breeding pair involved in nesting and rearing of young. One is a sea bird (the Southern Skua, *Stercorarius lonbergi*), others are small forest insectivores (the Whitehead, *Mohoua albicilla*; the Yellowhead, *Mohoua ochrocephala*; the Brown Creeper, *Finschia novaeseelandiae*; and the Rifleman, *Acanthisitta chloris*), while the remainder are rails (Takahe, *Notornis mantelli*; Weka, *Gallirallus australis*; and the Pukeko, *Porphyrio porphyrio*). The range of environments occupied (from sub-Antarctic to subtropical, from oceanic islands to dense evergreen forests and from alpine grasslands to lowland swamps), the range in size (from our smallest to some of our largest species), and the range in demography (from short-lived to long-lived birds that rarely breed before the age of nine years) make the quest for a universal explanation of cooperative breeding a difficult task indeed.

Pukeko have the most complex social system of all these species. (Pukeko is the Maori name for this gallinule, which is known as the Swamphen in other places.) Most Pukeko are found in communal groups of up to a dozen birds. All participate to a greater or lesser extent in the rearing of chicks and the defense of the territory. Birds are capable of breeding from one year of age but most do not until at least three years of age. Among the active breeders, mate-sharing is total and one to three females contribute eggs to a common nest. All birds in the group brood and provision the precocial chicks and, when a second clutch is laid, young of the first brood will help to feed the younger chicks. Offspring, with the exception of a few reared by pairs, remain in their natal territory as non-breeders before eventually becoming breeders.

The overall social system of Pukeko raises a plethora of challenging questions. Not only can the subtle interactions and behavior that allow individuals to act as a coordinated group be investigated but the opportunity to study the development of parental behavior also exists. Usually researchers can observe individuals only with their parents or as parents, but with Pukeko we can observe the gradual transition between the two. Being able to observe Pukeko breeding as pairs and in groups allows speculation on the reasons why groups form and persist. Potential problems of incest also require consideration. Finally, communal breeders represent a paradox for classical evolutionary theory. Why do animals share mates and assist in raising offspring that are not their own when it is argued that as a result of natural selection individuals will tend to maximize the representation of their own genotype in subsequent generations? In attempting to answer this question researchers must always be mindful of the difficulties of separating cause and effect and of distinguishing between alternative explanations. Most of all, a firm base of natural history observations is needed to ensure that the questions asked are biologically relevant.

General description and habits

The plumage of female and male Pukeko is identical. Each has a large red shield on its head, which tends to be larger in birds that are breeding. Young birds are indistinguishable from older non-breeding adults by nine months of age. Body size varies with latitude but, on average, females are 80–90% of the size of males and birds can be sexed by measurement with 96% accuracy (Craig *et al*. 1980).

Pukeko are primarily herbivorous, feeding on the shoots of grasses and sedges but also on roots, leaves and seeds. All gizzards investigated by Carrol (1966) contained a predominance of vegetable material. Animal foods are taken occasionally and include earthworms, insect larvae, insects and spiders, but more rarely small fish, eggs and young birds. Larger animals are eaten as carrion. Carrol (1966) reported animal material in 25% of gizzards but said that it was present in 'minute amounts'. Chicks are fed large quantities of invertebrates but no quantitative data are available.

The conversion of forest to open pasture has greatly increased the amount of suitable habitat used by Pukeko, although they are most commonly associated with swamps, bogs or drains. Areas with water and cover are chosen as nest sites, although on offshore islands, which lack mammalian predators, nests are often built away from water. The association of Pukeko with open habitat such as pasture, their colorful plumage and their reluctance to fly make them ideal for observation. They are easily conditioned to associate food with colored flags and so can be easily manipulated for observation from hides.

We and our colleagues have studied the cooperative behavior of Pukeko in three different habitats in New Zealand for a total of 10 years. (Pukeko are also found in southern Europe, Africa, Asia, the Phillipines, Australia and many islands in the Atlantic, Indian and Pacific Oceans.) An initial three year study was made in two areas in the Manawatu district of New Zealand; one at Linton was an area of small ribbon swamps in valleys of pastural land while the other, Pukepuke, was on the swampy edge of coastal dune lake and surrounding pasture (Craig 1979). Our more recent study, begun in 1978, is on a coastal farm at Shakespear Regional Park, 25 km north of Auckland (Fig. 13.1).

Social structure

Social groupings of Pukeko include non-territorial flocks, and territorial pairs and groups of all sizes and the proportion and persistence of these varied with locality. All the pairs and groups we have studied defended territories year round. Two of our areas, Pukepuke and

Fig. 13.1. One of the swampy valleys surrounded by pasture in Shakespear Regional Park. This valley was the major study area for a seven year study.

Shakespear Park, were invaded by birds from other areas during late summer–autumn of each year. These birds moved as a single unit of up to 70 birds and did not attempt to defend their area. Relationships within the group were mediated by a social hierarchy (Craig 1979). Each of these units was called a flock and only one flock was known in any area. The origin of these birds was unknown but appeared to include family groups. Another study elsewhere in New Zealand (G. Tunnicliffe, personal communication) showed that some birds that defended breeding territories in spring and summer moved to join flocks when these areas became flooded. It was suspected that flocks in our areas were also made up of birds that held territories in marginal areas during the breeding season. Flock birds in all areas dispersed immediately prior to breeding. Banding of flock birds at Pukepuke showed a predominance of males (60+ %) and young birds (60+ %). Some banded birds rejoined the flock year after year.

At Pukepuke, where an attempt was made to follow flock birds in detail, groups of flock birds attempted to set up new territories in areas little used by the resident territorial groups. Four territories (totalling 14 birds) were established by flock members over a two-year period. All nested but breeding success was extremely low, with most failing to produce any surviving young. Membership of these new groups was not constant from year to year, most had an excess of males and aggression within the group was common (Craig 1979). Six flock birds were admitted to established territories during two years of study.

A minority of birds remained non-territorial all year round. These birds were seen regularly in all study areas and banding showed that they had fairly restricted areas of use. Some tried to gain admittance to existing groups and the behavior used at this time was characteristic. When attacked by members of the resident group they would run away but always in circles until the aggressor stopped. Tolerance was often short lived and, once attacked by more than one resident, the intruder had little choice but to retreat from the territory.

Size and composition of territorial groups varied in different areas (Fig. 13.2). At Linton, pairs were common and although they retained their young on the territory through the winter, the offspring left or were aggressively expelled prior to egg-laying the following season. At Pukepuke, groups of four to six birds were common, comprising two female and two male breeders and up to two non-breeders, who were always offspring of previous years. Overall sex ratios were 1:1 ($n = 32$), although there were more breeding males than breeding females in groups. At Shakespear, group sizes were larger, many had an excess of males (up to seven) and most had non-breeders, who again were offspring of previous years. The greatest number of non-breeders was seven. Overall, the sex ratio among breeders at Pukepuke and Shakespear favored males (Table 13.1) with only one of 19 groups having as many females as males. Females predominated as non-breeders (Table 13.1).

Groups defended their territories year-round against neighbors and a few non-territorial floaters. Interactions between neighboring groups at boundaries were observed frequently and adult breeding males accounted

Fig. 13.2. Frequency histogram of group sizes of Pukeko found in three different areas for all years combined. Means ($\pm$ s.e.m.) are given in parentheses.

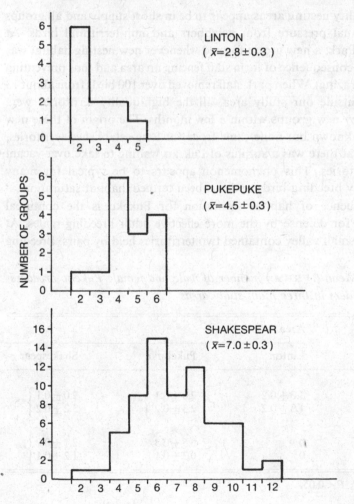

for the vast majority of these defensive interactions (>70%, Craig 1979). Not only did they do more, but they were also more effective.

Recording the position where territorial disputes occurred showed that the status of individuals within a group was a good measure of their aggressive abilities (ability to make an intruder retreat) at territorial boundaries. The more dominant adult males were able to defend to the greatest distance from the center of the territory, while the most subordinate non-breeders defended the least area (Craig 1976). This knowledge of the aggressive abilities of individuals, along with a study of the dynamics of territorial boundary disputes by groups of different size, gave us an insight into why Pukeko live typically in groups rather than in pairs.

Good-quality nesting areas appear to be in short supply and all groups suffer continual pressure from neighbors and non-territorial birds. At Shakespear Park, a new group formed whenever new nesting habitat was created (as a consequence of farm staff fencing an area and thus preventing stock from grazing). When park staff removed over 100 birds from about 13 territories outside our study area, all the high-quality territories were reoccupied by new groups within a few months. The origin of these new birds was unknown but none came from our long-established territories, indicating that there was a surplus of Pukeko waiting to take over vacant quality territories. This phenomenon appears to be typical for many cooperatively breeding birds and has been termed 'habitat saturation'.

A consequence of habitat saturation for Pukeko is the continual requirement for defense by the more effective adult breeding males. At Linton, one small valley contained two territories held by pairs. Breeding

Table 13.1. *Mean (±s.e.m.) number of male and female Pukeko breeders and non-breeders in three main study areas*

	Area		
	Linton	Pukepuke	Shakespear
Breeders			
Female	1.3 ± 0.2	2.0 ± 0.1 }*	2.0 ± 0.1 }**
Male	1.3 ± 0.2	2.3 ± 0.2	3.2 ± 0.2
Non-breeders			
Female	0	0.5 ± 0.3	2.1 ± 0.2 }**
Male	0	0.3 ± 0.1	1.2 ± 0.1

*, $P < 0.05$. **, $P < 0.05$.

was always asynchronous and hence the male in the territory with a current nest was only a part-time defender as he was also involved in incubation. The incubating male always lost part of his territory to his neighbor, although some was regained when the encroaching bird began nesting while the other was feeding chicks. Once chicks began to reach full size they helped with defense and the boundary moved back to its original position. This see-saw battle over the boundary went on each year and involved an area half the size of either territory (Craig 1979). Neither bird had another neighbor and with only one male opponent, each was able to persist in a pair for the length of the study.

At Pukepuke, one pair set up a territory at a time when neighboring groups were nesting and actively defended boundaries were short. By the time this pair had hatched their chicks, they had lost most of their territory due to the inability of the single male to defend two boundaries against groups with two males each. The following year this area was defended successfully by two new groups, each with more than one male (Craig 1979).

Each year in autumn and winter, birds from the flock encroached on the territory of a particular group which had two adult males. The flock was predominantly male and rapidly overran part of the area held by the resident group. Each year this group allowed one of the more dominant flock males to join their group and so they gained an extra defender and maintained the rest of the territory against the flock. As breeding approached and the flock dispersed, trespass declined and the group expelled this extra male.

These examples illustrate the general trend of a positive relationship between the number of males in a group and (1) the length of territorial boundaries $(215 \pm 69.4\,\text{m}:47.5 \pm 15.4\,\text{m}$, multimale:single male; $t = 4.8$, 13 d.f., $P < 0.001$) and (2) the number of males in adjacent groups $(6.1 \pm 1.9:1.6 \pm 1.2$ neighboring males, multimale:single male; $t = 5.5$, 3 d.f., $P < 0.001$) (Craig 1984). These patterns suggested to us that the number of adult males in a group determines a group's ability to retain a breeding territory. Why then do not all groups have many males to ensure possession of a territory and the opportunity to breed?

Adult males appear to be effective in maintaining a breeding territory but a consequence of this coactive defense is that reproductive output must be shared among the males. This would not be a problem in terms of individual fitness of the males if larger groups were more successful in chick production. However, there as a negative relationship between group size and reproductive success in all areas (see below). Thus, the risk to a lone male of losing his territory with the concomitant chance to breed increases as population density increases, but his alternative of admitting an additional male means his genetic contribution to following generation probably declines. If he is the first to admit an additional male and he can then take over part of a neighbor's territory, overall success may increase. If his neighbor counters by also admitting an additional male, both the original males are left with the same size territory but their own reproductive success has declined as they must now share the area and females with additional males. This scenario is similar to the Prisomer's Dilemma game put forward by Axelrod and Hamilton (1981). This game demonstrates how 'cooperative' or coactive behaviour between two individuals may be the only *stable* solution to a problem, even though they may be at a disadvantage compared to when acting alone (Craig 1984).

Breeding success

Reproductive success of Pukeko was highly variable. Environmental conditions, especially the presence of water and green grass, were positively correlated with reproductive success. Cover was also important for survival of eggs and chicks (Craig 1980a). The effects of these factors are best demonstrated by comparing within and between habits.

At Linton, water levels were always high and Pukeko managed to raise chicks successfully from two clutches in all years. At Pukepuke, the first clutches of most groups were successful. During late first clutches and second clutches, water levels in the lake decreased and none of the later clutches was successful (Craig 1980a). At Shakespear, there was little water by the time the second clutches would have been laid and many groups laid only one clutch. In 1984, farm staff fenced off one area of swamp and cover within this area increased dramatically over previous years. The five groups using this area had markedly higher chick survival to independence at four months than did the eight groups that used other swamps still grazed by stock (3.8:1.4 chicks/group; $P < 0.01$, Mann–Whitney U-test). Some groups in territories which lacked water and cover produced no chicks in any year. These results further demonstrate the degree of habitat saturation; not only is there a lack of vacant area where Pukeko can establish territories, but some areas that are occupied appear to be poor breeding habitat.

In addition to these environmental effects on reproductive success, there was also a negative trend with regard to group size. Territories were all of similar size, but the more birds in a territory the smaller the clutch size and the fewer chicks raised. Per territory, per breeder or per territory member, productivity was higher for pairs than it was for groups (Craig 1980*a*). Thus, unlike cooperative breeders, Pukeko that bred as pairs had the highest reproductive success.

Adult mortality varied from year to year (mean of 0.3–1.7 adults lost per year per group) but averaged one bird per group per year. A few groups lost as many as five adults in one year and the remaining group members were subsequently displaced from their territory by neighboring groups. In spite of these fluctuations, mean group size varied only by one bird over the seven years of study at Shakespear and did not vary greatly at the other study areas either (Table 13.2).

Dispersal and inbreeding

The majority of chicks (75%) produced in any breeding effort die within the first three months after hatching. In most pair territories, the young that survived were expelled as yearlings prior to the next breeding season. Young males and females from pair territories at Linton dispersed and were often seen in adjacent territories and surrounding areas. At times they returned to their natal territory after being away for a number of days, suggesting that they used it as a base until they became established in a new territory. Flock birds also dispersed widely from where we banded them and one was found 90 km away. Such long-distance dispersal has been noted elsewhere (Sutton 1967) and may explain the presence of this species on some small islands distant from large land masses. Few birds moved large distances and most never left their natal territory. Of 16 birds known to have dispersed, the mean distance moved was 11.2 (±6.0) km.

Data from our long-term study at Shakespear Park shows that, of the birds for which we have three or more years of sightings, only eight (9%) actively dispersed from year-round territories and all were males. Four of these were breeding adults that emigrated to an adjacent territory at the start of the breeding season. They maintained membership to both their old and new territories briefly for a period of weeks. They were able to do this presumably because they were the dominant male in their old territory and became the dominant male in their newly adopted territory as well. Where known, these movements by breeding males were to territories with higher previous reproductive success or with fewer breeding males. The other four males that dispersed were subadults leaving their natal territories. Such male-biased dispersal contrasts with the pattern for other cooperative breeders (see other chapters in this volume).

Almost all birds remained in their natal group until at least two years of

Table 13.2. Pukeko population structure by year for the three study areas

	Linton			Pukepuke			Shakespear		
	No. of territories	$\bar{x}$ no. of breeders	$\bar{x}$ group size	No. of territories	$\bar{x}$ no. of breeders	$\bar{x}$ group size	No. of territories	$\bar{x}$ no. of breeders	$\bar{x}$ group size
1970–1	4	2.5	2.8	5	4.3	6.3			
1971–2	4	2.8	2.8	8	4.0	4.6			
1972–3	4	2.3	2.7	10	4.6	5.0			
1979–80							13	4.4	6.8
1980–1							13	4.6	7.5
1981–2							13	6.2	6.9
1982–3							13	4.5	7.8
1983–4							15	5.6	6.7
1984–5							15	5.4	6.6
1985–6							15	5.3	7.8

age and most (71%: 28 males, 31 females) remain permanently. (A minority of birds (20%: 9 males, 7 females) lost possession of their territory but did not leave the area. They established a new territory, joined a neighboring group or became non-territorial 'floaters'). Marked groups were known to admit unrelated birds in only three group-years, whereas no external recruitment occurred in the remaining 23 group-years. Generally, non-breeders become breeders in their own natal group and often end up copulating incestuously. The diagrammatic representation of three years of group PG-2 demonstrates this (Fig. 13.3). Gradual death of breeders led to young of this group attaining reproductive status. Thus, by the third year the only breeding female is the likely daughter of one of the males and the likely brother of another. Both copulated with her and produced four normal chicks.

The dynamics of group membership where young are retained in their natal territories and eventually become breeders, along with minimal intergroup movement, means that the probability of incestuous mating must be high. Multiple paternity and maternity does mean that few chicks are likely to be full siblings. However, if only offspring of previous years become breeders in most territories then the levels of relatedness among group members must be extremely high, although the exact relatedness is impossible to calculate. Futhermore, as the few observations of dispersal were predominantly to adjacent territories, there must be considerable genetic relatedness among neighboring groups as well.

We observed a few examples of what might be considered incest avoidance behavior. Four subadult males dispersed from group territories containing their probable mother. Two of these males were observed attempting to copulate with their mother but were stopped by her aggressive behavior. Three additional males were observed to have reduced sexual activity when their mother was present in their territory (Craig and Jamieson 1988). Male Pukeko normally dominate females during interactions over food (see below), but breeding females tend to dominate young males. Therefore these examples of possible inbreeding avoidance are confounded by parental dominance. That males never dispersed or showed reduced sexual activity when sisters, but not mothers, were present and that females showed neither response in the presence of fathers or brothers is suggestive that early parental dominance may affect the reproductive behavior of young males. Older males were observed copulating with their likely mother ($n = 12$). In total, we recorded 34 cases of incest as compared to only seven of possible incest avoidance and 17 of

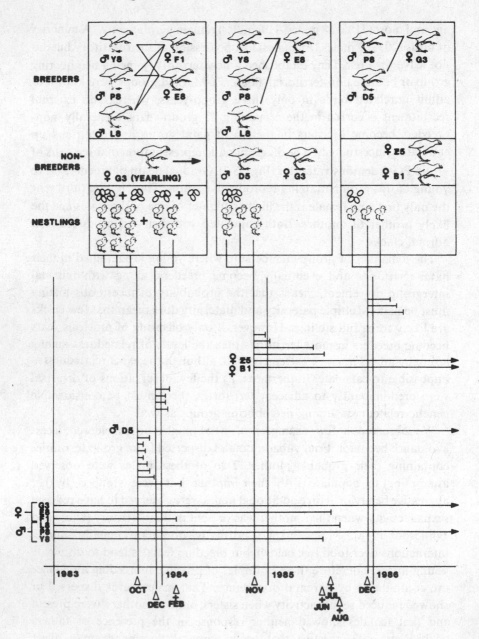

Fig. 13.3. Group membership of a Pukeko group (PG-2) over a $3\frac{1}{2}$ year period. In 1984 the group had a typical number of breeding members: three males and two females that all copulated with each other. One yearling from the brood of the previous year was also present and assisted in feeding the young. Two clutches were

likely outbreeding. We have observed no examples of deformed young or anything that could be considered inbreeding depression.

Dominance hierarchies

Social patterns within groups are easily followed. Provision of small amounts of grain causes birds to congregate in a small area and interact with each other (Fig. 13.4). Within a 2h observation session, sufficient interactions among all members of the group can be recorded to allow us to rank them in a dominance hierarchy. Status is absolute and few reversals are seen. Observations away from artificial sources of food confirm the hierarchical order obtained there (Craig 1979).

Pukeko act in accordance with their status at all times. Subordinates give way to dominants and frequently signal their submissiveness by *Bowing* (a posture where they tilt their body forward and put their head on the ground) of by *Headflicking* (a posture where the beak is pointed upwards). Aggressive postures of dominants tend to involve holding the head high and pointing it downwards. These postures (Craig 1977) allow us to suggest that the red beak is the focus for aggressive and dominance actions. Presenting the beak in a position for which a peck can be delivered signals aggression, whereas putting the beak in a position that makes pecking unlikely, such as a bow or a headflick, signals submissiveness. Extreme submissiveness is communicated by raising the tail and wings to expose and frame the white undertail coverts as the bird retreats. Avoidance and the adoption of appropriate displays ensures that overt aggression within a group is rare. However, even though status is so readily observed, apart from the consequences of ordered group interaction and priority of access to food, the functional importance of status is unknown. This is particularly true when we look at the role dominance plays in mating success.

Caption for Fig. 13.3 (*cont.*)

produced and each had eggs from the two females, although the subordinate female (E8) laid fewer eggs (4 + 3) than the dominant (F1) (7 + 5 eggs). Most chicks hatched but few survived for more than a few months. One, a male (D5) survived to the following year when he along with his two year old sister (G3) were non-breeders that assisted in provisioning the chicks, two of which survived (Z5 and B1).

In 1984 the dominant breeding female died but was not replaced. In 1985 the only breeding female and two of the breeding males died and the two non-breeders became breeders. Thus, G3 became the only breeding female and copulated with her brother and father. These three breeders were assisted, in feeding the chicks, by the two yearling female non-breeders.

Breeding behavior

Courtship and breeding are truly communal events for Pukeko. Allopreening and allofeeding were seen between all sexually active members of a group but both were infrequent compared to attempted and completed copulations which begin up to two months before egg-laying. Not only do males copulate with females but intrasexual mountings by both sexes are common. Mating calls given prior to a copulation are noisy and as a result attract other breeding males from within the group. Loud calls even brought neighbors to the boundaries of their own territories on several occasions. The frequent and noisy mating behavior of the Pukeko made our job of recording who was participating all the easier.

Surprisingly, we found no evidence of mate-guarding behavior by dominant males (Jamieson and Craig 1987a) as has been found in other cooperative breeders (see Chapter 14). The dominant male can displace subordinates when they attempt copulation but he is inconsistent in this behavior. Sometimes he interrupts, sometimes he responds to copulations

Fig. 13.4. A group of seven Pukeko around an artificial food pile. Note the patagial wing tags for individual identification. The three outer birds are avoiding the more dominants around the food. The second bird from the left in the foreground has just grabbed food from under a more dominant and is running away. The most dominant bird, fourth from the left is about to peck a subordinate that is feeding.

but merely observes, while on other occasions, he will totally ignore a copulation attempt, even though he is in view and has been alerted by the precopulatory calls of other males. Particularly during the egg-laying period, females can be hounded by males and may copulate with up to three different males within the space of a few minutes. Multiple paternity of a clutch would be expected from such behavior and preliminary work using electrophoretic analysis of proteins has confirmed this.

Optimality arguments suggest that intense competition for paternity should result in dominant males interacting in such a way that they gain the most fertilizations. Our observations, however, suggest that dominant males are indifferent towards their potential mating competitors (Jamieson and Craig 1987a). Furthermore, when several males court a female simultaneously, they do not behave cooperatively in any strict sense of the word, but behave coactively in that they direct their behavior towards the female and do not appear to respond to the actions of other males.

For females, the alpha breeder is involved in more copulations (57%, $n = 126$; Craig 1980b) than is the beta, but dominants do not behave aggressively towards, or inhibit copulations involving, less dominant females. Alpha females generally lay more eggs $(4.1 \pm 0.3 : 3.6 \pm 0.5$ (means $\pm$ s.e.m.); Craig 1980a), but this appears to be related to age and experience rather than dominance. All the breeding females in a group lay eggs in the same nest at the same time. Where the clutch size is too large to be covered by a single bird, a second bowl is constructed adjacent to the first (Fig. 13.5) and two birds incubate at the same time. Unlike the observations for Acorn Woodpeckers (*Melanerpes formicivorus*) (see Chapter 14), Groove-billed Anis (*Crotophaga sulcirostris*) (Vehrencamp 1977; see Chapter 11) and Mexican jays (*Aphelocoma ultramarina*) (Trail *et al.* 1981; J. L. Craig, personal observation; see Chapter 9), we have no evidence that females attempt to reduce the reproductive contribution of other females within their group. Each female's eggs are of a distinctive color and pattern (Fig. 13.5), but the number of eggs laid is the number incubated. These patterns of minimal or no competition amongst breeding females as well as males contrast markedly with the observations of overt reproductive competition in other cooperative breeders.

Parental care behavior

Usually, only Pukeko which have previously copulated sub-sequently incubate the eggs. However, previous breeders who fail to copulate or lay eggs for one nesting attempt will still incubate any eggs. Even a few non-breeders are attracted to the nest before the eggs hatch and

Fig. 13.5. A twin-bowl Pukeko nest with a combined clutch of 19 eggs. Note the differences in patterning on the eggs in the nearest bowl, showing that at least two females had contributed to the nest. Two breeders would incubate at the same time at such a nest.

these birds have also been observed incubating. In general, breeders are much more likely than non-breeders to visit nests and so encounter eggs, for a variety of reasons, and hence breeders show a much higher incidence of incubation. These observations suggest to us that participation in incubation is dependent not on certainty of parentage (Craig and Jamieson 1985) but on likelihood of association with the nest and its contents.

Pukeko chicks are precocial and can leave the nest within hours of hatching. All of their food is fed directly to them by other group members for the first six weeks, after which they begin to feed themselves. Initially, chicks are kept close to cover and in areas of deep water, other group members cut vegetation to form a floating platform for the chicks. At night, young chicks are brooded in brood nests which are built in cover near current feeding areas.

Birds which participate in incubation are the first to care for the chicks soon after they hatch. However, as they are led from the vicinity of the nest to food, chicks are brooded and fed by all members of the group, even juveniles of broods from earlier in the season.

This outline of the natural history of the Pukeko shows similarities to and differences from, other cooperative breeders. As in all others, habitat saturation and a shortage of prime breeding territories appears to be important, and all birds help to feed chicks. Why non-breeders feed chicks appears open to a number of interpretations and will be considered in more detail. Like Acorn Woodpeckers (see Chapter 14), Pukeko have a polygynandrous mating system which involves mate-sharing by both males and females. Unlike Acorn Woodpeckers and other cooperative breeders, Pukeko appear to breed without competition or attempts to interrupt the efforts of other group members. Finally, although a high degree of philopatry is not unusual for cooperatively breeding birds, frequent incidents of incest may be. These questions and theoretical problems are the subject of the second part of this chapter.

Explanations and speculations

Cooperative or 'communal' breeding has become a major area of investigation within the field of behavioral ecology and sociobiology. One important reason for this interest is that helping behavior on one level of analysis appears to represent an evolutionary dilemma. Here is an example where individuals are apparently not looking after their own interest – the perpetuation of copies of their own genes – but are assisting the reproductive efforts of others. Not only has this been the major justification for the research on cooperative breeders, but the use of 'loaded' terminology such as altruism, helping, cooperation etc. has ensured that new and ongoing studies have been directed into the same mold. Thus the majority of studies to date have concentrated on trying to determine what is the adaptive significance (in terms of costs and benefits) of the behaviors which are associated with cooperative breeding. We do not deny that the results of these studies support current evolutionary hypotheses, but we wish to point out that this approach is not the only level of investigation of biological phenomena. Niko Tinbergen (1963) once suggested that the study of animal behavior should be subdivided into four areas: causation, development, function and evolution. In the following section, we would like to stress the need for a more integrated approach by employing all four of these aspects in order to obtain a broader understanding of cooperative breeding.

Does 'helping' need a special explanation?

In the course of normal breeding behavior by most mammals and birds, as well as some reptiles, amphibians and fish, offspring are fed and cared for by their parents. By so doing, parents increase their offsprings' chances of survival and subsequent reproduction. We accept that this standard stimulus–response interaction between parents and offspring is a product of natural selection.

In these same animal groups, there are a few species with populations that have individuals in addition to the parents that care for young. The behavior of these so-called altruistic individuals is explained by separate and often *ad hoc* adaptive reasoning. Such reasoning is most highly developed in the communal-bird literature. For example, we are told that helping behavior in the form of feeding nestlings has evolved in cooperative breeders because it improves an individual's lifetime reproductive success by improving breeding experience, serving as a payment to resident breeders for continued access to limited resources on their territory, or improving the helper's own chances of obtaining a reproductive position. Helping may also improve the kin component of inclusive fitness by improving the breeding success of close relatives. However, rather than leap solely for explanations of evolution and adaptive significance (i.e. function) and rather than treating 'helping' as a new and unique behavior, why do not we also consider changes in the social environment of cooperative breeders and their effect on the development and expression of parental care behavior (Jamieson and Craig 1987b)?

Recent research by Jay Rosenblatt and his colleagues (see e.g. Rosenblatt *et al.*, 1979) has shown that the expression and maintenance of parental care in rats is largely the result of continuous stimulation from the young. Male and non-reproductive female rats who do not normally show parental care behavior can be induced to do so through association with newly born young. Even lactation in breeding females is enhanced by prior stimulation and is only maintained through contact with the pups. Adaptive reasoning would predict such a strong link between parenting and the presence of young. However, that male and non-breeding female rats do not normally exhibit parenting behavior is largely the consequence of a social system that limits the opportunities of encountering newborn young. Do most birds also have a latent potential to exhibit parental care?

Typically, birds are only exposed to young nestlings when they are parents and opportunities to care for young other than their own would be expected to be rare. However, in artificial situations where many birds are retained in aviaries, feeding by birds additional to the parents is well documented. There is also a growing number of examples of birds in the wild that suddenly adopt the young of a neighboring bird (frequently not even of the same species) when these young hatch before their own, or of non-breeding birds showing up at a nest and feeding the young (Shy 1982). Similarly, we are all familar with example of brood parasitism, where a nestling is raised by a host species. As with mammals, it does not seem to matter whether an individual is a breeder or not, as researchers have shown that reproductive or hormonal preconditioning is not a necessary prerequisite for exhibiting parental care behavior in birds (Lehrman 1964; Lofts and Murton 1973).

All of these examples presumably demonstrate the strong selection pressure acting on the neurological system of birds so that they react to the stimuli of active, gaping nestlings. The constraints on the operation of this stimulus–response system would make it difficult for a bird to react in some situations, but not in others. This is best illustrated by the ability of some species to discriminate and reject foreign eggs added to their nest by parasitic breeders, but the apparent inability of these same birds to discriminate and reject a foreign nestling added artificially to their broods. Lack of discrimination occurs even though the nestling in no way resembles its own offspring. Only in some colonial species does the ability to recognize young from their own nest develop rapidly and parents in these species tend to react indifferently or aggressively toward unfamilar chicks.

The stimulus–response system would also be operating in young immature birds that remain in their natal territories and that come in contact with begging nestlings. The only real difference then between parental behavior and helping behavior is the *context* in which we see the same sterotyped behavioral pattern. Therefore, rather than using the loaded word 'helping', which implies that the behavior is somehow different from that shown by the true parents, we should use a more descriptive term such as provisioning behavior. It can then be acknowledged that this behavior can be seen in a variety of contexts, including those involving breeders (or parents) and non-breeders (or auxiliaries).

This change in perspective alters the nature of the questions that we ask about cooperative breeders. The familiar questions such as 'Why has helping behavior evolved?' or 'Why do helpers help?' could be replaced by a more appropriate question 'Why do non-breeders or auxiliaries in cooperative breeders regularly provision nestlings?'. The key to answering this question can no doubt be found in every study presented in this book; the unique social environment of cooperative breeders, which involves young birds associating in a bonded, cohesive group with breeders, results in auxiliaries being exposed to offspring and the concomitant chance to provision. The standard cost–benefit measurements used in most studies are used to support the hypothesis that such behavior is adaptive in cooperative breeders. Because workers define helping behavior as a separate behavioral unit and because they use adaptive arguments, they are then either implicitly or explicitly supporting the hypothesis that helping behavior (as distinct from parental behavior) is the result of selection acting on genes for helping (as distinct from genes for typical parental behavior).

We cannot deny that such a hypothesis is supported because of the way in which the hypothesis is formulated in the first place. We can say, however, that this strictly functional approach is misleading because not only does it treat provisioning behavior by non-breeders as being a behavioral trait distinct from that of provisioning by breeders but it also treats the evolution of cooperative breeding and the evolution of 'helping' behavior as conceptionally separate (historical) events. So we find that Emlen (1982a) divided the topic into two separate parts; one dealing with a model for the evolution of social groups and the other dealing with the evolution of helping behavior once social groups have evolved. He justifies this separation by remarking that the ecological constraints model is only useful in predicting when extended family social organizations will be found in nature, but by itself it does not explain why non-breeding auxiliaries will actively contribute to the reproductive efforts of others. For him, and for others (e.g. Brown and Brown 1980), helping behavior will only evolve if it increases the inclusive fitness of the helper.

In contrast, we see an organism's behavior as the product of a dynamic interaction with its environment, and thus the behavior of auxiliaries in cooperatively breeding populations has 'evolved' without any necessary genetic and selective changes in the control of provisioning behavior. In other words, changes in life history parameters such as retention of non-breeders on breeding territories result in a significant change in the social environment. This *in itself* can lead to the early development and expression of inherent behavioral patterns which are not normally exhibited until later stages of ontogeny. Thus, provisioning by auxiliaries may represent an adaptive behavioral response of an unaltered genotype to a persistent (social) environmental condition.

Conceptualizing organism–environment interactions in this manner should allow us to make predictions about when provisioning by auxiliaries should occur among cooperatively breeding species. We would expect provisioning behavior to be expressed in direct proportion to the likelihood of non-breeders encountering and associating with dependent offspring. Thus, in cooperative nidifugous species such as Pukeko, where all group members are likely to encounter the precocial chicks, we would expect provisioning behavior by non-breeders, including chicks of earlier broods. In nidicolous species, with altricial young restricted to a nest, only individuals which are likely to have access to the nest will provision. Hence, species like Australian Black-backed Magpies (*Gymnorhina tibicen*), where parents usually exclude all other group members from the vicinity of the nest (C. Veltman, personal communication), are unlikely to exhibit a communal habit which includes provisioning by non-breeders. This argument, presented here in its simplest form, can only attempt to explain whether or not non-breeders will provision; separate arguments are needed to explain differences in frequency of provisioning among individuals of cooperative groups (see below).

Should breeders invest equally?

One interpretation of parental investment theory suggests that if an individual forgoes any chance of future reproduction it should invest according to the probability of parentage (see e.g. Trivers 1972; Joste *et al.* 1982). In Pukeko, incubation always incurs a cost of missing copulations for males, or reducing time available for females to accumulate sufficient resources needed to lay eggs in a subsequent clutch. In spite of this, breeding Pukeko always invest in incubation irrespective of probability of parentage (Craig and Jamieson 1985). For example, only one female will lay in some clutches whereas two will lay in others. However, regardless of whether or not females lay, all participate in incubation and chick care, a finding that appears to contradict the probability of parentage hypothesis but not the ideas of access and association to offspring. The latter hypothesis gains support when we observe non-breeding birds, where certainty of parentage is undeniable, incubating eggs and caring for young. The only way to make sense out of these observations under a functional framework is to perceive parental care behavior by non-breeders as having a different evolutionary origin. Yet this ignores the fact that most immature animals are capable of expressing many behaviors which are not normally exhibited until adulthood.

Why do some birds provision more than others?

Although all members of a cooperatively breeding group have the potential to provision nestlings, this does not mean that all will react equally to nestlings. Differences in physiology and prior experience between breeders and non-breeders, for instance, could result in differences in the level of responsiveness to young in a nest.

Observations of Pukeko (Jamieson 1988) show that breeders which

Fig. 13.6. The percentage of time that different classes of non-breeding Pukeko fed chicks. Dashed lines represent inexperienced yearlings while solid lines represent older, more experienced non-breeders.

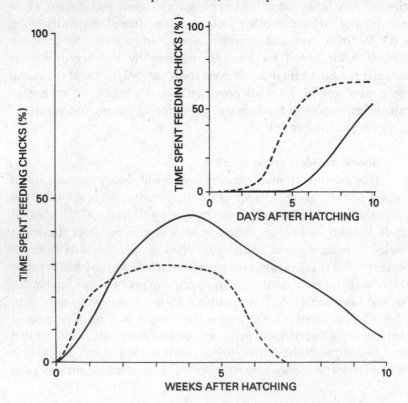

incubate are the first to brood and provision chicks when they hatch. This occurs as a result of incubators being sensitized to stimulations from chicks prior to hatching. Sensitization occurs through movement and vocalization of embryos, 'pipping' of egg shells and at hatching itself. Non-breeders that incubated also showed an early response to newly hatched chicks. However, from the first week till mid-way through the provisioning period, there is no significant difference in the percentage of time spent provisioning by breeders and non-breeders (Sign test, $n = 105$, $P > 0.10$). This is in agreement with the outcome of detailed studies on parental care in rodents. It appears that the maintenance of provisioning is not due to increased hormone levels attained by breeding or during incubation but is due primarily to contact with chicks.

Inexperienced yearling Pukeko provision at a higher rate during the first 10 days after hatching than older, more experienced non-breeders (Fig. 13.6). This heightened response appears to be due to naive birds being exposed to the novel stimulus of young hatchlings, as the same birds showed a reduced response during this 10 day period when they were two-year-olds. However, the novelty effect waned quickly in yearlings and older birds usually assumed the majority of chick care (Fig. 13.6).

Many functional explanations have been given to account for asymmetries in provisioning in other species. For example, older individuals of the sex most likely to inherit all or part of the natal territory is expected to gain more by aiding younger siblings, who could help to defend the territory until the older helper gained breeding status (Woolfenden and Fitzpatrick 1978). Or, the sex more likely to disperse with other siblings should gain more by helping to raise them (Ligon and Ligon 1978).

The above observations on Pukeko hint at the importance of viewing provisioning behavior as a basic stimulus–response interaction, subject to positive feedback effects. The observed patterns described here would not have been detected if the number of provisioning incidents for experienced and inexperienced non-breeders were simply summed. Therefore instead of presenting data on asymmetries in behavioral patterns in the form of absolute or relative differences in frequency, and then attempting to explain these differences by solely adaptive reasoning, it may be more profitable to devote more attention to the dynamics of how asymmetries develop.

This alternative approach to studying cooperative breeding can also address questions of whether only relatives should provision. A functional approach requires arguments of kin selection or reciprocal altruism to explain who participates. We would suggest that if, as for Pukeko, only chicks of previous seasons are present in a group as non-breeders, then only kin will provision. This occurs purely as a consequence of the social structure of Pukeko where outsiders join only as breeders. In other species such as Pied Kingfishers (*Ceryle rudis*) (see Chapter 17) and Green Woodhoopoes (*Phoeniculus purpureus*) (see Chapter 2), non-relatives have the opportunity to encounter young and so participate in provisioning. The point is, provisioning behavior occurs in cooperative breeders regardless of coefficents of relatedness or opportunites for future reciprocations. This is why we recommend a departure from these types of argument as the only explanations for the occurrence of so-called helping behavior.

Incest: a consequence of kin social systems?

A final example of the types of problem that are inherent in functional thinking in biology today is our outlook on incest. Most workers on cooperatively breeding species have at some stage addressed the problem of incest and the likelihood of selection for inbreeding avoidance mechanisms. This appears to be based on an almost universal acceptance that inbreeding is always deleterious.

It is tacitly assumed that potential disadvantages of inbreeding resulting from segregational load (e.g. expression of recessive lethals and the loss of heterozygosity) far outweigh any advantages resulting from lowered recombinational load (e.g. preservation of parental gene complexes). Also it is assumed that the advantages of avoiding extreme inbreeding are greater than the costs of migration and exposure to differing habitat variables.

A growing number of researchers are questioning the assumption that inbreeding is *always* detrimental (see e.g. Shields 1983). Recent work has suggested that extreme outbreeding can be equally or more disadvantageous than extreme inbreeding and that many organisms typically achieve a mixture between the two. However, several factors can affect the likelihood of the relative cost–benefits of any position along the inbreeding–outbreeding continuum. A long history of inbreeding may result in a loss or marked reduction in the frequency of deleterious recessive alleles and so initial selection may have reduced the disadvantageous genetic consequences of inbreeding. Also, high fecundity coupled with high ecological mortality may make the loss of offspring due to genetic effects inconsequential. Furthermore, differential fusion of gametes according to genotype, as occurs in some plant species, could reduce the chances of harmful genetic effects prior to fertilization. The first two of these factors could have direct relevance to the breeding system of Pukeko; the third is an unknown quantity in animals and needs further investigation (for more details, see Craig and Jamieson 1988).

Usually we have no accurate way of accessing the various costs and benefits associated with the inbreeding–outbreeding continuum and thus we should not presume any advantage or disadvantage from inbreeding. Nor can we draw conclusions about the existence of inbreeding avoidance mechanisms from observations of birds dispersing from their natal territory without first eliminating confounding variables such as intrasexual competition and parental dominance. Recent results from the cooperatively breeding Splendid Fairy-wren (*Malurus splendens*) (see Chapter 1) found that inbreeding was no more deleterious than outbreeding and suggest that our results of high occurrence of inbreeding among Pukeko may not be unique.

What should be the directions for future research?

Research on cooperative breeders provides some of the best long-term information on the behavioral ecology of social animals. The database of natural history is immense. The most unfortunate aspect of this avenue of study has been the preoccupation with, or over emphasis of, a single level of functional explanations. The importance of developmental factors and aspects of social learning have only been acknowledged in passing.

We hope that our work with Pukeko will stimulate others to broaden the scope of their investigations. No one can predict what will be the most fruitful avenues of research, but unless a more objective and integrated approach is attempted we will always be left with only one answer. That there is only one answer can in no way be used to confirm whether it is correct or not, it will be purely a demonstration that only one form of questioning was used. If Tinbergen (1963) was correct, there must be more than one level of question and answer.

Bibliography

Axelrod, R. and Hamilton, W. D. (1981). The evolution of cooperation. *Science* **211**: 1390–6.

Brown, J. L. and Brown, E. R. (1980). Kin selection and individual fitness in babblers. In *Natural Selection and Social Behaviour: Recent Results and New Theory*, ed. R. D. Alexander and D. Tinkle, pp. 244–56. Chiron Press: New York.

Carrol, A. L. K. (1966). Food habits of pukeko (*Porphyrio melanotus*, Temminck). *Notornis* **13**: 133–41.

Craig, J. L. (1976). An interterritorial hierarchy: an advantage for a subordinate in a communal territory. *Z. Tierpsychol.* **42**: 200–5.

Craig, J. L. (1977). The behaviour of the Pukeko, *Porphyrio porphyrio melanotus*. *N.Z. J. Zool.* **4**: 413–33.

Craig, J. L. (1979). Habitat variation in the social organization of a communal gallinule, the Pukeko *Porphyrio porphyrio melanotus*. *Behav. Ecol. Sociobiol.* **5**: 331–58.

Craig, J. L. (1980a). Breeding success of a communal gallinule. *Behav. Ecol. Sociobiol.* **6**: 289–95.

Craig, J. L. (1980b). Pair and group breeding behaviour of a communal gallinule, the Pukeko *Porphyrio porphyrio melanotus. Anim. Behav.* **32**: 23–32.

Craig, J. L. (1984). Are communal Pukeko caught in the prisoner's dilemma? *Behav. Ecol. Sociobiol.* **14**: 147–50.

Craig, J. L. and Jamieson, I. G. (1985). The relationship between presumed gamete contribution and parental investment in a communal breeder. *Behav. Ecol. Sociobiol.* **17**: 207–11.

Craig, J. L. and Jamieson, I. G. (1988). Incestuous mating in a communal bird: a family affair. *Amer. Nat.* **131**: 58–70.

Craig, J. L., McArdle, B. F. and Whettin, P. D. (1980). Sex determination of the pukeko or swamphen. *Notornis* **27**: 287–91.

Emlen, S. T. (1982a). The evolution of helping. I. An ecological constraints model. *Amer. Nat.* **119**: 29–39.

Emlen, S. T. (1982b). The evolution of helping. II. The role of behavioral conflict. *Amer. Nat.* **119**: 40–53.

Jamieson, I. G. (1988). Provisioning behaviour in a communal breeder: an epigenetic approach to the study of behavioural asymmetries. *Behavior* **104**: 262–80.

Jamieson, I. G. and Craig, J. L. (1987a). Dominance and mating in a communal polygynandrous bird: cooperation or indifference toward mating competitors? *Ethology* **75**: 317–27.

Jamieson, I. G. and Craig, J. L. (1987b). Critique of helping behaviour in birds: a departure from functional explanations. In *Perspectives in Ethology, vol. 7,* ed. P. Bateson and P. Klopfer, pp. 79–98. Plenum Press: New York.

Jamieson, I. G. and Craig, J. L. (1987c). Male–male and female–female courtship and copulation behaviour in a communally breeding bird. *Anim. Behav.* **35**: 1251–3.

Joste, N. E., Koenig, W. D., Mumme, R. L. and Pitelka, F. A. (1982). Intragroup dynamics of a cooperative breeder: an analysis of reproductive roles in the acorn woodpecker. *Behav. Ecol. Sociobiol.* **11**: 195–201.

Lehrman, D. S. (1964). Control of behavior cycles in reproduction. In: W. Etkin (ed.) *Social behavior and organization among vertebrates,* ed. W. Etkin, pp. 143–66. University of Chicago Press: Chicago.

Ligon, J. D. and Ligon, S. H. (1978). Communal breeding in green woodhoopoes as a case for reciprocity. *Nature (Lond.)* **276**: 496–8.

Lofts, B. and Murton, R. K. (1973). Reproduction in birds. In *Avian Biology,* vol. III, ed. D. S. Farmer and J. R. King, pp. 1–107. Academic Press: New York.

Rosenblatt, J. S., Siegel, H. I. and Mayer, A. D. (1979). Progress in the study of maternal behavior in the rat: hormonal, nonhormonal, sensory, and developmental aspects. In *Advances in the Study of Animal Behaviour,* vol. 10, ed. J. S. Rosenblatt, R. A. Hinde, C. Beer and M. C. Busnel, pp. 226–311. Academic Press: New York.

Shields, W. M. (1983). Optimal outbreeding and the evolution of philopatry. In *The ecology of animal movement,* ed. I. R. Swingland and P. J. Greenwood, pp. 132–59. Oxford University Press: London.

Shy, M. M. (1982). Interspecific feeding among birds: a review. *J. Field Ornith.* **53**: 370–93.

Sutton, R. R. (1967). Strong homing instinct in a pukeko. *Notornis* **14**: 221.

Tinbergen, N. (1963). On aims and methods of ethology. *Z. Tierpsychol.* **20**: 410–33.

Trail, P. W., Strahl, S. D. and Brown, J. L. (1981). Infanticide and selfish behaviour in a communally breeding bird, the Mexican Jay (*Aphelocoma ultramarina*). *Amer. Nat.* **118**: 72–82.

Trivers, R. L. (1972). Parental investment and sexual selection. In *Sexual Selection and the Descent of Man,* ed. B. Campbell, pp. 136–79. Aldine: Chicago.

Vehrencamp, S. L. (1977). Relative fecundity and parental effort in a communally nesting anis, *Crotophaga sulcirostris. Science* **197**: 403–5.

Wolfenden, G. E. and Fitzpatrick, J. W. (1978). The inheritance of territory in group-breeding birds. *BioScience* **28**: 104–8.

14 ACORN WOODPECKERS: GROUP-LIVING AND FOOD STORAGE UNDER CONTRASTING ECOLOGICAL CONDITIONS

14

Acorn Woodpeckers: group-living and food storage under contrasting ecological conditions

W. D. KOENIG AND P. B. STACEY

Acorn Woodpeckers (*Melanerpes formicivorus*) are distributed in foothill and montane habitats in Oregon, California, the American south-west, western Mexico, and southward through Central America to Colombia. Throughout their range, they are closely associated with oaks (genus *Quercus*), most commonly being found in pine–oak woodlands. They are generally quite common and conspicuous, and are well known for their unique habit of storing acorns, often by the thousands, in specialized trees known as storage trees or granaries. Acorn storage is characteristic of many, although not all, populations of Acorn Woodpeckers throughout their range.

Acorn Woodpeckers are also cooperative breeders. Within the family Picidae, they share this habit with the Red-cockaded Woodpecker (Chapter 3) and several tropical and Caribbean forms which, like the Acorn Woodpecker, belong to the melanerpine line. Birds in social units store and defend acorns and other mast communally. Although acorns constitute a major portion of their diet, particularly during the winter, Acorn Woodpeckers also engage in a wide variety of other foraging techniques, including sapsucking, flycatching, bark gleaning and seed eating.

We studied Acorn Woodpeckers at three sites (Fig. 14.1): Hastings Reservation (HR) in California (W.D.K.; this study, begun by MacRoberts and MacRoberts (1976), has been ongoing since 1971), Water Canyon (WC), New Mexico (P.B.S.; primarily between 1975 and 1984), and the Research Ranch (RR), Arizona (P.B.S. and C. Bock; conducted from 1975 to 1978). The Hastings study site (mean annual rainfall 56 cm), located near Monterey in central coastal California, is coastal oak woodland, and contains five common, and an additional two uncommon, species of oak. Water Canyon, located in the Magdalena Mountains of central New

Mexico (mean annual rainfall 29 cm), consists of riparian canyon bottom and pine–oak woodland and contains two species of oak. The Research Ranch in south-eastern Arizona (mean annual rainfall 43 cm), is primarily oak savannah and oak woodland, and contains two species of oak. All three studies involved exhaustive color-banding and extensive behavioral observations. We are thus able to provide an overview of cooperative breeding in the Acorn Woodpecker based on comparative data from three geographically distinct populations living under very different ecological conditions. An earlier demographic comparison of the HR and WC populations was made by Stacey (1979*a*).

Territories and mast storage

Acorn Woodpeckers are highly territorial, and in most parts of their range are permanent residents (Stacey and Koenig 1984). As mentioned above, territories in temperate regions usually contain one or more granaries, which the birds defend aggressively from both intra- and interspecific competitors. Granaries are filled each year with mast, usually

Fig. 14.1. Photos of representative parts of the study sites. (*a*) The Research Ranch, at the edge of the Huachuca Mountains in south-eastern Arizona near Elgin. (*b*) Water Canyon, in the Magdalena Mountains of central New Mexico. (*c*) Hastings Reservation, in the northern Santa Lucia Mountains of central coastal California.

acorns, which are typically produced in the autumn. Food storage appears to be more variable in tropical areas: granaries exist in some areas (see e.g. Miller 1963), but the available evidence suggests that birds in many tropical areas do not store food (e.g. Kattan 1988), or that, if they do, they often store in bromeliads and other natural locations rather than in granaries (Stacey 1981).

Mast stores play a critical role in allowing groups in temperate areas to remain on territories during the winter, when few alternative food sources are available, and to breed successfully the subsequent spring. Mast harvest and use is a communal activity. That is, individuals do not store or defend their own cache of acorns. Instead, all group members work together to harvest nuts, which become a communal resource freely available to all group members. Granaries and the acorns stored in them have important influences on virtually all aspects of Acorn Woodpecker biology, and almost certainly are related to the birds' group-living habits (see also Trail 1980). Indeed, one benefit of group-living in this species may be the collective defense of storage facilities.

Mast storage allows Acorn Woodpeckers to extend the availability of a highly seasonal resource. Groups that eventually exhaust their stores may be forced to leave and to move considerable distances in search of food. In WC, for example, 41 of 55 groups (75%) that exhausted their stores before 1 May subsequently abandoned their territories. Depending on the magnitude and extent of acorn crop failures, birds that abandon their territories may find space and acorns close by (see e.g. MacRoberts and MacRoberts 1976). Other birds are forced to leave the immediate vicinity, but remain close enough to return the following spring; of the 41 groups that abandoned territories at WC, for example, birds from 14 (34%) returned to reoccupy their original territories at a later date. Similar patterns of abandonment and reoccupation occur during years of local crop failure at HR (Hannon *et al.* 1987).

In some years, widespread failure of the mast crop can lead to permanent disappearance of a large proportion of the population. For example, at WC a virtually complete mast crop failure in the autumn of 1978 led to the abandonment of 17 of 19 territories (89%); at HR, a similarly poor crop in autumn 1983 resulted in the abandonment of 24 of 34 territories (71%). The fate of most birds forced to abandon their territories in years of widespread crop failure is unknown, but almost certainly most die.

Mast failures can lead to precipitous population declines. A crop failure at HR in autumn 1978 reduced the breeding population from 109 in spring 1978 to only 48 birds in spring 1979, a decline of 56% in one year. Crop failures also result in widespread changes in territory ownership and thus profoundly influence the continuity of genetic lineages occupying different territories. At WC, for example, 10 of the 17 territories (59%) abandoned in 1978 were later recolonized by an entirely different set of individuals, and 70% of the 37 previous group members on territories which were not permanently abandoned also disappeared. Current evidence indicates that such events occur fairly frequently, on the order of every four to five years (that is, approximately once per woodpecker generation), even at HR, where oak diversity is high (Hannon *et al.* 1987). Largely as a result, there is a significant degree of annual variation in population size (Fig. 14.2). This finding is at odds with the conventional view that cooperative breeders tend

Fig. 14.2. Annual variation in Acorn Woodpecker population size at Water Canyon (1975–84) and at Hastings Reservation (1973–86). Both graphs show the number of birds present at the start of each breeding season and the number of young fledged by the population during the breeding season, plotted cumulatively. Major population declines are due to acorn crop failures. At WC, a crop failure took place in the autumn of 1978, while at HR, crop failures occurred during the autumns of 1978 and 1983.

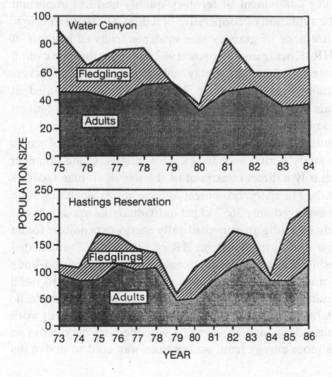

to live in relatively stable environments that lack such dramatic ecological shifts in conditions (see e.g. Brown 1974).

The ability of a group to store sufficient food to last through the winter depends in part on mast production and group size, but more so on the size of the storage facilities available. Thus, large granaries result in a direct, immediate benefit in terms of overwinter food availability. In WC, 58% of groups ($n = 55$) that had large storage facilities (defined as >3000 holes) were able to accumulate enough stores to last through the winter and into the following breeding season, whereas those with medium facilities (1000 to 3000 holes; $n = 47$) and small facilities (<1000 holes; $n = 62$) had only a 23% and 9% chance, respectively, of their stores lasting through the winter ($G = 34.7$, d.f. $= 2$, $P < 0.001$). This advantage feeds back into a cycle of increasing survivorship and reproductive success, leading to increased group size and increased ability to drill yet more holes and create even larger granaries. A limit on granary size is set by physical characteristics of the granaries themselves, including their size, resistance and longevity. Ultimately, granary size is probably set by the long-term availability of acorns in an area, since birds make new holes in winter only when they already have mast stores (MacRoberts and MacRoberts 1976). In any case, granary size is a key component of territory quality and has important implications for the evolution of cooperative breeding, as discussed below.

Despite the importance of granary size, evidence from an analysis of acorn storage at HR (Koenig and Mumme 1987) suggests that the overall contribution of stored acorns to the energy budget of Acorn Woodpeckers is surprisingly small. Stored acorns in 24 granaries were counted at bimonthly intervals for seven years and their total energetic content estimated from knowledge of their size, species composition, degree of insect damage, and nutritional composition. The maximum number of acorns stored during the study period was 36 846. This relatively small number appears to be primarily a direct constraint of the size of storage facilities available to the birds. The energetic content of these stores when they were at their maximum averaged only 3657 kJ per individual (the maximum was 6366 kJ per individual). Using an estimated daily energy expenditure for an Acorn Woodpecker during the winter at HR of 186 kJ day^{-1} (calculated from the relationship between total daily energy expenditure and body mass; see Koenig and Mumme 1987), these values are sufficient to fulfill only 11.3% (maximum) and 6.5% (average) of the total estimated energetic needs of the population between 1 December and 1 June. More recent work (Koenig, unpublished data) suggests that the woodpeckers may derive as much as 2.5 times more energy from acorns than was used to derive the above values. Using this revised estimate, stored acorns may fulfill up to 28.3% (maximum) or 16.3% (average) of their energetic needs between 1 December and 1 June, still a small fraction of their needs.

Besides storing them, Acorn Woodpeckers also eat acorns directly off the trees in the autumn. Furthermore, the energetic role of acorns varies considerably from group to group, year to year, and in particular may be quite different in other populations. For example, there are areas in which granaries are considerably larger, containing 50 000 or more holes, than those at either HR or WC (e.g. Ritter 1938). A group containing such vast storage facilities has yet to be studied. None the less, the proportion of the energy budget of Acorn Woodpeckers at HR provided by stored mast is certainly far smaller than other apparently comparable systems. For example, Vander Wall and Balda (1977) estimated that a Clark's Nutcracker (*Nucifraga columbianus*) stores 2.2 to 3.3 times the number of pinyon pine seeds (*Pinus edulis*) than it needs to survive the six months of winter during which they are dependent on food stores. This comparison renders it all the more surprising that acorns are so important to the survivorship and reproductive success of Acorn Woodpeckers. Apparently, they provide a crucial backup source of 'fast food', quickly and easily eaten when alternative food resources are temporarily unavailable.

The central role of mast stores and granaries in the social life of Acorn Woodpeckers is indicated by the behavior of these birds at the RR in south-eastern Arizona (Stacey and Bock 1978). In this population, the birds breed almost exclusively in pairs (Fig. 14.3). Habitat is open oak savannah forming a transition between grasslands at lower altitudes and pine–oak woodlands at higher elevations. Only two species of oak are present, and mast production is highly variable from year to year. The birds in this population usually do not make storage holes, and most territories lack storage facilities. Acorn stores are generally placed in natural cracks and crevices of tree bark, and are usually exhausted soon after acorn production ceases in the autumn. The woodpeckers then individually leave their territories, wander briefly in the vicinity, and eventually disappear for the winter. Most likely, the birds migrate to the extensive oak woodland habitat in the Sierra Madre Occidental of northern Mexico, but this has not been confirmed.

Following abandonment in the autumn, territories at the RR remain unoccupied until the following spring, when the woodpeckers return. Of banded birds that migrated, 14 of 42 breeding adults (33%) and 1 of 18 juveniles (6%) came back to the study area; 10 of the adults and none of the juveniles returned to the territory that they had inhabited the previous autumn. Birds return independently, not as a group, and there is little fidelity to mates: of four cases where both members of the pair migrated and returned to breed on the study area the following spring, none re-paired with the same partner. The remaining birds disappeared and were not seen again; some undoubtedly dispersed and bred elsewhere. Overall, turnover is high: of 65 adults and juveniles banded between 1975 and 1977, only one was present in 1981.

This behavior contrasts with that observed in Acorn Woodpeckers at either WC or HR, where group size is generally much larger. Instead, it resembles that described for the Red-headed Woodpecker (*Melanerpes erythrocephalus*) and Lewis' Woodpecker (*M. lewis*), both close relatives of the Acorn Woodpecker. Both these latter species typically emigrate from

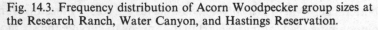

Fig. 14.3. Frequency distribution of Acorn Woodpecker group sizes at the Research Ranch, Water Canyon, and Hastings Reservation.

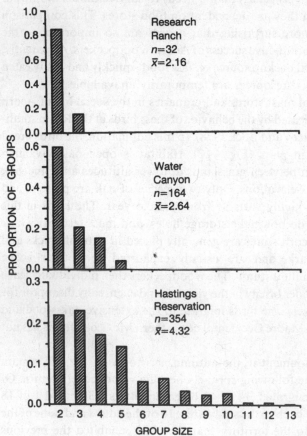

their breeding areas in the autumn, moving in search of regions of high mast production (Bock 1970; Moskovits 1978; Stacey, unpublished data). Both species defend winter territories individually rather than in groups, and store mast in natural locations rather than in self-constructed, individual holes. As for Acorn Woodpeckers at the RR, breeding occurs in temporary male–female pairs and there are few helpers.

Although most Acorn Woodpeckers at the RR are migratory and breed non-cooperatively, exceptions occur both in the study site and in nearby areas. On the RR itself, birds living in territories with access to artificial feeders were not only year-round residents but frequently bred in trios containing two males and a female. Failure to migrate in the autumn also occurred in 1977, when both species of oak produced a large crop. Five of the seven groups (pairs plus young of the year) present in November 1977 remained resident throughout the winter, although all of the young dispersed or disappeared prior to the spring.

Outside the study area proper, storage trees and cooperatively breeding groups are also found in the surrounding mountains, usually in well-watered canyons with many oaks. Thus, the capacity for acorn storage, permanent residency and cooperative breeding exists in this population, even though it is apparently expressed only when conditions are particularly favorable. This behavioral plasticity suggests that acorns play a key role in allowing populations to be resident, and, at least indirectly, may lead to group-living and cooperative breeding.

Social organization

Average group size of Acorn Woodpeckers varies considerably among the three study sites, ranging from a mean of over four individuals (maximum of 13) at HR, where birds are generally permanent residents, to marginally greater than a pair (maximum of three) at the RR, where they are usually migratory. However, even though there are considerable differences among populations in mean group size and composition, small groups of two or three individuals are most common in all three populations. The distribution of group sizes is presented in Fig. 14.3; population and group characteristics for the three populations are compared in Table 14.1.

The mating system of Acorn Woodpeckers can be described as being opportunistically polygynandrous. Within a group, up to four male co-breeders may compete for matings with one to three female co-breeders, who lay their eggs within the same nest cavity. Superimposed on this 'breeding core' of individuals may be up to 10 non-breeding helpers, usually

Table 14.1. *Group and population characteristics of Acorn Woodpeckers*

	Hastings Reservation	Water Canyon	Research Ranch
Mean breeding group size	4.39 ± 2.41 (2–12) (55%)	2.64 ± 0.89 (2–5) (11%)	2.15 ± 0.36 (2–3) (4%)
♂♂ breeders group⁻¹	1.84 ± 0.99 (1–4) (54%)	1.23 ± 0.53 (1–3) (11%)	1.05 ± 0.22 (1–2) (5%)
♀♀ breeders group⁻¹	1.19 ± 0.44 (1–3) (37%)	1.00	1.00
♂♂ non-breeders group⁻¹	0.70 ± 1.26 (0–5) (180%)	0.24 ± 0.53 (0–2) (79%)	0.10 ± 0.30 (0–1) (86%)
♀♀ non-breeders group⁻¹	0.49 ± 0.86 (0–5) (176%)	0.08 ± 0.28 (0–1) (104%)	0.00
Percentage of groups that are pair only	23.5 (35%)	59.8 (29%)	85.4 (8%)
Percentage adult non-breeders	25.0 (25%)	12.0 (64%)	5.0 (85%)

Data given are $\bar{x}$ ± s.d. (range), or $\bar{x}$ %; the lower value is the coefficient of variation based on annual values.

offspring from prior years. Thus, Acorn Woodpeckers exhibit two major types of cooperative breeding: 'helping at the nest' by non-breeding offspring and the much rarer phenomenon of 'mate-sharing' and 'joint-nesting' (referring to males and females, respectively) by individuals sharing breeding status within groups.

Although we are able to propose this outline for the mating system of Acorn Woodpeckers with reasonable certainty, the precise determination of the mating system of Acorn Woodpeckers poses considerable difficulties and is controversial. There are at least three reasons for this. First, the pairing behavior often seen in apparently monogamous cooperative breeders is absent. Second, as in many cooperative breeders (e.g. Florida Scrub Jays), copulations are only observed rarely; when they are seen, often more than one male is involved. Third, unlike many apparently monogamous cooperative breeders, all group members typically share in incubation, brooding and feeding of young. These characteristics are consistent with, but fail to prove, the hypothesis that Acorn Woodpeckers are not monogamous. We will first summarize the data concerning females, followed by the less conclusive evidence concerning males.

Evidence for joint-nesting is lacking from either WC or the RR, where groups containing more than one female are rare, but is certain at HR. In the latter population, three females have been observed to lay eggs in the same nest on two occasions and joint nesting by two females has been directly observed (based on checking the nest contents immediately following visits by females or by the laying of at least two eggs within a 24 h period) in at least 18 cases (see Mumme *et al.* 1983*b*). Indirect evidence for joint-nesting, on the basis of the presence of exceptionally large clutches, is even more extensive. Overall, about 22% of groups at HR are estimated to contain two or more breeding females, and nearly 36% of females nest jointly.

In 16 cases where the relatedness of (definitively) joint-nesting females was known, 13 were half or full siblings, two were of no known relationship, and one was a joint-nesting trio involving a mother and two of her daughters. Thus, 88% of this sample of joint-nesting females are close relatives, usually siblings. The reproductive bias between joint-nesting females is relatively small and not statistically significant; thus, each female is the mother of, on average, about half the offspring fledged from joint nests (Mumme *et al.* 1983*b*).

In addition to these joint-nesting females, groups at HR also may contain a variable number of additional females that do not lay eggs, even though they may participate in incubation and feeding of young. These birds are generally offspring of the breeders, and in particular the breeding male(s). Following the death and replacement of the males by a new, unrelated set of breeders, young females may subsequently nest jointly with their mothers, such as occurred in the case of three joint-nesting females mentioned above. This pattern of reproductive suppression and joint-nesting is consistent with the hypothesis that an 'incest taboo' exists within Acorn Woodpecker groups such that young remaining in their natal group act as non-breeding helpers as long as a parent of the opposite sex is in the group, but may breed once that parent is replaced by birds from elsewhere (Stacey 1979b; Koenig and Pitelka 1979; Koenig et al. 1984). However, these data are based on relatively few definitive cases, and many, although we believe not all, can at least in part be explained by an alternative hypothesis that reproductive competition, rather than incest avoidance, is driving the pattern of reproductive suppression observed in this species (Shields 1987; Craig and Jamieson 1988).

Although the data concerning the precise factors influencing the reproductive roles of females within groups are few, data for males are considerably worse. To begin with, the problems associated with proving the very existence of multiple paternity, much less its extent and reproductive consequences, have proved considerable. Extensive studies at both WC (Joste et al. 1985) and HR (Mumme et al. 1985), using starch-gel electrophoresis of blood and muscle proteins have yielded but one definitive case of multiple paternity within a clutch (at WC) and one additional highly probable case (at HR). Consequently, we are reduced to relying on more indirect evidence of the reproductive roles of males within groups.

There are four types of indirect evidence suggesting that multiple paternity occurs. First is the lack of typical pairing behavior and communal sharing of nesting activities described above. Particularly notable for males is that more than one male frequently incubates and broods at night (as far as is known, breeding males perform nocturnal incubation in all species of woodpeckers). Second, in the WC population, apparently unrelated males frequently join groups already containing a potentially breeding male (Stacey 1979b). When such males enter the group prior to egg-laying, they invariably participate in the feeding of young, whereas they do not feed, and may even kill the young, when they enter the group following egg-laying (Stacey 1979b; Stacey and Edwards 1983). This suggests that, at least under these circumstances, male feeding-behavior is directly linked to breeding, and that more than one male will reproduce and subsequently help to raise young. Third, studies of mate-guarding by males at HR demonstrate that several males may guard a single breeding female in a group (Mumme *et al.* 1983*a*). The incidence of guarding generally follows the converse of the situation described above for females; that is, male siblings that have filled a reproductive vacancy and joined a group guard the breeding female, while male offspring do not guard their mothers, but will guard new, unrelated breeding females simultaneously with their father. Fourth, recent experiments at HR suggest the possibility that males temporarily removed during egg-laying, and thus whose confidence of paternity is forced to zero, may destroy the nest when returned to their group and force a renest in which they may participate. This suggests that, similar to female egg destruction (see below), males may have to share paternity fairly evenly in order that all group members participate in raising young.

These indirect lines of evidence, combined with the electrophoretic data, are all consistent with, but again do not prove, the hypothesis that multiple paternity occurs regularly and that male reproductive behavior is constrained by the same factors important to females; that is, that male offspring act as 'non-breeding helpers' while their mother (or other closely related breeding females) is present in the group, but may inherit and breed in their natal group along with their father once a new, unrelated famale is present.

A primary source of difficulty is that females behave during the breeding season in ways that make observations of copulations almost impossible. For example, only 26 copulations were recorded in over 1400 h of observations at WC devoted specifically to observing copulations in multi-male groups. Those copulations that were observed occurred almost exclusively at dusk just prior to roosting, when identification of individuals was extremely difficult (for humans and probably the woodpeckers as well).

Just before matings, females often fly away when a male approaches, and may lead males on extended chases throughout the territory. Copulation itself is very brief, lasting only 3 to 5 s, and is not accompanied by vocalizations. When two or more males land together with the female at a copulation perch, interference between the males may or may not take place. These behaviors contrast with those of the Acorn Woodpecker's two close relatives, the Red-headed and Lewis' Woodpeckers. In both these latter species, copulations usually occur on exposed perches during the morning hours, are frequent and prolonged, and are easily observed (Stacey, personal observation).

By making mating cryptic and difficult to detect, and by mating multiply with several males in a group, females increase the parental uncertainty of each male and make it more difficult for one male to monopolize group reproduction. Because males that copulate are more likely to feed young and less likely to disperse (Stacey 1979*b*, 1982), these behaviors probably result in increased total parental care by the co-breeding males and thus contribute to the greater reproductive success of groups containing more than one male (see below).

An unfortunate side effect of these behaviors is that it has been impossible to unravel the actual genetic situation within groups. We have tended to respond to this difficulty in slightly different ways. One of us (P.B.S.) has taken the conservative view, and is uncomfortable assigning breeding or non-breeding status to males when no data exist on whether they actually sire offspring. Thus, data from WC are usually presented only by overall group size or total number of adult males or females. Statistical conclusions are not altered by either lumping or separating categories of potential breeders and non-breeding helpers in this population.

The other (W.D.K.) has chosen the more sanguine approach, categorizing individuals in the population based on their *potential* breeding status, given their origin and the apparent incest taboo described above. Thus, data from HR are given for both breeding and non-breeding individuals, based not necessarily on known reproductive success but rather by assigning breeding status to individuals whose history indicates that they had both the opportunity and motive to breed, whether or not they were in fact successful in doing so. This approach is similar to that adopted by many avian biologists studying non-cooperative breeders who, because of kleptogamy, must deal with the fact that the 'genetic' mating system is not necessarily congruent with the 'apparent' mating system (e.g. Gowaty and Karlin 1984).

Data with which to resolve this problem are likely to be found in the near future with currently developing genetic methods such as DNA fingerprinting. Such techniques represent a promising area of research in this, as well as other, species of cooperative breeders (see Chapter 15). For now, however, lack of parentage information, particularly among males, remains a basic weakness of all studies of Acorn Woodpeckers.

Allowing for the differences in group size and the different approaches of the authors toward the problem of multiple parentage, the basic social organizations of the study populations appear to be similar. However, there is a striking difference in the patterns of relatedness in the HR and WC populations. At HR, male co-breeders are generally brothers, a father and one or more of his sons, or both, and female co-breeders are either sisters or a mother and her daughters. During the study, only one set of two unrelated male co-breeders (1% of male co-breeding units) and two sets of two unrelated female co-breeders (7% of female co-breeding units) were found. Non-breeders are invariably offspring fledged by the group in prior years, and may include birds up to at least five years old.

In contrast, male co-breeders within groups in the WC population frequently consist of unrelated individuals (50% of 28 co-breeding units; see also Stacey 1979b). Thus, the pattern of close relatedness among co-breeders observed at HR is not consistent throughout the range of this species. Unrelated co-breeding units at WC typically form when previously abandoned territories are reoccupied, either by the same or different birds. Resettlement often occurred at different times, and, since many of the birds were unbanded, it is possible that a few of the birds were actually relatives. However, only a single case in WC has been recorded in which a bird left its group, lived elsewhere, and then rejoined relatives. Dispersal and independent colonization of territories also occurs at HR, but almost invariably involves closely related individuals.

Stacey (1979b) has hypothesized that the reason why an unrelated bird at WC might be allowed to join and co-breed in a new group is because of the considerable reproductive advantage of breeding in trios or larger groups in this population: groups with three adults, for example, fledge on average significantly more young per adult than do groups consisting of a simple pair (0.48 ($n = 77$) vs 0.76 ($n = 30$) young adult^{-1} year^{-1} for pairs and trios, respectively; Kruskal–Wallis 1-way ANOVA, $\chi^2 = 5.45$, d.f. $= 1$, $P < 0.05$; see Fig. 14.4).

Demography

Reproductive success

Table 14.2 compares the basic reproduction rates of the three populations. Although groups fledged more young at HR than at WC or the RR, this can reasonably be explained primarily because of larger average group size: there were no significant differences in the number of fledglings per bird between HR and WC (Mann–Whitney U-test, $z = 0.40$; $P > 0.5$) or in the proportion of groups breeding successfully among the three populations ($\chi^2 = 3.1$; d.f. $= 2$, $P > 0.20$). Thus, differences in group size, and other demographic differences directly or indirectly resulting from differences in group size, cannot be explained by an average difference in productivity among the populations.

A variety of factors influence reproductive success in Acorn Wood-peckers. The most important of these is the availability of acorns stored in the autumn. At HR, groups with stores remaining in the spring initiate breeding earlier (7 May $\pm$ 23.7 days ($n = 157$ groups) vs 30 May $\pm$ 36.4 days ($n = 20$), Mann–Whitney U-test, $P < 0.001$), have larger clutch sizes (4.55 $\pm$ 1.03 ($n = 93$ clutches) versus 3.57 $\pm$ 0.65 ($n = 14$), $P < 0.001$, based on clutches of single females only), and fledge over five times as many young than groups with no stores remaining (2.89 $\pm$ 1.87 ($n = 137$ groups) versus 0.51 $\pm$ 1.11 ($n = 43$), $P < 0.001$; all data given are $\bar{x} \pm$ s.d.). A similar, although not as dramatic, enhancement of reproduction is shown at WC, where groups with stored acorns remaining fledge over twice as many young as groups without stores (2.88 $\pm$ 1.30 ($n = 43$) versus 1.12 $\pm$ 1.40 ($n = 95$), $P < 0.001$). Overall, the proportion of groups with stores remaining in the spring that successfully fledge young is 83% at HR and 93% at WC, while only 23% and 46% of groups that do not have stores fledge young in

Fig. 14.4. Reproductive success per group and per capita as a function of Acorn Woodpecker group size in Water Canyon and Hastings Reservation. Plotted as $\bar{x} \pm$ s.e.m. Sample size (n, number of groups) in each category are listed in parentheses.

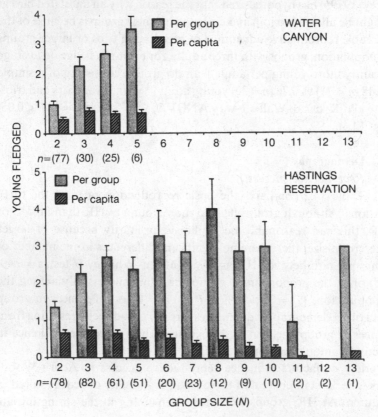

the two sites, respectively; differences in success between groups with and without stores are highly significant ($P < 0.001$) in both sites.

Stored acorns are sometimes fed directly to older nestlings, but the primary means by which stored acorns enhance reproduction is apparently by allowing adults to feed a greater proportion of the insects they catch to nestlings. The availability of acorns during the winter also results in adults in better physical condition for breeding, as shown by the larger and earlier clutches described above in birds with access to stores, as well as an increased probability that birds without stores will fail to initiate any breeding attempt whatsoever during the season. Acorns also enhance reproductive success at HR by increasing the probability that groups will attempt a second nest in the spring and by permitting autumn nesting.

At HR, other variables examined for an effect on reproductive success include various weather variables, flying insect abundance (insects are the primary food source fed to nestlings), territory quality, population size, and whether the breeders have had prior breeding experience with each other. Of these variables, prior breeding experience is most important: groups in which no turnover in breeders had occurred within the past year fledged an average of 2.69 ± 2.03 ($n = 134$ groups) young compared to only 1.04 ± 1.57 ($n = 76$) young for groups in which a turnover had taken place (Mann–Whitney U-test, $P < 0.001$). There is also a suggestion of density dependence in reproductive success, at least when years following poor acorn crops are excluded, i.e. when both population size and subsequent reproductive success are low. If the two poor crop years at HR that occurred between 1975 and 1985 are ignored, there is a significant inverse correlation

Table 14.2. *Reproductive success of Acorn Woodpeckers*

	Hastings Reservation	Water Canyon	Research Ranch	P-value
Fledglings per group	2.38 ± 2.30	1.67 ± 1.62	1.06 ± 1.17	**
Fledglings per bird	0.62	0.58	0.49	n.s.
% groups not attempting to breed	15.4	26.2	37.1	**
% nest failures	27.3	12.6	20.0	**
% groups with second nests	4.5	8.8	0.0	—
% groups breeding successfully	64.3	60.1	48.5	n.s.
n group-years	347	164	33	—

n.s., not significant; **, $P < 0.01$.

between population size and young fledged ($r_s = -0.82, n = 9, P < 0.05$). A similar density dependence does not emerge at WC, even when the two poorest crop years are excluded ($r_s = -0.14, n = 8, P > 0.5$).

Territory quality, as reflected by the amount of mast and storage facilities, is also important to some aspects of reproductive success, but apparently much more so at WC than at HR (Table 14.3). In particular, there is a highly significant difference in reproductive success among groups with access to different-sized granaries at WC. This is not true at HR, even though data from both sites indicate that storage facilities significantly influence group size and the probability of a group having stored enough acorns to last through the winter.

One consequence of the importance of acorns is that the major source of mortality of nests and nestlings appears to be related to starvation, at least at HR. (Fates of eggs were generally not determined at WC.) Of 693 eggs laid in 180 nests, 16.0% starved as nestlings and 10.4% failed to hatch. Only 2.0% of eggs were almost certainly lost to predation, but even including all eggs lost to unknown causes, a maximum of only 9.8% of eggs were subject to depredation as either eggs or nestlings. Thus, a maximum of only 17% of

Table 14.3. *Influence of territory quality (measured by mast storage facilities) on Acorn Woodpecker group demography*

	Hastings Reservation	Water Canyon[a]
Adult breeding group size		
Small (<1000 holes)	3.21 ± 1.23 (100)	2.18 ± 0.47 (62)
Medium (1000–3000 holes)	4.04 ± 2.02 (138)	2.62 ± 0.89 (47)
Large (>3000 holes)	5.51 ± 2.68 (116)	3.18 ± 0.89 (55)
χ^2 (Kruskal–Wallis test)	54.4***	33.0***
Probability of stores during breeding		
Small	0.58 (87)	0.09 (62)
Medium	0.74 (118)	0.23 (47)
Large	0.78 (99)	0.58 (55)
χ^2 (contingency test)	10.2**	34.7***
Young fledged per group		
Small	2.03 ± 2.07 (100)	0.96 ± 1.42 (62)
Medium	2.22 ± 2.16 (138)	1.76 ± 1.92 (47)
Large	2.75 ± 2.62 (116)	2.33 ± 1.71 (55)
χ^2 (Kruskal–Wallis test)	4.28	16.5***

Data presented are $\bar{x} \pm$ s.d. (n groups). **, $P < 0.01$; ***, $P < 0.001$.
[a] Water Canyon data from Stacey and Ligon (1987).

all losses were from predation, far less than the 75% of all eggs which are lost to predation in the Florida Scrub Jay (see Chapter 8). Interestingly, egg predation in both Scrub Jays and Acorn Woodpeckers (at both HR and WC) is believed to be primarily by snakes. Presumably cavity-nesting is the major reason that predation losses in the woodpeckers are so much lower than in the jays.

Group size and composition also have important effects on reproductive success (Fig. 14.4). At HR, the number of young fledged increases with the numbers of non-breeding helpers, breeding females and breeding males. However, in an ANOVA including these three aspects of group composition, and whether a turnover in breeders had occurred or not, only turnovers and the number of breeding males is significant (Koenig and Mumme 1987).

Earlier analyses of HR data (Koenig and Mumme 1987) indicated that there was considerable enhancement of reproductive success by helpers in poor years, when the overall success of the population was low, but that there was little or no effect of helpers in good years. Results through 1986 dividing years into good and poor years still support the hypothesis that the relative influence of non-breeders is greater in poor years, but the enhancement is now significant for both sets of data. Similarly, at WC, non-breeders had a significant positive effect on reproductive success in both good and poor years, but the enhancement was relatively greater in poor years (Table 14.4).

In both the WC and HR populations, mean group reproductive success increases significantly with group size (Fig. 14.4); similarly, reproductive success increases with the number of (breeding) males or (breeding) females in both populations. (As discussed above, breeders were distinguished at HR but not at WC; thus, these results, and others presented below, are based on the number of presumed breeding birds at HR and on all birds of the appropriate sex at WC.) However, an important difference between the two populations is that per capita reproductive success at WC, but not at HR, is significantly greater in groups than in pairs (Fig. 14.4). Similarly, per (breeding) male and per (breeding) female, reproductive success decreases with group size at HR (significantly so for females) but not at WC (Stacey 1979a; Koenig and Mumme 1987).

This latter difference is critical and may form the basis for several important behavioral, as well as demographic, differences between the two populations. For example, the pattern of young fledged per male with increasing number of males per group at WC means that, on average, males do not suffer a loss in reproductive success with additional co-breeders, and

Table 14.4. *Effect of non-breeding Acorn Woodpecker helpers on reproductive success in good and bad years*

	No non-breeders	Non-breeders present	Percentage change with non-breeders	P-value
Hastings Reservation[a]				
Good years (n = 7)	2.90 ± 2.31 (99)	3.96 ± 2.24 (85)	+37	<0.01
Poor years (n = 8)	0.99 ± 1.40 (90)	1.61 ± 1.84 (75)	+63	<0.05
Water Canyon[a]				
Good years (n = 5)	2.14 ± 1.78 (25)	3.26 ± 1.43 (30)	+52	<0.005
Poor years (n = 5)	0.78 ± 1.56 (18)	1.72 ± 1.27 (65)	+121	<0.01

Data presented are $\bar{x} \pm$ s.D. (*n* groups). Statistical tests are by ANOVA.
[a] For both populations, 'good' years were those in which the average number of young fledged per group was > 2.0, and 'poor' years were those in which an average of < 2.0 young were fledged per group.

may explain in part why unrelated males are frequently allowed to join groups in that population.

Adult survivorship

At HR, overall annual survivorship is 82.4% for breeding males and 71.2% for breeding females, while values for WC are an average of over 20% per year lower for adult males and females (61.3% and 51.5%, respectively; Table 14.5). These differences are in part a function of differences in the samples included in the two sites. In particular, 'non-breeding helpers', who have a considerably higher probability of dispersing than breeders, are not included in the figures for HR, but are, at least potentially, in those for WC (see discussion above). None the less, these figures suggest substantially higher overall survivorship rates at HR, and lower annual survivorship for females than males at both study sites.

As with reproductive success, survivorship is significantly influenced by the acorn crop, especially among breeding females. During the winter following the poor 1978–9 crop at HR, for example, disappearance and probable mortality of breeding females was 46.2% ($n = 13$) compared to 5.8% ($n = 137$) in other winters ($G = 12.4$, $P < 0.001$). Age, territory quality, and prior breeding experience do not significantly influence apparent survivorship at HR (Koenig and Mumme 1987). In contrast, territory quality significantly affects apparent survivorship at WC, with survivorship

Table 14.5. *Annual percentage survivorship of Acorn Woodpecker adults depending on adult group composition*

	Hastings Reservation	Water Canyon
Males		
Overall	83.4 (273)	61.3 (155)
Groups with 1 male	70.3 (64)	55.4 (56)
Groups with 2 males	86.6 (112)	72.7 (55)
Groups with 3 or 4 males	85.9 (99)	68.2 (44)
Females		
Overall	72.5 (302)	51.5 (103)
Groups with 1 female	70.0 (200)	50.6 (73)
Groups with 2 females	77.5 (102)	63.3 (30)

Data are mean annual survivorship (*n* bird-years). For HR, only birds known or believed to be breeders are included; for WC, all adults of the appropriate sex are included (see the text).

being significantly enhanced in birds living on territories with larger storage facilities (Fig. 14.5).

Survivorship in both study areas is significantly higher in groups with additional adults (Table 14.5). At HR, survivorship of breeding males is enhanced by the presence of co-breeders or the presence of non-breeding helpers, even when controlling for territory quality, breeding experience, and group composition in a logit analysis, but female survivorship is not significantly enhanced when controlling for these other variables (Koenig and Mumme 1987). Furthermore, examination of the seasonality of mortality indicates that females are particularly at risk during the spring and summer. Although the reasons for this are not known, it suggests that reproduction entails a greater cost or increased risk to females than males. Breeding male survivorship is constant among seasons but significantly enhanced in larger groups, suggesting that predation may be decreased by the presence of increased numbers of vigilant group members.

Fate of offspring

Except in poor acorn years, when many birds are forced to abandon their territories, offspring only rarely disperse from their natal

Fig. 14.5. Survivorship curves for Acorn Woodpeckers at Water Canyon banded as adults and assigned an age of one year when captured. Median lifespans for birds in small-, medium- and large-category territories were 1.65 ($n = 65$), 1.73 ($n = 45$), and 2.31 ($n = 46$) years, respectively (Lee–Desu test, $D = 12.4$, $P < 0.005$). (Redrawn from Stacey and Ligon 1987.)

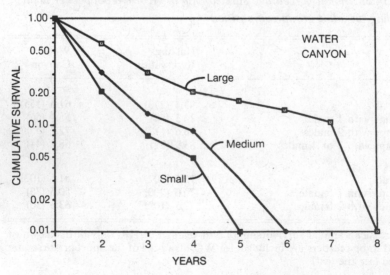

groups prior to their first spring: only 1.1% of 361 nestlings at HR and 3.0% of 133 nestlings at WC are known to have left their natal groups and moved elsewhere in the study area before their first spring for reasons other than exhaustion of stores. Consequently, we can in general be confident that disappearing first-year birds have died.

Survivorship to the first spring is 56.8% at HR and 36.6% at WC; this 20% difference again parallels the differences in survivorship shown by adults (males and females cannot be sexed in the field until their first autumn, and thus sex-specific first-year survivorship values are not available). During their first year, one of four fates may befall offspring: (1) they may disappear from the study area, (2) they may move locally to another group within the study area, (3) they may inherit and become a breeder in their natal group, or (4) they may remain in their natal group as a non-breeding helper. The average proportions of males and females undergoing these fates each year at HR and WC are presented in Table 14.6. Also presented is the propensity of offspring to disperse with siblings of the same sex in these populations.

The proportion of non-breeding helpers disappearing is higher for females than for males in both populations. Birds at WC are more likely to disappear, move locally and especially to inherit their natal territory than at

Table 14.6. *Fate of offspring of Acorn Woodpeckers*

	Hastings Reservation		Water Canyon	
	♂♂	♀♀	♂♂	♀♀
Percentage of first-year birds surviving the winter	56.8 (447)		36.6 (123)	
Percentage non-breeding helpers[a]				
Disappearing each year	19.1	30.1	28.8	38.6
Moving locally each year	15.6	11.7	19.7	22.7
Inheriting natal territory	9.5	3.4	28.8	11.4
Remaining as non-breeder	55.7	54.9	19.7	27.3
n non-breeder-years	262	206	66	44
Percentage birds dispersing with at least one other sibling	68.4	42.9	8.7	0.0
Mean number of siblings dispersing together	2.10	1.48	1.05	1.00

[a] Includes first-year birds that survived to February (HR) or to be sexed (WC) and older non-breeding helpers.

HR. Conversely, considerably more birds (both males and females) remain as non-breeders at HR than at WC. When birds at HR do disperse, they are much more likely to do so with other siblings. These differences are consistent with the hypothesis that dispersal is more difficult at HR, leading to a higher proportion of first-year birds remaining as non-breeding helpers and an increased frequency of joint dispersal. A likely basis for these differences is the lower adult survival rates at WC, leading to an increased incidence of breeding vacancies and reproductive opportunities for young birds.

Lifetime reproductive success

At HR, estimates of lifetime reproductive success have been made for: (1) birds acting as non-breeding helpers versus those which disperse in their first year, and (2) birds co-breeding with relatives versus those dispersing and breeding on their own. These estimates can be made both directly (using the observed success of individuals during their lives) and indirectly (on the basis of simulations using observed average demographic parameters for birds pursuing specific behavioral options).

Direct estimates of lifetime reproductive success yield no significant differences between birds which breed in their first year compared to those remaining in their natal groups as non-breeding helpers (for males, number of young fledged $= 2.04 \pm 2.20$ ($n = 8$ birds) for those that bred in their first year versus 3.26 ± 4.40 ($n = 19$) for those that stayed; for females, 3.33 ± 2.89 ($n = 3$) versus 2.92 ± 3.94 ($n = 12$); both $P > 0.05$). These results are not changed if the analysis is restricted to birds which did not inherit and breed in their natal territory.

Indirect approaches also indicate that there is at best only a small fitness difference between birds helping in their first year compared to dispersing and breeding on their own. From a demographic model, Koenig and Mumme (1987) estimated that, on average, a non-breeding helper gains 0.263 offspring equivalents in indirect fitness by helping for one year (Table 14.7). Approximately two-thirds of the benefits are through the 'present' component of indirect fitness; that is, by increasing the reproductive success of the breeders. The remainder accrues by increasing the survivorship of male breeders (the 'future' component of indirect fitness). Using similar methods, breeding males are estimated to gain 0.39 offspring during their lifetimes by being aided by a single non-breeding helper, while breeding females gain only 0.05 offspring.

These models suggest that the fitness benefits of remaining as a non-breeding helper at HR are on average small, at least in absolute terms. The relative importance of indirect benefits is harder to estimate. However, on the basis of these and other models, and assuming that dispersal of offspring is constrained (see below), an estimate can be made of the proportion of total fitness benefits derived indirectly (I_k, the index of kin selection; Vehrencamp 1979). Estimates of I_k range between 24% (for males when no other non-breeding helpers are present) and 57% (for females when two other non-breeding helpers are present). Apparently the relative importance of indirect selection to the maintenance of remaining as a non-breeding helper at HR is great, contributing up to half of the inclusive fitness benefits. Interestingly, one-third of these benefits are a consequence of increased longevity of male breeders (the 'future' component of indirect fitness; see Brown 1987), and thus are unlikely to be related to the act of feeding young usually considered to be the critical feature of cooperative breeding systems.

Analogous models can be made, comparing males that share mates and females that nest jointly with individuals breeding by themselves. The results of these models are variable and depend on the specific assumptions involved (Koenig and Mumme 1987; Mumme *et al.* 1988). In general, however, they suggest that males derive considerable fitness benefits by sharing breeding status with other birds, primarily through increased survivorship. For example, the estimated lifetime reproductive success of a dominant male initially sharing breeding status with a subordinate male is 7.48 young compared to only 6.78 young for the same male initially breeding on his own. Even considering indirect effects, a male still has higher fitness sharing breeding status with one or two other males, even if those males are unrelated to himself, and even if there is relatively strong reproductive bias. Thus, there appear to be considerable fitness benefits to males sharing breeding status.

Table 14.7. *Average annual values of present and future components of indirect fitness received by non-breeder Acorn Woodpeckers in offspring equivalents at HR*

	Via male breeders	Via female breeders	Total	Percentage of toal
Present	0.083	0.092	0.176	66.9
Future	0.157	−0.070	0.087	33.1
Total indirect fitness	0.240	0.022	0.263	100.0

From Koenig and Mumme (1987).

The situation for breeding females is less clear. Joint-nesting females do not appear to reap direct fitness benefits compared to females nesting singly. Inclusive fitness benefits accrue only if bias is relatively weak and if competition for breeding vacancies is intense, a situation which renders an advantage to females that compete for vacancies in large sibling units (Hannon *et al.* 1985).

Population size and stability

Population size, like virtually all other aspects of the demography of Acorn Woodpeckers, is strongly tied to the acorn crop. In good acorn years, reproductive success can be very high, and groups at HR may breed successfully in the autumn as well as twice during the spring and early summer. In poor years, groups may be forced to abandon their territories in the autumn, and those few birds remaining the following spring have very poor reproductive success. Survivorship of adults is also lower, leading to small group and population sizes which may take several years to recover (see Fig. 14.2).

This critical effect of acorns on demography also has ramifications for their geographical ecology: Acorn Woodpeckers are more abundant in areas containing at least three species of oak, suggesting a diversity threshold resulting from the presumed inverse relationship between the probability of total crop failures and oak species diversity (Bock and Bock 1974; see also Roberts 1979).

All factions of Acorn Woodpecker populations vary annually, but breeders do so much less than do non-breeding helpers (Fig. 14.2; Table 14.1). Non-breeders occur in direct proportion to the overall population size; thus, in years when there are many birds, relatively more young remain at home as non-breeders (r_s between breeding population size and percentage group offspring remaining as non-breeders $= 0.76$, $P < 0.05$). In contrast, there is relatively little variation in the proportion of territories occupied. These characteristics suggest that non-breeders have the effect of buffering the population, but particularly so with respect to the number of groups in an area, and less so with respect to group size. Overall sex ratios in all populations favor males. For WC, mean $\pm$ s.D. (male:female) ratio for adults at the start of the breeding season was 1.32 ± 0.17 ($n = 10$ years), while at HR the ratio was 1.43 ± 0.19 ($n = 16$ years).

Behavior in a saturated environment: Hastings Reservation

Acorn Woodpeckers exhibit a wide range of cooperative and competitive behaviors within and between groups. Within groups, birds cooperatively store acorns, maintain the granary, defend the territory, and help to raise offspring (MacRoberts and MacRoberts 1976; Mumme and de Queiroz 1985). As a result of the additional group members, feeding visits per hour at HR nests increases with group size up to about eight individuals, paralleling the relationship (Fig. 14.4) between group size and reproductive success (Mumme 1984). In general, breeders deliver more care than non-breeding helpers, and females more than males (Fig. 14.6).

This high degree of cooperation is presumably facilitated, at least in part, by the close genetic relatedness among group members at HR, with the exception of the breeders of opposite sex. As discussed earlier, this pattern appears to be maintained by a combination of an incest taboo and reproductive competition. Consequently, there is a relatively low rate of inbreeding and an overall mean coefficient of relatedness (F) among mated sets of birds (following the criteria discussed earlier) equal to 0.022 ($n = 82$ breeding units).

Despite the close relatedness among co-breeders at HR, there is considerable reproductive competition. For breeding males, competition takes the form of mate-guarding behavior, the propensity for males to maintain close association with breeding females during their fertile period.

Fig. 14.6. Individual feeding rates (feedings per nestling per hour) at Hastings Reservation in relation to Acorn Woodpecker reproductive status and sex. Data presented are $\bar{x} + 95\%$ confidence interval; numbers above bars are the number of individuals used in the analysis. Differences between breeders and non-breeders, and between males and females, are significant by a 2-way ANOVA (both $P < 0.05$). (Redrawn from Mumme 1984.)

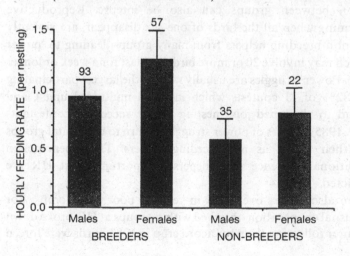

At HR, guarding is invariably shown by males when more than one potential co-breeder is present within a group. Females are not guarded when only a single male is present, a result consistent with the hypothesis that mate-guarding is a response to within-group reproductive competition (Mumme *et al.* 1983*b*). Apparently the possibility of kleptogamy by a male outside the group is not enough of a threat to induce mate-guarding, even though kleptogamy has been demonstrated with electrophoretic studies of paternity (Mumme *et al.* 1985).

Among joint-nesting females, there is considerable competition to lay eggs in the communal nest: as in the egg-tossing described in the Groove-billed Ani (see Chapter 11), females laying first invariably have their eggs destroyed by their co-breeders in an apparent attempt to bias maternity of the clutch (Mumme *et al.* 1983*a*). Such egg destruction occurs despite the close relatedness between the females and the eggs they destroy. Recent experimental evidence (our unpublished results) suggests the possibility that males may similarly destroy eggs when they are denied the opportunity to parent the young (see also Stacey and Edwards 1983). Infanticide has not been observed at HR, in contrast to WC (see below), but it is possible that this difference reflects a lack of opportunity rather than a behavioral difference between the populations.

As a result of egg destruction by breeding females, reproduction is synchronized between joint-nesting females, and maternity is split fairly evenly (Mumme *et al.* 1983*b*). Another result of egg destruction is to eliminate the possibility of groups having more than one active nest simultaneously; females laying in separate nest cavities will reciprocally destroy each other's eggs until they are laid in the same hole.

Competition between groups can also be intense. Reproductive vacancies, forming when all the birds of one sex disappear, are violently contested by non-breeding helpers from many groups, leading to 'power struggles' which may involve 20 or more birds and last for a week or longer (Koenig 1981). Power struggles are usually won by the largest participating sibling unit (82% of 11 contests which included multi-bird units), and frequently lead to presumed joint-nesting by the successful contestants (Hannon *et al.* 1985). Losers of power struggles return to their natal groups and resume their status as non-breeding helpers. This phenomenon provides additional evidence that dispersal opportunities at HR are severely restricted.

Competition also occurs over food in years of poor acorn crops. For example, the usual cooperation observed within groups at HR broke down during the winter following the poor acorn crop in 1983. Birds were forced to leave groups, apparently in inverse order of their dominance, i.e. first-year birds followed by non-breeding females, non-breeding males and breeding females (Hannon *et al.* 1987). Thus, cooperation within groups is dependent on the availability of sufficient resources to support the group.

Competition and cooperation are intimately entwined in the social fabric of Acorn Woodpeckers. Cooperation within groups and sharing of resources in times of plenty degenerates into competition during times of food shortage. Similarly, egg destruction by joint-nesting females and competition for matings by co-breeding males abruptly switches to cooperation in incubation, brooding, and feeding of nestlings upon the completion of the clutch.

Options in an unsaturated environment: Water Canyon

Acorn Woodpeckers at WC exhibit a similar range of cooperative and competitive behaviors, with several interesting exceptions. First, joint-nesting by two or more females has not been observed, although the reason for this may be simply that groups containing two females, unrelated to the breeding male(s), are rare at WC. Second, although many groups exhaust their acorn stores during the winter, birds share equally until the food is depleted, and there is no consistent pattern of dispersal as a function of dominance, as described above for HR. Third, joint dispersal by siblings or parent–offspring units is rare. Although a newly created reproductive vacancy, particularly one in a high-quality territory, may attract many birds, each woodpecker appears to act independently during the ensuing contest (Stacey and Edwards 1983). Other than when entire groups moved together to a vacant territory, only three cases have been observed of two relatives immigrating together into an established group. Thus, birds do not appear to gain from acting together as a unit during competition for reproductive vacancies, and the intense power struggles observed at HR are rare.

There is, however, one form of reproductive competition that is particularly strong at WC. Immigrants joining a group after a breeder has disappeared and while there are young in the nest (eggs or nestlings) will regularly kill those young. Infanticide was first observed in female immigrants (Stacey and Edwards 1983); since that report, the same behavior has been seen also in males. For example, in 1983 the breeding male of a pair was killed by a bull snake (*Pituophis melanoleucus*) while the bird was brooding recently hatched nestlings. Although the female continued to feed the young, within one day a new male had joined the group and killed the nestlings.

Infanticide occurs only during the early part of the breeding season while a group still has a chance to re-nest. Birds joining groups late in the season following egg-laying neither help to raise nor kill the young (Stacey 1979*b*). This suggests that killing young in Acorn Woodpeckers is primarily an adaptation allowing immigrants to begin breeding in their new group as quickly as possible. The absolute frequency of infanticide at WC has been difficult to determine, but, once it was discovered, it has been observed at least once each year. We therefore suspect that it is an important phenomenon in this population.

Similar cases of infanticide may have been missed at HR, but are almost certainly uncommon. A possible reason for this difference is the relatively high mortality in WC (Table 14.5), which may make immediate breeding more important there than at HR.

Many of the more interesting differences between WC and HR reflect differing levels of competition for breeding opportunities. For example, the smaller group size (Fig. 14.3) and high annual mortality at WC results in a relatively high rate of reproductive vacancies in established groups. Further, all territories are not continuously occupied. Each year some groups exhaust their mast stores and abandon their territories. These are then available for recolonization the following spring. Some of these previously occupied territories are not recolonized and remain vacant throughout the next breeding season. Both factors create opportunities for potential helpers to disperse from their natal groups and breed independently. Stacey (1979*a*) provided a careful examination of the effects of these differences between WC and HR on various group and population-level demographic parameters based on the first three years of both studies.

One critical fact which has emerged from WC is that all territories are not equally likely to be occupied. The key determinant appears to be the size of the storage facilities. Territories that have the largest amount of mast storage facilities (LFT) were always occupied during the summer ($n = 49$ territory-summers), while those with a medium amount (MFT) were occupied for 92% of summers ($n = 61$) and those with small granary facilities (SFT) were occupied for only 46% of summers ($n = 127$; $G = 86.9$, $P < 0.001$). Storage facilities also have a profound effect on the fitness of birds living on the territory (Stacey and Ligon 1987). Not only is reproductive success twice as high for groups breeding on LFT than on SFT (Table 14.3), but annual survivorship is also significantly higher (adults: 0.44, 0.51 and 0.67 year^{-1} on SFT, MFT, and LFT, respectively, $G = 10.3$, $P < 0.01$; juveniles: 0.19, 0.32 and 0.50 year^{-1}, respectively, $G = 8.5$, $P < 0.025$).

Each year in WC there are territories that could be colonized and are not. Some birds are philopatric and remain as non-breeding helpers even when breeding opportunities exist nearby. To examine how the variance in territory quality affects dispersal decisions, a demographic model was developed to estimate lifetime reproductive success for birds as a function of (1) the number of years they serve as non-breeding helpers, and (2) the quality of the territory on which a bird eventually breeds, either through dispersal or inheritance (Stacey and Ligon 1987). Some of the results, including the number of offspring produced and the indirect inclusive fitness benefits of helping on each territory type, are given in Table 14.8.

These analyses assume that there is no habitat saturation or other constraint to independent breeding: any bird that leaves its group is assumed to be able to breed immediately. The results indicate that remaining as a helper in a high-quality territory with many storage holes, even for several years, is a strategy superior to dispersing and breeding immediately on a territory of lesser quality. For example, a young bird that remains as a helper for two years on a high-quality territory and does not breed until its third year on a territory of similar quality will leave nearly twice as many offspring equivalents (OEs) than if it dispersed and began breeding in its first year on a low-quality territory (2.77 versus 1.41 OEs), and about the same number of OEs as a bird that disperses immediately and breeds on a medium-quality territory (2.77 versus 2.66 OEs). The same pattern emerges if one considers either only the number of actual offspring produced or the number of those offspring that themselves become breeders (Stacey and Ligon 1987).

Table 14.8. *Expected lifetime inclusive fitness[a] of Acorn Woodpeckers that help for varying numbers of years at Water Canyon according to the size of granary facilities (from Stacey and Ligon 1987)*

Territory storage facility	Annual indirect fitness benefits[b]	Number of years helping		
		0	1	2	3
Small (< 1000 holes)	0.94	1.41	1.33	1.32	1.29
Medium (1000–3000 holes)	0.77	2.66	1.84	1.60	1.46
Large (> 3000 holes)	0.79	4.88	3.30	2.77	2.39

[a] Includes both direct and indirect fitness benefits.
[b] Estimated average net effect of adding one additional group member; assumes that non-breeders feed full siblings.

Thus, philopatry can be a superior strategy in WC even when there is no limitation of breeding opportunities, at least for birds on high-quality territories. As predicted, young birds in WC frequently become helpers on high-quality territories and almost never do so on low-quality territories (27% ($n = 56$) of fledglings remained as helpers on LFT, 15% ($n = 41$) on MFT, and only 2% ($n = 36$) on SFT; $G = 11.0$; $P < 0.05$).

These results are predicated on the considerable range in quality of territories at WC (as measured by mast storage facilities) and the strong effect that territory quality exerts on reproductive success and survivorship in this population. If the comparisons in Table 14.8 were restricted to only a single territory quality category, or if all territories were combined, the results would suggest that immediate reproduction would always be superior to remaining as a non-breeding helper, assuming that territories are available.

Evolution of group-living

Sociality in Acorn Woodpeckers is more than usually complex, and shows considerable geographic variation. Thus, it is not surprising that there is considerable evidence for, and against, more than one hypothesis for why this species usually lives in groups, and for the evolution of helping behavior once groups have formed.

In general, it appears that options for young birds are limited. Evidence supporting this hypothesis comes from numerous sources, including a comparison of HR with WC (Stacey 1979b), the correlation between the dispersal of young birds and the number of vacancies opening in the population (Stacey 1979b; Emlen and Vehrencamp 1983), the severe competition for breeding vacancies at HR, and analyses of the lifetime fitness benefits of offspring by remaining and helping compared to dispersing (Koenig and Mumme 1987). The primary resource constraining the options available to young birds is certainly access to granary facilities (Koenig and Mumme 1987; Stacey and Ligon 1987).

One interpretation of these results is that severe habitat limitation forces young to remain in their natal territories until they can find a breeding vacancy elsewhere in the population, or until they are able to inherit and breed in their natal territory. In support of this hypothesis, Koenig and Mumme (1987) have recently calculated, using survivorship and fecundity information from HR and the equation of Fitzpatrick and Woolfenden (1986), that there are on average nine times more offspring present in the population than vacancies for them to fill. Indeed, the hypothesis that ecological constraints are the critical factor in the evolution of group-living in Acorn Woodpeckers has been a dominant theme among workers considering this species.

However, there are several parts of the story which do not readily fit into the classical interpretation of habitat saturation. At WC, some territories are vacant each summer, and some non-breeding helpers remain in their natal groups even when breeding opportunities exist nearby (Stacey and Ligon 1987, 1990). At HR, Koenig and Mumme (1987), although acknowledging the shortage of available territories relative to non-breeders available to fill them, none the less conclude, on the basis of a comparison of immigrants versus young disappearing from the population, that virtually all offspring eventually attain breeding status somewhere. This result conflicts with the expectation of most habitat saturation models that there should be a high cost of dispersal. Instead, the cost appears to be small, other than perhaps the relatively high probability of new groups failing in their first year.

There are several possible solutions to the dilemma posed by these and similar observations. One, suggested by Koenig and Pitelka (1981), is that the shortage of reasonable territories to which non-breeders can disperse and breed, rather than the population surplus itself, is the key to group-living. This hypothesis predicts (1) a relatively high ratio of 'optimal' versus 'marginal' territories, and (2) that young are constrained as a consequence of their overproduction relative to good territories.

An alternative, recently suggested by Stacey and Ligon (1987) based on their work with the WC population, also begins with the fact that good territories are in relatively short supply compared to birds which could possibly fill them. However, rather than suggest a limited proportion of marginal territories leading to a difficulty of dispersal by young, they propose that the key is the advantage offspring gain by remaining in good territories. By dividing territory quality into different categories based on storage facilities, these authors show that, at WC, young are better off remaining as non-breeders and then breeding later on than dispersing to low-quality territories (see Table 14.8), which do not appear to be in short supply. Thus, this hypothesis predicts a relatively large amount of variance in territory quality, rather than the smaller amount predicted by the hypothesis of Koenig and Pitelka (1981). Group-living is not forced on young as a result of a limited availability of marginal habitat, but rather because remaining on good territories as non-breeders provides the best route toward obtaining a high-quality territory (Fig. 14.7).

These contrasting hypotheses for the ecological factor leading to group living in cooperative breeders provide a variety of differing predictions about the importance of non-breeding helpers, the dispersal of young, and the pattern of territory occupancy (Stacey and Ligon 1990). Support for either hypothesis can be garnered for both the HR and WC populations; at present, however, the evidence (or at least the biases of the authors) suggest that the benefits of philopatry are more important at WC and that habitat saturation is more important at HR. We believe that detailed investigation of these hypotheses will be an important means by which future work will further our knowledge concerning the evolution of cooperative breeding.

There is similarly contradicting evidence concerning the evolution of the much rarer phenomenon of mate-sharing in the Acorn Woodpecker. Part of the difficulty involves the current lack of definitive data concerning this phenomenon, especially in males. We believe that the evidence for polygynandry (multiple breeders of both sexes within a social unit) in the Acorn Woodpecker, combined with the apparent suppression of reproduction by offspring when living with their parent of the opposite sex, is good. However, this conclusion is but the beginning of the solution to the riddle provided by the mating system of this species. What are particularly desired are data on the extent of multiple paternity, the bias in paternity that exists in groups containing different numbers of reproductive males, and the correlation between paternity and subsequent parental-like behavior.

For example, consider a group containing two male siblings which filled a reproductive vacancy and are thus unrelated to the breeding female. What proportion of the offspring are sired by each of the males? And, equally importantly, does the subsequent parental-like behavior of the males depend on the *actual* number of young sired in a nest, on the *average*

Fig. 14.7. Two hypothetical scenarios for the evolution of cooperative breeding. Under the benefits-of-philopatry hypothesis, cooperative breeding may or may not lead to habitat saturation (dashed line). (Redrawn from Stacey and Ligon 1987.)

A. Habitat saturation hypothesis:

Habitat saturation and ⟶ Imposed ⟶ Cooperative
absence of marginal habitat   philopatry   breeding

B. Benefits of philopatry hypothesis:

Access to critical resource ⟶ Philopatry ⟶ Cooperative --→ Habitat
or group benefit   breeding   saturation

probability of each male having sired an offspring, or on the average probability of each male having sired *any* offspring in the nest? Even the possibility that males can recognize their own offspring and behave differentially toward nestlings on the basis of their relatedness to them can as yet not be excluded, although it seems unlikely.

Beyond these initial difficulties loom problems concerning the fitness consequences of mate-sharing and joint-nesting. Specifically, how, and under what circumstances, does shared breeding increase the fitness of individuals? At both WC and HR, demographic data indicate that males benefit directly from living and breeding in mate-sharing duos and trios compared to breeding solitarily, primarily because of a substantial increase in survivorship among males sharing mates. Thus, these data indicate that a mutualistic increase in fitness, not habitat saturation, may be responsible for mate-sharing by males (Stacey 1979*a*, 1982; Koenig and Mumme 1987).

For females, which have only been observed to nest jointly at HR, the situation is more complex, and analyses indicate that a variety of ecological and social factors influence the relative advantages and disadvantages of joint-nesting. In general, the fitness of females nesting as duos appears to be comparable to females nesting singly. Females nesting jointly experience slightly higher survivorship and an increased probability of obtaining a territory, while those nesting singly experience considerably higher annual reproductive success (Koenig and Mumme 1987; Mumme *et al.* 1988).

Interestingly, the apparent differences in the fitness effects of mate-sharing in males and joint-nesting in females parallel two different theoretical expectations. Stacey (1979*b*, 1982) hypothesized that mate-sharing occurs only when both parties mutually benefit from the behavior: for example, when two males together raise more young than can a single male breeding in a pair. Conversely, if a second male in the group has little effect on reproductive success, dominant males should monopolize matings and keep other males from copulating; that is, dominants should enforce a mating system which is basically monogamous.

In contrast, Vehrencamp (1979, 1983) on the basis of work done by herself and her colleagues on joint-nesting in Groove-billed Anis (see Chapter 11), predicted that mate-sharing and a lack of mating bias should occur only when fitness is independent of the number of cooperating individuals, and hence when birds that breed in pairs outside the group have the same fitness as individuals breeding as a group. Bias between co-breeders is predicted to increase as the fitness of individuals in groups exceeds that of singletons, in theory reaching complete monopolization (i.e. monogamy) as the fitness of solitary individuals approaches zero.

Thus, Stacey and Vehrencamp provide opposing predictions for the conditions under which a lack of reproductive bias among mate-sharing or joint-nesting individuals should occur. Stacey's model predicts equitable breeding when the fitness benefits of living in groups is greater than living singly, while Vehrencamp's model predicts equitable breeding when there is little or no benefit to living in groups.

As yet, the evidence available from Acorn Woodpeckers is unable to discriminate between these two models. On the one hand, the fitness of males does appear to benefit directly by mate-sharing in both populations, and, in support of Stacey's (1982) hypothesis, mate-sharing is relatively common among males. However, the fitness of females does not appear to be affected much by group-living and, consistent with Vehrencamp's model, bias among joint-nesting females at HR is small. More precise testing of these models, as well as answers to more basic questions concerning mate-sharing and joint-nesting, will have to await the emergence of unequivocal data on paternity and maternity, data which the currently developing techniques of DNA fingerprinting now seem likely to provide within the next several years.

Conclusion

A combination of factors appears to be responsible for group-living in Acorn Woodpeckers. Ecological constraints to dispersal and independent reproduction are most apparent in the HR population, where reproductive vacancies are few, a high proportion of territories are continuously occupied, and the fitness benefits to be gained by remaining as a non-breeding helper are in general small. Constraints also exist at WC, at least in the form of a limited number of high-quality territories available for colonization by potential breeders. However, the benefits to remaining in the natal territory, even as a non-breeder, are relatively greater in WC, at least as long as birds have the option of remaining on a high-quality territory. The WC data thus suggest that individuals remain as non-breeding helpers, not because of ecological constraints but rather because of direct benefits to remaining philopatric. Such benefits also exist at HR, at least for breeding males, for which sharing reproduction with other co-breeders appears considerably to enhance estimated lifetime fitness. Thus, both ecological constraints and direct benefits to group-living appear to be necessary to explain cooperative breeding in Acorn Woodpeckers, although the relative importance of these factors are different in the two sites.

Our work emphasizes some of the advantages of parallel studies of the same species. The behaviors of each of the populations that we have studied have turned out to be surprisingly distinct, and each has yielded quite different insights into the evolution of cooperative breeding. We fully expect that future studies of other Acorn Woodpecker populations, particularly those in Central and South America, will reveal further unexpected and bizarre behaviors, and will force additional re-evaluation of the causes of group-living and cooperative breeding in this unusual species.

We thank our coworkers Carl Bock, Tom Edwards, Phil Hooge, Roxana Jansma, Nancy Joste, Susan Hannon, Dave Ligon, Ron Mumme, Frank Pitelka, Mark Stanback and Bob Zink, whose extensive efforts helped to provide much of the material summarized here. Ron Mumme graciously allowed us to use unpublished results. We also thank our field assistants for their help and the National Science Foundation for support.

Bibliography

Bock, C. E. (1970). The ecology and behavior of the Lewis Woodpecker (*Asyndesmus lewis*). *Univ. Calif. Publ. Zool.* **92**: 1–91.

Bock, C. E. and Bock, J. H. (1974). Geographical ecology of the Acorn Woodpecker: diversity versus abundance of resources. *Amer. Nat.* **108**: 694–698.

Brown, J. L. (1974). Alternate routes to sociality in jays – with a theory for the evolution of altruism and communal breeding. *Amer. Zool.* **14**: 63–80.

Brown, J. L. (1987). *Helping and Communal Breeding in Birds: Ecology and Evolution.* Princeton University Press: Princeton.

Craig, J. L. and Jamieson, I. G. (1988). Incestuous mating in a communal bird: a family affair. *Amer. Nat.* **131**: 58–70.

Emlen, S. T. and Vehrencamp, S. H. (1983). Cooperative breeding strategies among birds. In *Perspectives in Ornithology*, ed. A. H. Brush and G. H. Clark, Jr, pp. 93–120. Cambridge University Press: New York.

Fitzpatrick, J. W. and Woolfenden, G. E. (1986). Demographic routes to cooperative breeding in some New World jays. In *Evolution of Animal Behavior*, ed. M. H. Nitecki and J. A. Kitchell, pp. 137–60. University of Chicago Press: Chicago.

Gowaty, P. A. and Karlin, A. A. (1984). Multiple paternity and maternity in single broods of apparently monogamous Eastern Bluebirds. *Behav. Ecol. Sociobiol.* **15**: 91–95.

Hannon, S. J., Mumme, R. L., Koenig, W. D. and Pitelka, F. A. (1985). Replacement of breeders and within-group conflict in the cooperatively breeding Acorn Woodpecker. *Behav. Ecol. Sociobiol.* **17**: 303–12.

Hannon, S. J., Mumme, R. L., Koenig, W. D., Spon, S. and Pitelka, F. A. (1987). Acorn crop failure, dominance, and a decline in numbers in the cooperatively breeding Acorn Woodpecker. *J. Anim. Ecol.* **56**: 197–207.

Joste, N. E., Koenig, W. D., Mumme, R. L. and Pitelka, F. A. (1982). Intragroup dynamics of a cooperative breeder: an analysis of reproductive roles in the Acorn Woodpecker. *Behav. Ecol. Sociobiol.* **11**: 195–201.

Joste, N. E., Ligon, J. D. and Stacey, P. B. (1985). Shared paternity in the Acorn Woodpecker. *Behav. Ecol. Sociobiol.* **17**: 39–41.

Kattan, G. (1988). Food habitats and social organization of Acorn Woodpeckers in Colombia. *Condor* **90**: 100–6.

Koenig, W. D. (1980). Variation and age determination in a population of Acorn Woodpeckers. *J. Field Ornith.* **51**: 10–16.

Koenig, W. D. (1981a). Space competition in the Acorn Woodpecker: power struggles in a cooperative breeder. *Anim. Behav.* **29**: 396–409.

Koenig, W. D. (1981b). Reproductive success, group size, and the evolution of cooperative breeding in the Acorn Woodpecker. *Amer. Nat.* **117**: 421–43.

Koenig, W. D., Hannon, S. J., Mumme, R. L. and Pitelka, F. A. (1988). Parent–offspring conflict in the cooperatively breeding Acorn Woodpecker. *Proc. 19th Intern. Ornith. Congr.*: 1220–30.

Koenig, W. D. and Mumme, R. L. (1987). *Population Ecology of the Cooperatively Breeding Acorn Woodpecker.* Princeton University Press: Princeton, N.J.

Koenig, W. D., Mumme, R. L. and Pitelka, F. A. (1983). Female roles in cooperatively breeding Acorn Woodpeckers. In *Social Behavior of Female Vertebrates*, ed. S. K. Wasser, pp. 235–61. Academic Press: New York.

Koenig, W. D., Mumme, R. L. and Pitelka, F. A. (1984). The breeding system of the Acorn Woodpecker in central coastal California. *Z. Tierpsychol.* **65**: 289–308.

Koenig, W. D. and Pitelka, F. A. (1979). Relatedness and inbreeding avoidance: counterploys in the communally nesting Acorn Woodpecker. *Science*, **206**: 1103–1105.

Koenig, W. D. and Pitelka, F. A. (1981). Ecological factors and kin selection in the evolution of cooperative breeding in birds. In *Natural Selection and Social Behavior: Recent Research and New Theory*, ed. R. D. Alexander and D. W. Tinkle, pp. 261–280. Chiron Press: New York.

Ligon, J. D. and Stacey, P. B. (1989). On the significance of helping behavior in birds. *Auk* (in press).

MacRoberts, M. H. and MacRoberts, B. R. (1976). Social organization and behavior of the Acorn Woodpecker in central coastal California. *Ornith. Monogr.* **21**: 1–115.

Miller, A. H. (1963). Seasonal activity and ecology of the avifauna of an American equatorial cloud forest. *Univ. Calif. Publ. Zool.* **66**: 1–78.

Moskovits, D. (1978). Winter territorial and foraging behavior of Red-headed Woodpeckers in Florida. *Wilson Bull.* **90**: 521–35.

Mumme, R. L. (1984). *Competition and cooperation in the communally breeding Acorn Woodpecker.* Ph.D. thesis, University of California, Berkeley.

Mumme, R. L. and de Queiroz, A. (1985). Individual contributions to cooperative behaviour in the Acorn Woodpecker: effects of reproductive status, sex, and group size. *Behaviour* **95**: 290–313.

Mumme, R. L., Koenig, W. D. and Pitelka, F. A. (1983a). Mate guarding in the Acorn Woodpecker: within-group reproductive competition in a cooperative breeder. *Anim. Behav.* **31**: 1094–106.

Mumme, R. L., Koenig, W. D. and Pitelka, F. A. (1983b). Reproductive competition in the communal Acorn Woodpecker: sisters destroy each other's eggs. *Nature (Lond.)* **306**: 583–4.

Mumme, R. L., Koenig, W. D. and Pitelka, F. A. (1983c). Are Acorn Woodpecker territories aggregated? *Ecology* **64**: 1305–7.

Mumme, R. L., Koenig, W. D. and Pitelka, F. A. (1988). Costs and benefits of joint nesting in the Acorn Woodpecker. *Amer. Nat.* **31**: 654–77.

Mumme, R. L., Koenig, W. D., Zink, R. M. and Marten, J. A. (1985). Genetic variation and parentage in a California population of Acorn Woodpeckers. *Auk* **102**: 312–20.

Ritter, W. E. (1938). *The California Woodpecker and I.* University of California Press: Berkeley, CA.

Roberts, R. C. (1979). Habitat and resource relationships in Acorn Woodpeckers. *Condor* **81**: 1–8.

Shields, W. M. (1987). Dispersal and mating systems: investigating their causal connections. In *Mammalian Dispersal Patterns: The Effects of Social Structure on Population Genetics*, ed. B. D. Chepko-Sade and Z. T. Halpin, pp. 3–24. University of Chicago Press: Chicago.

Stacey, P. B. (1979*a*). Habitat saturation and communal breeding in the Acorn Woodpecker. *Anim. Behav.* **27**: 1153–66.

Stacey, P. B. (1979*b*). Kinship, promiscuity, and communal breeding in the Acorn Woodpecker. *Behav. Ecol. Sociobiol.* **6**: 53–66.

Stacey, P. B. (1981). Foraging behavior of the Acorn Woodpecker in Belize, Central America. *Condor* **83**: 336–9.

Stacey, P. B. (1982). Female promiscuity and male reproductive success in birds and mammals. *Amer. Nat.* **120**: 51–64.

Stacey, P. B. and Bock, C. E. (1978). Social plasticity in the Acorn Woodpecker. *Science* **202**: 1298–300.

Stacey, P. B. and Edwards, T. C. (1983). Possible cases of infanticide by immigrant females in a group-breeding bird. *Auk* **100**: 731–3.

Stacey, P. B. and Jansma, R. (1977). Storage of piñon nuts by the Acorn Woodpecker in New Mexico. *Wilson Bull.* **89**: 150–1.

Stacey, P. B. and Koenig, W. D. (1984). Cooperative breeding in the Acorn Woodpecker. *Sci. Amer.* **251**: 114–21.

Stacey, P. B. and Ligon, J. D. (1987). Territory quality and dispersal options in the Acorn Woodpecker, and a challenge to the habitat saturation model of cooperative breeding. *Amer. Nat.* **130**: 654–76.

Stacey, P. B. and Ligon, J. D. (1990). The benefits-of-philopatry hypothesis for the evolution of cooperative breeding: variance in territory quality and group size effect. *Amer. Nat.* (in press).

Trail, P. W. (1980). Ecological correlates of social organization in a communally breeding bird, the Acorn Woodpecker, *Melanerpes formicivorus. Behav. Ecol. Sociobiol.* **7**: 83–92.

Vander Wall, S. B. and Balda, R. P. (1977). Coadaptations of the Clark's Nutcracker and the piñon pine for efficient seed harvest and dispersal. *Ecol. Monogr.* **47**: 89–111.

Vehrencamp, S. L. (1979). The roles of individual, kin, and group selection in the evolution of sociality. In *Social Behavior and Communication*, ed. P. Marler and J. Vandenburgh, pp. 351–394. Plenum Press: New York.

Vehrencamp, S. L. (1983). A model for the evolution of despotic versus egalitarian societies. *Anim. Behav.* **31**: 667–82.

15 DUNNOCKS: COOPERATION AND CONFLICT AMONG MALES AND FEMALES IN A VARIABLE MATING SYSTEM

15

Dunnocks: cooperation and conflict among males and females in a variable mating system

N. B. DAVIES

In Old English 'dun' means dull brown and 'ock' signifies little and, true to its name, the Dunnock (*Prunella modularis*) is the archetypal little brown bird. Although sparrow-sized and commonly called the Hedge Sparrow, it has a thin bill and is not a true sparrow but rather one of 12 species of accentors of the family Prunellidae, mostly montane birds which occur throughout Europe and Asia. Accentors often inhabit dense vegetation, where they forage solitarily, creeping about with mouse-like action in their search for small insects and seeds. The sexes are similar in appearance, with streaked plumage and the males have short, simple, warbling songs.

The Dunnock favors habitats with dense undergrowth. In Britain it is one of the 11 most common woodland birds and is the second most abundant bird of farmland hedgerows. It also inhabits downland, marshland and coastal scrub, and is common in suburban parks and gardens. Despite its abundance, this species' modest appearance and skulking habits did not, until recently, attract any detailed study. Indeed, the Reverend F. O. Morris, writing in his *History of British Birds* in 1856, encouraged his parishioners to emulate the humble life of the Dunnock: 'Unobtrusive, quiet and retiring, without being shy, humble and homely in its deportment and habits, sober and unpretending in its dress, while still neat and graceful, the dunnock exhibits a pattern which many of a higher grade might imitate, with advantage to themselves and benefit to others through an improved example'. This recommendation turns out be unfortunate. We now know that the Dunnock belies its dull appearance, having extraordinary sexual behaviour and an extremely variable mating system.

In the Reverend Morris's day, when animals were often viewed as behaving for the good of their species, or for the glory of God, the

457

observation of cooperation among individuals in the raising of offspring did not present a problem for interpretation. Any squabbles between members of a pair would, perhaps, have been regarded as brief mistakes in what was otherwise marital harmony. In sharp contrast, our emphasis today is on the expectation that individuals will behave so as to promote their own lifetime reproductive success. Detailed studies of how individuals behave have shown the prevalence of conflicts of interest in animal populations. Indeed, our wonder now may be how individuals ever come to cooperate sufficiently to rear any offspring at all!

The variable mating system of the Dunnock provides an excellent opportunity for observing conflict and cooperation among breeding adults. In this review, I begin by describing the variability in mating combinations. Then I examine the behavior of individuals and ask whether it makes adaptive sense, given the reproductive outcomes which individuals gain from their behavior. Finally I use the Dunnock study to raise the general problem of how cooperation can be stable, given the conflicts of interest between individuals.

A variable mating system

The first detailed studies of the Dunnock were of populations in a woodland in southern England (Snow and Snow 1982), and in a large garden near Edinburgh, Scotland (Birkhead 1981). Both studies showed that mating combinations varied considerably, even within a small population, including pairs (monogamy), a male with two females (polygyny), a female with two or even three males (polyandry) and more complex combinations involving two males sharing several females (polygynandry). In October 1980, I began a long-term study on a larger population with the aim of discovering the reasons behind this great variability. Even what textbooks describe as 'monogamous species' have some variability, with occasional polygyny for example, but the degree of variability shown by the Dunnock is unusual.

My study site is the Cambridge University Botanic Garden (Fig. 15.1), an area of 16 ha with a mosaic of different habitats including woodland with dense undergrowth, open woodland with little undergrowth, hedgerows (mainly hawthorn (*Crataegus*), and evergreens (*Taxus* and *Thuja*)), flower beds, shrubs, and areas of open lawn. This patchwork of different vegetation types and densities provides an ideal site for observing on a small scale how habitat influences behavior. There are 80 to 90 breeding Dunnocks in the garden each summer and most, if not all, are color-ringed for individual recognition. The nests are built by the female in hedges or evergreen shrubs. Each female has two to three broods per year, usually laying three to five eggs per clutch (Davies and Lundberg 1985). The eggs are incubated by the female alone for about 11 days and the young are then fed by both sexes on small insects for a period of 11–12 days in the nest. Once young leave the nest, they are fed for a further two to three weeks before reaching independence, soon after which they leave their natal territory.

Many young disperse outside the study area. Females are more likely to leave than males, and, of those young which do remain, females tend to disperse further than males. Both males and females breed when one year old. Adults are sedentary, usually remaining on or near their breeding territory throughout the year. The breeding population is more or less stable from year to year (46 males and 31 females in 1982, 48 males and 38 females in 1983, and 42 males and 40 females in 1984). Adult survival from the start of the breeding season in one year to the start of the next (1 April) was 56.0% 1981–2, 70.8% 1982–3 and 53.2% 1983–4 (Davies and Houston 1986). Of the disappearing adults, roughly 25–50% are replaced by young birds born on the study area the previous year and the remainder by

Fig. 15.1. Part of the Cambridge University Botanic Garden, where there is a breeding population of 80–90 Dunnocks. The birds feed mainly on the ground among the rockeries, in dense vegetation and on the lawns, and nests are built in hedges and shrubs, such as those in the background in the middle of the photograph.

immigrants. Ringing recoveries show that Dunnocks rarely disperse far, so most immigrants probably come from less than 5 km away.

The various mating combinations are summarized in Table 15.1. Close relatives were not involved in any of the more complex mating systems. For example, the origins of the two males were known in 23 cases of polyandry and polygynandry and in none of these were they relatives. Likewise, females involved in polygynous and polygynandrous combinations were never close relatives (origins known in 16 cases). Throughout the first year of the study, I found the variability bewildering, but then it became clear from watching the behavior of individuals that these different mating combinations were just variations on a single theme, namely an individual's ability to monopolize mates. A male's access to females increases, reading down the list of mating combinations in Table 15.1, from no females to shared access to one female (polyandry), sole access to one female (monogamy), shared access to several females (polygynandry) and finally, best of all in terms of mating success, sole access to several females (polygyny). From a female's point of view, however, mating success increases from the bottom of the list to the top. Females have least access to males with polygyny, where they have to share one male, and most with polyandry, where they have sole access to several males.

Observations showed that males were behaving so as to increase their access to females, while females were behaving so as to increase their access

Table 15.1. *The mating combinations observed in a population of Dunnocks in the Cambridge University Botanic Garden, with the number of breeding attempts for each case during the years 1981 to 1984*

| Mating combination | No. cases | Mating success | |
		For a male	For a female
Unpaired ♂	5	No females	—
Polyandry 3♂ 1♀	4	Share one	Sole access
2♂ 1♀	77	female	to several males
Monogamy 1♂ 1♀	62	Sole access to one female	Sole access to one male
Polygynandry 2♂ 2♀	47	Share several	Share several
2♂ 3♀	16	females	males
2♂ 4♀	1		
3♂ 2♀	1		
Polygyny 1♂ 2♀	21	Sole access to several females	Share one male

to males. This suggested that the variable mating system may be a reflection of different outcomes of sexual conflict. To examine this idea we first need to describe individual behavior in more detail and see how the different mating combinations come about (from Davies 1985; Davies and Lundberg 1984).

Behavior of males and females

In the spring, females defend exclusive territories against other females. The males then compete with each other to defend the female territories. The simplest case, shown in Fig. 15.2, is where one male defends one female territory, giving rise to monogamy. The stability of the pair is constantly threatened, however, because the male attempts to expand his territory to include a second female. The female, for her part, will drive off other females in an apparent attempt to keep the male to herself. In

Fig. 15.2. Diagrams to show how male (dashed lines) and female (solid lines) territories of Dunnocks overlap in the various mating combinations. The arrows indicate the directions in which alpha male or female behavior encourages changes in the mating system. (From Davies 1989.)

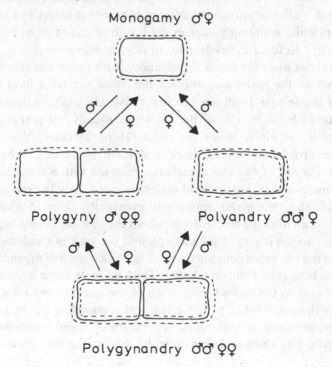

polygyny, where a male defends a territory large enough to encompass two female territories, the females often chase each other and the male hops about in between them as if to keep the peace! Sometimes females desert their eggs as a result of harassment by other females. The aggression between the females makes good adaptive sense because the cost of polygyny to a female is shared male parental care. Each female builds a nest and lays eggs and then, if the young hatch synchronously, the male usually helps each female part of the time. If one female has many more young than the other, then the male usually helps her full time and leaves the other female to rear young alone (Davies 1986).

The stability of monogamy is threatened in another way (see Fig. 15.2). First, a female may wander over a large territory so that one male is unable to defend it economically. In these cases two males begin by occupying adjacent territories, with the female wandering over both. The two males then compete with each other to monopolize the whole of the female's territory. After a while, they usually come to what looks like an agreement to share it. Their squabbling decreases to a low frequency and they coalesce their territories, share song posts and defend the whole female territory as a team against neighboring males. Second, males who are unpaired attempt to force themselves on to monogamous pairs and settle as a second male. The resident male attempts to evict the newcomer but sometimes he fails. This occurs quite commonly later in the breeding season when bereaved males attempt to force themselves on to neighboring pairs.

In both these cases the result is polyandry, with two males sharing the female. One of the males can displace the other one from food or the vicinity of the female. I call the dominant male the 'alpha male' and the subordinate male the 'beta male'. Beta males are usually first-year males (20 of 29 cases) and alpha males are older (21 of 23 cases), though the dominance order may reverse back and forth in successive breeding attempts (in seven of 45 cases dominance reversed within a season), and there is some evidence that very old males are more likely to drop in rank.

Although the two males apparently eventually agree to share the territory, it is an uneasy alliance. From about the time the female begins to line the nest to the laying of the last egg, and the start of incubation, the alpha male tries to monopolize her, guarding her closely and attempting to prevent the beta male from copulating. The beta male keeps approaching the female and so the males spend much of the time chasing each other. Sometimes there are fights, with the two males grappling on the ground. These fights can cause serious injury, often to the eyes, and sometimes even lead to death. The alpha male has difficulty in guarding the female, partly because he may lose her in the dense vegetation, but mainly because the female actively encourages the beta male to copulate with her! She often flies off suddenly through the undergrowth, as if to escape the alpha male's close attentions, and approaches the beta male to solicit a copulation. Once the alpha male has lost the female, it is often 10–20 min (once 3 h!) before he finds her again and in the meantime she may have been hiding away with the beta male. Even with close guarding, the beta male may be able to copulate while the female is hidden from the alpha male's view, behind a rock or some vegetation.

The more complex mating combinations are simply extensions of the three basic systems illustrated in Fig. 15.2. Sometimes in polyandry there are three males sharing a female's range (alpha, beta and gamma). Polygynandry arises when one male is unable to defend two female territories alone, so that two males (or sometimes three) come to share the defense. Once again, one of the males is alpha and attempts to monopolize both females. In many cases this is impossible simply because the females have exclusive territories and, if they are laying eggs synchronously, then whenever the alpha male is guarding one of them, the beta male can copulate freely with the other.

Relating behavior to paternity by DNA fingerprinting

In 1988, Ben Hatchwell and I observed some females from the time they finished nest building until the start of incubation, a period of six to 10 days during which males compete for copulations (mean observation time per female = 4.02 h, $n = 34$). Monogamous males guarded their females closely, spending on average 81.1 (± 4.5)% of their time within 10 m of the female (range 62.1–97.3%, $n = 10$) and intruders gained exclusive access (defined as the presence of only one male within 10 m) in only three of the 10 cases, a total of just 3 min in 23.98 h of observation (mean for the 10 cases = 0.17 (± 0.11) % of total observation time).

By contrast, in polyandry and polygynandry, where two males shared a female, competition for matings was intense. In eight out of 11 cases of polyandry and 10 out of 13 cases of polygynandry, the beta male enjoyed at least some period of exclusive access to the female. We recorded 85% of the cases where beta males had some access within the first hour of observation, so our observation times of *c*. 4 h per female were well above those needed to give a reliable measure. The mean copulation rate during periods of exclusive access was no different for alpha males (mean 1.98 $\pm$ 1.07 per hour) and beta males (mean 2.15 $\pm$ 0.44 per hour; $n = 16$, Wilcoxon matched pairs test, not significant (n.s.) and, furthermore, there was no significant variation in the proportion of exclusive access gained by alpha or beta males at different stages of the mating period. Exclusive access time therefore gave a good measure of mating success and was easier to record than copulations, especially on territories where males remained hidden with females for long periods of time. As in monogamy, intruding males gained very little exclusive access (only three out of 24 cases, a total of 5 min in 112.7 h, mean of 0.05 ($\pm$0.04)% of observation time).

We took blood samples, from the brachial vein, from all adults (caught in mist nets) and from 6–7-day old nestlings. Terry Burke and Mike Bruford (Leicester University) then produced DNA fingerprints for each individual. As expected from our observations during the mating period, the fingerprinting showed that intruding males gained little success; of the 133 young (45 broods), all but one were fathered by a male resident on the territory. The exception was a case of polyandry, where one chick from a brood of four was fathered by an alpha male from a neighboring territory (Burke *et al.*, 1989). All chicks from monogamous mating systems were sired by the monogamous male, while in polyandry and polygynandry paternity was often shared between alpha and beta males (Fig. 15.3; Table 15.2). In all cases maternity was assigned to the resident female on the territory, so there was no intraspecific brood parasitism.

In all 15 cases of monogamy, the young were fed only by the monogamous male and female. In the 30 cases of polyandry and polygynandry where we recorded chick feeding, the female fed the young at all 30 nests, helped by both alpha and beta males in 14 cases, by the alpha male only in 11 cases, by the beta male only in four cases (in one case an alpha male died during incubation, and in one case a beta male died), and in one case the female received no male help. We recorded 90% of the cases where beta males fed the young within the first hour of observation, so our mean observation time for chick feeding (4.35 h per nest, $n = 45$) gave a reliable measure of who brought food. In cases where both alpha and beta males helped to feed, there was a dramatic change in their behaviour from conflict during the mating period to cooperation once the chicks hatched; sometimes they now collected food side by side, whereas two weeks previously there would have been a serious fight.

From the DNA fingerprinting analysis the most striking result was that males were more likely to help to feed the young if they had paternity than if they did not (Table 15.3). Two results suggest that this association does not arise because males can recognize their own offspring. First, in two cases the alpha male fed nestlings, even though he had no paternity and in four cases the beta male did so (Table 15.3). In two of these cases both males helped to

Fig. 15.3. Analysis of paternity in Dunnock broods by DNA fingerprinting (from Burke *et al.* 1989). Two broods are shown from territories where there were two males and a female (polyandry). The spread of different-sized DNA fragments, shown as bands on the gel, provides a unique 'genetic fingerprint' for each individual. Analysis of a large family showed that each parental band had a 0.5 probability of being transmitted to an offspring. Paternity can be assigned by the presence/absence of diagnostic (paternal male-specific) bands in nestlings. Markers indicate diagnostic alpha (filled arrow heads) and beta (open arrowheads) male bands. In the left-hand example, all the bands in the three offspring A, B and C can be traced to either the female or the alpha male; so the alpha male sired all three young. In the right-hand example, however, there is mixed paternity, with the beta male siring offspring D, E and F and the alpha male siring offspring G.

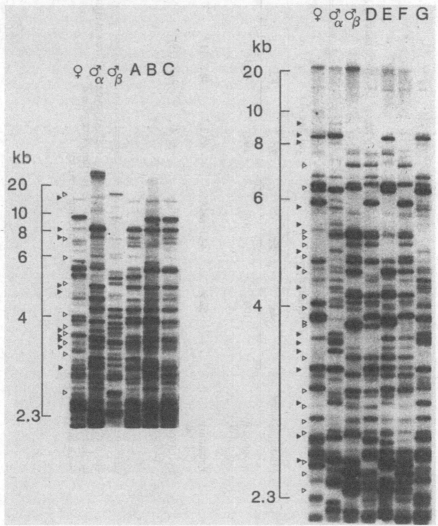

465

Table 15.2. Analysis of paternity by DNA fingerprinting for three Dunnock mating systems (from Burke et al. 1989).

Mating system	% total young (n) fathered by		Mean % paternity per nest (n)		Details for each nest: proportion fathered by α♂
	α♂	β♂	α♂	β♂	
Monogamy (1♂ 1♀)	100	— (49)	100	0 (15)	4/4, 4/4, 4/4, 4/4, 4/4, 4/4, 3/3, 3/3, 3/3, 3/3, 3/3, 3/3, 3/3, 2/2, 2/2
Polyandry (2♂ 1♀)	52.9	44.1 (34)[a]	49.2	48.5 (11)[a]	4/4, 3/4,[a] 3/3, 2/2, 2/4, 1/4, 1/4, 1/3, 1/3, 0/2, 0/1
Polygynandry (2♂ 2♀; 2♂ 3♀)	72.0	28.0 (50)	68.4	31.6 (19)	3/3, 3/3, 3/3, 3/3, 3/3, 2/2, 2/2, 2/2, 1/1, 3/4, 3/4, 2/4, 2/4, 2/3, 1/3, 1/2, 0/2, 0/1, 0/1

The proportions of young sired by the two males in polyandry and polygynandry are not significantly different. All but one of the 133 young were fathered by a male resident on the territory (see note a).
[a] One chick was fathered by a male on a neighboring territory

Table 15.3. *Association between paternity, feeding of nestlings and access to the female during the mating period for alpha and beta male Dunnocks in polyandry and polygynandry (from Burke et al. 1989).*

Paternity (from DNA fingerprinting)	Nest watches showed male		Observations of female during mating period showed male	
	Fed young	Did not feed	Had some exclusive access	Did not
Alpha male				
Broods where he had paternity	23	1	13	0
Broods where he had no paternity	2	3	4	1
	$P = 0.011^a$		$P = ^b$	
Beta male				
Broods where he had paternity	14	3	10	0
Broods where he had no paternity	4	8	2	6
	$P = 0.011^a$		$P = 0.005^a$	

Chick feeding by males was recorded for 29 out of the 30 nests in Table 15.2, and there were observations during the mating period for 18 of these cases.
[a] Fisher Exact probability test.
[b] Sample size too small to test.

feed a single chick. Second, when the young left the nest they were often divided among the parents, with each adult taking sole care of part of the brood and feeding them for a further two weeks until they became independent (Byle 1987). There was no tendency for males to pick out their own young. We observed 28 young as dependent fledglings, from 12 broods where both alpha and beta male had helped to feed the nestlings; 11 young were sired by the alpha male, of which five were cared for by their father and six by the beta male, while 17 were sired by the beta male, of which nine were cared for by their father and eight by the alpha male (Burke et al. 1989).

How, then, did the association between chick feeding and paternity come about? A male's access to the female gave a good prediction of his chances of paternity (Table 15.3), and males apparently used their access to the female to determine whether they fed the young. Including data from all polyandrous and polygynandrous mating systems observed in 1981-8, beta males fed the young in 22 of the 27 cases where they had some exclusive access to the female during the mating period, compared to only one of 11 cases where they were not observed to have access ($P < 0.001$; Burke et al. 1989). Furthermore, beta males brought a greater proportion of the male feeds where they had enjoyed a greater proportion of the exclusive access time during the mating period (Fig. 15.4).

The apparent inability of Dunnocks to recognize their own young is, perhaps, not surprising given that they will feed young cuckoos (*Cuculus canorus*) and the young of other species introduced into their nests (Davies and Brooke 1989). Nevertheless, a simple rule of chick feeding in relation to access to the female during the mating period, which is a good predictor of paternity, leads the males to behave adaptively.

Mating display and interference

There is one other factor which may influence paternity in the Dunnock, namely the extraordinary precopulatory display first described in 1902 by Edmund Selous and recorded in his book *Evolution of Habit in Birds* (1933). The male hops behind the female, she shivers her wings and raises her tail so as to expose her cloaca. The male then pecks her cloaca for on average 50 s (maximum 2 min) before he copulates (Fig. 15.5). The pecking display must be very important because during this long period there is an increased chance that another male will interrupt before copulation itself occurs. In fact, 40% of attempts are interrupted at the pecking stage, which makes one wonder why the male does not, like most birds, simply copulate quickly straight away. During the pecking, the female's cloaca becomes pink and distended and makes strong pumping movements, during which the female dips her abdomen down suddenly. When performing some of these dips, the female ejects feces but on other occasions she ejects a small droplet of fluid. There is little doubt that this is what the male has been waiting for during the pecking because he looks at the ejected droplet and, as soon as it is produced, he copulates. Copulation itself is brief and the male jumps at the female, cloacal contact lasting for only a fraction of a second.

Fig. 15.4. Relationship, for polyandry and polygynandry in Dunnocks, between percentage exclusive access time with the female enjoyed by the beta male during the mating period, and percentage male feeds brought by the beta male to the nestlings. Exclusive access time is the total time that either the alpha or beta male was the only male within 10 m of the female. Data from 1981–8 for cases where beta males gained some exclusive access ($y = 5.75 + 0.798x$: analysis on log-transformed data: $r = 0.775$, 15 d.f., $P < 0.01$; the slope does not differ significantly from 1, $t = 1.33$, 15 d.f.). In polygynandry, a male's provisioning of the young at one nest may be influenced by the activities of other females in the mating system so cases where other females were laying or had young simultaneously were excluded. (From Burke *et al.* 1989.)

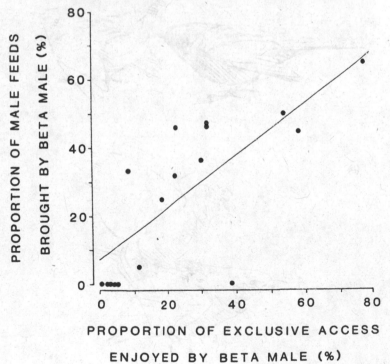

I have spent a great deal of time crawling about searching for these droplets and so far have found only three. In each case they have proved, under the microscope, to be a mass of sperm (Davies 1983). I do not know how often these droplets are ejected and so the interpretation of this display must be speculative. It is tempting to suggest that with frequent copulations from two males the female's sperm stores become full. The copulation rate in Dunnocks is higher than that described for other small birds, averaging once or twice per hour throughout the 10 day mating period. If a female's sperm store did become full, then the only way a male could put sperm in would be to first stimulate the female to eject some. Furthermore, sperm in

Fig. 15.5. Copulation in Dunnocks. (*a*) The male first stands behind the female and pecks her cloaca. Mean duration of display was 50.2 s, range 5–120 s; mean number of pecks was 27.9, range 0–118 pecks; 74 displays observed. (*b*) Then he copulates. Copulation itself is very brief; the male appears to jump over the female, cloacal contact lasting for a fraction of a second. (Drawings by John Busby. From Davies 1983.)

(*a*)

(*b*)

the store may be another male's, whereas sperm a male is about to put in is certainly his own. We can speculate still further that it may pay the female to eject sperm in front of a male to try to convince him that he will have paternity and so encourage him to feed the young.

What happens if the alpha male guards the female successfully and prevents the beta male from copulating? Sometimes the beta male leaves the territory to try elsewhere (six of 19 cases). In other cases he remains on the territory showing no interest in the female until she begins to build a new nest for the next breeding attempt. Sometimes, however, the beta male chases the female during incubation and visits the nest. The alpha male chases him away. In some cases this behavior has been correlated with cracked eggs in the nest, desertion by the female or the disappearance of young chicks (Davies 1986). I have no direct evidence, but I strongly suspect that this interference is directly due to the beta male. It would certainly pay a beta male who had not copulated to interfere if he could; the result is that the female lays a replacement clutch within one to two weeks, compared with the four to five weeks that the beta male would have to wait if he simply stood aside while the alpha male and female raised the young to independence. Interference by a beta male, therefore, would hasten the day that he has another chance to father young himself.

There is therefore a dual incentive for the female to copulate with the beta male; first she gains a second male's help with chick feeding, and second she reduces the risk of interference. I now consider whether this behavior of individuals, a mixture of cooperation and conflict, makes adaptive sense by looking at the outcomes of behavior in terms of reproductive success.

Reproductive success of males and females in the various mating combinations

The young are fed on billfulls of very small invertebrates, especially spiders, beetles, midges and collembola, with only occasional larger items such as earthworms and caterpillars (see also Bishton 1985; Tomek 1988). The provisioning rate to the brood increases with brood size but it also varies with the mating system (Fig. 15.6). Nestlings obtain most food where two males and a female feed them (polyandry), less where only one male and a female provision (monogamy), or where a female has the part-time assistance of two males (polygynandry), and least of all where a female has only the part-time help of one male (polygyny) (for an analysis of how hard males and females work in the different mating systems, see Houston and Davies 1985). These differences in provisioning rate are reflected in nestling weight (Davies 1986) and also the number of young fledged from a nest, which is proportional to the amount of male help which a female gets with chick feeding (Fig. 15.7).

Are these differences in reproductive success caused by differences in the mating system? It is possible, for example, that greatest success occurs with cooperative polyandry because this mating combination occurs on the best-quality territories, or because the best-quality individuals are involved in this mating system. The greater reproductive success could then be attributed to greater food abundance and better parental ability, rather than to the fact that three adults provisioned the brood. Two lines of evidence suggest that it is the mating combination which is the main determinant of reproductive success, rather than these or other possible confounding variables. First, when one of the males dies in polyandry, the number of young produced decreases (Fig. 15.7). Second, we can compare

Fig. 15.6. The provisioning rate to nestlings increases with brood size in Dunnocks but varies markedly across the different mating systems. Each point refers to a different nest, observed when the chicks were 7 to 11 days old. (From Davies 1986.)

(a) Where the female has full-time help from a male in monogamy (♂♀), provisioning rate is greater than where she only has the part-time help of a polygynous male (♂♀♀) (analysis of covariance; slope n.s., elevation $p < 0.02$). (b) Where a female has the full-time help of two males in a polyandrous mating system (♂♂♀), the provisioning rate is greater than in polygynandry (♂♂♀♀), where on average a female has only the part-time help of two males (analysis of covariance; slope n.s., elevation $p < 0.01$). The top line in (b), cooperative polyandry, is significantly higher in elevation than the top line in (a), monogamy ($P < 0.001$), but does not differ in slope. There is no significant difference between monogamy and polygynandry.

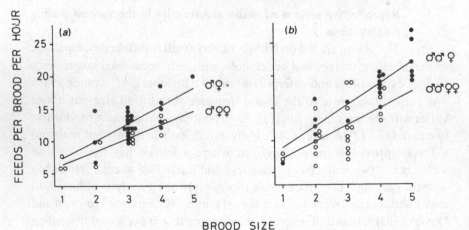

the reproductive success of the same female on the same territory but in different mating combinations. This is possible because mating combinations vary between broods and between years, so that a particular female may have breeding attempts with each of the three combinations in Fig. 15.7. These 'matched comparisons' show the same significant differences as the overall result in Fig. 15.7 (for details, see Davies 1986).

What do these differences mean for male and female reproductive success in the various mating combinations? Lifetime reproductive success of individuals involved in different mating systems is impossible to obtain because individuals frequently change mating systems during their lives, even within the same breeding season. If mating systems are divided into four types (monogamy, polygyny, polyandry and polygynandry), then for females observed to have two breeding attempts within a season, 11 out of 33 changed mating system and for females who had three breeding attempts, eight out of 14 changed mating system at least once during the season. Considering females observed in two different years, 10 out of 13 were observed in two or more different mating systems (four of the 10 were observed in three different systems) and all four females observed over three years took part in at least two different mating combinations. (This analysis excludes changes in mating system induced experimentally by the provision of feeders, described below.) Many of these changes in mating system occurred because of mortality. For example: one male in a polyandrous trio died, leaving a monogamous pair; a beta male left a polyandrous trio to

Fig. 15.7. The number of Dunnock young fledged from a nest is proportional to the amount of help which a female gets in chick feeding. In polygyny the female gets the part-time help of one male. In monogamy she gets the full-time help of one male. In cooperative polyandry she gets the full-time help of two males. *, $P < 0.05$; ***, $P < 0.001$. (From Davies 1986.)

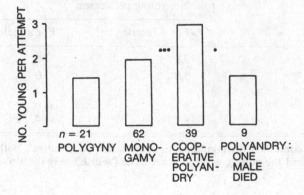

take over a widowed female on a neighboring territory; a polygynandrous female died, leaving polyandry; a monogamous male died and the neighboring polyandrous males took over the female, thus creating polygynandry; and so on.

It would also be preferable to have a measure of the number of young produced who survive to breed; this is impossible because most of the young disperse from the study area soon after independence. The best measure of reproductive success that I can obtain from the data available is a calculation of the number of fledged young an individual would produce per breeding season from the different mating combinations. This calculation takes into account the probability that a breeding attempt will be successful, the number of fledged young produced per successful attempt and the time taken for successful and unsuccessful attempts (for details, see Davies and Houston 1986). Table 15.4 presents these calculations separately for males and females.

A female has least success with polygyny, where she has to share the help of one male. Her aggression to the other female makes good sense; if she drove her away she could claim the male's full-time help and hence the greater success of monogamy. A female has greatest success of all with cooperative polyandry, where two males help her to raise young. Her encouragement of copulations by the beta male therefore also makes good sense. If the beta male fails to copulate, then the outcome, labeled selfish polyandry in Table 15.4, is less successful than with monogamy.

For a male, the mating combinations have exactly the reverse effects in

Table 15.4. *The number of fledged Dunnock young produced per breeding season by males and females in the different mating combinations (from Davies and Houston 1986)*

Mating system	No. young per season	
	For a female	For a male
Polygyny 1♂ 2♀	3.8	7.6
Monogamy 1♂ 1♀	5.0	5.0
Polyandry 2♂ 1♀		
Cooperative	6.7	α4.0, β2.7
Selfish	4.4	α4.4, β0

In polyandry, two outcomes are distinguished; in some cases ('cooperative'), both alpha and beta male feed the young, while in other cases ('selfish') only the alpha males does so.

terms of reproductive success. A male does best with polygyny, the combination in which a female does worst! Although, because of less help, each female produces fewer young, the total production of young by two females in polygyny exceeds that of a monogamous female. A male's attempts to monopolize a second female therefore pays, as does his attempts to prevent disputes between his two females.

To calculate male reproductive success in cooperative polyandry, we need to know how paternity is shared. Of 17 cases of polyandry and polygynandry observed in 1981–4, where both males were seen to copulate, alpha males performed 78 of the total of 126 copulations (61.9%), a mean proportion of 58.3% (Davies 1985), and it was assumed therefore that the relative success of alpha and beta males was probably about 60%:40% (Davies and Houston 1986). The DNA fingerprinting results in 1988 showed that this assumption was reasonable. For the 20 nests in Table 15.2 where both males in polyandry and polygynandry had access to the female during the mating period, or both fed the young (and so, by implication, both had enjoyed access), the mean paternity split was 55% alpha, 45% beta (total of 58 young, 56.9% sired by the alpha male; Burke *et al.* 1989). In Table 15.4, I have therefore followed an earlier analysis and assumed the paternity split is 60% alpha:40% beta. Under this assumption, it is better for an alpha male to try to evict the beta male if he can, just as is observed, so as to claim the greater reproductive success from monogamy. If he cannot drive off the beta male, then it is best at least to try to prevent him from copulating because full paternity in 'selfish polyandry' has greater payoff for an alpha male than shared paternity in 'cooperative polyandry'. This is again exactly what we observe from the male's behavior. An alpha male never agrees to share the female; he only comes to cooperate by default when he is unable to prevent the beta male from copulating. From the beta male's point of view, monogamy is clearly much better than the share he gets in cooperative polyandry. As predicted, whenever a neighboring female becomes available through the death of a male, a beta male is always quick to leave a polyandrous trio to claim her.

The way paternity is shared is critical for this interpretation. If, for example, the alpha male could claim more than 75% of the paternity, then he would do better with cooperative polyandry than monogamy, assuming the same number of young are raised as in Table 15.4. Under these circumstances it would pay the alpha male to agree to share the female. However, it is unlikely that the alpha male could ever impose a strict limit on the beta male's access to the female. First, the alpha male could not simply sit on a perch, watch the beta male copulate and then chase him off once he had performed say 25% of the matings. Obviously it would pay the beta male to go for a larger share than the alpha male wanted to give him! Second, from the female's point of view a 75%:25% paternity split, say, in favor of the alpha male may not maximize her reproductive success. Given that each male is likely to work harder the greater the proportion of matings he achieves, the female will probably maximize the total work that males put into her brood if she gives each male 50% of the matings. It would be very nice to be able to tether the two males to see how a female preferred to distribute the copulations! With a 50:50 paternity split, a male does worse than he would with monogamy (Table 15.4). The conflicts between alpha and beta males are, therefore, not hard to understand. The main barrier to the evolution of cooperation is the difficulty of reaching a stable agreement on how the female should be shared.

Polygynandry has been omitted from Table 15.4 because the way in which two males share two females is very variable and not yet understood (Davies 1986). In the commonest polygynandrous combination, of two males with two females, the number of fledged young produced by each female is calculated to be 3.65 per breeding season (Davies and Houston 1986). This is slightly less than the 3.8 young produced per season by a polygynous female (Table 15.4). It is not known why a polygynandrous female, who on average gains the part-time help of two males, does not do better than a polygynous female, who gains only the part-time help of one male. In polygynandry, however, there was often much interference between females, which sometimes led to desertion of clutches (Davies 1986). From a male's point of view, with polygyny he gets full paternity of both females' broods, whereas with polygynandry paternity is shared. Therefore the attempts of the alpha male to drive away the beta male makes good sense.

The general conclusion is that the behavioral conflicts observed are what would be expected given the reproductive payoffs for males versus females in Table 15.4. I suggest that the various mating systems reflect different outcomes of conflicts of interest. Where a female is able to gain her optimum at the expense of males, we observe cooperative polyandry. Where a male is able to gain his optimum at the expense of females, we observe polygyny. In monogamy the female has not been able to gain another male and the male has not been able to gain another female. Polygynandry (e.g. two males with two females) can be viewed as a kind of 'stalemate'; the alpha male is unable to drive the beta male off and hence claim both females for himself, and neither female is able to evict the other and claim both males for herself.

What determines who wins when there are conflicts of interest?
Three factors are probably important.

(*a*) *Population structure.* The sex ratio in the breeding season will influence
the degree of competition for mates. In Dunnocks, the female is smaller
than the male and is subordinate at feeding sites in the winter. In a hard
winter, female mortality is higher than that of males (Davies and Lundberg
1984) and, with a greater male-biased sex ratio in the spring, polyandry is
more common.

(*b*) *An individual's competitive ability.* Male Dunnocks with larger territories
can encompass more female territories and so enjoy greater mating success
(Fig. 15.8). One of the factors influencing a male's competitive ability
appears to be age, with alpha males in polyandrous trios tending to be old
males and beta males first-time breeders. It is not yet known exactly why a
male should become a better competitor as he gets older; other factors may
be important as well as age.

Fig. 15.8. Dunnock males with larger territories have greater mating
success (Spearman rank correlation = 0.641, $P < 0.01$). The different
mating combinations are ranked up the y-axis in increasing order of
male mating success. Territory size is measured by the size of a male's
song polygon in early spring. In polyandry and polygynandry,
neighboring males coalesce their singing territories and eventually
come to share the whole of one or two females' territories (see the text
and Fig. 15.2), with one male becoming alpha and the other beta. The
graph shows the size of song polygons before any coalescence of male
territories had occurred. (From Davies and Lundberg 1984.)

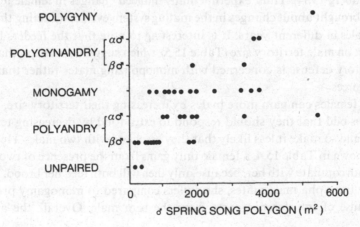

More long-term data are needed to discover what influences a female's ability to gain access to males. One factor determining the number of males associating with a female is her territory size; the larger a female's territory the more likely she is to have two males because one male is less able to defend a large territory economically. What, then, determines a female's territory size? Dunnocks spend most of their time foraging in amongst dense vegetation. In some parts of the Botanic Garden these good feeding areas are densely distributed, whereas in other parts they occur as more widely scattered patches. Female territory size is larger in places where food patches are more widely dispersed (Fig. 15.9(*a*)). The consequence of a larger female territory size is that she is more likely to be associated with more than one male (Fig. 15.9(*b*)). These observations suggest that a female's territory size is influenced by food distribution and the size of her territory then influences the mating system through her ease of monopolization by males.

Arne Lundberg and I investigated this further by means of an experiment. We put out extra food (protein-rich softbill food) on some randomly chosen female territories. As predicted, females with feeders had smaller territories than did control females and, as a result, they had fewer males monopolizing them (Table 15.5). This experiment provides direct evidence that food distribution can influence the mating system via changes in female territory size. We did the feeder experiment in two years and were able to do matched comparisons for 10 females who were present in both years but who had extra food in only one of the years. Nine of the ten females had a smaller territory in the year they had a feeder; females whose ranges increased from one year to the next had more males associated with them, while females whose range decreased had fewer males (Davies and Lundberg 1984). Thus, experimentally induced changes in female territory size brought about changes in the mating system even considering the same females in different years. It is interesting to note that the feeders had no effect on male territory size (Table 15.5), which suggests, perhaps, that male territory defense is concerned with monopolizing mates rather than food resources.

If females can gain more males by increasing their territory size, then it seems odd that they should respond to extra food by decreasing territory size and so make it less likely that they associate with two males. However, as shown in Table 15.4, a female only gains from the presence of two males if both copulate with her, because only then will both feed her brood. Where only the alpha male mates, she suffers compared to monogamy probably because of interference costs from the beta male. Overall, the average

Fig. 15.9. (*a*) Female Dunnocks have larger territories in poorer
feeding areas. The two symbols refer to different years. The curve
drawn through the points is the expected curve if all females had the
same average total feeding area within their territories. (*b*) The larger
a female's territory, the more likely she is to associate with more than
one male. The top graph plots cases where one female associated with
one male (monogamy) or two or three males (polyandry). Territories
of monogamous females are smaller than those of polyandrous
females ($P < 0.002$). The bottom graph plots the territory size of each
female where the mating system was two females associated with one
male (polygyny) or two or three males (polygynandry). ((*a*) and (*b*)
from Davies and Lundberg 1984.)

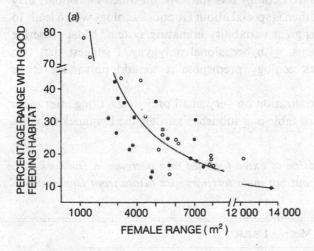

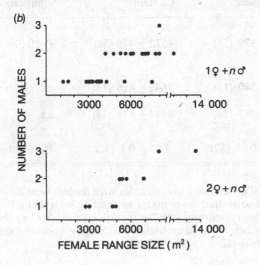

reproductive success of a female in polyandry (including 'cooperative' and 'selfish' polyandry) is no greater than for monogamy (2.09 young per breeding attempt, $n = 77$, compared with 1.95 for monogamy, $n = 62$, n.s.; Davies 1986). Thus there is no incentive for a female to increase her territory so as to gain two males unless she can be confident of getting them both to copulate.

With each sex out to exploit the other, it is clear that many factors will influence who 'wins' the conflict and many more data are required before this will be understood.

(*c*) *Ecological conditions.* Ecology sets the stage on which individuals play their behavior. What then is special about Dunnock ecology which leads to the expression of this great variability in mating system? Most passerine birds are monogamous, with occasional polygyny. I suggest that two features of Dunnock ecology predispose it to add polyandry to its repertoire.

The first is its specialization on very small prey items. Long after other species have left a bird table in a suburban garden, the Dunnock remains,

Table 15.5. *The provision of extra food led to a decrease in Dunnock female territory size, but not male territory size (Data from Davies and Lundberg 1984)*

	Mean $\pm$ 1 S.E.M.		
	With feeder	Control	Significance
Territory size (m²)			
Females	2776 ± 379 (28)	4572 ± 476 (39)	$P < 0.01$
Males			
Defended by one male	2864 ± 340 (11)	2642 ± 416 (13)	n.s.
Defended by two males	5276 ± 797 (14)	6614 ± 674 (17)	n.s.
No. males defending a female	1.05 ± 0.07 (37)	1.59 ± 0.1 (32)	$P < 0.001$

n.s., not significant.
As a consequence of their smaller territory size, females with feeders were more easily monopolized by males and so had fewer males associating with them. For each female the number of males associating with her is calculated by dividing the number of males by the number of females within the mating combination. Significance levels shown for two-tailed *t*-tests.

picking up invisible scraps which other birds find unprofitable. Even when feeding its young, tiny insects are the main prey. The major rate-limiting factor in chick feeding is the long time needed to collect a billfull of small prey. It seems likely, therefore, that the rate at which food is delivered to the nest is limited not primarily by food abundance but rather by the work force available to collect it. It may pay a female Dunnock, more so than a species that exploits easily collected bonanzas of large prey (e.g. caterpillars), to have more than one male to help to care for the young.

The second ecological factor of importance may be the Dunnock's habit of foraging in dense cover. This means that it is relatively easy for intruding males to trespass undetected and so makes it difficult for one male to monopolize a female. The best way to test these ideas would be to use the comparative approach to see whether species with similar ecology have the same variability of mating system.

The evolution of stable cooperation

Although the measures of reproductive success in Table 15.4 make reasonable sense of individual behavior, all we have really done is to push the problem back one step. Why have the reproductive payoffs ended up at these particular values? Why, for example, does the beta male not help more with chick-feeding so that it pays the alpha male to cooperate? Why does a polygynous male not help his females more so that they agree to mate polygynously?

To answer questions like these we need to think in more general terms about the evolution of life histories, in particular the trade-offs between survival and reproductive success. For example, how hard should an individual work at feeding its young? The harder it works (e.g. the more food it brings) the more likely the young are to survive, but harder work is likely to decrease the adult's own chances of survival. Some intermediate level of effort would therefore seem to be the optimum. However, the problem is not as simple as this. Where two or more individuals are cooperating to raise offspring, the optimum effort for one individual depends on how hard the others are prepared to work.

There is evidence from a number of species, including the Dunnock, that individuals can work harder at chick feeding than they actually do (Chase 1980; Houston and Davies 1985). For example, when one parent dies or deserts, the other increases its effort, though not usually sufficiently to compensate for the loss of the other. Alternatively, where parents have helpers at the nest they often work less hard themselves than they would if alone. Given that an individual works harder if others do less, and less hard if others do more, how do two or more individuals come to a stable agreement on how much work each should put into the cooperative venture of raising offspring? The problem is that of cheating; each individual will be tempted to get away with less than its fair share of work. This problem is, of course, not restricted to chick-feeding but arises whenever two or more individuals are engaged in a cooperative task.

Alasdair Houston and I have used a model to help us think about this problem (Houston and Davies 1985), developing a model first suggested by Chase (1980) (Fig. 15.10). Consider a pair of birds. The male will have a 'best response', for example in terms of chick-feeding effort, to a given effort put in by the female. If the female works harder, it will pay the male to do less and vice versa (Fig. 15.10(*a*)). The female will likewise have a 'best response' to what the male is prepared to do (Fig. 15.10(*b*)). In both

Fig. 15.10. Stable cooperation in Dunnock chick feeding. (From Houston and Davies 1985). (*a*) It pays the male to work harder the less the female does. (*b*) It pays the female to work harder the less the male does. For simplicity, these 'reaction curves' have been drawn as straight lines. (*c*) The male and female reaction curves plotted on the same graph. The intersection point is the stable solution to the conflict; see the text. ESS, evolutionarily stable strategy.

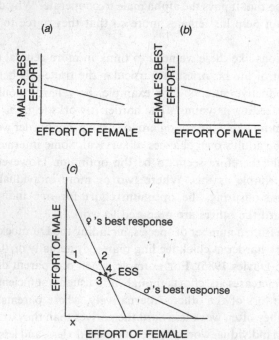

cases, these lines have been drawn with a slope of less than -1, which means that if one parent does less work the other responds by doing more but not sufficient to compensate fully for the loss, which is the pattern generally observed in real life.

Given that each member of the pair has a reaction curve, what will be the result? To see this, we simply plot both responses on the same graph (Fig. 15.10(c)). The intersection point turns out to be the stable solution to the conflict. For example, if the female played effort x, then the male's best response would be 1. The female would then reply with 2, the male with 3 and so on, reactions proceeding by smaller and smaller amounts until the intersection point was reached. This is the stable effort for male and female because at this point it does not then pay either sex to change. Other outcomes are possible depending on how the reaction curves meet. For example, if the curves have a slope greater than -1, the intersection point is unstable. Here, if one parent reduces its effort, the other reacts by increasing its effort to more than compensate. The first parent will then be tempted to reduce its effort further and reactions will proceed by larger and larger amounts until one parent ends up doing all the work. Which parent cares will depend on the starting conditions.

In theory, the stable solution to the conflict could be reached by male and female reacting to each other's bids on a short time scale. Much of the complex courtship seen between members of a pair may reflect this kind of bargaining. Alternatively, we may not actually see any conflict now because the game may have already been played over evolutionary time, with individuals now designed by natural selection to play the stable effort.

We can do the same kind of analysis for three cooperating individuals, to represent cooperative polyandry for the Dunnock. In this case the optimum effort for one adult is a function of the efforts of the other two. Once again we can calculate the stable equilibrium of effort for each of the three parents (Houston and Davies 1985). How can we test this model? To do so we need to know two things, first how a parent's effort influences both its own survival and that of the young, and, second, the shape of the reaction curves, namely how one individual's best effort varies as a function of what others are doing. Neither of these are well known for any animal, so at the very least the model is useful in identifying gaps in our knowledge.

Although we cannot yet test the model quantitatively, we can use it to explore possible outcomes. One important result is as follows. In Dunnocks the young are fed more when three adults provision them rather than just two (Fig. 15.6). This means that a female benefits from cooperative polyandry and much of the observed conflict followed from this. The model shows, however, that there are survival functions for which the total effort is not expected to increase with an increase in work force. As more and more individuals help, all that happens is each does less work and the total effort remains constant. This pattern is seen in some species of birds which have helpers, for example Gray-crowned Babblers (*Pomatostomus temporalis*) (Brown *et al.* 1978), and Moorhens (*Gallinula chloropus*) (Gibbons 1987). In both these species the total feeding rate to the young does not increase with an increased number of helpers. To understand why in other species, such as the Dunnock, total effort does increase, we need to quantify how effort influences adult and chick survival.

Two general conclusions

Two general messages can be learned from the Dunnocks. The first is that conflicts of interest are important and so we need to calculate payoffs for each individual in a breeding group and compare these with its alternative options. Different mating systems may often reflect different resolutions to conflicts of interest (Davies 1989). The interesting question to ask is: 'under what circumstances can some individuals gain at the expense of others?' Second, even when we see individuals engaged in a cooperative task such as feeding nestlings, with no visible squabbles among them, nevertheless conflicts of interest will have influenced the degree of cooperation which we now see. Twenty years ago we had many descriptive data on the breeding behaviour of birds, with rather little theory to help us make sense of it. Now we have some useful theory, but it is clear that we need better-quality data to test it.

Bibliography

Birkhead, M. E. (1981). The social behaviour of the Dunnock, *Prunella modularis*. *Ibis* **123**: 75–84.

Bishton, G. (1985). The diet of nestling Dunnocks *Prunella modularis*. *Bird Study* **32**: 59–62.

Brown, J. L., Dow, D. D., Brown, E. R. and Brown, S. D. (1978). Effects of helpers on feeding of nestlings in the Grey-crowned Babbler, *Pomatostomus temporalis*. *Behav. Ecol. Sociobiol.* **4**: 43–59.

Burke, T., Davies, N. B., Bruford, M. W. and Hatchwell, B. J. (1989). Parental care and mating behaviour of polyandrous dunnocks *Prunella modularis* related to paternity by DNA fingerprinting. *Nature (Lond.)* **338**: 249–51.

Byle, P. A. (1987). Behavior and ecology of the dunnock, *Prunella modularis*. Ph.D. thesis, University of Cambridge.

Chase, I. D. (1980). Cooperative and non-cooperative behaviour in animals. *Amer. Nat.* **115**: 827–57.

Davies, N. B. (1983). Polyandry, cloaca-pecking and sperm competition in Dunnocks. *Nature (Lond.)* **302**: 334–6.

Davies, N. B. (1985). Cooperation and conflict among Dunnocks, *Prunella modularis*, in a variable mating system. *Anim. Behav.* **33**: 628–48.

Davies, N. B. (1986). Reproductive success of Dunnocks, *Prunella modularis*, in a variable mating system. I. Factors influencing provisioning rate, nestling weight and fledging success. *J. Anim. Ecol.* **55**: 123–38.

Davies, N. B. (1989). Sexual conflict and the polygamy threshold. *Anim. Behav.* **37**: (in press).

Davies, N. B. and Brooke, M. de L. (1989). An experimental study of co-evolution between the cuckoo *Cuculus canorus* and its hosts. II. Host egg markings, chick discrimination and general discussion. *J. Anim. Ecol.* **58**: 225–36.

Davies, N. B. and Houston, A. I. (1986). Reproductive success of Dunnocks, *Prunella modularis*, in a variable mating system. II. Conflicts of interest among breeding adults. *J. Anim. Ecol.* **55**: 139–54.

Davies, N. B. and Lundberg, A. (1984). Food distribution and a variable mating system in the Dunnock, *Prunella modularis*. *J. Anim. Ecol.* **53**: 895–912.

Davies, N. B. and Lundberg, A. (1985). The influence of food on time budgets and timing of breeding of the Dunnock, *Prunella modularis*. *Ibis* **127**: 100–10.

Gibbons, D. W. (1987). Juvenile helping in the Moorhen *Gallinula chloropus*. *Anim. Behav.* **35**: 170–81.

Houston, A. I. and Davies, N. B. (1985). Evolution of cooperation and life history in Dunnocks. In *Behavioural Ecology: The Ecological Consequences of Adaptive Behaviour*, ed. R. Sibly and R. H. Smith, pp. 471–87. British Ecological Society Symposium. Blackwell Scientific Publications: Oxford.

Morris, F. O. (1856). *A History of British Birds*. Groombridge: London.

Selous, E. (1933). *Evolution of Habit in Birds*. Constable: London.

Snow, B. and Snow, D. (1982). Territory and social organisation in a population of Dunnocks, *Prunella modularis*. *J. Yamashina Inst. Ornith.* **14**: 281–92.

Tomek, T. (1988). The breeding biology of the Dunnock, *Prunella modularis*, in the Ojcow National Park (South Poland). *Acta Zool. Cracov.* **31**: 115–66.

16. WHITE-FRONTED BEE-EATERS: HELPING IN A COLONIALLY NESTING SPECIES

16

White-fronted Bee-eaters: helping in a colonially nesting species

S. T. EMLEN

The evolution of cooperative breeding is a topic of considerable interest among evolutionary biologists. The reason is that helping behavior at first glance appears to be altruistic – that is, the behavior benefits other individuals (the recipient breeders and their offspring), while entailing a cost to the donor (the helper). As such, the forgoing of breeding and participation in helping poses an evolutionary paradox, since it appears to contradict the fundamental theorem of individual natural selection.

The production of this volume attests to the collective interest of ornithologists in this topic. Some contributors were drawn to their studies by an inherent interest in complex avian societies; others were drawn by the challenge of finding an answer to the puzzle of altruism. Whatever the reasons, the last decade has seen a surge of empirical studies on cooperatively breeding species. And, along with the increase in field data, have come the beginnings of a general unified theory for the evolution of avian helping.

The emerging solution to the paradox of helping behavior is best viewed as a two-step process (Brown, 1974, 1987; Emlen 1982*a*, 1984; Emlen and Vehrencamp 1983). We should ask two questions: 'why don't all individuals breed on their own?', and 'why do non-breeders often become helpers?'. The answer to the first in many cases seems to be that they cannot – that various constraints exist which either prevent them from attaining breeding status or which raise the costs of independent breeding to prohibitive levels, thereby favoring a postponement of reproduction. The exact form of the constraint varies from species to species, but the end result is the same – grown offspring delay dispersal and remain with their natal group into the succeeding breeding season.

If the option of independent breeding is closed, the second question 'why help?' becomes less formidable. Numerous gains could accrue to a non-breeder that provided help to a natal nest. The critical point is that, for helping to evolve, the inclusive fitness of a helper need not be greater than that of a breeder but only greater than that of an individual that remains with its natal group and does *not* help. Much current work is attempting to identify the various benefits of helping to helpers, as well as attempting to quantify the relative importance of direct as opposed to indirect benefits in different species. (For reviews of this topic, see Brown 1983, 1987; Emlen 1984).

The archetypical cooperative breeder lives in nuclear family groups, resides on all-purpose group territories, and has as its primary ecological constraint a severe shortage of breeding territories. The Florida Scrub Jay, *Aphelocoma c. coerulescens*, provides a classic example of such a species (Woolfenden and Fitzpatrick 1984, and see Chapter 8). But many variants occur, as the species accounts in this volume attest. Before the constraints model outlined above can be accepted as having general value, it must be tested in as diverse an array of types of cooperative breeders as possible. This chapter describes the biology of one such variant species, the White-fronted Bee-eater, *Merops bullockoides*.

The bee-eaters: family Meropidae

The bee-eaters, family Meropidae, comprise a group of 24 species in the order Coraciiformes. They are entirely Old World in distribution, being found throughout tropical Asia and Africa, with single species having reached Australia and southern Europe.

All species are insectivorous, flycatching for aerial insects from semi-exposed perches; most include stinging hymenopterans as a regular part of their diet (Fry 1984).

In Africa, some 18 species of bee-eater have radiated into a variety of habitats ranging from evergreen forest (*Merops mulleri*; *M. gularis*), through various types of savannah and grassland, to arid, semi-desert steppe (*M. revoilii*; *M. albicollis*). Paralleling their diversity in habitats, bee-eaters also exhibit a wide spectrum of social organizations. Some species are solitary and widely dispersed, while others are highly colonial. In some, the mated pair alone raises the young; in others non-breeding individuals routinely help at the nest. Furthermore, these two 'axes of sociality' (degree of coloniality and degree of cooperative breeding) are uncoupled in the Meropidae. Species can be found that are colonial but not cooperative (e.g. *Merops nubicus nubicoides* in Zimbabwe; S. T. Emlen and N. J. Demong, unpublished observations), while others are solitary yet have numerous helpers at the nest (e.g. *Merops albicollis*, Fry 1972; Dyer and Crick 1983). The family thus offers a rich potential for comparative studies of social organization.

Many years ago, when I first became interested in the topic of cooperative breeding, I decided to study a species that was far from the archetype mentioned in the introduction, one that exhibited simultaneously coloniality and helping behavior. Only a handful of avian species are currently known to satisfy both of these criteria. One such is the White-fronted Bee-eater.

History of the study

The range of the White-fronted Bee-eater extends across the entirety of central Africa, closely adhering to the belt of wooded savannah habitat to which it is largely restricted. It is found from the Democratic Republic of Congo and Angola in the west, to Kenya, Tanzania, Mozambique and Natal (South Africa) in the east. Within this vast region, it reaches its greatest densities in the wooded savannahs of the Great Rift Valley.

Together with Natalie Demong, I initiated pilot studies of bee-eaters in the Rift Valley of Kenya in 1973 and 1975. Then, beginning in 1977, I began a longitudinal study of a color-marked population of birds, located in Lake Nakuru National Park, which continued until 1987 (Fig. 16.1). Over the years I have been assisted by numerous field assistants and collaborators. Although these persons are thanked in the acknowledgements, I wish to single out four for special mention: my wife, Natalie J. Demong; my former graduate student, Robert E. Hegner; my first field supervisor, Carolyn E. Miller, and my current collaborator, Peter H. Wrege. Each has played a major role in the research described herein.

During the first three years of the longitudinal study, we closely monitored two populations of bee-eaters residing in the southern part of the park (the Makalia and the Badlands populations). However, during the autumn of 1979 an infectious disease struck the Makalia birds and over 90% of the population died within a period of six weeks. Shortly thereafter we shifted our attention to the Baharini population located in the northeast corner of the park and we monitored this population until 1987.

Our basic field methods have been described elsewhere (Emlen 1981; Hegner 1981; Emlen and Wrege 1986). They include capturing and marking birds for individual identification, making intensive behavioral observations on these individuals, and collecting demographic data on fecundity (reproductive success), dispersal and survival.

In this report I attempt to provide a broad outline of the basic features of the breeding biology and social organization of the White-fronted Bee-eater. The results are partitioned into sections dealing with the species' spacing system, social system, breeding system and reproductive success. I end with a brief speculative section relating helping behavior in these bee-eaters to the general two-step conceptual framework discussed in the introduction.

Spacing system

Population structure

White-fronted Bee-eaters are found throughout the grassland and open scrub savannah habitats of Lake Nakuru National Park. They are year-round residents whose basic habitat requirement appears to be access to a suitable foraging area that is located within easy travel distance of the vertical cliffs that serve as colony sites. During the day, bee-eaters travel routinely to foraging areas 4 km from the active colony (and some groups travel as far as 7 km; Hegner *et al.* 1982). In the late afternoon, all birds return to the colonies to roost. The colonies also serve as social foci during

Fig. 16.1. Natalie Demong and Carrie Miller overlooking a large White-fronted Bee-eater colony near Nakuru, Kenya. Photo taken in April, during the long rains.

the breeding season, when all birds nest gregariously and relatively synchronously.

On the basis of roosting and breeding colony locations, we have determined that six different populations of bee-eaters are present in the park, two each along the Nderit and Makalia river systems, one along the Naishi river, and one amongst a series of geological fault cuts and man-made muram pits located in the park's north-east corner. These populations, and the foraging areas they encompass, are shown in Fig. 16.2. The sizes of the populations ranged from 80 to 194 individuals at the beginning of the study (see Table 16.1).

Operationally, each bird can be assigned to membership in a specific population on the basis of the location and composition of its roosting and breeding colonies. For example, all birds with foraging territories located within the area marked MAK in Fig. 16.2 would converge each evening to roost together at a colony site along the Makalia river. Similarly, all those with foraging areas in the BL area would converge at some colony site further to the south. Individual bee-eaters occupy the same specific foraging territories (see p. 497) for several years and perhaps for their lifetimes, but they generally shift the location of their active colony in the months prior to each breeding. When the birds move to a new colony location, they *all* do so, keeping precisely the same foraging territories and population composition as before. The term colony therefore refers to the spatial location where members of a population are currently aggregating, either for the purpose of roosting or breeding.

Table 16.1. *Basic demographic and breeding information for the six populations of White fronted Bee-eaters residing within Lake Nakuru National Park, Kenya*

Population name	Initial population size	Initial sex ratio (M:F)	Breeding season
Makalia	194	1.03:1.00	Long rains
Badlands	95	1.16:1.00	Short rains
Enderit	178	1.14:1.00	Long rains
Naishi[a]	80[a]	1.00:1.00[b]	Long rains
Pitz	94	1.00:1.00	Short rains
Baharini	174	0.96:1.00	Short rains

M, male; F, female.
[a] Population size is estimate (from counts at roosting).
[b] Based on sample of only 38 laparotomized individuals.

White-fronted Bee-eaters are also highly sedentary. At the start of our study we marked over 400 individuals with leg bands and patagial wing tags. This represented virtually all members of the Makalia, Badlands and Enderit populations. During the succeeding $2\frac{1}{2}$ years, we recorded all cases of immigration into the Makalia population by either untagged birds or

Fig. 16.2. Map of Lake Nakuru National Park showing the locations of White-fronted Bee-eater populations. Shown on the map is the main road from Nairobi to Kampala (heavy dashed line), the boundary of the national park (dotted line), and the boundaries of the foraging areas used by members of each of the six populations of bee-eaters inhabiting the park. All colonies' sites used for breeding have been located within the respective population's foraging area. Lake Nakuru is a shallow alkaline lake bordered on the west by the Rift Valley escarpment and on the north and east by several volcanic hills. Population designations are: BAH, Baharini; PITZ, pitz; END, Enderit; MAK, Makalia; BL, Badlands; NSI, Naishi.

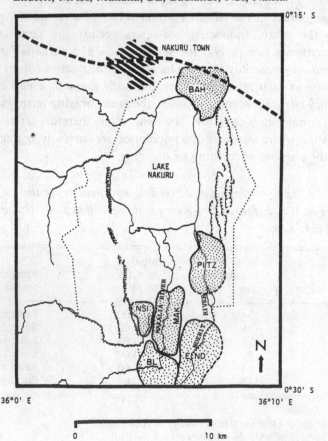

birds tagged in adjacent populations. On this basis we calculated that the annual immigration rate was 15%. More recent study of the Baharini population gives a similar value of 16%. If we can assume that immigration and emigration rates are equal, we must conclude that the majority of White-fronted Bee-eaters live their entire lifetimes within the spatial and social confines of the population into which they are born.

One additional aspect of population structure deserves special mention. Like many avian species, White-fronted Bee-eaters appear to time their breeding to periods when the probability of food (insect) abundance is greatest. In savannah habitats marked by alternating wet and dry seasons, this generally means breeding with the rains. The rains in Kenya, however, are under the meteorological influence of the inter-tropical frontal system which brings two, not one, rainy seasons per year (Sinclair 1978, p. 484). White-fronted Bee-eaters in Kenya may initiate breeding *either* with the long rains of March–May *or* with the short rains of October–November (Brown and Britton 1980). Any given individual, however, breeds in only one season each year. The result is that different populations can be characterized as either long-rains or short-rains breeders. We have discovered that both types co-exist in the Rift Valley of Kenya, and any spatial map would reveal a checkerboard patchwork of populations of each type. The breeding seasons for each of the six populations in Lake Nakuru National Park are given in Table 16.1.

Coloniality

White-fronted Bee-eaters are obligatorily colonial. The requirements for a colony site appear to be a nearly vertical, dirt cliff-face that is clear of vegetation. In our study area these have ranged from eroded banks of rivers, to geological fault-cuts in grasslands, to muram pits and earthen sections of gravel quarries created by human road crews. They have ranged in height from 2 to 12 m, and in horizontal width from 6 m upward. Into such areas the birds excavate horizontal tunnels averaging 1.2 m in length which end in an enlarged roosting or nesting chamber. Burrows are dug in close proximity, with neighboring entrances typically 15–20 cm apart. As a result of this high nest density, the cliff of an average sized bee-eater colony is riddled with tunnel entrances and looks not unlike a colony of bank swallows (sand martins, *Riparia riparia*) of more northern latitudes.

Suitable colony locations are abundant in the Nakuru area. The banks of the Makalia and Nderit rivers contained numerous cliff-faces that either were used during our study or showed signs of former activity (large numbers of abandoned, excavated chambers). Oxbow sand-cliffs further to the south and fault cuts and muram pits in the north-east provided similar evidence of multiple colony sites for the Badlands and Baharini populations. The availability of numerous colony locations may be a permissive factor in our finding that bee-eater populations generally change colony site in the few months prior to each breeding. At this time the population moves together to the new location and new chambers are excavated. Of 32 instances in which a population was monitored in successive breeding seasons, the colony location changed in 26 (81%). In those instances when a pre-existing colony cliff face was re-used, the birds typically dug new chambers or cleaned and lengthened existing ones prior to nesting.

Fig. 16.3. Top: frequency distribution of breeding colony sizes of White-fronted Bee-eaters from the Rift Valley of Kenya. Bottom: proportion of White-fronted Bee-eaters nesting in colonies of size n.

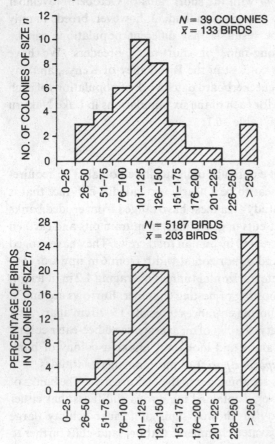

The adaptive explanation for these colony shifts remains to be studied, but one plausible hypothesis is that, by moving and excavating new colonies, the birds leave their fouled chambers behind and thus reduce the likelihood of ectoparasite infestation and disease transmission. During their many months of use, excrement and regurgitated fragments of non-digestible insect parts accumulate on the chamber floor and create an environment in which feather lice and bird ticks thrive. Although the evidence was inconclusive, a viral disease transmitted by *Ornithodoros erraticus* ticks was strongly implicated in the epidemic die off of the Makalia population in late 1979 (Emlen *et al.*, unpublished data; H. Hoogstraal, personal communication).

All members of a population generally roost and breed together at the same colony; consequently colony size typically is the same as population size. However, on four occasions, a population has temporarily divided and nested in two colonies. Once the Makalia birds split into two breeding colonies located 2.2 km apart along the river. Similarly, during three successive years, the Baharini population split in a similar fashion and nested simultaneously at two locations 0.4 to 1.1 km apart. In each case, the full population roosted together prior to breeding, and reassembled to continue roosting in aggregate after nesting was completed.

Breeding colony sizes for the six Nakuru populations have ranged from a low of 31 birds to a high of 169. A frequency histogram of the distribution of sizes of 39 colonies censused in the larger area of Kenya's Rift valley is shown in Fig. 16.3 (top). Average colony size was 133 birds. Viewed somewhat differently, the average bee-eater bred in a colony of 203 individuals (Fig. 16.3 (bottom)). If the three largest colonies are excluded, the adjusted average colony size becomes 107 birds and the modal bee-eater resides in a colony of 127 individuals.

Clan foraging territories

During the day, members of a bee-eater colony disperse to forage over an area of many square kilometers. However, the birds do not wander nomadically over this large area. Instead, each individual or pair of bee-eaters consistently returns to the same geographic area of 1.5 to 2 ha day after day, throughout the year. Furthermore, members of the same social unit (or clan, see p. 499) are foraging neighbors that occupy adjacent and partially overlapping home ranges. The resident forager will tolerate most intrusions by other members of *its* clan, but will aggressively chase off birds belonging to *other* clans. Thus the clan has a spatial representation in the form of a clustered neighborhood of foraging home ranges (FHRs) located some distance from the colony. We call these areas clan foraging territories (Hegner *et al.* 1982; Hegner and Emlen 1987).

Every young bee-eater begins occupying a FHR of its own when roughly five months old. This initial FHR overlaps heavily with areas used by other clan members but includes previously unoccupied areas as well. When a bird takes a mate, the new pair settle on a FHR of their own (see p. 500). This new FHR generally represents either an enlargement of one of the mate's former feeding areas (44% of cases), or a shift to an area near the periphery of the foraging territory of the clan of one of the mates (56% of cases).

The clan territories of White-fronted Bee-eaters are feeding areas only and do not contain other resources critical for reproduction (such as nesting sites). We have speculated that the adaptive value of clan territoriality lies in the enhanced foraging success and the decreased risk of predation that accompany long-term familiarity with a specific area (Hegner and Emlen 1987).

Birds occupying clan foraging territories located more than 1.5–2 km from a breeding colony temporarily cease using them when provisioning young in the nest. The time and energy required to make the numerous commuting flights to and from distant territories reaches the point where continued use of such areas is no longer economically profitable. Birds that abandon their foraging territories provision nestlings by feeding nomadically in the vicinity of the colony itself. Such birds have a lower foraging success, and suffer a lower reproductive success, than do birds that occupy (and continue using) their foraging territories near the colony (see Tables 8 and 9 of Hegner and Emlen 1987).

One might conclude from this that clan territories located near an active colony site would be of higher quality than territories located more distantly. Recall, however, that colony locations change between successive breeding seasons, while territory locations do not. The result is that a clan foraging territory located close to the active breeding colony in one year may be far from it the next, and vice versa. This unpredictability of future colony locations places bee-eaters in a difficult situation. Since foraging efficiency increases with long-term use of a familiar area (Hegner and Emlen 1987), short-term abandonment of a foraging territory is perhaps best viewed as an attempt by the occupants to make the best of a temporarily bad situation.

There are two other important differences between bee-eater foraging territories and the group territories of many other cooperatively breeding species. First, foraging areas are not limiting in the sense of restricting dispersal and 'forcing' offspring to remain with their natal clans. Unoccupied areas of seemingly suitable habitat (judged by occupancy at some time in the study) existed throughout the Makalia and Badlands areas in all years of the study. In the Baharini area, fully 40% of the foraging areas were unoccupied. Perhaps for this reason, we have never had any bee-eater, single or paired, that did not obtain its own FHR at the time of independence, and again later at pairing. If foraging space was limiting, one would expect birds to expand their feeding areas when neighboring clans decreased in size or died out. Yet this has not occurred; instead, the vacated foraging areas of such clans remain largely unused.

Second, breeding status and territory ownership are not linked in White-fronted Bee-eaters. Most individuals are paired long before the start of a breeding season and every pair occupies and defends its own FHR. Yet not all pairs breed in any given season (see p. 506). Similarly, many pairs that do breed temporarily abandon their FHRs. An analysis of the probability of breeding showed no significant difference between pairs occupying 'high-quality' clan foraging territories (defined by proximity to the current breeding colony) and those occupying 'low quality' (distant) areas (see Table 10 of Hegner and Emlen 1987). Territory ownership thus is neither a necessary nor a sufficient condition for becoming a breeder. Robert Hegner and I concluded that foraging territories are not a critical ecological constraining factor for White-fronted Bee-eaters in Kenya. (For a more detailed discussion of the habitat saturation hypothesis as it applies to bee-eaters, see Hegner and Emlen 1987).

Social system
The basic social unit

During our initial field seasons, Natalie Demong and I were struck by what we interpreted to be the openness and fluidity of group memberships among White-fronted Bee-eaters. We erroneously assumed that the primary social unit of these birds was the cooperative breeding group tending a given nest. In those seasons, 'group' sizes ranged from two to seven individuals, with a mean of 3.0 (Emlen 1981). But we also noticed that if a nesting attempt failed, the 'group' often disbanded, with some individuals shifting to roost communally with other birds in the colony, and other individuals joining foreign 'groups' at different nests at which they now became helpers! In 1981, I wrote: 'Bee-eater social organization combines a large measure of openness and fluidity of group membership and frequent changes between breeding and helping status with a stability imposed by the reestablishment and fidelity of certain social bonds.

'...whatever the nature of the formation of such bonds, it is the faithfulness of their reestablishment that gives the stability and the pattern to the fabric of bee-eater society' (Emlen 1981, p. 229).

We now know that the basic social unit in White-fronted Bee-eater society is really a multi-generation, extended family containing anywhere from three to 17 members. We refer to these family units as clans, and a typical clan will include from one to five mated pairs plus a small assortment of single birds that either have not yet paired or that have been widowed and are not yet re-paired. The distribution of clan sizes from five well-studied populations in the Nakuru area is presented in Fig. 16.4.

During a breeding season a variable number of these pairs will breed, and the clan becomes temporarily subdivided into a number of smaller nesting associations (the 'groups' described above). Many of the non-breeding individuals act as helpers at one of the active nests of the clan. Each such non-breeding helper associates with only one nest for the duration of the breeding attempt. If the nest is successful, the helper continues to aid in feeding the fledglings until the latter become independent, roughly six weeks later. If the nest fails, the helper *as well as the failed breeders* may join at another active nest of the clan, where they become full contributing helpers. We refer to these birds as *redirected helpers*. It was these exchanges which led us initially to interpret membership in bee-eater groups as open and fluid.

By knowing the identity of each breeding pair and banding the young in each nest, we can assign (behavioral) parentage to all young produced in the population. Slowly, across the years, we have been able to construct genealogies for most members of the study populations. These behavioral genealogies show clearly that clans are actually extended family units comprising three, and sometimes four, overlapping generations of kin. Individual bee-eaters regularly interact not only with their parents and siblings but also with grandparents, aunts and uncles, nieces and nephews, and cousins. When birds reproduce after re-pairing (after the death or divorce of a mate), clans will include step-parents, half-siblings, and half-cousins as well. The complexity of the genealogical structure of two large clans is shown in Fig. 16.5.

Pair formation in bee-eaters

When pair formation occurs in *Merops bullockoides*, one member of each newly forming pair leaves its natal clan and joins the clan of the other. This dispersing individual greatly diminishes (but does not totally end) its interactions with members of its own natal group and becomes integrated into the clan of its partner. One result of this pairing system is that one (the resident) member of each pair continues living in a network of close genetic kin, while the other (dispersing) member of the pair forms social bonds with members of its new clan, but kinship links are lacking. As will be discussed below, this distinction is important, since natal birds and their unrelated mates ('in-laws') show very different tendencies to become helpers.

Greenwood (1980), reviewing sexual asymmetries in dispersal patterns,

Fig. 16.4. Top: frequency distribution of clan sizes from five populations of White-fronted Bee-eater. Clan sizes were censused just prior to the second breeding season that each population was under study; clan sizes are at their lowest annual level at this time, since young of the year are absent. Data include all clans from the Baharini, Pitz, Makalia and Glanjoro populations (the latter being located outside Lake Nakuru National Park to the northwest) and a partial (unbiased) sample from the Badlands population. Bottom: proportion of White-fronted Bee-eaters living in clans of size *n*.

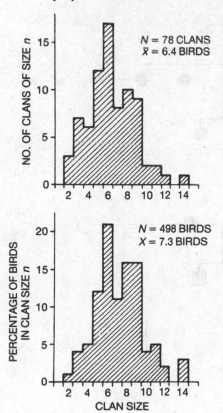

Fig. 16.5. Genealogies of two clans from the Baharini population of White-fronted Bee-eaters. Genealogies include all individuals present in early 1983 (shortly after the autumn 1982 breeding). Selected ancestors no longer alive are indicated by symbols with an X and the word DEATH written below. Square symbols indicate males; circles indicate females; and symbols labeled JUV indicate juveniles of unknown sex. Shading identifies natal clan members, hatching identifies natal members that are 'step-kin' to the shaded members, and clear symbols identify unrelated mates ('in-laws'). The direction of the arrow connecting a mated pair indicates which partner immigrated into the clan of the other at the time of pairing.

(a) Clan IXB. An example of a relatively simple clan structure consisting of the nuclear families of two, different-aged, brothers together with their surviving mother.

(b) Clan I. An example of a more complex genealogy formed when breeders remate following the deaths of their original mates. The result is a clan consisting of different sublineages involving step as well as full relatives.

(a) BAHARINI CLAN IXB (1983)

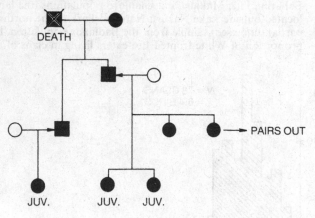

(b) BAHARINI CLAN I (1983)

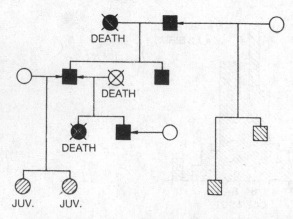

502

concluded that females generally were the dispersing sex in birds, while males were more philopatric. Our data on the sexes of bee-eaters that left their natal clans support this trend. Of 24 pairings in the Makalia population, 21 involved females moving to join the clans of their male mates while in three the reverse was true. In the Baharini population, 95 pairings have been monitored closely. In 84 of these it was the female that changed clans. When pooled, these data show a strong female-biased pattern of pairing dispersal. Both sexes do disperse, however, with the overall dispersal ratio being 88% females and 12% males.

When choosing mates, White-fronted Bee-eaters invariably avoid selecting close genetic relatives as partners. In 81 recorded pairings, we have never witnessed an incestuous bond (defined when relatedness to the mate is more than 0.06; Emlen and Demong 1984). Adopting the rule of thumb 'pair with a partner from outside your natal clan' may be one mechanism by which such inbreeding is avoided.

Once a pair breeds together, they generally remain monogamously mated for much of the remainder of their lifetimes. Data from 62 'pair-years' in which both members of a pair survived until the following breeding season show that in fully 87% of the cases the pair bond remained intact. Given that the expected lifespan of a young bee-eater reaching independence is 2.5–3 years (Emlen, unpublished data), and barring cases involving the death of a mate, we can calculate that two-thirds of bee-eaters will remain mated to the same partner for their entire lives.

The importance of the colony

As temperate-zone ornithologists, most of our experiences with colonially breeding birds come from marine seabirds, gulls and terns, herons and egrets, and perhaps from various gregarious swallows, swifts and blackbirds. Among these more 'typical' colonists, the only permanent social bonds are between members of each mated pair. Temporary bonds form between parents and their young, but terminate when the young become independent. Parents apparently do not recognize their grown offspring in later years, nor do adult siblings seem to remember one another. As far as is known, these birds show no differential treatment to specific individuals apart from mates and current offspring. Indeed, most breeding bird colonies seem to be no more than anonymous assemblages of nesting pairs.

The social networks of White-fronted Bee-eater colonies are very different from those of the taxa mentioned above. Each bee-eater 'knows' and behaves differentially toward a large number of specific individuals far beyond its mate and nestlings. 'Known' individuals include, at a minimum, all members of the current clan and, in the case of unrelated mates ('in-laws'), many members of the former natal clan as well.

At non-breeding times of the year, each clan occupies and defends a set of burrows at the colony in which they communally roost. All members of the clan generally have access to *any* of the clan's chambers and much greeting and visiting between the burrows occurs each evening. The actual composition of the groups roosting in any given chamber may vary considerably from night to night or week to week, but access is almost always restricted to clan members only.

During a breeding season, a variable number of pairs in each clan initiate nesting attempts. Each such pair, together with its helpers, excavates a new chamber and begins defending it, often against other clan members as well as members of the colony at large. As a result, clans temporarily become subdivided into several smaller groups, each tending a specific nesting burrow. However, these subdivisions are far from absolute. Current clan members continue to interact regularly on the foraging territory, and to visit one another's nesting burrows at the colony, on an almost daily basis. Furthermore, any change of activity at one particular burrow can lead to a coalescence of clan members at another. For example, when a bird loses its mate, it generally moves into the chamber of another subunit of the clan. If a breeding attempt fails, the individuals shift over to roost with, and often to help at, another ongoing nest of the clan (see p. 507). When a nest successfully fledges young, all members of the successful subunit increase their frequency of visiting and roosting in the other active chambers of the clan. The result is a clan structure that combines tremendous fluidity of subgroup associations with a stability imposed by the membership limitations of the clan itself.

In the case of unrelated mates that have paired into new clans ('in-laws'), the social bonds can encompass members of their former (natal), as well as their current, clan. Because White-fronted Bee-eaters nest in dense colonies, an 'in-law's' own natal family is typically nesting and roosting nearby. Through visits and greetings at pre-roost time, unrelated mates maintain loose links with their former clan. Such birds (if they are not nesting during a given breeding season) may even temporarily abandon their mates in order to return 'home' and serve as helpers at the nest of one or both of their genetic parents (Emlen and Wrege 1988). Upon the completion of that nesting attempt, the helping 'in-law' returns to its mate.

Maintenance of social links with members of one's natal family after dispersing from home and pairing into a new social group is a rare phenomenon in animal societies. In White-fronted Bee-eaters, it is made possible by the combination of sedentary, year-round residency and obligate coloniality.

Breeding system

Frequency of helping and activities of helpers

Helping-at-the-nest is a common feature of the breeding system of White-fronted Bee-eaters. Helpers are typically adults that are not currently breeding themselves. Juvenile helpers occasionally aid sporadically at late nests in the season of their hatching. But the consistency and the magnitude of their contributions are much lower than those of adults; consequently juvenile helpers have been omitted from analyses in this chapter.

During the eight years of this study, 50% of all nesting attempts have included helpers. A composite histogram of the frequency of occurrence of nests tended by breeding pairs alone and by pairs aided by various numbers of helpers is shown in Fig. 16.6. This figure includes data from four populations: Makalia, in which 50% of 159 nests had helpers; Badlands,

Fig. 16.6. Frequency distribution of group sizes tending 611 nests of White-fronted Bee-eaters. Data include the 1977-9 nestings of the Makalia and Badlands populations and the 1980-4 nestings (both primary nestings and re-nestings) of the Baharini population. Numbers above histograms give sample sizes of nests in each category.

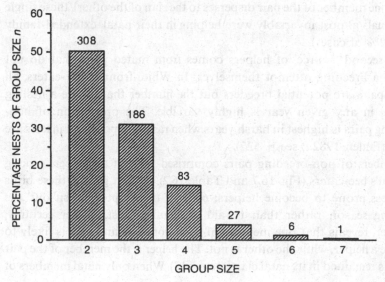

where the figures were 46% of 54 nests; Enderit, with 54% of 56 nests, and Baharini with 49% of 342 nests.

While being a helper, a bee-eater affiliates with one particular nest where it remains for the duration of that breeding attempt. It assists in virtually all aspects of nesting, including digging and defending the nest chamber, feeding the breeding female prior to her egg-laying, incubating the eggs, feeding the nestlings, and continuing to care for and feed the young until they become independent some six weeks after fledging.

Origins of helpers

Bee-eater helpers come from several 'sources', from paired as well as unpaired individuals, and from breeders as well as non-breeders. The data presented below were compiled from the 1980-4 breedings of the Baharini study population.

Most pairing in White-fronted Bee-eaters occurs before the breeding season, so as nesting activity approaches, most birds have already established mating partners. A typical clan at this time consists of several mated pairs plus any unpaired yearlings (many birds do not pair until their second year) and the occasional recently widowed individual.

These unmated individuals form the first and largest source of bee-eater helpers. Of all helpers, 59% came from the ranks of unpaired birds (Fig. 16.7 and Table 16.2). Such individuals also showed the greatest likelihood of becoming helpers. Fully 78% of unpaired birds became helpers (102 of 130). Since bee-eaters remain in their natal clans until such time as they pair (when one member of the pair disperses to the clan of the other), these single individuals almost invariably were helping in their natal, extended-family clan (92% of cases).

The second 'source' of helpers comes from mated pairs that do not initiate a breeding attempt themselves. In White-fronted Bee-eaters, all mated pairs are potential breeders but the number that initate a nesting attempt in any given year is highly variable. The proportion of non-breeding pairs is highest in harsh years when rains are poor and insects are scarce (Emlen 1982a; see p. 522).

Members of non-breeding pairs comprised 14% of all helpers among Nakuru's bee-eaters (Fig. 16.7 and Table 16.2). At first glance, these birds seem less prone to become helpers; 66% of them opted to 'sit out' the breeding season rather than to aid at another nest. Closer scrutiny, however, reveals that one member of each non-breeding pair is likely to become a helper, while the other is not. The helper is the member of the pair that has remained in its natal clan (see p. 513). When only natal members of non-breeding pairs are examined, 51% became helpers (compared to 12% of the unrelated mates).

The final source of helpers comes from the ranks of breeders whose own nesting attempts have failed. Failure rates were high at Nakuru, with 57% of all nests started during the eight years of our study being lost prior to fledging (see Table 16.5). Predators such as spitting cobras and mongooses may enter a nest and eat the contents, or a portion of the cliff face may collapse in heavy rains, carrying the nest contents with it. More frequently, nestlings starve, one by one (see p. 516). When a nesting attempt fails, one or more breeders may shift allegiances and join up as a helper at another active nest of their clan. This *redirected helping* (Emlen 1981) is a common feature of the breeding system of White-fronted Bee-eaters. Fully 27% of all helpers came from this source (Fig. 16.7 and Table 16.2). Failed breeders were intermediate between unpaired birds and non-breeding pairs in their likelihood of becoming helpers. Natal members of failed breeding pairs helped in 69% of all possible opportunities, while their unrelated mates helped in 20%.

Fig. 16.7. Origins of White-fronted Bee-eater helpers. The histogram shows the percentage of helpers originating as (1) unpaired individuals, (2) paired birds that do not initiate a breeding of their own, and (3) paired birds that do breed on their own but become redirected helpers at another nest after their own breeding attempt fails ('current season former breeders'). These three origins total 100%. The fourth bar gives the frequency of helpers that have *ever* been breeders prior to helping ('all seasons former breeders'). Numbers above histograms refer to sample sizes of helpers in each category. Data include all helpers from the 1980–4 breedings of the Baharini population.

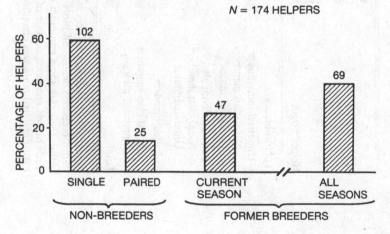

Table 16.2. *The origins of helpers among White-fronted Bee-eaters*

Helper origin	First breeding (seasons C1, D1, E1, F1 and G1)		Re-nesting (seasons D2, E2 and F2)		Total	
	Potential helpers	Actual helpers	Potential helpers	Actual helpers	Potential helpers	Actual helpers
(1) Non-breeders						
Unpaired individuals						
Natal clan members	88	80	27	14	115	94
Unrelated mates[a]	10	7	5	1	15	8
Paired individuals						
Natal clan members	21	16	20	5	41	21
Unrelated mates	13	3	19	1	32	4
(2) Failed breeders						
Natal clan members	39	29	15	8	54	37
Unrelated mates	37	10	14	0	51	10
Total	208	145	100	29	308	174

Potential non-breeding helpers include all birds that did not initiate breeding themselves; potential redirected helpers (failed breeders) include all birds that bred and failed in their own attempt and which had another active nest available within their clan for a minimum of one week following the date of their own nest failure. Data include 10 cases where birds helped recipients outside of their current clan. Data are from the Baharini population, 1980–4.
[a] An unrelated mate was classified as unpaired if it was recently widowed and still living in the clan of its former mate.

These shifts, where a breeder becomes a helper at another nest, demonstrate a fundamental point concerning White-fronted Bee-eater social organization. In most species of avian cooperative breeders, helping is a role played by young individuals that have not yet attained the social status or dominance necessary to attract a mate or to take over a breeding territory of their own. In *Merops bullockoides*, by contrast, close to half of the helpers are already paired (41%), and all (100%) occupy their own foraging home ranges and can excavate their own roosting (potential nesting) chambers. In other species of cooperative breeders, helping typically is a transitory phase in the life cycle. Once an individual achieves breeding status, it remains a breeder for the remainder of its lifetime, rarely if ever reverting back to helping status again. But in White-fronted Bee-eaters, 27% of all helpers are recently failed breeders. If we count all helpers that formerly have been breeders (meaning they bred in *any* former season, not just the current one), this value increases to 40% (69 of 174 cases). Role reversals from breeder to helper and vice versa are commonplace in White-fronted Bee-eaters and our data indicate that most individuals act repeatedly in both breeder and helper capacities during their lifetimes.

Sex and age of helpers

Both sexes act as helpers among White-fronted Bee-eaters. Data from five years show that in some years female helpers predominated; in others males did (Table 16.3). The stronger female bias in 1983 and 1984 was due to a male-biased mortality in the Baharini population which accompanied the Sahel drought of 1983. When all years are pooled, the frequency of male and female helpers does not differ from equity (53% females; $\chi^2 = 0.76$; $P > 0.25$).

Table 16.3. *Sex of helpers among White-fronted Bee-eaters. Data are from the Baharini population (1980–4)*

Breeding season	No. of males	No. of females	Total
1980 (C)	21	16	37
1981 (D)	19	17	36
1982 (E)	26	21	47
1983 (F)	14	27	41
1984 (G)	1	9	10
Five-year total	81	90	171[a]

[a] An additional three helpers were of unknown sex.

Further, the likelihood that any given bird would become a helper was independent of its gender. There was no sexual bias in the probability of helping for any category of natal potential helper ($G = 0.64$, not significant (n.s.)). Analyses could not be performed on unrelated mates because of the small sample size of males).

The age distribution of bee-eater helpers is more difficult to determine. Exact ages are known only for individuals hatched after our studies began. However, if we make the oversimplifying assumption that all helpers known to be at least n years of age are n years old and no older, we can obtain a first approximation of the age distribution of helpers. Such a distribution for the Baharini population is shown in Fig. 16.8. Because of the above assumption, Fig. 16.8 greatly underestimates the true ages of many helpers. Nevertheless, it is apparent that while yearlings constitute the largest group of helpers, older birds continue to help throughout their lifetimes. This is due primarily to the phenomenon of redirected helping – the tendency for older breeding individuals to become helpers at another nest if their own reproductive attempt fails.

Kinship of helpers

White-fronted Bee-eaters are an excellent species in which to study the importance of kinship in helping behavior. They live in extended family

Fig. 16.8. Minimum age distribution of White-fronted Bee-eater helpers. Data are from the 1980–4 breedings of the Baharini population. Sample sizes are given above each histogram. Note that figure underestimates the true ages of helpers (see the text).

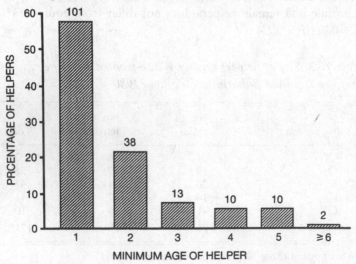

clans within which several pairs may breed in any given season. Further, many clans (15 to 25) aggregate to breed in dense colonies. From the prospective of a would-be helper, this means that large numbers of non-kin, as well as an assortment of kin of varying degrees of relatedness, are available as potential recipients of its aid. Given this range of choices, whom do bee-eater helpers help?

Peter Wrege and I recently completed an analysis of this question and much of this section is drawn from that study (Emlen and Wrege 1988). We based our kinship calculations upon genealogical data, using the assumption that behavioral parents (breeders) were the genetic parents of the young they reared. In actuality, we know that both extra pair copulations and intraspecific parasitism do occur in our study population, with a combined frequency of between 9% and 12% (Emlen and Wrege 1986; Wrege and Emlen 1987). Such behaviors will cause some kin assignments to be overestimates (where a helper aids its parents, but the parents have been cuckolded/parasitized); others will be underestimates (where the helper is not the genetic offspring of its putative parents, or where the helper aids a nest at which *it* was the cuckolder/parasite). The first two biases partially cancel out and the third is uncommon (Emlen and Wrege 1986); consequently our calculations should closely approximate the true average relatedness between helpers and recipient nestlings.

Table 16.4 summarizes our data from the Baharini population on the putative family relationships between 174 helpers and the breeding pairs that they helped. The average relatedness between helper and nestlings being helped was 0.33. Some 88% of the cases involved kin recipients and the most frequent associations were among the closest kin. Grown offspring helping their parents to rear full siblings accounted for 44.8% of all cases. This was followed by grown offspring helping a parent and step-parent to rear half-siblings (19.0%). The next most common association involved an age reversal, with a parent helping its son and daughter-in-law in their breeding effort (10.3%). Although a wide spectrum of relatives received aid, 11.5% of cases involved helpers aiding breeders for whom no genealogical link was known.

To test whether bee-eaters exhibited kin discrimination, that is whether they *preferentially chose* to aid close kin, we contrasted the observed distribution of helping associations with that expected if recipient nests had been selected at random (Fig. 16.9). Only nests potentially available *within the helper's current clan* were included in this analysis; thus the expected values represent the proportion of available potential recipient nests belonging to the different kin-classes. The null hypothesis of randomly

Table 16.4. *Whom do bee-eaters help? Family relationships between helpers and the breeders that they help*

Breeders		No. of cases	% of cases
Father × Mother	0.50	78	44.8
Father × step-mother	0.25	17	9.8
Mother × step-father	0.25	16	9.2
Son × non-relative	0.25	18	10.3
Brother × non-relative	0.25	12	6.9
Grandfather × grandmother	0.25	5	2.9
Half-brother × non-relative	0.13	3	1.7
Uncle × non-relative	0.13	2	1.1
Grandmother × non-relative	0.13	1	0.6
Grandson × non-relative	0.13	1	0.6
Great-grandfather × non-relative	0.06	1	0.6
Non-relative × non-relative	0.00	20	11.5
Total		174	100

r values refer to coefficients of relatedness between the helper and the nestlings receiving help. Data include 10 cases where birds helped recipients outside of their current clan. Modified from Emlen and Wrege (1988).

Fig. 16.9. Expected and observed frequencies of helping to recipients of various kin-classes in White-fronted Bee-eaters. Expected values show the proportion of recipient nests available belonging to the different kin-classes (r). Data include 174 cases of helping and 535 potential recipient nests, all from the 1980–4 breedings of the Baharini population. The null hypothesis of randomly directed helping is rejected, providing strong support for the alternative hypothesis that bee-eaters show kin discrimination in choosing the recipients for their aid (for statistics see the text) (from Emlen and Wrege 1988).

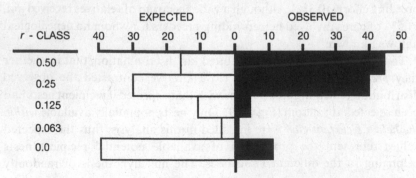

directed helping is rejected ($G = 126$; $P \ll 0.001$). White-fronted Bee-eaters are far more likely to aid nestlings to whom they are closely related and far less likely to help to rear non-kin than is predicted by chance ($P < 0.01$ for r-classes of 0.5, 0.25 and 0.00; $P < 0.05$ for r-class 0.125; binomial test).

Recall that only 56% of potential helpers became actual helpers in this population of birds. To examine more closely the importance of kinship in the decision of whether or not to help, we partitioned all potential helpers into categories based upon (1) their relatedness to the most closely related recipients available within their clan, and (2) the presumed cost of their providing aid. This latter division was based upon each bird's activities immediately prior to the helping opportunity. We assumed that high-effort activities (feeding nestlings at a prior nesting attempt) would cause a bird to be more energetically stressed and thus raise the cost of its becoming a helper again.

We then calculated the conditional probability that a potential helper would aid a recipient of relatedness r, given that no more closely related recipient was available. These conditional probabilities of helping are plotted as a function of kinship and prior effort in Fig. 16.10. Both factors influenced the likelihood of helping. The presumed 'high-cost' birds were less likely to become helpers than 'low-cost' birds. This effect was significant for kin classes of $r = 0.5$, 0.25 and marginally for $r = 0.00$ ($P < 0.05$ and $P < 0.06$, respectively, Fisher Exact test). Within each presumed cost category, the likelihood of helping decreased with decreasing r values ($G = 70$ and 46 for low- and high-cost birds, respectively; both $P \ll 0.01$). Among low-cost birds, this was true across all four relatedness values. For high-cost birds, the probability of helping seemed to reach a lower threshold at $r = 0.125$.

How are helping choices made? Three simple rules of thumb go a long way in explaining both *whether* a given bird will become a helper and *whom* it will help.

First, only help potential recipients that are members of your current clan. Helping is not randomly distributed throughout the colony. Fully 164 of the 174 cases of helping (94%) were directed to current clan members.

Second, if you are an unrelated mate ('in-law'), do *not* help potential recipients in your current clan until you have grown offspring of your own that have become breeders within the clan (by our operational definition, a bird ceased being called an 'in-law' at this time). When potential helpers are partitioned according to natal or 'in-law' status within their clan, a major dichotomy in the likelihood of helping emerges. Natal members show a *much* greater tendency to help when an opportunity is available than do unrelated mates. For non-breeding potential helpers the probabilities of helping were 72% versus 19% ($G = 55.1$, $P \ll 0.01$); for redirected potential helpers (failed breeders), the values were 69% versus 10% ($G = 41.3$; $P \ll 0.01$; Emlen and Wrege 1988).

The third rule of thumb is to select among the actively breeding pairs in your clan and help the most closely related one. Emlen and Wrege (1988) analyzed 115 instances in which a natal clan member had a *choice* between two or more nests with young of differing degrees of relatedness. In 108 of the 115 cases (94%), the helper chose to aid the nest with the most closely related young.

This ability to discriminate among different classes of potential recipients suggests that bee-eaters possess a fairly sophisticated system of kin

Fig. 16.10. The conditional probability that a potential helper becomes an actual helper, plotted as a function of its coefficient of relatedness to the recipient nestlings. Data are partitioned according to the activities of the potential helper immediately prior to the helping opportunity (presumed 'high cost' versus 'low cost' of providing aid; for definitions see the text). This analysis is restricted to within-clan potential recipients; consequently the 10 cases of helpers aiding recipients outside their clans are classified here as not helping. Numbers above histograms refer to sample sizes of potential helpers in each category. Both the presumed cost of helping and the degree of relatedness to the recipient strongly influence the likelihood that a bee-eater will act as a helper (modified from Emlen and Wrege 1988).

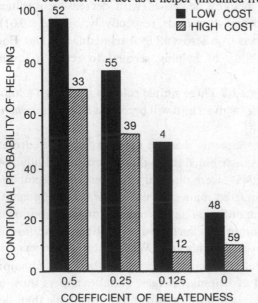

recognition. Together with Peter Wrege, I am currently investigating possible mechanisms that might underlie such kin recognition through a series of cross-fostering experiments in which recently hatched chicks are transferred to 'foster' nests and reared by breeders that are not their true parents. The later behavior of these individuals toward their foster clan as well as toward their true natal clan (both of which will be present at the colony) should help us to understand the proximate mechanisms that bee-eaters use to make their choices.

Reproductive success and the role of helpers

White-fronted Bee-eaters typically breed once each year, timed either to the long rains of March–May or to the short rains of October–November (see p. 495). On three occasions, however, numerous members of the Baharini population have re-nested during the same season. Data from both primary nestings and re-nestings are pooled in the analyses that follow.

Patterns of egg and nestling mortality

Throughout the eight years of our study, bee-eater breeding has been characterized by a highly variable, but generally low, level of nesting success. Many nests are abandoned early; and many of those not abandoned suffer high rates of nestling starvation.

Table 16.5. *Fledging success of White-fronted Bee-eaters in the Nakuru area*

Population	Number of nests	Number of nests producing n fledglings					% nests successful
		$N=0$	$N=1$	$N=2$	$N=3$	$N=4$	
Baharini	295	146	91	48	10	0	54
Makalia	173	143	20	10	0	0	17
Badlands	52	17	16	14	3	2	63
Grand total	520	306	127	72	13	2	43

Both the percentage of successful nests (defined as producing at least one fledgling) and the frequency distribution of nests producing different numbers of fledglings are presented. Five years' data from the Baharini population (1980–4; both primary nestings and re-nestings) and three years' data each from the Makalia and Badlands populations (1977–9) are included. Sample sizes for successful nestings are slightly larger than for number of fledglings because some nests were known to fledge young successfully but the number of fledglings was not known with certainty.

Table 16.5 summarizes the success rate and average number of fledglings produced in each of our three study populations. Summing the data from eight breeding seasons (totalling 520 nests), White-fronted Bee-eaters in Nakuru successfully fledged young from only 43% of their nesting attempts. Of the young that successfully fledge, fully one-half will die before reaching independence five months later. Assuming that the probability of death is independent of the number of young fledging from a given nest, this means that only one nest in four produces independent juveniles. The importance of this high failure rate for understanding the evolution of redirected helping and the fluid nature of bee-eater helping associations has already been mentioned.

A detailed breakdown of the sources of pre-fledging mortality in the Baharini population is shown in Fig. 16.11. Approximately one-quarter of all eggs laid are lost prior to hatching. Another third are lost after hatching, as the nestlings die of starvation. After additional losses to predation, catastrophies, and unknown causes, only 26% survive actually to fledge from the nest.

Fig. 16.11. Sources of egg and nestling mortality in the Baharini population of White-fronted Bee-eaters (1980–4 including re-nesting seasons). Catastrophic losses include flooding, cave-ins, and mortality due to safari ants (sources where group size could play no role in influencing success). Starvation refers to nestlings that showed signs of morphological retardation and disappeared from nests singly. Swifts refers to cases where horus swifts (*Apus horus*) physically removed nest contents and took over use of the nest themselves; predation refers to all non-swift cases where the entire nest contents were lost between successive nest checks.

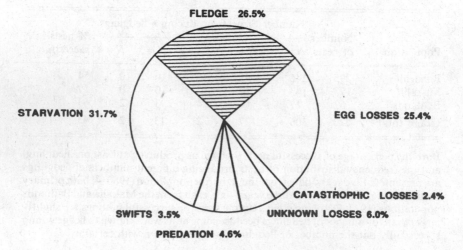

Effects of helpers on fledging success

I mentioned previously that bee-eater helpers assist in most aspects of nesting, from excavating and defending the nesting chamber, to incubating the eggs, to feeding and tending the young both as nestlings and as fledglings. What is the effect of this helping upon the survival of the young?

The standard way to test the hypothesis that helpers really 'help' is to contrast the production of young from nests tended by breeding parents alone with that from nests tended by breeders plus various numbers of helpers. The average production of fledglings from nests tended by differing numbers of adults is shown in Fig. 16.12. Included are data from five breeding seasons (1980–4) of the Baharini population. (Similar results from the Makalia population have been published by Emlen (1981).) Fledgling production increases dramatically with the presence of helpers. In a multiple regression analysis controlling for the effects of seasonality and clutch size, helper number was a highly significant predictor of number of fledged young ($F_s = 65.8$; $P \ll 0.01$). The magnitude of the helper effect is

Fig. 16.12. The effect of group size of fledging success in White-fronted Bee-eaters. Values presented are least-square means ± 1 S.E.M. from an analysis of covariance in which the effects of season and clutch size were controlled ($F_s = 65.8$, $P \ll 0.01$). Data are from the 1980–4 nestings (including re-nestings) of the Baharini population.

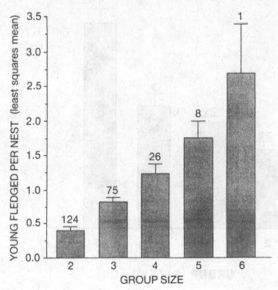

large, with each helper accounting for an average of 0.44 additional offspring. Furthermore, the trend shows no decrease as helper number increases. In other words, there is no diminishing return on helper investment across the range of group sizes recorded in nature (Emlen and Wrege 1988).

How do helpers help?

The major effect of bee-eater helpers comes through their provisioning of food to the nestlings. Fig. 16.11 showed that starvation was the major cause of nestling mortality. Consequently Peter Wrege and I performed a detailed study of provisioning by quantifying the rate at which food was brought to 61 nests. For each nest, we made observations on two consecutive mornings and afternoons for a total of 8 h per nest. To minimize variance in feeding rate due to poor weather and differing energy demands of the young, we restricted sampling to nests with young between 16 and 25 days of age and to days with good weather. Our findings are graphed in Fig. 16.13 and show a steady increase in total feeding rate as group size increases. An analysis of covariance controlling for the effects of

Fig. 16.13. The effect of bee-eater group size on rate of feeding to the nest. Values are least-square means from an analysis of covariance in which the effects of brood size and year were controlled ($F_s = 18.0$, $P \ll 0.01$). Data are from the 1981 and 1982 nestings of the Baharini population. (Feeding data were not collected systematically in other years due to other research priorities.)

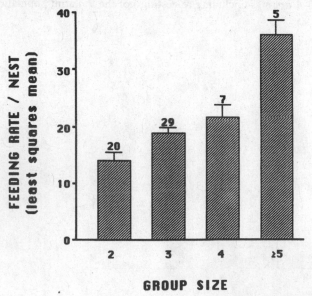

brood size and year showed that group size was a highly significant predictor of total feedrate ($F_s = 18.0$; $P \ll 0.01$), accounting for 55% of the model sum of squares (model $r^2 = 0.62$; $P \ll 0.01$). A similar analysis for starvation rate versus group size demonstrated that starvation losses decreased as helper number increased ($F_s = 43.2$; $P \ll 0.01$; model $r^2 = 0.62$; $P \ll 0.01$).

These analyses do not rule out other ways in which the presence of helpers might be beneficial to breeders and/or to their dependent young. But they do indicate that one major way in which helpers increase the production of fledgings is by reducing starvation losses caused by severe food limitation.

Speculation: do bee-eaters fit the basic model?

White-fronted Bee-eaters do not fit the archetypical image of a cooperative breeder. For starters, they are colonial, they do not inhabit all-purpose territories, and the territories which they do occupy (foraging territories) are not in limited supply. Add to this the facts that helpers come in both sexes and all ages, and that fully 40% of helpers are themselves former breeders and you have an unusual, albeit fascinating, exception to Brown's (1974) early portrait of a cooperative breeder.

A critical question thus is whether White-fronted Bee-eaters fit the emerging two-step conceptual model of the evolution of helping behavior. Are bee-eaters ecologically constrained from becoming independent breeders? And, if so, does the benefit of helping exceed that of remaining integrated with one's natal clan but *not* helping?

In earlier papers (Emlen 1982a, 1984; Emlen and Vehrencamp 1983) I have argued for the generality of ecological constraints in causing grown offspring to delay reproduction and to remain with their natal (parental) groups. I proposed that such constraints often are a necessary – but not a sufficient – first step for the evolution of helping behaviors. I listed three factors that could constrain the option of personal reproduction: (1) a shortage of breeding territories, (2) a shortage of mating partners, and (3) a prohibitively high cost of breeding. In the first two, constrained individuals are unable to become breeders; in the third, breeding is a possibility, but an unfavorable benefit to cost ratio selects against its occurrence. As correctly pointed out by Stacey and Ligon (1987), this is an important distinction. So long as breeding is possible, the expected benefits of helping should be compared against those of breeding (since helping might be the optimal strategy to pursue). When breeding is not possible, the appropriate contrast is between helping and not helping.

The first two factors do not seem to be applicable to White-fronted Bee-eaters at Nakuru. Foraging territories are not in short supply (see pp. 498–9). Neither does territory occupancy appear to be related to the likelihood of breeding. Instead breeding opportunities and territory ownership are decoupled in this species. Many pairs residing on presumedly high-quality foraging territories (located near the active colony) did not breed, while many others residing farther away bred but abandoned their territories while provisioning dependent young (see Table 10 of Hegner and Emlen, 1987). As for a shortage of mates, many helpers (41%) are already paired; and those that are single are equally likely to be female as male (Table 16.3).

I believe that the factor limiting breeding attempts among White-fronted Bee-eaters is the low probability of successfully reproducing at all. Success in breeding depends both upon environmental harshness (the availability of food) and group size tending the nest (presence and numbers of helpers). Concerning environmental harshness, I wrote in 1984:

> In variable and unpredictable environments, erratic changes in the carrying capacity create the functional equivalents of breeding openings and closures. As environmental conditions change from year to year, so too does the degree of difficulty associated with successful breeding. In benign seasons, abundant food and cover decreases the costs to younger, less experienced individuals of... breeding independently. In harsher seasons, the costs associated with such reproductive ventures increase, eventually reaching prohibitive levels. As conditions deteriorate, breeding options become more constrained and the constraints first hit the younger, more subordinate individuals.
>
> (Emlen 1984, p. 321)

The effects of low food abundance would be ameliorated if the breeders could increase the number of adults provisioning the nest. As a breeder ages, its likelihood of having grown offspring as helpers of its own increases (unpaired offspring almost invariably help their parents). We also know that older males frequently harass or otherwise interfere with the reproductive attempts of their paired offspring, a behavior that often leads offspring to abandon their own reproductive efforts and join as helpers at the nests of their fathers (Emlen 1981; S. T. Emlen, P. H. Wrege and N. J. Demong, unpublished data). As a result, older individuals are more likely to gain helpers of their own. Since provisioning rate is directly related to group size (Fig. 16.13), older breeders may have a high probability of successfully breeding even in harsh years. The same is not true for young breeders who rarely have helpers of their own and whose nesting attempts often are actively disrupted.

Figure 16.14 (which is modified slightly from Fig. 5.4 of Brown 1987), graphically depicts the effects of environmental harshness on the breeding success of a hypothetical species of cooperative breeder living in a variable environment. Following Brown (1987), I assume that there is some threshold level (B) of total provisioning to the nest required to fledge young successfully. (Provisioning is an appropriate measure for bee-eaters, since food stress on nestlings is a major factor in their reproductive success; one could equally well plot total 'parental' (= breeder plus helper) effort.) I further assume (following Brown) that the total rate of provisioning to the nest increases as some monotonic function of the age of the breeder(s). This may be because of an age-related improvement in the foraging skills of the breeder(s) themselves (as emphasized by Brown) or because older breeders are more likely to gain helpers to aid them in defending or provisioning their young. For the reasons discussed above, I feel that the latter is the important factor for bee-eaters.

The effect of a deterioration in environmental conditions is to make

Fig. 16.14. Brown's (1987) 'learning skills' model, showing how severe environmental harshness (depicted by the 'bad year' curve) can function as a constraining factor leading to postponement of breeding among younger members of a population. The horizontal line, B, represents a minimum threshold level of provisioning necessary for successful reproduction. I modify Brown's interpretation to emphasize age-dependent gaining of helpers rather than age-dependent learning of foraging skills; for a fuller discussion, see the text.

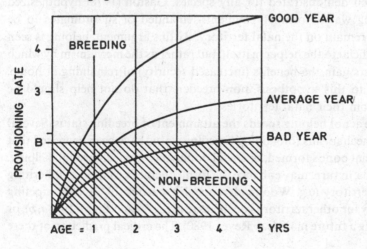

provisioning more difficult, in essence to lower the slope of the provisioning curve (compare the 'good year' with the 'bad year' curve in Fig. 16.14). Since B is fixed, the result is an increase in the number of younger (unaided) pairs that will be unable to reproduce successfully.

The model predicts that the frequency of non-breeders will increase as environmental conditions become increasingly harsh. Data supporting this prediction for bee-eaters have been published previously (Emlen 1981, 1982a).

White-fronted Bee-eaters thus appear to fit the first step of the conceptual model. Given that many individuals do not attempt to breed, the next question is: 'Why should they help?'

Several possible advantages have been hypothesized to accrue to helpers by virtue of their helping (for reviews, see Brown 1983, 1987; Emlen and Vehrencamp 1983; Emlen 1984; Woolfenden and Fitzpatrick 1984). These hypotheses fall into four basic categories:

(1) enhancing one's probability of survival while waiting to become a breeder,

(2) increasing one's probability of becoming a breeder in the future,

(3) increasing one's fecundity when one does become a breeder,

(4) increasing the production of non-descendant kin.

The first three provide direct fitness gains to the helper; in the fourth, the benefit is indirect (*sensu* Brown 1980).

In (1), it is assumed that helping somehow reduces the risk of mortality. For increased survival to be a benefit of helping *per se*, rather than a passive effect of delayed dispersal, it must be shown that non-helpers suffer a higher mortality than do helpers. This is counter-intuitive and, to my knowledge, has not been demonstrated for any species. Gaston (1978) hypothesized that helping was a form of payment demanded of subordinates to be allowed to remain on the natal territory. By this argument, helping is seen not as beneficial to the helper in itself, but rather as the mechanism by which non-breeders gain the benefits (increased security) of remaining at home. According to this hypothesis, non-breeders that do not help should be expelled from their natal territories.

In (2) the act of helping speeds the attainment of breeding status. Several proximal mechanisms for gaining breeding status have been proposed that involve social bonds formed, or social dominance gained, through helping. These bonds, in turn, may enhance the former helper's chances of inheriting the natal territory (e.g. Woolfenden and Fitzpatrick 1978), of competing successfully for other territory vacancies (e.g. Ligon and Ligon 1978a, b), or of obtaining a future mate (e.g. Reyer 1980). The critical prediction of these hypotheses is that the occurrence (or magnitude) of helping is correlated with an increased probability of becoming a breeder in the near future.

In (3), an individual gains both nesting and social 'experience' by being a helper. Nesting experience refers to the learning or practicing of skills associated with tending eggs and rearing young. Social experience refers to dominance gained via social interactions performed while a helper. When the individual later becomes a breeder itself, its early nesting experience makes it more skilled at parenting. Its social experience enhances its likelihood of gaining helpers of its own. Both could translate into enhanced reproductive success. The critical prediction of this category of hypotheses is that a positive correlation exists between the amount of experience gained as a helper (including both nesting and social experience) and later reproductive success.

Finally, in (4), the helper benefits by increasing its inclusive fitness through the production of non-descendant genetic relatives. Two critical predictions here are that helping enhances fledging success and that the recipients of helping are close kin.

Which of these proposed benefits are important for White-fronted Bee-eaters in Kenya? In recently completed analyses, Peter Wrege and I have calculated that survival of helpers is not greater than that of non-helpers; nor are non-helpers expelled from clan membership.

Since a shortage neither of mates nor of foraging territories appears to limit breeding, helping cannot easily be viewed as a stratagem for attaining reproductive status. In all probability, the primary direct benefit of helping among these bee-eaters comes from the experience so gained. The magnitude of this benefit has yet to be analyzed quantitatively.

We do know, however, that the magnitude of the *indirect* gain to a bee-eater helper is large (category (4) benefits). This gain can be thought of as the product of two terms, the increase in the number of young successfully produced by the breeding pair as a result of the activities of the helper, and the coefficient of relatedness between the helper and the young it helps to rear. The effect of helping in enhancing fledging success was shown dramatically in Fig. 16.12. The genetic closeness between donor and recipient was documented in Table 16.4. The average coefficient of relatedness between a bee-eater helper and the nestlings it helps is 0.33. By living in a social unit comprised of extended family members, a non-breeder is virtually assured of the availability of close genetic relatives as potential recipients of its aid.

White-fronted Bee-eaters satisfy the critical predictions for category (4) benefits. The fact that bee-eaters show active kin discrimination in their choice of recipients further indicates that indirect benefits are important for the maintenance of helping behavior in this species. By preferentially choosing to aid close, as opposed to distant, kin (documented on pp. 510–14), a bee-eater can gain almost as much in indirect fitness as a helper as it could in direct fitness as a breeder.

The four categories of potential benefits listed above are not mutually exclusive; in fact they are additive. This point is often overlooked in debates over the importance of kin selection for the evolution and maintenance of helping. I do not yet know the relative importance of direct versus indirect benefits in tipping the balance of the cost/benefit ratio to favor helping over non-helping among non-breeding bee-eaters. But since the editors urged authors to be speculative, I will hazard a guess that kin selection will prove to be very important for this species. I base this prediction on the simple fact that non-breeders for whom relatedness values to potential recipients are small (unrelated mates, or 'in-laws') usually chose not to become helpers at all (Table 16.2). This would not be predicted if direct benefits were important to the helper, since direct gains would be accrued regardless of relatedness to the recipient.

I conclude with a plea for the importance of comparative studies in the development and testing of basic social theory. Many early reviews of helping behavior concluded that cooperative breeding occurred in such a diverse array of species that no one set of circumstances could explain its evolution (e.g. Fry 1972; Rowley 1976). But I feel that the beginnings of a unified general theory for the evolution of helping behavior are at hand (Brown 1974, 1983, 1987; Emlen 1982a, b, 1984; Emlen and Vehrencamp 1983). Studying a species as different from the archetypical cooperative breeder as the White-fronted Bee-eater has reinforced my optimism. Bee-eaters fit a modified version of the two-step, ecological constraints model outlined in the introduction to this chapter. It will be exciting to see if the model's applicability continues to hold as an increasingly diverse array of cooperatively breeding species are studied in the years ahead.

This work would not have been possible without the help and encouragement of many persons. Foremost among these are the individuals who assisted in the collection of field data, I. Brown, N. Demong, D. Emlen, G. Farley, R. Hegner, E. Kellogg, M. Kinniard, M. Kippelian, C. Miller, S. Moore, M. Read, G. Tabor, R. Wagner and P. Wrege. For initial hospitality and logistical support in Kenya I thank J. Hopcraft, R. Terry and C. van Someren; later field work was made possible by authorization from the Office of the President and academic sponsorship from the National Museums of Kenya. Long-term financial support has been provided by the National Science Foundation and the National Geographic Society.

Additional financing has come from the John Simon Guggenheim Foundation, the Chapman Fund of the American Ornithologists' Union, and personal funds. To all of these persons and institutions, I owe an enduring debt of gratitude. The manuscript itself has benefited from editorial comments by R. Curry, W. Koenig, P. Stacey, G. Woolfenden, and P. Wrege.

Bibliography

Brown, J. L. (1974). Alternate routes to sociality in jays with a theory for the evolution of altruism and communal breeding. *Amer. Zool.* **14**: 63–80.

Brown, J. L. (1980). Fitness in complex avian social systems. In *Evolution of Social Behavior: Hypotheses and Empirical Tests*, ed. H. Markl, Dahlem Konferenzen, pp. 115–28. Verlag Chemie, Weinheim.

Brown, J. L. (1983). Cooperation: a biologist's dilemma. In *Advances in Behavior*, vol. 13, ed. J. S. Rosenblatt, pp. 1–37. Academic Press, New York.

Brown, J. L. (1987). *Helping and Communal Breeding in Birds*. Princeton University Press: Princeton, NJ.

Brown, L. and Britton, P. (1980). *The Breeding Seasons of East African Birds*. East African Natural History Society: Nairobi.

Dyer, M. and Crick, H. Q. P. (1983). Observations on White throated Bee-eaters breeding in Nigeria. *Ostrich* **54**: 52–5.

Emlen, S. T. (1981). Altruism, kinship, and reciprocity in the white-fronted bee-eater. In *Natural Selection and Social Behavior: Recent Research and New Theory*, ed. R. D. Alexander and D. Tinkle, pp. 217–30. Chiron Press: New York.

Emlen, S. T. (1982a). The evolution of helping. I. An ecological constraints model. *Amer. Nat.* **119**: 29–39.

Emlen, S. T. (1982b). The evolution of helping. II. The role of behavioral conflict. *Amer. Nat.* **119**: 40–53.

Emlen, S. T. (1984). Cooperative breeding in birds and mammals. In *Behavioural Ecology: An Evolutionary Approach*, ed. J. R. Krebs and N. B. Davies, pp. 305–39. Blackwell Scientific Publications: Oxford.

Emlen, S. T. and Demong, N. J. (1984). The bee-eaters of Baharini. *Nat. Hist.* October: 50–9.

Emlen, S. T. and Vehrencamp, S. L. (1983). Cooperative breeding strategies among birds. In *Perspectives in Ornithology*, ed. A. H. Brush and G. A. Clark Jr, pp. 93–120. Cambridge University Press: Cambridge.

Emlen, S. T. and Wrege, P. H. (1986). Forced copulations and intraspecific parasitism: two costs of social living in the White-fronted Bee-eater. *Ethology* **71**: 2–29.

Emlen, S. T. and Wrege, P. H. (1988). The role of kinship in helping decisions among white-fronted bee-eaters. *Behav. Ecol. Sociobiol.* **23**: 305–15.

Fry, C. H. (1972). The social organization of bee-eaters (Meropidae) and cooperative breeding in hot-climate birds. *Ibis* **114**: 1–14.

Fry, C. H. (1984). *The Bee-eaters*. T. and A. D. Poyser: Calton, Staffs.

Gaston, A. J. (1978). The evolution of group territorial behavior and cooperative breeding. *Amer. Nat.* **112**: 1091–100.

Greenwood, P. J. (1980). Mating systems, philopatry and dispersal in birds and mammals. *Anim. Behav.* **28**: 1140–62.

Hegner, R. E. (1981). Territoriality, foraging behavior, and breeding energetics of the white-fronted bee-eater in Kenya. Ph.D. thesis, Cornell University: Ithaca, NY.

Hegner, R. E. and Emlen, S. T. (1987). Territorial organization of the White-fronted Bee-eater in Kenya. *Ethology* **76**: 189–222.

Hegner, R. E., Emlen, S. T. and Demong, N. J. (1982). Spatial organization of the White-fronted Bee-eater. *Nature (Lond.)* **298**: 264–6.

Ligon, J. D. and Ligon, S. H. (1978a). The communal social system of the Green Woodhoopoe in Kenya. *Living Bird* **17**: 159–98.

Ligon, J. D. and Ligon, S. H. (1978b). Communal breeding in Green Woodhoopoes as a case for reciprocity. *Nature (Lond.)* **276**: 496–8.

Reyer, H.-U. (1980). Flexible helper structure as an ecological adaptation in the Pied Kingfisher (*Ceryle rudis*). *Behav. Ecol. Sociobiol.* **6**: 219–27.

Rowley, I. (1976). Cooperative breeding in Australian birds. In *Proc. 16th Intern. Ornith. Congr.*: 657–66.

Sinclair, A. R. E. (1978). Factors affecting the food supply and breeding season of resident birds and movements of palearctic migrants in a tropical African savannah. *Ibis* **120**: 480–97.

Stacey, P. B. and Ligon, J. D. (1987). Territory quality and dispersal in the Acorn Woodpecker, and a challenge to the habitat-saturation model of cooperative breeding. *Amer. Nat.* **130**: 654–76.

Woolfenden, G. E. and Fitzpatrick, J. W. (1978). The inheritance of territory in group-breeding birds. *BioScience* **28**: 104–8.

Woolfenden, G. E. and Fitzpatrick, J. W. (1984). *The Florida Scrub Jay: Demography of a Cooperative-breeding Bird.* Princeton University Press: Princeton, NJ.

Wrege, P. H. and Emlen, S. T. (1987). Biochemical determination of parental uncertainty in White-fronted Bee-eaters. *Behav. Ecol. Sociobiol.* **20**: 153–60.

17 PIED KINGFISHERS: ECOLOGICAL CAUSES AND REPRODUCTIVE CONSEQUENCES OF COOPERATIVE BREEDING

17

Pied Kingfishers: ecological causes and reproductive consequences of cooperative breeding

H.-U. REYER*

Studies on the adaptive significance of behavioral strategies are most promising when there is high variability, both in the behavioral trait and the ecological conditions under which it occurs. This allows comparison of the costs and benefits of pursuing different strategies under the same conditions and of pursuing the same strategies under different conditions. The Pied Kingfisher (*Ceryle rudis*) shows such variability in its helper system. I have studied this species over eight breeding seasons (1976–83) at Lake Naivasha and Lake Victoria in Kenya. The following account of the results consists of four sections: (1) a description of the life history and the helper structure; (2) a functional interpretation of the cooperative breeding from the helpers' and from the breeders' points of view; (3) an analysis of the causal mechanisms which allow the birds to choose the strategy which maximizes their fitness under the prevailing ecological conditions; and (4) some speculations as to the origin of this helper system.

Life history and helper structure
General biology

Pied Kingfishers range, in three subspecies, from eastern Asia through Asia Minor to South Africa. They occur along many rivers, but are particularly frequent in the marginal regions of big freshwater lakes. They feed almost exclusively on fish. To catch their prey, they either dive from papyrus stems, dead trees or other perches along the shore, or – more often – fly over the water searching, sometimes hovering above the surface, and plunging swiftly when they see a fish. Although individual birds have preferential perches and hunting areas, there are no defended territories. A

---

* Present address: Zoologisches Institut der Universität Zürich, Winterthurerstrasse 190, CH-8057, Zürich, Switzerland.

529

suitable perch may be used by as many as 10–15 birds simultaneously. The foraging areas not only overlap considerably but also shift during the day with changing conditions such as velocity and direction of wind.

Outside the breeding season, Pied Kingfishers can be seen singly, in pairs or in small groups along the whole shore area. In the breeding season (April–August), the beginning of which coincides with the rains, they concentrate at rivers, canals, road embankments, and other places having sandy or clay banks, not too far from the lake (Fig. 17.1). Here, several birds excavate and defend nesting-holes, which can be as close as half a meter to each other. In this way breeding colonies are formed. Both males and females of the breeding pair, who can be distinguished by plumage differences, take turns digging the nest hole and incubating the four to six eggs. After 18 days the young hatch, naked and blind. From the first day they are fed with fish brought from the lake, mainly by the male and possible helpers; later, when brooding declines, the female joins

Fig. 17.1. Section of the river where the Lake Victoria Pied Kingfishers breed during the rainy season. A positive effect of the rain is that the banks become soft enough for digging nest holes; a negative effect is that the river may become a torrent and make parts of the banks collapse. For the birds this often results in loss of a nesthole and clutch, while the native Luo people welcome the sand and clay as building material.

increasingly in feeding the young. Nestlings are fully fledged after about 26 days and can fish for themselves roughly two weeks later, but stay with their parents away from the breeding site for several months.

Demographic data

Using only clutches in which all young could be sexed, the male: female ratio among nestlings was 0.9:1 ($n = 38$) at Lake Victoria and 1.1:1 ($n = 104$) at Lake Naivasha. Neither is significantly different from each other ($\chi^2 = 0.3$, d.f. $= 1$, not significant (n.s.)) or from a 1:1 ratio (both $z < 0.5$, n.s., Binomial test, two-tailed). However, out of 36 marked juveniles returning as yearlings only one was a female while 35 were males. This 3%:97% ratio differs significantly from expectation based on the even sex ratio among the nestlings ($z = 5.5$, $P < 0.001$). Thus, young females disperse before the end of their first year, while males return (e.g. male 298, born in 1980, and males 594 and 9245, born 1982; Fig. 17.2).

After the first year, there seems to be no dispersal of either sex – unless breeding conditions become unfavorable. This is exemplified by a comparison of the two lakes. At Lake Naivasha, fluctuations in the lake level and fortification of banks by men made some previously occupied breeding sites unavailable (colonies A and B in Fig. 17.3), while digging of ditches opened new possibilities (colony C in Fig. 17.3). Thus, birds were forced (and able) to move, which precluded any reliable calculation of return and mortality rates.

At Lake Victoria, conditions were more stable (Fig. 17.3). Here, in 1979 and 1982, a total of 351 adult Pied Kingfishers were checked in three neighboring colonies lying 2, 4 and 9.5 km away from the main study area where in the previous years (1978 and 1981) 110 marked adults had been present. Only one of these marked adults, a female, was found in the nearest separate colony. This limited extent of colony change (0.9%) allows the use of return rates as a measure of survival. The resulting figures are 64.9% ($n = 185$) for males and 56.5% ($n = 106$) for females (breeding and non-breeding birds combined). In addition, the year-to-year stability in colony size, plus the fact that reproduction is seasonal, allows the calculation of the mean adult survival of males from the number of breeding adults surviving into the next year, divided by the number of surviving male yearlings plus adults (see e.g. Vehrencamp 1978). The resulting figure is 53.8% ($n = 52$), similar to the fractions of male breeders returning the year after ringing, which is 57.1% ($n = 49$). Thus, under stable conditions, male mortality seems to be compensated for by juvenile males from the same colony.

For females the latter method of calculating survival could not be used, as juvenile females do not return to their natal area. Consequently, mortality of females must be completely compensated for by females immigrating from other areas. As these immigrating females were not ringed and could not be aged by plumage characteristics, their precise age is

Fig. 17.2. Life histories of some Pied Kingfishers at Lake Victoria from 1980 to 1983. White boxes at the top of the broken lines denote breeding groups, gray boxes below broken lines show the young produced by the respective groups. Members of the breeding groups are separated into male breeders ($\male$), female breeders ($\female$), primary helpers (p) and secondary helpers (s). For further explanations, see the text.

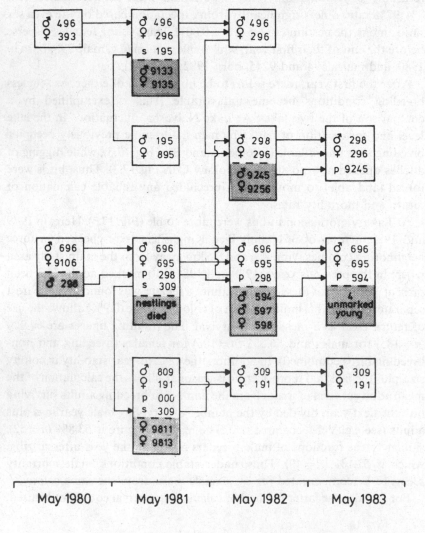

May 1980   May 1981   May 1982   May 1983

unknown. But with almost no dispersal of adults, they can be assumed to be yearlings. These females bred in the year of their arrival and hand-reared birds of both sexes bred in captivity at the age of 11 months. Thus, sexual maturity and breeding *can* be reached within the first year and probably *is* in all females. Males, however, rarely breed in their first year (see p. 542), but act as helpers for one to two years. In some of these helpers, even sexual maturity appears to be delayed until the age of two or three years (see p. 543).

Two types of helper

During all early stages of the breeding season (i.e. before hatching) several pairs in both colonies were accompanied by one or two additional adults. These extra birds do not participate in tunneling, incubating and brooding, nor do they copulate. But they feed the male of the breeding pair, support it in feeding the female, and assist the pair in chasing off rivals and nest predators such as the monitor lizard (*Varanus niloticus*), cobras (*Naja* spp.), and the ichneumon (*Herpestes* spp.). Finally, after the young have

Fig. 17.3. Population sizes from 1977 to 1983 for one colony of Pied Kingfishers at Lake Victoria (filled symbols) and three colonies (A–C) at Lake Naivasha (open symbols). ——, Colony A; ·····, colony B; ----, colony C; ● and ○, total mated birds; ▲ and △, total population size.

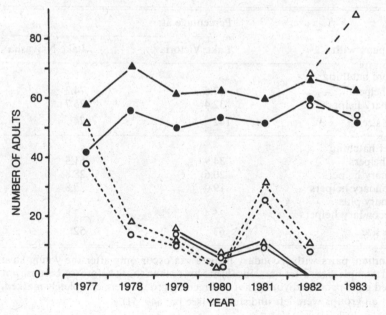

hatched, they help the parents to feed the nestlings. Such birds which accompany pairs from the very beginning of the breeding season I call 'primary helpers'.

About one week after the young hatch, the number of breeding pairs with helpers and the number of helpers per pair increases. These additional helpers, although present in the colony from the beginning of the breeding season, are firmly associated with pairs only after hatching and are called 'secondary helpers'. Secondary helpers join breeders with, and breeders without, primary helpers about equally often (Table 17.1, rows 5 and 6) and there can be as many as one primary plus three secondary helpers per pair. The proportion of pairs with primary helpers was similar at the two lakes, both before and after hatching (Table 17.1, rows 2 and 4). Pairs with secondary helpers, however, were relatively more frequent at Lake Victoria than at Lake Naivasha, where a higher proportion of breeding pairs remained unassisted ($\chi^2 = 17.4$, d.f. $= 3$, $P = 0.0006$; χ^2-test, applied to rows 3–6 of Table 17.1). This makes the average breeding group size at Lake Victoria higher than at Lake Naivasha (Fig. 17.4; $z = 2.635$, $P < 0.01$, Mann–Whitney U-test, two-tailed).

Table 17.1. *Percentage of mated pairs of Pied Kingfishers at Lake Victoria and Lake Naivasha with no helpers, with primary helpers only, with secondary helpers only and with primary plus secondary helpers*

	Percentage at:	
Mated pairs with	Lake Victoria	Lake Naivasha
(a) Before hatching		
(1) No helpers	67.6	74.3
(2) Primary helpers	32.4	25.7
Sample size	142	74
(b) After hatching		
(3) No helpers	34.9	61.5
(4) Primary helpers	20.6	28.8
(5) Secondary helpers	19.0	5.8
(6) Primary plus secondary helpers	25.4	3.8
Sample size	63	52

By definition, pairs with secondary helpers can occur only after the young have hatched. Sample sizes after hatching (b) are lower than those before (a), because not all mated pairs produced young, not all breeding groups were completely marked, and not all groups were left undisturbed (see pp. 549–51).

Among the 71 primary and the 53 secondary helpers that I have recorded all were males. Eight of the secondary helpers were mated males who did not breed in the year in which they helped (e.g. s 195 in 1981, Fig. 17.2), the remaining 93.5% of all helpers were unmated. The high number of unmated males results from a highly biased sex ratio. At Lake Victoria, adult males outnumbered adult females by 1.58:1 (S.D. = 0.44, $n = 8$ years), and at Lake Naivasha by 1.57:1 (S.D. = 0.36, $n = 7$ years). The reason for, and the significance of, this bias will be discussed on pp. 551–5.

The date at which helpers associate with breeders is one characteristic in which primary and secondary helpers differ. The way in which they associate with breeders is a second characteristic (Fig. 17.5). From the very beginning of the season, primary helpers restrict their activities to one mated pair only, e.g. helper A to pair 1, helper B to pair 2. Color-banding over eight years has revealed that most primary helpers are the sons of at least one mate of that pair. This means that they usually assist in raising younger siblings. Of all cases, 38% were full siblings (e.g. p. 594 in 1983, Fig. 17.2), 54% were half-siblings (e.g. p. 298 in 1981), and for only 8% were primary helpers not related to the young ($n = 24$). The resulting average coefficient of genetic relationship between primary helpers and nestlings is $r = 0.32$.

Secondary helpers, on the other hand, do not appear to be closely related

Fig. 17.4. Relative frequency of breeding group sizes of Pied Kingfishers at Lake Victoria (open blocks; $n = 63$ groups) and Lake Naivasha (filled blocks; $n = 52$ groups). Group sizes were measured after the young hatched but include adults only.

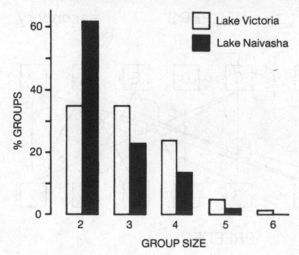

to the young that they raise. They were certainly not the young of the breeding pairs that they joined; and color-banding showed no evidence for any other close genetic relatedness down to the level of 1/8. As inbreeding seems to be low (no incestuous matings were observed and young females disperse before breeding), the average coefficient of relatedness between a secondary helper and the random young he helps to rear can be calculated to be $r \leq 0.05$. This is more than six times lower than with primary helpers. The 'random choice' of secondary helpers is also supported by the way they join the breeders (Fig. 17.5). Unlike the primary helper, a potential secondary helper, such as C, initially approaches various pairs, e.g. 3, 6 and 7; and one pair, such as 6, may be visited by different potential helpers, e.g. by C, E and F. But by and by the secondary helpers focus on one particular pair e.g. helper C on pair 7, helper E on pair 6. If that pair is not successful, secondary helpers may switch again (e.g. s 309 in 1981, Fig. 17.2). But normally they finally restrict their activities to this pair and help in warding off predators and feeding young in the same way as the primary helpers do.

Quantitatively, however, there are large differences between the two helper categories (Fig. 17.6). Primary helpers invest as much as male and female breeders do, no matter whether we consider nest-guarding against predators or food contributions to young in terms of numbers of fish/day,

Fig. 17.5. Schematic representation of associations between primary helpers (A–B), secondary helpers (C–?) and breeding pairs (1–7) at the beginning of the Pied Kingfisher breeding season. ?, Unmarked bird. Secondary helpers C and D are bracketed to indicate that they are brothers joining different pairs.

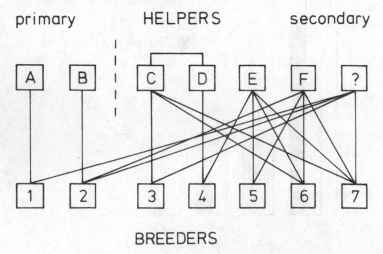

primary　　　　HELPERS　　　secondary

BREEDERS

Fig. 17.6. Investment of Pied Kingfisher breeders (♂, ♀), primary helpers (ph) and secondary helpers (sh) in nest-guarding (*a*), and in feeding, with respect to number of fish per day (*b*), energy content per fish (*c*) and total energy per day (*d*) (1 cal = 4.184 J). Shaded bars and thin vertical lines, pairs with ≤2 helpers; filled bars and thick lines, pairs with >2 helpers. Bars represent means, vertical lines 95% confidence limits. Bars are connected by horizontal lines if the difference between them is significant (continuous line $P \le 0.01$; dashed line $P \le 0.05$), or tends to be so (dotted line $P \le 0.10$; Mann–Whitney *U*-test). Numbers under the graphs are sample sizes for pairs with ≤2 helpers and those with >2 helpers (in parentheses).

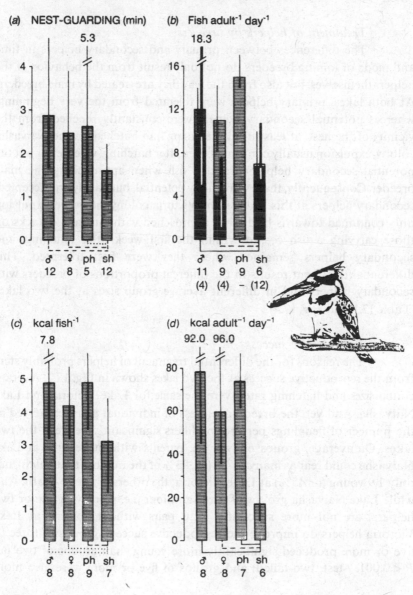

average size of fish and total energy/day. Secondary helpers, on the other hand, take it much more easy: they spend less time nest-guarding (Fig. 17.6(a)), feed both fewer fish (Fig. 17.6(b)) and smaller ones (Fig. 17.6(c)). In total they provide the young with only about a quarter of the energy which the other birds provide (Fig. 17.6(d)). Thus, the investment in nestlings is a third characteristic in which primary and secondary helpers differ.

Treatment of helpers by breeders

The differences between primary and secondary helpers in time and mode of joining breeders do not only result from the behavior of the helpers themselves, but also from the way they are treated by male breeders. At both lakes, primary helpers were tolerated from the very beginning, whereas potential secondary helpers were constantly repelled from the vicinity of the nest, at least until the young had hatched. In the Naivasha colony, expulsion usually continued even after hatching, whether or not the potential secondary helper carried a fish when approaching the male breeder. Consequently, there were many potential, but hardly any accepted, secondary helpers at this lake. In the Victoria colony, however, expulsion only continued towards birds that approached without prey. Attacks on those carrying a fish ceased within the first week after hatching, and secondary helpers remained where they were first tolerated. This differential treatment results in the different proportions of breeders with secondary helpers and in different average group sizes at the two lakes (Table 17.1 and Fig. 17.4).

Reproductive success

The reasons for the differential treatment of helpers probably stem from the reproductive success at the two lakes shown in Fig. 17.7. Average clutch sizes and hatching rates were the same for Lake Victoria and Lake Naivasha. And yet, the breeding success of individual pairs, expressed as the number of fledglings per group, differs significantly between the two lakes. On average, groups of two (i.e. parents without helpers) at Lake Naivasha could rear as many as 4.0 (i.e. 83% of the number that hatch), but only 1.9 young (or 45%) at Lake Victoria; the others starve to death. And while Lake Naivasha groups of three and four (i.e. pairs with one or two helpers) are not more successful than pairs without helpers, at Lake Victoria helpers do improve the reproductive success: groups of three to five or more produced significantly more young than groups of two (all $P < 0.001$, t-test, two-tailed) and groups of five or more were also more successful than groups of three ($P < 0.02$). If Lake Victoria pairs are split into those having only primary and those having only secondary helpers, it turns out that the number of additionally surviving nestlings per primary helper is 45% higher than that per secondary helper (Fig. 17.7). This very likely results from the higher feeding contribution of primary helpers (Fig. 17.6).

Ecological reasons

There are two main ecological reasons for the differences in reproductive success between the two lakes: (1) the type of prey and (2) its availability.

(1) At Lake Victoria, the main prey item is the cyprinid fish *Engraulicypris argenteus*. At Lake Naivasha the birds feed mainly on cichlid species. Because of its slender body shape, an *Engraulicypris* of about 5 cm length yields less energy than a bulky cichlid of the same length. Moreover, *Engraulicypris* does not grow

Fig. 17.7. Clutch size, number of hatched, and number of fledged young for different group sizes of Pied Kingfisher. Bars give average values for Lake Victoria (open blocks) and Lake Naivasha (filled blocks), vertical lines represent one standard deviation. For Lake Victoria the numbers of additionally surviving young per primary (stippled) and per secondary (hatched) helper are also shown.

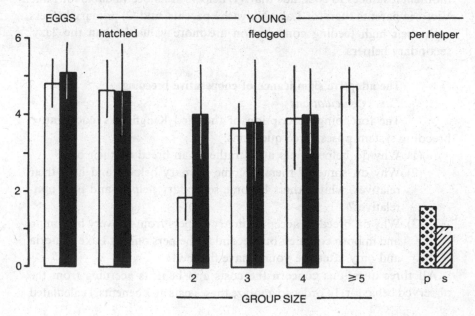

as big as cichlids. To compensate for that, Lake Victoria parents have to catch more fish for their young than do Lake Naivasha parents.

(2) To catch and feed one fish demands more time in hovering and flying from a Lake Victoria bird than from a Lake Naivasha bird. This is because the distance between the fishing grounds and the colony at Lake Victoria is about twice that at Lake Naivasha (c. 600 m versus 300 m). It is also because of differences in water conditions. At Lake Victoria, one of the largest freshwater lakes in the world, there is always a strong wind sweeping from across the lake, roughening the surface and forming breakers along the shore. Under these conditions, the Pied Kingfishers have to fly out far and only 24% of their dives are successful ($n = 107$). In contrast, at the much smaller Lake Naivasha the wind comes from the land and there is a coastal zone of calm, relatively clear water, protected by a high papyrus belt. Here 79% of all dives are successful ($n = 52$), a highly significant difference ($\chi^2 = 42.704$, $P < 0.001$).

These two environmental conditions taken together have the effect that Lake Victoria parents must invest more time and energy than Lake Naivasha parents to feed their young – apparently so much more that they cannot do it alone, but need helpers to prevent their young from starving. This energy aspect will be dealt with in more detail on pp. 546–9). For the moment it suffices to conclude that: (1) helpers are more valuable for Lake Victoria parents than for Lake Naivasha parents, and (2) primary helpers with their high feeding contribution are more valuable than the 'lazy' secondary helpers.

The adaptive significance of cooperative breeding
Fitness calculations

The foregoing description of the Pied Kingfishers' cooperative breeding system poses three questions:

(1) Why do helpers help at all rather than breed on their own?
(2) Why do some of them become primary helpers and help their relatives, while others become secondary helpers and help non-relatives?
(3) Why do breeders accept primary helpers from the very beginning and in both colonies, but secondary helpers only at Lake Victoria and only after the young have hatched?

All three questions concern the costs and benefits accruing from the observed behavior. In order to analyze these costs and benefits, I calculated the genetic fitness of birds following different behavioral strategies (e.g. helper versus breeder, primary versus secondary helper, rejecting helpers versus accepting them) which is basically the product of the following three factors:

(1) The probability P, that a bird will obtain a certain status (e.g. become a breeder or a helper).

(2) The number of young which are produced as a consequence of the bird's contribution. If he is breeder, this is simply the number of his own young. If he is a helper, it is the difference between the number of young surviving with his contribution (N^+) and the number of young which would have survived without his contribution (N^-), divided by the total number of helpers at this nest (H).

(3) The coefficient of genetic relationship between the bird and the young, estimated as $r = 0.50$ for breeders, $r = 0.32$ for primary and $r \leq 0.05$ for secondary helpers (pp. 535–6).

Figure 17.8 shows the result of such calculations for young Lake Victoria

Fig. 17.8. Sequence of choices a young male Pied Kingfisher has to make between four different strategies (dendrogram) and inclusive fitness after two years if he starts as a breeder, a primary helper, a secondary helper or a delayer (stippled bars). The highest inclusive fitness value from Table 17.2, that of breeders, has been set to 1, values of the three other categories have expressed relative to it.

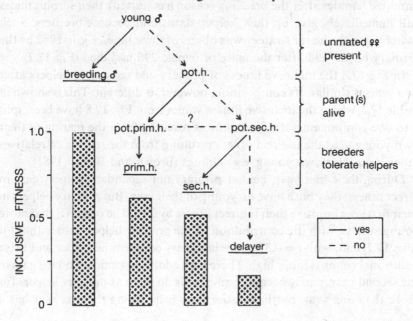

males starting as breeders, primary helpers, secondary helpers or 'delayers'. Delayers are birds, which – when not mated in their first year – do not help, but wait for a breeding opportunity to occur the next year. I have restricted the calculations to the first two years of life, partly because this is the period in which males differ in their behavioral strategies, partly because it is the period for which reasonably large sample sizes are available. Details about the calculation and the data on which it is based have been given by Reyer (1984b). Where results differ from those in previous publications (Reyer 1984b, 1986) new data have been added.

Costs and benefits for helpers

According to the histograms in Fig. 17.8, helping is superior to delaying, but inferior to breeding. Why then do not all young males breed? The answer is simple: they cannot. For their own reproduction they need females, and with sex ratios of 1.6:1 females are a limited resource at both lakes. Under these conditions of high competition for females, it is mainly the young, unexperienced males who do not get a mate. Only three of 64 males of known age (4.7%) obtained a female in their first year. Hence, helping becomes the second best strategy (Fig. 17.8) and will be favored by selection over delaying. This answers the first functional question, why helpers help rather than breed on their own: they choose the best *available* strategy. If breeding becomes possible (e.g. by the occasional appearance of unmated females after the breeding season has started), then surplus males will immediately give up their helper status and become breeders. Such switching to the better strategy was observed three times, e.g. in 1982 by the primary helper 298 after the mate of female 296 had died (Fig. 17.2).

In Fig. 17.8 the inclusive fitness of primary and secondary helpers after two years is similar. Its composition, however, is different. This is shown in Table 17.2, where the inclusive fitness values from Fig. 17.8 have been split into two components: the direct fitness resulting from the rearing of their own young, and the indirect fitness resulting from the rearing of relatives other than their own young, e.g. siblings (Brown and Brown 1981).

During their first year, neither primary nor secondary helpers gain in direct fitness; they both have no young of their own. But primary helpers in their first year improve their indirect fitness by 0.45. The reason is that more young survive with the contribution of the primary helper than without it (Fig. 17.7) and, with $r = 0.32$, the relatedness between the helper and these additional young is fairly high. There is an additional indirect fitness gain in the second year, partly because some birds do serve as primary helpers for more than one year, partly because their help during the first year has a long-term effect on their parents' survival and future reproduction of siblings. In total, the primary helpers' indirect fitness gain after two years is 0.65.

For secondary helpers it is only 0.05, primarily because of the low relatedness between secondary helpers and the young they help to rear ($r \leq 0.05$). Because of this low probability that secondary helpers and young will share genes identical by descent, it would not pay the secondary helpers to invest as much as primary helpers and parents do. This probably explains why they guard less and feed fewer and smaller fish (Fig. 17.6).

But being a secondary helper offers other advantages: it markedly improves the chances of breeding in the following season, hence boosting the direct fitness component. When a male breeder dies, it is usually his former secondary helper who takes over the female (e.g. s 309 in 1982, Fig. 17.2). There even occur prolonged fights between male breeders and their secondary helpers with a 9% probability that the helper will displace the breeder. Overall, 91% of the surviving marked secondary helpers were mated in the next season ($n = 23$), 47.6% of them to the female they had helped the year before (for details see Table 3 of Reyer 1986). This, and a high survival rate of 74%, results in a high breeding probability. From these

Table 17.2. *Direct, indirect and inclusive fitness values for primary and secondary helpers of Pied Kingfishers during their first two years of life*

Status	Year	Gain in fitness		
		Direct	Indirect	Inclusive
First-year breeder	1	0.96	0	0.96
	2	0.80	0	0.80
Total		1.76	0	**1.76**
Primary helper	1	0	0.45	0.45
	2	0.42	0.20	0.62
Total		0.42	0.65	**1.09**
Secondary helper	1	0	0.04	0.04
	2	0.87	0.01	0.87
Total		0.87	0.05	**0.92**
Delayer	1	0	0	0
	2	0.30	0	0.30
Total		0.30	0	**0.30**

Bold figures are the overall sums.

data it follows that in his second year a secondary helper can expect 0.87 in direct fitness (Table 17.2). This is more than for all other male categories.

Delayers, who have an equally high survival rate (70%), gain only 0.30 because they have much lower mating chances (33%). Breeders, with high chances to be mated again (98%), gain 0.80 due to lower survival rates (57%), while primary helpers gain only 0.42 because of their low survival rates (54%) and medium probability of obtaining a mate (60%). The lower survival of primary helpers and breeders probably results from their higher feeding effort (Fig. 17.6). The poorer mating chances of primary helpers are due to the fact that, even if they do survive, they rarely take over the female that they have helped the year before. In many cases this would lead to a son mating with his mother, which may be prevented by incest avoidance.

Thus, in terms of direct fitness, being a primary helper is worse than being a secondary helper, whereas in terms of indirect fitness it is the other way round. Adding both components together results in similar inclusive fitness values for the two strategies (1.09 versus 0.92; Table 17.2 and Fig. 17.8). Why then do some surplus males become primary helpers to their parents, while others become secondary helpers to strangers?

The answer to this question is in the probability that a young male will attain a particular status. Remember that breeding is the best strategy but is only achieved by 4.7% of yearling males. Potential secondary helpers face a similar problem: not always can they become secondary helpers. Under certain ecological conditions, such as those prevailing at Lake Naivasha, helpers will not be tolerated by breeding males. Then, they may be forced into the inferior delayer strategy. For primary helpers, who experience much greater tolerance, the likelihood of being forced to become a delayer is much lower. It is therefore not surprising that unmated males prefer the primary over the secondary helper status. If at least one of his parents is still breeding, an unmated male will usually remain as a primary helper rather than become a secondary helper elsewhere. Thus, surplus birds choose the best *available* strategy, which – depending on conditions – for some is the primary, for others the secondary, helper strategy. There were only two exceptions to this rule. Both concerned breeding pairs which arrived in the colony with two primary helpers each. Only one helper of each pair remained, while the others left and became secondary helpers before their siblings hatched.

The above benefits accrue only to males. Females, as the dispersing sex, find themselves among unrelated birds so that raising young of others does not improve their indirect fitness. It would not increase their direct fitness either, because, being the limiting sex, females can always be sure to get a mate. With breeding being highly superior to helping, it is therefore not surprising that we have never witnessed a helping female.

Costs and benefits for breeders

With the above information in mind, we can also answer the third question as to why the two types of male helpers are treated differently, both within and between colonies. One only has to assume that potential helpers are accepted when the benefits they contribute to the breeders' fitness override the fitness costs, but are rejected when the reverse is true. Benefits derive from improved survival of present young through additional feeding; costs result from lowered chances of having young in the future as the helper may displace the breeder in the next season.

Because of the male surplus, these costs accrue only to male breeders. Therefore it is not surprising that female breeders are much more tolerant towards potential secondary helpers. With equal sex ratios at both lakes (1.6:1), the likelihood that a mated male will be displaced is probably similar at the two lakes. We can thus focus on differences in benefits in order to understand the cost/benefit ratios and the resulting behavior of male breeders.

At Lake Victoria, with its poor feeding conditions, the reproductive benefits of having helpers are high (Fig. 17.7). Therefore, both primary and secondary helpers are accepted. At Lake Naivasha, however, with its good feeding conditions, having helpers is not crucial for high reproductive success. Under such conditions, a male breeder should reject those helpers which feed little and threaten his status. These are the secondary helpers. Primary helpers, on the other hand, pose no such threat and can be tolerated even when the benefits they offer are low as at Lake Naivasha.

The same argument holds for the time at which helpers are accepted by breeders. Before the young hatch, helpers are of little value at either lake, because no feeding is required. Therefore, they should be rejected when the probability for competition is high (as in secondary helpers), but can be accepted when this probability is low (as in primary helpers). On a proximate level, these differential probabilities are reflected by differences in blood plasma levels of testosterone (Reyer *et al.* 1986). While average titers of potential secondary helpers (0.58 ng/ml) were found to be similar to those of mated males (0.51 ng/ml), those of primary helpers (0.17 ng/ml) were significantly lower ($P < 0.01$; Mann–Whitney U-test, two-tailed). As low titers were paralleled by small gonad sizes and no sperm production, primary helpers, in contrast to potential secondary helpers, may not be able to fertilize eggs and therefore do not threaten the mated male's paternity.

Thus, primary helpers are tolerated from the beginning of the breeding season because of the lower competitive threat that they pose to breeding males, while secondary helpers are only accepted when the need for help is high, as it is late in the season at Lake Victoria.

Proximate mechanisms of decision-making

Thus, Pied Kingfishers seem to choose the strategy which, under the prevailing ecological and demographic conditions, yields the highest inclusive fitness. But what are the proximate mechanisms for this decision-making? A bird cannot calculate inclusive fitness values – and even if he could, the information would be available to him only at the end of his life, and then it is too late to choose a better strategy. Therefore, we must look for proximate mechanisms which are (a) constantly available and (b) good predictors of inclusive fitness.

Decision mechanisms for young males

For young males, these mechanisms follow immediately from the foregoing discussion about the availability and profitability of the various strategies (Fig. 17.8). The first thing a male has to do at the beginning of a breeding season is to check whether or not unmated females are present. If yes, he should join her and breed, if not, he should try to become a helper. In the latter case the next question is whether or not his parent(s) are still alive. If yes, he should join them as a primary helper, if not, he should approach various other pairs as a potential secondary helper. The final decision is not one of the young male, but one of the breeders with whom he wants to associate. If they tolerate him, the potential helper turns into an actual helper, if not, he will be forced into the delayer strategy. In our study, this happened only to potential secondary helpers, but there may be conditions where primary helpers are not tolerated either. Then the rejected bird perhaps would try to become a secondary helper (? in Fig. 17.8). A behavioral program which tells the young male to follow this hierarchy of decisions and to react according to the actual situation will automatically lead to the best available strategy. This program is effective regardless of the ecological conditions at the colony.

Decision mechanisms for breeders

For breeders, the marked differences between Lakes Victoria and Naivasha in food availability and reproductive success suggested that their decision whether or not to accept secondary helpers could be based on need, i.e. on the food requirements of the young and the feeding capacities of the parents. To test this hypothesis, we first measured the daily energy expenditure (DEE) of feeding adult Pied Kingfishers with the doubly-labeled water technique (Reyer and Westerterp 1985). We then related the DEE of parents to their reproductive success, their behavior towards potential secondary helpers and the food requirements of their young. Finally, we manipulated clutch size in order to change these food requirements and hence the energetic stress of parents (Reyer and Westerterp 1985).

Energy expenditure of parents

As the daily energy expenditure of adults increases, the amount of food delivered to nestlings rises in a linear fashion at both lakes, but with significantly different slopes (Fig. 17.9). Thus, a Lake Victoria bird will achieve a lower feeding contribution than one from Lake Naivasha for the same amount of energy expended (see e.g. dashed line at $210 \, \text{kJ day}^{-1}$ in Fig. 17.9). The ecological reasons for this have been mentioned before: (1) less profitable fish and (2) poorer hunting conditions at Lake Victoria than at Lake Naivasha. The line at 210 kJ was not chosen arbitrarily. Pied Kingfishers expending less than $210 \, \text{kJ day}^{-1}$, on average maintained or even increased their body weight, those expending more than $210 \, \text{kJ day}^{-1}$ lost an average of $3 \, \text{g day}^{-1}$, which is 3.8% of the mean body weight of 78 g. Thus, 210 kJ seems to represent a physiologically determined energy threshold. Although a bird may exceed this threshold for a day or two, it apparently cannot maintain such a high performance for prolonged periods without a decline in body condition that is large enough to threaten survival. The threshold, 210 kJ, is about $4 \times \text{BMR}$ (basic metabolic rate), an upper limit which seems to hold for many more bird species. By combining this threshold with the regression lines of Fig. 17.9, one can predict the average feeding capacities for the two lakes: at Lake Victoria a parent Pied Kingfisher can bring a maximum of $102 \, \text{kJ day}^{-1}$, at Lake Naivasha he can deliver as much as 267 kJ, or 2.6 times more.

Growth and survival of young

The effect of these feeding capacities on nestling development can be deduced from Fig. 17.10, which plots the daily weight change per young against the amount of food he receives. The resulting solid regression line intersects the line of constant weight at about 90 kJ. Provided with that much energy, a nestling will neither gain nor lose weight. At Lake Victoria with a maximum food contribution of 102 kJ through one parent and with an average clutch size of 4.6 at hatching, each nestling will receive an average of 44 kJ day^{-1} ($= 2 \times 102/4.6$; black arrow on *x*-axes) if parents try to raise their offspring without helpers. This is insufficient and would lead to an average daily body mass loss of 5.6 g (black arrow on *y*-axes) if it were not for competition between nestlings; some will get enough and survive at the expense of others. At Lake Naivasha, with a maximum food contribution of 267 kJ per parent, each nestling will receive an average of 111 kJ day^{-1}, which is sufficient for all of them to add weight (white arrows), even if their parents have no helpers. These energy calculations support the data on reproductive success in Fig. 17.7.

The information for the parents on whether or not their young have enough food very likely comes from the begging of the nestlings, which has a clear influence on the adults' feeding patterns. Parents resting in the

Fig. 17.9. Amount of food delivered to Pied Kingfisher nestlings (kJ adult^{-1} day^{-1}) in relation to daily energy expenditure of feeding adults (kJ day^{-1}). ●——●, Lake Victoria; ○——○, Lake Naivasha; ·····, upper limit of energy expenditure and resulting feeding capacities, respectively.

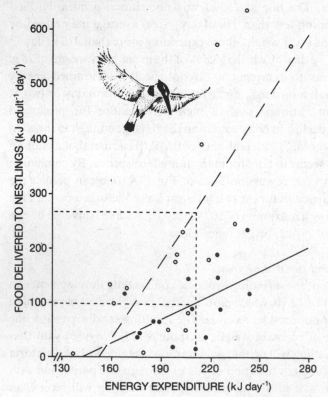

colony were regularly observed approaching the nest entrance and either resuming rest when the begging response was soft or immediately flying to the lake when it was intense. Also, begging duration increases with decreasing food supply (Fig. 17.10). Consequently, nestlings at Lake Victoria can be expected to beg more than nestlings at Lake Naivasha.

Manipulation experiments

According to these results, the different demands of the nestlings (as communicated in their begging) plus the different energetic stress on parents (Fig. 17.9) could indeed provide the proximate mechanisms responsible for the different treatment of secondary helpers at Lake Victoria and Lake Naivasha. In order to test this hypothesis further, energetic stress and begging duration in the two colonies were reversed through manipulation of clutch size and by comparing the treatment of helpers under normal and manipulated conditions (Reyer and Westerterp 1985). Under normal conditions, with a clutch size of four to six nestlings,

Fig. 17.10. Body mass change of Pied Kingfisher nestlings (●———●) and begging duration (+-----+) in relation to the amount of food received. Arrows show the average amount of food per day and the average daily change in body mass for a nestling at Lake Victoria (filled arrows) and Lake Naivasha (open arrows).

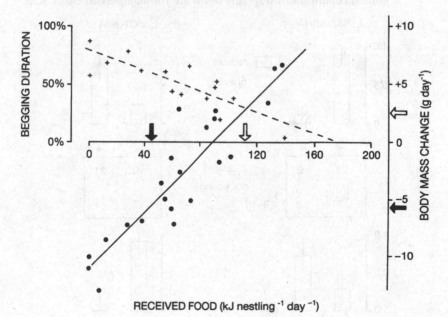

RECEIVED FOOD (kJ nestling $^{-1}$ day $^{-1}$)

each parent in a pair without helpers has to care for two to three young. Under these conditions, the vast majority of the breeders rejected potential secondary helpers at Lake Naivasha, whereas at Lake Victoria all accepted them (Fig. 17.11; $\chi^2 = 48.398$, $P < 0.001$). Some clutches at Lake Naivasha were then experimentally increased to eight to ten young, or four to five young per parent. According to the energy measurements, this put the parents into a position similar to that of Lake Victoria birds: they could no longer provide enough food. And indeed, in 10 such experiments eight pairs accepted secondary helpers (Fig. 17.11). This differs significantly from normal conditions at this lake ($P < 0.001$; Fisher's Exact probability test, one-tailed). The reverse experiment was equally conclusive. When at Lake Victoria clutch size was reduced to one to two nestlings, so that each parent had to care for no more than one, potential secondary helpers were not

Fig. 17.11. Treatment of potential Pied Kingfisher secondary helpers by breeders under normal conditions (four to six young) and when clutch size was increased to eight to ten young or reduced to one to two young. Lake Naivasha left, Lake Victoria right. The 2 × 2 contingency tables in the center of the graph give the number of pairs which rejected (−) or accepted (+) secondary helpers. The histograms show the proportion of encounters in which male breeders either attacked (stippled) or greeted (open) potential secondary helpers when meeting them near the nest. Histograms above the tables are for normal conditions; histograms below are for manipulated clutch sizes.

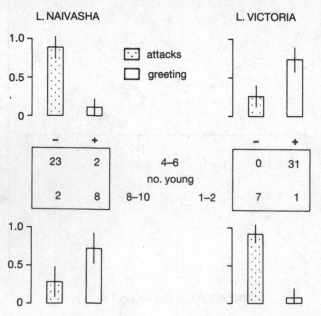

allowed in all but one case; again significantly different from normal conditions ($P < 0.001$).

It could be argued that this result is not very surprising, because breeders spending more time in getting food for more young, of course, have less time to chase away potential secondary helpers, which then may sneak in undetected. Such an explanation is ruled out by looking at the actual encounters between male breeders and potential secondary helpers shown in Fig. 17.11. When clutch size was normal, attacks prevailed at Lake Naivasha, and greeting prevailed at Lake Victoria. When clutch size was increased, Naivasha birds switched from attacking helpers to greeting them, whereas Victoria birds switched from greeting to attacking when clutch size was reduced. Thus, the tolerance towards helpers is, indeed, determined by the demands of young and the parents' energetic abilities, which in turn depend on the ecological conditions, particularly the type of food and its availability.

The basis for the cooperative breeding system

From the foregoing scenario, two features emerge as the most crucial components for the Pied Kingfishers' cooperative breeding system.

(1) Under *severe ecological conditions* breeders alone cannot success-fully raise all their young.

(2) A *strongly biased sex ratio* among adults leads to a surplus of unmated males.

Condition (1) creates the need for helpers, while condition (2) creates a supply of potential helpers. Breeding in colonies allows the potential helpers to choose among various breeding pairs, and allows the needy breeders to choose among various helpers, including related (= primary) and unrelated (= secondary) ones. What, however, is the origin of the parents' limitation and the reason for the biased sex ratio?

(1) The feeding capacities of parents seem to be limited by physiological mechanisms constraining the maximum daily energy expenditure to about $4 \times BMR$, a threshold which has also been found in other birds (Drent and Daan 1980). Recruiting helpers when food conditions are poor is one way to guarantee sufficient feeding rates despite such parental limitation. Apparently, this is true not only for Pied Kingfishers. Using published data on 15 species of other cooperative breeders, I found a significant positive correlation between hourly feeding rates of unassisted pairs and the percentage increase in feeding rates when helpers are present ($r = 0.820$, $P < 0.001$). In other words, in species with low parental feeding rates (i.e.

< 10 trips h^{-1}) helpers did not increase the overall food contribution to the young; they only reduced the parents' contribution. But in species with high (> 20 trips h^{-1}) and in some species with medium (10–20 trips h^{-1}) parental feeding rates the overall contribution went up and parents did not reduce their effort when helpers were present. This suggests that in harder working species parental feeding capacities are limited and helpers are needed (recruited?) for better food supply (Reyer, unpublished results).

An alternative answer to parental feeding limitations would be to select better areas, where one can have a higher reproductive success without risking competition from helpers. So, why do Lake Victoria Pied Kingfishers not move to places such as Lake Naivasha? The main limiting resource seems to be suitable nesting sites (= sandy banks) close to the fishing grounds. This is indicated by the changing availability of breeding sites at Lake Naivasha and by the birds' readiness to accept man-made features (p. 530), which is also known from other areas (Douthwaite 1970). Even where suitable nesting sites *appear* to be superabundant, they may not be, due to the properties of the soil. If the soil is too loose, burrows may collapse; if it is too hard digging may become difficult or even impossible, as in 1980 when at Lake Victoria the rains failed. Out of 27 pairs which congregated in the colony in that year, only three found places soft enough for excavating nesting-holes. Also, the tunnel and the nest chamber must be constructed and oriented in such a way that the young are buffered against stressful temperature fluctuations and are protected from a harmful build-up of CO_2 and NH_3 and a diminution of O_2 caused by the metabolism of the birds and of microorganisms living on the accumulating feces and pellets (White *et al.* 1978).

There are probably few places which fulfil all these breeding site requirements and, in addition, offer good feeding conditions. Because of such resource limitations, Pied Kingfishers will not only be forced to breed colonially – sometimes even in unsafe areas where high predation pressure and human disturbances lead to a very poor reproductive success (Douthwaite 1978). Many of them will also be driven to places such as Lake Victoria, where obtaining food is time and energy consuming and reproductive success is low, unless helpers are recruited.

(2) A biased sex ratio among adults with a surplus of unmated males exists in most cooperative breeders and has often been invoked as an important component for the evolution of cooperative breeding (Brown 1978; Emlen 1984). Alternatively, some authors have argued that the biased sex ratio may be a result rather than a source of cooperative breeding (Koenig and Pitelka 1981). Basically, the first hypothesis assumes that the

bias is produced by the parents and, therefore, already present at the end of parental care, while the second assumes that the bias occurs after parental care has terminated and is caused by differential dispersal and mortality resulting from the differing breeding options for males and females.

The first idea is exemplified by the 'repayment model' proposed by Emlen *et al.* (1986) and extended by Lessels and Avery (1987). If grown offspring of one sex remain with their parents and help them to rear successive young, they 'repay' part of the costs of their production. Consequently, they become the 'cheaper' sex. With parents investing equally in the production of sons and daughters (Fisher 1930), the evolutionarily stable sex ratio will deviate from 1:1, because selection will favor females who produce more offspring of the helping than of the non-helping sex.

For testing this hypothesis, it is essential to know the sex ratio among offspring at the termination of parental care (TPC), which in Pied Kingfishers is three to four weeks after fledging. Unfortunately, the young follow their parents to the lake within a few days after leaving the nest and I am thus unable to compile these data. Yet, there are two reasons why the repayment model is unlikely to apply to Pied Kingfishers.

(*a*) Any 'intended' shift from a 1:1 sex ratio among nestlings to an excess of males could only be produced by selectively starving females as the more expensive sex. (More expensive because they do not repay through helping and not because they require more food; there is no sex difference in fledgling size.) The benefits of such selective starvation should increase with increasing stress on parents. Consequently, the bias in sex ratios should occur before the parents engage in maximum effort (Clutton-Brock *et al.* 1985), which is around the end of the second week after hatching. My data, however, give absolutely no evidence to support these predictions. The 1:1 ratio is still present in nestlings of three weeks and older.

(b) Although the 1:1 ratio among nestlings may shift *slightly* toward a surplus of males until the end of parental care, it is unlikely that a few weeks will produce a *marked* bias toward males. Thus, assuming that the TPC and nestling sex ratios are the same, I fitted my data into equations (12) and (16) of Lessels and Avery (1987). The resulting predictions for the evolutionary stable male:female ratios were very different for the two lakes, namely 1.44:1 for Lake Victoria and 0.97:1 for Lake Naivasha. Yet, neither the nestling nor the adult sex ratio differed significantly between the two colonies.

Therefore, the surplus of males in adult Pied Kingfishers cannot be satisfactorily explained through an overproduction of a cheaper and more valuable sex. Differential mortality after the termination of parental care seems to offer a better explanation for the shift from a 1:1 ratio among nestlings to the 1.6:1 ratio among adults. In Pied Kingfishers there are two sources for higher female than male mortality.

(1) Juvenile females disperse while juvenile males remain near and return to their natal colony (p. 531). Although quantitative data are lacking because of the difficulty of finding dispersing birds, it seems likely that juvenile females on their way into unknown areas suffer higher mortality than juvenile males remaining at home.

(2) Female breeders take a bigger share in incubating the eggs and brooding the young than do male breeders; consequently females are more endangered by nest predators or when nesting-holes cave in or are flooded. This results in different male and female mortality rates among adults (p. 531).

The notion that higher female than male mortality creates a surplus of males, who are potential helpers, does not automatically mean that cooperative breeding originates from differential mortality. It also could be the other way round: if cooperative breeding (or any other behavioral trait) makes it worthwhile for females to disperse and take the greater share in reproduction, then differential mortality would be a result rather than a cause of helping (Koenig and Pitelka 1981). In both cases, however, the question arises: why do young females take the greater risk?

The reasons for dispersal in general and sexual differences in particular are still poorly understood. According to Greenwood (1980) the following link between dispersal patterns and mating systems exists: if males invest little in the offspring, and if the acquisition of females as the limiting resource, mainly depends on the distribution of the females themselves, then males can enhance their reproductive success by moving around in search of mates. This is the typical pattern in mammals, in which males usually disperse more than females. If, however, the acquisition of females requires the defense of limited resources such as feeding territories or breeding sites, and if such defense is facilitated through familiarity with the locality, then philopatry of males should result. This is the typical pattern in birds, including Pied Kingfishers. At the beginning of the breeding season, male-biased groups of three to nine Pied Kingfishers perform noisy aerial chases and ground displays which seem to attract females to a suitable breeding site where pair formation takes place (Douthwaite 1970). Suitable breeding sites, however, are limited at least in some areas (see above). Those males who have found one stick to it very closely year after year. At Lake Victoria, 83% of the helpers nested within a few meters of the place where they had helped the year before ($n = 30$) although the colony stretched over more than 500 m. In eight cases concerning breeders, the male and female of a newly formed pair had bred the previous year more than 50 m apart. In seven of these cases the female joined the male in his previous site and in one case the reverse was true. This differs from an equal probability of site change ($P = 0.035$, binomial test, one-tailed). Thus, for males but not for females, there seems to exist a 'home advantage' effect, which favours philopatry of male Pied Kingfishers.

This, however, does not explain why females disperse in Pied Kingfishers and most other bird species. Greenwood (1980) suggests that dispersal has evolved as a mechanism against inbreeding depressions. We do not know whether or not such depressions would occur in Pied Kingfishers, but inbreeding is definitely avoided. Even when their fathers die, primary helpers do not mate with their widowed mothers (Reyer 1986). The costs of such possible fitness depressions, which have to be weighed against the costs of dispersal, would affect both sexes equally. The costs of dispersal, however, would be higher for males, who would lose the benefits they derive from philopatry. Consequently, 'which sex disperses may be the outcome of a conflict between the sexes, where the relative costs and benefits of dispersal and philopatry to the sexes determine the outcome' (Greenwood 1980, p. 1155).

Similarly, the answer to the question 'who takes the bigger share and risk in reproductive effort' will also depend on the relative costs and benefits to the sexes. Although reproduction, through its effects on mortality, imposes higher costs on female than on male Pied Kingfishers, these may be offset by higher benefits for females. Females breed from their first year on, whereas young males usually have to go through the less profitable helper and delayer strategies before they can reproduce. Overall, the fitness gain for the two sexes may turn out to be the same, despite the marked differences in dispersal and reproductive patterns. This, however, cannot be proved verbally, but requires some modelling.

Although such models can test whether or not the observed sex ratios are adaptive under the present system, they will not enable us to say that the entire system is more adaptive than others. The pattern of female-biased dispersal, incubation and mortality and of male-biased sex ratios is so widespread among birds that it will be very difficult to circumvent the 'phylogenetic inertia' argument, i.e. that the pattern, the function of which we do not yet know, may have nothing to do with cooperative breeding but be merely something intrinsic to birds.

I am grateful to the many helpers in this project, all of them unrelated and yet of 'primary' importance. My particular thanks go: to W. Wickler for giving me the opportunity to work in Kenya and for his continued interest in and support of the study; to D. Schmidl for his untiring and competent help during many years in the field and the Institute; to K. Westerterp, J. Dittami and M. Hall, who cooperated with me on energy expenditure and hormone titers; and – last but not least – to my wife, H. Reyer, for encouraging my work and patiently tolerating my repeated absence from home. The research was financed by the Max Planck-Gesellschaft with additional funds from the DFG (Re 553/1-1). E. K. Ruchiami (Office of the President) and M. L. Modha (Wildlife Conservation and Management Department) issued and extended my research permit (OP. 13/001/C1891/14). Permission to live and work on their properties was kindly given by C. Clause, J. Geoffrey, M. and S. Higgins, E. Sketch and J. Ndolo. Finally, I am grateful to my colleagues in Seewiesen for discussing with me various stages of the study and to W. Koenig and P. Stacey for their helpful comments on an earlier version of this article.

Bibliography

Brown, J. L. (1978). Avian communal breeding systems. *Ann. Rev. Ecol. Syst.* **9**, 123–55.

Brown, J. L. and Brown, E. R. (1981). Kin selection and individual selection in babblers. In *Natural Selection and Social Behavior: Recent Results and New Theory,* ed. R. D. Alexander and D. Tinkle, pp. 244–56. Chiron Press: New York.

Clutton-Brock, T. H., Albon, S. D. and Guinness, F. E. (1985). Parental investment and sex differences in juvenile mortality in birds and mammals. *Nature (Lond.)* **313**: 131–3.

Douthwaite, R. J. (1970). Some aspects of the biology of the Lesser Pied Kingfisher *Ceryle rudis* Linn. Ph.D. thesis, Makerere College, University of East Africa.

Douthwaite, R. J. (1978). Breeding biology of the Pied Kingfisher on Lake Victoria. *J. E. Afr. Nat. Hist. Mus.* **31**: 1–12.

Drent, R. and Daan, S. (1980). The prudent parent: energetic adjustments in avian breeding. *Ardea* **68**: 225–52.

Emlen, S. T. (1984). Cooperative breeding in birds and mammals. In *Behavioral Ecology. An evolutionary Approach*, 2nd edn, ed. J. R. Krebs and N. B. Davies, pp. 305–39. Blackwell: Oxford.

Emlen, S. T., Emlen, J. M. and Levin, S. A. (1986). Sex-ratio selection in species with helpers-at-the-nest. *Amer. Nat.* **127**: 1–8.

Fisher, R. A. (1930). *The Genetical Theory of Natural Selection*, 2nd edn. Clarendon, Oxford.

Greenwood, P. J. (1980). Mating systems, philopatry and dispersal in birds and mammals. *Anim. Behav.* **28**: 1140–62.

Koenig, W. D. and Pitelka, F. A. (1981). Ecological factors and kin selection in the evolution of cooperative breeding in birds. In *Natural Selection and Social Behavior: Recent Results and New Theory*, ed. R. D. Alexander and D. Tinkle, pp. 261–80. Chiron Press: New York.

Lessels, C. M. and Avery, M. I. (1987). Sex-ratio selection in species with helpers at the nest: some extensions of the repayment model. *Amer. Nat.* **129**: 610–20.

Reyer, H. U. (1980). Flexible helper structure as an ecological adaptation in the Pied Kingfisher, *Ceryle rudis. Behav. Ecol. Sociobiol.* **6**: 219–27.

Reyer, H. U. (1982). Nutzen und Kosten bei helfenden und brütenden Graufischern. *Naturw. Rundschau* **35**: 31–2.

Reyer, H. U. (1984*a*). The adaptive significance of cooperative breeding in the Pied Kingfisher (Japanese). *Anima* **6**: 50–5.

Reyer, H. U. (1984*b*). Investment and relatedness: a cost/benefit analysis of breeding and helping in the Pied Kingfisher (*Ceryle rudis*). *Anim. Behav.* **32**: 1163–78.

Reyer, H. U. (1985). Brutpflegehelfer beim Graufischer. In *Verhaltensbiologie*, 2. *Auflage*, ed. D. Francke, pp. 277–82. G. Thieme Verlag: Stuttgart.

Reyer, H. U. (1986). Breeder-helper-interactions in the Pied Kingfisher reflect the costs and benefits of cooperative breeding. *Behavior* **96**: 278–303.

Reyer, H.-U. (1988). Okologie und Evolution kooperativer Jungenaufzucht bei Vögeln. *Verh. Dtsch. Zool. Ges.* **81**: 169–82.

Reyer, H. U., Dittami, J. P., & Hall, M. R. (1986). Avian helpers at the nest: Are they psychologically castrated? *Ethology* **71**: 216–28.

Reyer, H. U. and Dunn, E. K. (1985). *Ceryle rudis*, Pied Kingfisher – voice. In *Handbook of the Birds of Europe, the Middle East and North Africa*, ed. S. Cramp, pp. 728–9. Oxford University Press: Oxford.

Reyer, H.-U., Migongo-Bake, W. & Schmidt, L. (1988). Field studies and experiments on distribution and foraging of pied and malachite kingfishers at Lake Nakuru (Kenya). *J. Anim. Ecol.* **57**: 595–610.

Reyer, H. U. and Westerterp, K. (1985). Parental energy expenditure: a proximate cause of helper recruitment in the Pied Kingfisher (*Ceryle rudis*). *Behav. Ecol. Sociobiol.* **17**: 363–9.

Vehrencamp, S. L. (1978). The adaptive significance of communal nesting in Groove-billed Anis (*Crotophaga sulcirostris*). *Behav. Ecol. Sociobiol.* **4**: 1–33.

White, F. N., Bartholomew, G. A. and Kinney, J. L. (1978). Physiological and ecological correlates of tunnel nesting in the European bee-eater, *Merops apiaster*. *Physiol. Zool.* **51**: 140–54.

18 NOISY MINERS: VARIATIONS ON THE THEME OF COMMUNALITY

18

Noisy Miners: variations on the theme of communality

D. D. DOW* AND M. J. WHITMORE*

Study species

The Noisy Miner, *Manorina melanocephala*, is one of the larger members of the Meliphagidae, the honeyeaters, which comprise about 160 species endemic to Australia, New Zealand, and the south-west Pacific region (Walters 1980). Of about 66 Australian meliphagids (Schodde *et al.* 1978), four are in the genus *Manorina*. The Yellow-throated Miner, *M. flavigula*, is most widespread and typical of the country's dry interior. Its range extends to the west coast. The Bell Miner, *M. melanophrys*, is found east of the Great Dividing Range, where it inhabits eucalypt forest and woodland with heavy undergrowth. In the south-east, it may live in gardens. The endangered Black-eared Miner, *M. melanotis*, is restricted to remnants of the mallee districts of Victoria, South Australia and New South Wales. The Noisy Miner ranges from tropical Queensland (16° S) over 3000 km south into temperate Tasmania (43° S), and from the coastal lowlands west across the Great Dividing Range, penetrating inland in places as much as 800 km. Its traditional habitat was probably dry sclerophyll forest and woodland but it has also successfully colonized, and now flourishes in, suburban gardens and city parks.

The Noisy Miner, averaging 60 g, is strikingly marked and not easily confused with any other species. Black crown and cheeks contrast sharply with a patch of naked yellow skin behind the eye. The rest of the plumage is mainly gray, save for a wash of yellow on the lateral edges of some retrices, remiges and wing coverts. White edges or tips evident on new primaries and retrices wear off quickly. The feet are strong and the claws sharp. There is a recognizable juvenile plumage (Dow 1973) but no obvious sexual

* Present address: The University of Michigan Biological Station, Pellston, MI 49769, U.S.A.

dichromatism or dimorphism. Males on average tend to be larger and their markings slightly more intense than the females', although generally these differences are apparent only in hand-held birds.

Noisy Miners are one of the most common species in eastern Australia, yet Dow (1970) was the first to discover that they breed cooperatively. Subsequent studies indicate that Noisy Miners are obligate rather than opportunistic communal breeders; i.e. populations show no tendency to vacillate between cooperative breeding and pair breeding from one year to the next. The few records of breeding by pairs are based mostly on observations of unmarked birds at a time when cooperative breeding was not a well-known phenomenon.

We studied Noisy Miners intensively at two sites during a long-term project in the state of Queensland. One site (1971–3) consisted of 7 ha of a much larger dry sclerophyll woodland near Laidley, 72 km west of Brisbane. Average annual rainfall is 825 mm and temperatures range from a mean minimum of 5.5 °C in July to a mean maximum of 30.9 °C in January. The vegetation is dominated by several species of *Eucalyptus*. Ground cover consists primarily of sedges and grasses, with exposed soil covered by leaf and bark litter. There is no well-developed subcanopy or shrub layer. Woody vegetation in the first 4 m from the ground is very sparse, consisting of saplings, shrubs and vines. Most trees have few lower branches, the foliage beginning about half way up. The canopy is open and averages about 15 m in height, with a few emergent trees.

The other study site (1979–81) included 240 ha of a 949 ha pastoral property near Meandarra, 310 km west of Brisbane (Fig. 18.1). Much of the land in the region had been cleared for grazing by sheep and cattle and only remnants of the original forest remained, usually in strips or scattered stands. It is in these remnants that Noisy Miners live. Brigalow, *Acacia harpophylla*, is the most common species of tree, although belah, *Casuarina cristata*, and poplar box, *Eucalyptus populnea*, are abundant in some places. Small trees and shrubs are found throughout the area. Ground cover consists mostly of grasses and herbs.

Meandarra is in a transitional zone between semiarid and sub-humid regions (Kalma 1974). The area receives an average annual rainfall of 585 mm, with most falling from December to March. However, Meandarra is known locally as occupying a dry belt with highly unpredictable rainfall. Droughts are common, possibly cyclical, and sometimes severe. The region was drought-stricken during part of the study period. Temperatures ranged from −3 °C to +47.5 °C and commonly varied by more that 20 deg. C within a 24 h period.

Fig. 18.1. Noisy Miners at the Meandarra site occupied remnants of the original vegetation, which had been left as shade strips or small patches. The dominant tree was brigalow, an acacia. These tended to be of fairly uniform size with only an occasional emergent. (*a*) The ground cover after good rains near the end of 1981. (*b*) Conditions prevailing during a drought. The picture was taken from the same place almost exactly four years earlier.

Generally, the climate at Laidley is more salubrious and predictable than at Meandarra. Trees at Laidley are larger and more evenly distributed than those at Meandarra, producing a higher and more open canopy. Woodland at Laidley is extensive; trees at Meandarra are smaller and found mostly in scattered, dense patches.

Although a honeyeater and possessing a brush-tipped tongue, a specialization for feeding on nectar, the Noisy Miner is an opportunistic feeder that takes a large proportion of its food in insects and other invertebrates gleaned from the foliage, flushed from the ground or hawked in the air. It frequently visits blossoms when these are available, presumably to take nectar, and may pollinate some species of trees (Paton and Ford 1977). It feeds on the white, waxy secretions of psyllids when these are abundant on eucalypt and acacia leaves. We have frequently seen miners capture and eat large caterpillars, moths, spiders, mantids and, once, taking a small tree frog.

Spacing system

Subtle differences in social organization and spacing of Miners may exist under different environmental conditions, particularly where man has altered the composition or configuration of native vegetation. Differences may also appear among populations of different densities, as we discovered by studying Miners at two localities.

Miners in our study areas did not live in small, discrete groups as do many cooperatively breeding species. Instead, they occurred in large year-round assemblages, which we have termed colonies (Dow 1979a), and obtained all their food there. Colonies occupied discrete areas. Some near our study areas were separated from others by expanses of vegetation apparently similar to that of occupied habitat, but virtually devoid of Miners. Other colonies were separated by barriers such as lakes, cultivated land or cleared suburban areas. Colonies can contain up to 500 individuals. At Laidley, the density varied from 7.7 to 9.0 birds ha^{-1}, while at Meandarra corresponding figures were 0.65 to 1.15.

Male Miners do not defend individual territories. They generally move in coalitions (temporary flocks) of two or more birds, usually about five to eight (Dow 1979b). Coalitions normally contain birds involved in the same activity: feeding, resting, roosting, bathing, chasing. The largest coalitions (40 to 50 birds) are found mobbing predators or other animals. The membership of coalitions is temporary and changes frequently with time and place because individual birds spend most time in relatively small activity spaces. Activity spaces are not equivalent to 'home ranges' (as commonly understood), for birds occasionally leave them for short periods. Rather, they are small areas with a high probability of occupancy by the same bird. At Laidley, activity spaces of males ranged in diameter from 45 to 212 m and averaged 129 m ($n = 20$) during the non-breeding season. Activity spaces were slightly smaller (average 114 m; $n = 17$) in the breeding period.

Males whose activity spaces are clustered in a particular area of the colony tend to associate with one another. These associations we termed 'coteries', after the analagous organization found by King (1955) among prairie dogs, *Cynomys ludovicianus*. At times, members of a coterie may be found in different coalitions, but these members regroup to show site-specific aggression to birds from adjacent or other coteries. Members of coteries readily leave their usual activity spaces to attack members of other coteries at what could be considered fairly loose boundaries. Behavior at this level could be considered territorial. In effect the social unit of greatest stability within the colony is the coterie, because there is little exchange of individuals. Coalitions, on the other hand, although likely to show some agonistic behavior when encountering other coalitions, have considerable freedom of movement and exchange of members within the area occupied by the coterie. The organization of male Noisy Miners is summarized schematically in Fig. 18.2. Note that aggression can be manifested at three levels: inter-individual, inter-coalition and inter-coterie.

Females occupy year-round activity spaces smaller than those of males. At Laidley, the diameter of their activity spaces ranged from 52 to 95 m and averaged 74 m ($n = 14$). The activity spaces of females rarely overlap. However, females show little agonistic behavior to other females and apparently maintain their almost exclusive female–female spacing by mutual avoidance. Thus, the spacing of females within the colony differs radically from the spacing of males.

Females may join coalitions that form near their activity spaces. Some females may occupy a portion of one coterie area, while others establish activity spaces at the boundary of adjacent coterie areas; their spaces therefore occupy a portion of two or more coterie areas.

Both males and females occupy the same parts of the colony from one year to the next, particularly after their first year of life. Although the shapes and sizes of activity spaces vary slightly from year to year, the dispersion of birds with respect to their neighbors changes little, suggesting a great potential for social stability.

The Noisy Miners' curious spacing behavior, involving as it does activity spaces rather than individual territories, leads to a paradox: the species with the highest reported number of auxiliaries (22) at nests has no discrete or even stable social group attending those nests. Aside from interactions between members of neighboring coteries, the frequency of interactions among birds within a colony probably depends largely on the size of the area it occupies and the density of the population that inhabits it. In large colonies such as the one at Laidley (approximately 400 birds), it is likely that some individuals would rarely if ever encounter one another.

Miners display remarkably unrestrained aggression toward members of their own and other species. While coterie areas are not always defended against intrusion by males from nearby coteries, aggression against strange miners from other colonies is collective and may be shared by members of different coteries. Birds of other species are virtually always attacked and sometimes killed (Dow 1977). Because miners are so successful at keeping other species out of their colony area, they have apparently been able to expand their feeding niche (Dow 1977). At Laidley they obtained food in seven different feeding zones, viz. tree trunks, limbs, branches, twigs, foliage, ground and air. The time spent in these zones differed in woodland

Fig. 18.2. A schematic summary of the social organization of male Noisy Miners. *Colony*: membership and location stable year round (dispersal and immigration possible. *Coterie*: greatest stability in non-breeding periods. When females are nesting, boundaries are more vague and individual males may penetrate further into areas occupied by other coteries. *Coalition*: transitory flocks largely unpredictable in membership, location and time. The coalition labeled 'M' represents the common observation of a mobbing coalition with members from more than one coterie. (From Dow 1979*b*.)

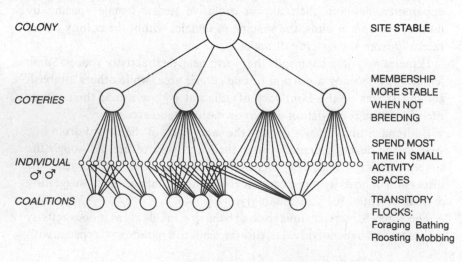

with dense understory versus more open ground. In the more dense woodland, more species of birds were present because they could escape the miners by retreating into the shrub layer.

The spatial organization of miners clearly differs from most cooperative breeders. Rather than living exclusively in small, social, territorial groups, birds may share in defending their part of a much larger area, the colony. Coteries fragment frequently into coalitions that may roost, forage, or rest together in the coterie area, but the membership of such coalitions is not always predictable.

Social system

We never found Noisy Miners as simple pairs in any of our study areas. The most stable social group below the level of the entire colony is the coterie, although this is not necessarily a closed system. A coterie, which usually contains in its area more than one female, may lose its social integrity when members of two or more coteries become mutually involved in caring for the young at the same nest. In areas of low density or highly fragmented vegetation, birds in a colony may behave as if belonging to a single coterie. Where habitat is continuous and more extensive, contiguous coteries form a colony. Although males from one coterie may visit a nesting female in an adjacent coterie, they are rarely found more than one coterie away from their own activity spaces.

After his first year of life, a male tends to dominate other males in only part of his activity space. In other parts, he is subordinate to other males. At high population densities (Laidley), obvious hierarchies do not emerge, probably because of the extensive overlap of activity spaces. At low population densities (Meandarra), site-specific hierarchies among males could usually be defined clearly, and reversals (switches in identity of dominant and subordinate birds) were very rare. It is important to understand that the positions or rankings of birds in a hierarchy depended on the location of the birds when they interacted. Thus, hierarchies could change dramatically from one location to the next.

Noisy Miners are among the most aggressive honeyeaters. Attacks occur commonly among adult males, but young birds are not exempt. We watched an adult male from a neighboring coterie at Meandarra kill a nestling. We suspected that infanticide occurred at three other nests (Whitmore 1986). Often, adult males begin to attack juveniles when the latter about 11 weeks old. Attackers may include males who have contributed most to the earlier feeding of the young birds. Attacks may coincide with the development of subsong, but we have no data to suggest that they are a consequence of it. Adult females are less aggressive toward juveniles, although mothers occasionally attack their offspring.

The development of agonistic behavior begins early in life. We frequently observed it among nestlings. Agonistic interactions may intensify after young fledge, resulting sometimes in the killing of a broodmate (two observations of 11-week old birds at Meandarra). At Meandarra, young birds that were socially subordinate early in life either died or disappeared from the colony before their eighth month.

Among newly fledged birds, the sex ratio was 1:1 at Laidley. Among adults in breeding populations, males outnumbered females (e.g. 3.3:1 at Laidley in June, 4.7:1 at Meandarra in July, 3.5:1 near Meandarra in November). It is difficult in this species to compare sex ratios of breeders and 'helpers' because there is not a class of helpers distinguishable from breeders. Often, in fact, a male breeder at one nest is simultaneously a helper at another. For example, two nests active at the same time at Laidley, each with 14 banded male visitors, had eight of these in common. Of these eight males, four visited yet other nests during the same period (see Dow 1979a).

The biased sex ratio seems to arise in two ways. Data from observations of 30 young birds at Meandarra suggest that aggression drives young females from their natal areas more frequently than it does young males. It is reasonable to assume that the females suffer greater mortality than do the young males that remain within the confines of the natal colony. Secondly, a dispersing male would have a greater chance of entering a new colony than would a female. This is because the spacing behavior of the sexes differs. Females are intolerant of other females. Their activity spaces tend not to overlap, so fewer can be accommodated in the colony. Males, on the other hand, customarily share overlapping activity spaces with other males and females. A new male will undoubtedly be harassed, but, if persistent, can remain. We have few data on dispersal of males, but think that emigration may not occur until the density within a colony reaches some critical level.

Breeding biology

Noisy Miners have an open system (*sensu* Emlen 1978) of cooperative breeding, rivalled only, it seems, by that of the White-fronted Bee-eater, *Merops bullockoides* (see Chapter 16). The closely knit family group, so often a trademark of cooperative breeding, is not readily apparent in miners. Some auxiliaries are not closely related to the breeders whose young they provision, and the allegiance of auxiliaries to particular nests or particular females is not always well developed. For example, at Laidley in a single breeding period, the most active male visited 14 nests. On average, a high-ranking male (defined as the male that visited a particular nest more than did other male attendants) attended the nests of 5.1 ($\pm$ 2.89) females ($n = 11$). Clearly, these high-ranking males could not be closely related to all females whose young they fed. At Meandarra, an 'average' male visited 1.29 ($\pm$ 0.61) nests in 1980 ($n = 14$) and 4.22 ($\pm$ 2.63) nests in 1981 ($n = 18$). An 'average' male visited the nests of 1.29 ($\pm$ 0.61) females in 1980 (no re-nesting occurred) and 2.72 ($\pm$ 1.53) females in 1981. Thus, the general lack of allegiance to any one breeding female and the pattern of attending many nests were apparent in both populations that we studied.

Miners have been recorded nesting in all months of the year, although most nests are built between July and November. The onset of nesting is not clearly related to rainfall, temperature, or other environmental factors. At Laidley, the initiation of nesting in three consecutive years differed by as much as 110 days. The duration of nesting there was 175, 115 and 190 days. At Meandarra, the onset of nesting in two years (1980 and 1981) differed by only 17 days, while the duration varied from 73 to 113 days. The degree of synchrony in nesting by females at both study areas varied considerably from year to year.

The number of nests begun also varies. In three years (1971–3) at Laidley, nests in a 7 ha study plot on the edge of the colony numbered 40, 53 and 18. Females may attempt to raise as many as four broods in one year. When all eggs in a nest are lost, the female generally builds a new nest, usually within 9 to 19 days. However, there is marked annual variation in the propensity to re-nest. In some years some females make no attempt whatever to breed, contributing to the variation in the numbers of nests found in an area.

One-year-old males and females may be physiologically capable of breeding. Regardless, social and environmental factors appear to limit their opportunities. Young males, which usually remain in their natal areas, are often socially subordinate to older males. If dominance is positively related to the probability of paternity, as we think it is, one-year-old males should rarely father offspring. Young females may be driven from their natal areas when only a few months old. While we never observed a nesting, color-banded, one-year-old female in our study areas, there were several unbanded nesting females and these could have been one-year-old immigrants.

The onset of breeding in the colony brings with it the performance of many fascinating vocal, flight, postural, and facial displays. (For a detailed description of these, see Dow 1975.) Some displays are given by solitary individuals, while others involve many birds. For example, one or more males may *Drive* a female, repeatedly supplanting and chasing her through a maze of branches and across open spaces. Between bouts of *Driving*, the female may perform a *Bowed wing display*, assuming a crouched posture and quivering her wings. Such courtship activity often becomes intensified into massive communal display and aggression.

The female selects a site and builds the nest. During the early stages, she advertises the nest's location, and perhaps her readiness to breed, by performing a *Head-up flight* when approaching the nest or nest site. She elongates her neck and holds her head up and back. Males, mostly those whose activity spaces are near the nest site, begin to visit the nest and also may approach it in a *Head-up flight*.

Such advertising display is unusual among passerines and certainly makes the nest site obvious to other miners, humans, and presumably to predators. It follows that the social advantages of advertising must outweigh the disadvantages of exposing the location of the nest and its contents. If females were conservative in recruiting males to assist at their nests, one would expect females to perform *Head-up flights* only at nests that contained young, because a male's most evident contribution is of food to the nestlings. (Males do not incubate eggs, nor do they feed the female while she does so.) That such a conspicuous display is used to attract males to the nest even as it is being built is evidence that females may compete for the help of males. Instead of advertising the nests's location only when it contains young, a female advertises it as soon as she selects its site, thereby establishing a pattern of visitation by males that may continue throughout the occupancy of the nest.

The performance of *Head-up flights* by males is more difficult to interpret. Because of the absence of sexual dimorphism and dichromatism in miners, a *Head-up flight* given by a male may appear to other males very similar to one performed by a nesting female. Males performing *Head-up flights* when approaching nests that they intend to visit and nests which contain their offspring may be actively recruiting (and deceiving) other males, who are potential visitors.

Given the miners' frenzied courtship activity, it is not surprising that we frequently observed the birds copulating, although copulation in many other communal breeders is notably secretive. One female may copulate with more than one male, and one male may copulate with more than one female, often on the same day. Promiscuity could be an important component in the evolution of cooperative breeding in this species, as we discuss later. If males that copulate with a female behave as though they have fathered her offspring, then a female could actively recruit auxiliary males through multiple copulations (Dow 1978a).

Once the nest is completed, the female lays from two to four eggs. At Laidley, the median was 3 and the average 2.6 (S.E.M. = ± 0.09, $n = 36$). At Meandarra in 1980, the mode was 3 and the average was 2.7 (S.E.M. = ± 0.11, $n = 20$). One or more males living nearby may visit the nest. At Laidley, as many as 13 banded males visited a nest containing eggs ($\bar{x} = 4.3 \pm 0.40$). The tongue is sometimes protruded to touch the eggs with a flicking motion, and the male may roll the eggs slightly with the bill. If more than one male is present at the nest, fights or chases may erupt. Sometimes, a group display occurs, with all males placing their heads into the nest cup and contacting the eggs. The structure of this display could be derived from nest-building actions. Through contact with the eggs, males may monitor the condition of the nest, with hatching stimulating the change of behavior from simply visiting to bringing food.

Patterns of attending nests differ for males and females. Females attend only their own nests, but a male may visit several nests (of several females) in one day and many nests during one season. One male at Laidley attended 14 nests during one season ($\bar{x} = 5.2 + 0.54$, $n = 33$). A male at Meandarra attended 11 nests during one season and five nests in one day. Further, more than one male usually attends any one nest; up to 10 males visited a nest containing young at Meandarra ($\bar{x} = 4.56 \pm 0.25$, $n = 42$). Twenty-two males attended one nest at Laidley.

No significant increase in the number of attendants through the nestling period was evident at Meandarra, where the mean numbers of attendants in the three periods we defined were 5.0 ($n = 3$), 4.5 ($n = 6$), and 4.6 ($n = 5$) in 1980 and 4.0 ($n = 8$), 4.5 ($n = 11$), and 5.1 ($n = 9$) in 1981. No significant increase in visiting rates was detected. In 1980, during a severe drought, hourly visiting rates in the three parts of nestling life averaged 20.0 ($n = 3$), 21.2 ($n = 6$), and 21.2 ($n = 5$), respectively. In 1981, when conditions improved, values were 12.2 ($n = 8$), 23.2 ($n = 11$), and 24.3 ($n = 9$).

At Laidley, the number of attendants and their rates of visiting increased significantly throughout the nestling period. Mean values for numbers of attendants were 6.1 ($n = 9$) in the first third of nestling life, 8.2 ($n = 9$) in the second third, and 12.0 ($n = 3$) in the final third, while average values for hourly visiting rates were 27.2, 35.8 and 56.9, respectively. Increases in the number of visitors and visiting rates at nests are effected solely by males as only one female attends a nest and her visiting rate (9.4, 8.5 and 10.1 visits h^{-1} at the nests just cited) tends to remain constant throughout the nestling period.

Several factors could produce an increase in the number of visitors and the visiting rates at nests. One relates to the conspicuousness of nests. Their locations are advertised throughout the nestling period by the performance of *Head-up flights*. Nests are not hidden in the foliage, but are placed in easily visible sites. The rim of the nest often contains wads of white or lightly colored material, which contrast sharply with the surrounding vegetation. Secondly, when the eggs hatch, a nest becomes the focus of activity for many birds. The comings and goings of these visitors may stimulate other males living nearby to join in social interactions at the nest and, perhaps, to feed the nestlings.

A third factor that might cause an increase in the number of visitors and the visiting rate at a nest is the vocal behavior of young Noisy Miners. Vocalizations of nestlings carry up to 1 km in open country and the structure of the calls makes them easily locatable. Such is not the case for the calls of most young passerines. Nestlings vary the rate of vocalizing, calling more rapidly prior to feeding and less rapidly following feeding (O'Brien and Dow 1979). This rise and fall in rate of vocalization could be monitored by males and could transmit information about both rate of feeding and the degree of hunger immediately after feeding. Thus, by using a variable rate of transmission rather than analog communication, nestling miners may communicate more effectively with all potential feeders, including birds near the nest and those far from it. Vocalizations become louder as the nestlings age, suggesting that young in different nests compete for visitors.

One important question posed by those who study cooperative breeders is whether there is a direct relation between reproductive success and the number of helpers at a nest. Analyses of data from a sample of 14 nests active in 1981 at Meandarra revealed that successful nests (those from which at least one bird fledged) had significantly greater feeding rates, visiting rates, number of feeders, and number of visitors than did unsuccessful nests ($F = 69.36, 66.35, 30.54, 7.07$, respectively; d.f. $= 1,108$). We performed a stepwise multiple regression to determine which of the four variables contributed most significantly to the probability of a nest producing at least one fledgling. In the first regression, visiting rate explained a significant amount of the variance in nest outcome: Outcome $= 0.362 + 0.016 \times$ Visiting rate. (Adjusted $r^2 = 0.192$; $\beta = 0.460$; $F = 10.48$; d.f. $= 1,39$). But rates of visiting, in themselves, are unlikely to increase the probability of success unless visits in some way deter, rather than attract, predators. More biologically important may be the amount of food delivered to nestlings and this is best reflected by the feeding rate.

Visiting rate and feeding rate were highly correlated ($r = 0.984$, d.f. $= 40$; part-whole correlation). When the analysis was repeated with feeding rate forced into the equation first, it explained a significant amount of the variance in nest outcome: Outcome $= 0.425 + 0.016 \times$ Feeding rate (Adjusted $r^2 = 0.170$; $\beta = 0.437$; $F = 9.20$; d.f. $= 1,39$). The number of feeders was also a significant variable if forced into the regression equation first.

These results, suggesting that a high level of activity by visitors at nests containing young positively influences reproductive success, lend support to our contention that breeding females may compete for the attention of males. Note, however, that many factors may influence visiting and feeding rates by males. Because most males visit more than one nest, the number of active nests in the area, the contents of these nests, the time they have been active and the distance between nests are all potentially important. Ideally, our analyses should consider these and other factors simultaneously. However, we were limited by the number of observations we could make concurrently at active nests in our study areas.

The ultimate benefits that birds derive from visiting nests are also difficult to determine. We analyzed data from nest watches at Meandarra to see if the age or sex of a bird influenced its pattern of attendance. Although our analysis included data from only four nests, the patterns it revealed seem to apply to other nests that we observed less intensively. Significant sets of mean values of visiting rate, feeding rate, and rate of removing fecal sacs could be distinguished statistically for attendants, allowing us to assign a bird to one of several levels of participation at nests (Table 18.1). The breeding female occupied the first (highest) or second level. Old males could be found in any level, although one or two of them almost always occupied the first level. One-year-old males usually occupied lower levels unless they were attending their mothers' nests. There was no apparent division of labor among the attendants: for any bird, rates of visiting, feeding, and removing fecal sacs were highly correlated.

Sex and age, then, were two important factors that influenced the contributions of attendants at a particular nest. Once we learned that not all males at a nest contributed equally, we asked whether a male at a nest contributed equally to other nests that he attended on that day or during that season. Could his contributions at one nest be predicted from his contributions at another? Alternatively, did his contributions vary significantly among the nests he visited? If so, why?

We analyzed data for 12 males that visited more than one nest in 1981. For those males, rates of visiting and feeding varied significantly among the nests they attended. The results of a model II ANOVA for each variable are summarized in Table 18.2.

Several examples of variation in visiting rates by one bird attending several nests illustrate the dynamic nature of communal behavior in miners. BRY's hourly visiting rates at five nests in 1981 were 0.2, 0.4, 0.5, 2.6 and 19.8. WOG's rates at seven nests were 0.1, 0.1, 0.9, 1.7, 2.9, 7.3 and 10.7. RGR's rates at six nests were 0.2, 0.3, 1.3, 3.0, 4.3 and 8.9.

To what can this variation be attributed? We analyzed nestwatch data from a subsample of nine males whose social status was well known. For each of the nine birds, we assessed the influence of (1) changing ecological conditions and the presence of dependent fledglings from previous nests; (2) his social and spatial relations with other attendants at a nest; (3) his probability of paternity, as reflected by physical attacks, supplanting behavior, and aggressive displays directed toward other males near the nest

Table18.1. *Summary of levels of participation at four Noisy Miner nests at Meandarra in 1981*

	M	M	M	M	M	M	M	M	F	M	M	
Nest 1					1	2	1	3	?	3	3	
												Visits
												Feeds
												Sacs
Nest 2	2	2	?	3	?	?	?	1	?	3		
												Visits
												Feeds
												Sacs
Nest 3					4	4	4	2	?	2	4	
												Visits
												Feeds
												Sacs
Nest 4					4	2	?	?	4			
												Visits
												Feeds
												Sacs
Lowest											Highest	

LEVEL →

Non-significant sets of mean values (determined using LSD method) for visiting rate (Visits), feeding rate (Feeds) and rate of removal of fecal sacs (Sacs) are connected by horizontal lines. Above the lines are the age (in years) and the sex (M = male, F = female) of the attendants.

Table 18.2. Summary of one-way ANOVAs for rates of visiting, feeding and removal of fecal sacs by 12 Noisy Miner males that attended more than one nest at Meandarra in 1981. Age is in years

Males	Age	Nests in sample	Visits				Feeds				Sacs			
			F	d.f.	P	$r_I{}^a$	F	d.f.	P	r_I	F	d.f.	P	r_I
BGB/W	?	6	27.21	5, 51	**	0.74	28.67	5, 51	**	0.75	9.05	5, 51	**	0.47
BYB/W	1	10	5.39	9, 74	**	0.35	5.03	9, 74	**	0.34	Not calculated			
GBW/W	?	5	3.14	4, 33	*	0.24	3.14	4, 33	*	0.24	5.02	4, 33	**	0.37
O/BGB	4	3	48.06	2, 32	**	0.82	45.82	2, 32	**	0.82	4.09	2, 32	*	0.24
O/BRY	3	5	28.54	4, 40	**	0.61	29.22	4, 40	**	0.77	8.77	4, 40	**	0.48
O/KWK	4	3	6.62	2, 29	**	0.36	8.77	2, 29	**	0.43	Not calculated			
O/OKR	2	4	7.93	3, 44	**	0.37	11.18	3, 44	**	0.47	Not calculated			
O/RGR	2	6	25.94	5, 40	**	0.78	21.49	5, 40	**	0.75	10.60	5, 40	**	0.58
O/WGY	4	2	7.19	1, 24	*	0.32	6.93	1, 24	*	0.32	1.69	1, 24	n.s.	not calc.
O/WOG	3	7	32.19	6, 50	**	0.80	34.00	6, 50	**	0.81	11.57	6, 50	**	0.58
O/WYW	3	4	20.15	3, 34	**	0.70	17.43	3, 34	**	0.66	10.98	3, 34	**	0.54
RYR/W	1	4	45.39	3, 44	**	0.80	43.67	3, 44	**	0.79	5.54	3, 44	*	0.29

$^a r_I$ = coefficient of intra-class correlation for model II ANOVA.
n.s., not significant; *, $P < 0.05$; **, $P < 0.01$.

site; (4) his genetic relationship to other adult attendants at the nest; and (5) the distance of the nest from his center of activity. The following results relate to each of these five factors.

(1) Ecological conditions within the breeding season and the presence of dependent fledglings did not significantly alter rates of visiting and feeding for three males, all of whom participated minimally at nests they attended. For five of the remaining six males, rates of visiting and feeding at first nests of the season were significantly greater than those at later nests. This could indicate that the supply of food declines as the season progresses, or that demands for food by recently fledged miners result in lower rates of feeding to nestlings in subsequent nests.

(2) All but one of nine males attended nests within their own coterie areas significantly more often than nests outside. Four of nine males attended nests only within their coteries' borders. Members of a coterie interact frequently with one another and collectively defend the area from intruders. Strong social ties among the males may result in their preferential attendance at nests within the coterie area, possibly owing to social facilitation. (We know that males may visit the nest together, and we suspect that single males may be attracted to a nest by the activity of other visiting males.)

(3) Three of nine males, including two one-year-old birds, did not dominate other males during interactions near any nest. For the other six males, mean contributions at nests built near or in places where they were dominant were significantly greater than at those built in areas where they were subordinate. In two instances, nests were located where the hierarchy among males was not clear cut and two males appeared to have equally high status. One case involved broodmates, while the other involved two males that shared the same mother but were of different ages. Closely related males are often neighbors or have overlapping activity spaces because male miners tend to settle in their natal areas. We did not know the genetic relationships among all males in the population at Meandarra, but our only observations of shared dominance were among close relatives.

(4) We examined the effect of genetic relationships among adults attending a nest by analyzing data collected on five first-year males. None of these birds, being about one year old and having low social status, was likely to have fathered any offspring. Three of the five males restricted their attendance to the nest(s) of a close relative – their mother's. One other male was seen once or twice at

two nests other than his mother's, but he made no substantial contribution at either of these. The remaining male's mother disappeared during his first year of life, and so he could not possibly have visited her nests. Rather, he attended 10 different nests of five females but contributed little.

(5) We expected an inverse relation between a male's visiting rate at a nest and the distance of that nest from his center of activity. A graph of average rates of visiting at a nest on distance to the nest from the center of activity for a sample of nine males at Meandarra is in Fig. 18.3. Points on the graph are not independent because the same males are represented more than once in the sample and the number of points for each male varies considerably. Because of this, we did not test the significance of the suggested inverse relation between visiting rate and distance. Such an inverse trend was evident among the denser population of males at Laidley.

These five factors are clearly interrelated. Independently, none satisfactorily explained all the variation in attendance by all males in the sample. We think that a model to explain the variation in a male miner's attendance at different nests must include at least these five factors. The model depicted in Figure 18.4 is based on intensive observation of eight males in the colony at Meandarra over a three-year period and continual

Fig. 18.3. Average visiting rates plotted against distance to the nest from center of activity for a sample of nine male Noisy Miners at Meandarra in 1981. For further details, see the text.

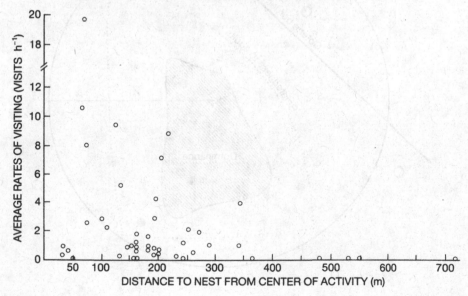

monitoring of the spatial and social relations of marked birds with whom they interacted.

In Fig. 18.4, X marks a male's center of activity; r_e, the radius of energetics, defines a circle of possibilities. Within this circle are nests that a male could conceivably attend regularly without disadvantageous expenditure of energy. (At Meandarra, r_e equalled about 300 m in 1981: males rarely travelled further to nests.) Males rarely attend nests outside the circle of possibilities, but attendance rates at nests within the circle vary greatly and without much regard to the distance of those nests from the males' centers of activity.

Often, males do not cross coterie borders to visit nests, even though these nests may lie well within the circle of possibilities. When males do visit nests

Fig. 18.4. A model depicting factors that influence contributions at nests by male Noisy Miners: distance to nests (r_e), coterie membership (heavy straight line), relatedness to adults (area bounded by dotted line) and social dominance (hatched area). Circles represent active nests and are shaded to indicate the frequency with which they are visited. Open circles are nests rarely, if ever, visited; filled circles nests visited often. For details, see the text.

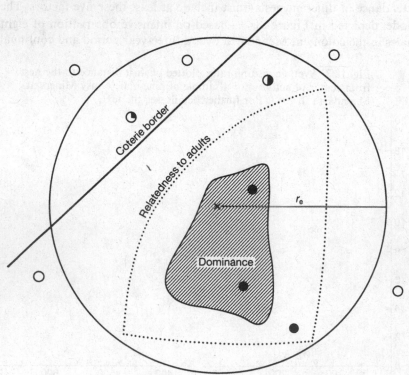

outside their own coteria areas, their rates of attendance are usually less than at nests within their coterie areas. All but one of nine males in the sample at Meandarra attended nests within their own coterie area significantly more often than nests outside those areas. Four of the nine attended nests only within their coteries' borders (Table 18.3).

Genetic relationships among adults also influence visiting rates. Because young males tend to settle in their natal areas, their activity spaces are more likely to be close to those of their mothers and their male siblings. This produces the spatial association of related birds shown in Fig. 18.4. Young males preferentially visit the nests of their mothers. Further, in areas where two adult males are co-dominant, high rates of attendance by both birds may reflect high probability of paternity for each. This is not to suggest that kin selection is a driving force in the evolution of communal breeding in Noisy Miners, but to illustrate that there may be evolutionary bonuses (*sensu* Ligon 1980) for male miners that attend nests.

Finally, a dominant male's presumed direct genetic investment in offspring is reflected in his substantial contributions at nests within his relatively small area of dominance.

It would be useful to rank the five factors in the model in terms of their importance. But their influence may change with time. The distance that a miner can travel affordably to nests may depend on local conditions, which may change both as the season progresses and from year to year. A male's genetic relationships to nearby adults may change owing to mortality,

Table 18.3. *Results of planned comparisons of rates of visiting and feeding among coteries for nine banded Noisy Miner males at Meandarra in 1981. Non-significant sets of means are connected by horizontal lines*

| | Visits | | Feeds | |
Males	Within coterie	Outside coterie	Within coterie	Outside coterie
O/BRY	Attended only nests within coterie area of residence			
O/KWK	Attended only nests within coterie area of residence			
O/RGR	Attended only nests within coterie area of residence			
O/WYW	Attended only nests within coterie area of residence			
BYB/W	0.90	1.17	0.57	0.67
O/BGB	5.21	0.06	5.09	0.03
O/OKR	0.79	0.08	0.45	0.00
O/WOG	4.25	0.11	4.09	0.00
RYR/W	7.64	0.09	7.22	0.04

natality, dipersal and immigration in the local population. Our observations indicate that social status is positively and strongly related to age, so the status of a male probably changes dramatically as it grows older. And finally in some seasons, females may build four nests, while in other seasons the same females may build only one or make no attempt to breed. Thus, the influence of dependent fledglings on the care of nestlings produced later in a season may change from year to year.

We have little information on the extent of inbreeding and, more generally, on the genetic structure of our study populations. The dispersal of young female miners would generally preclude the possibility of matings between siblings and between fathers and daughters. Sons often live near their mothers or share part of their activity spaces. However, matings between sons and mothers are probably limited initially by the social subordinance of these relatively young males. By the time sons have aged and become socially dominant, their mothers have often died or disappeared. Our observations indicate that the females replacing them are immigrants from other colonies, and hence not closely related to the resident males.

Emlen (1978) suggested that coloniality and the concomitant proximity of individuals might lead to extensive social and sexual bonding among birds. This bonding, in turn, could culminate in cuckoldry, intraspecific brood parasitism, and promiscuity. We have no evidence that intraspecific brood parasitism occurs among female Noisy Miners. The clutch size is small, eggs are added to the clutch at regular intervals, and the coloration, size and shape of eggs in a clutch varies little. The mating system, however, is highly promiscuous and the dispersion of birds is such that one female can interact (and copulate) with many males just prior to laying her eggs. This suggests that the genetic structuring in any population of this species is potentially more complex than that found in cooperatively breeding birds that live in family groups, where the number of potential and genetically compatable mates may be smaller.

Basic demography

Certain factors that make Noisy Miners interesting also make them difficult to study. And once studies are completed, it is sometimes difficult to compare the results to those obtained for other species. Breeding pairs, so often identified in other cooperative breeders, are not readily apparent in Noisy Miners, owing to a promiscuous mating system. Further, many people studying cooperative breeders look for correlations between the number of young fledged and the number of helpers in a group. In

Noisy Miners, this approach is made difficult by the absence of clearly defined groups. The number and identity of attendants (males) at nests may change during the nestling period, underscoring the dynamic social structure in this species.

Other parameters, such as reproductive success, longevity and age structure, are more easily compared with those of other species. For Miners at Laidley, nest success was remarkably low – 15.4%, 13.5% and 16.7% of nests known to contain eggs produced at least one fledgling over three years. At Meandarra, corresponding figures were higher, 37% and 44% of nests with eggs in 1980 and 1981, respectively; 27% and 31% of eggs produced fledglings.

Mortality of nestlings can be attributed to several factors, including starvation and infanticide. The within-brood effects (Ricklefs 1969) of infanticide may mimic those of starvation, in that both result in the loss of single nestlings. The relative influence of these two factors remains to be determined.

Predation was presumably the most common cause of death of adult miners, immatures and fledglings, and possibly of nestlings. Snakes and lizards, especially goannas, *Varanus varius*, were seen near nests, but we never actually observed predation in over ten years' experience with the species. The remains of six banded birds were recovered at Laidley. Owls were suspects as were brush-tailed possums, *Trichosurus vulpecula*.

The disappearance of individuals leads to difficulties of interpretation with any population study. The problem is exacerbated when the boundaries of the population are not discretely determined. At Laidley, it was difficult to check contiguous areas systematically for banded birds because of the high population density. If a banded bird left the immediate area, it did not have a great distance to travel to find another population nor did it necessarily leave the one in which it previously resided.

It cannot be concluded that all birds that disappeared died, but most miners at Laidley were highly sedentary. Of birds behaving like residents – as contrasted with those dwelling on the edge of the study area and occasionally entering it – several disappeared in 1972. Of birds known or presumed to be male, three disappeared prior to breeding, five disappeared during the breeding period, and one disappeared after the breeding period. Of two immatures born in 1971 and present at the end of that year, one disappeared before the onset of breeding. The other, a male, by then involved in reproductive activities, disappeared before the end of the breeding season. A banded adult female disappeared and returned later in the same season.

To estimate mortality rates in 1972, birds that disappeared could only be grouped with those definitely known to have died. No mortality occurred on the Laidley study site between 1 January and 1 July. Mortality of banded resident males in any of the remaining six months varied from 2.7% to 5.9% per month (2.9% to 11.8% if disappearances and deaths are combined). The average annual mortality rate was 11.0% (32.8% combined).

The number of banded adult females at Laidley was smaller, remaining fairly constant at about 15 over three years. Deaths of females occurred only in July (7.1%) and August (7.7%), coinciding with the peak of breeding. The annual mortality rate for females at Laidley was 14.3%. The mortality rate for males lies between 11% and 33% and thus may be considerably higher than that for females.

We knew of only one adult dying at Meandarra. It was found dead on the ground, pinned by a fallen tree, after a severe storm. Other birds that disappeared may have died or simply moved elsewhere. Combining deaths and disappearances yielded a mean annual mortality of 37.2% for adult males. Females may have shown a somewhat lower value, but our sample of banded females was too small for reliable estimation.

There are few records available on the longevity of Noisy Miners. Among all Australian banding records, the three oldest miners recovered were all found dead near the place of banding (D. Purchase, personal communication): a bird banded as a nestling, after 6 years 10 months; a bird banded as an adult, after 6 years 11 months; and another, after 8 years 1 month. One adult banded at Laidley was sighted in 1985, 9 years and 9 days later. On the basis of the estimated mortality rates, the mean life expectancy of adult miners at Laidley was 6.8 years for females and 2.7 to 8.7 years for males, high values for a passerine species. These values do not reflect mortality in the first year of life.

There are three main periods of mortality or disappearance of young miners. The first coincides with fledging. In a small population near Brisbane, a banded female successfully reared four broods of two young between September 1969 and June 1971. Three of these eight offspring disappeared within ten days of fledging and can be presumed to have died or been killed.

A second period of high mortality for young birds occurs two to three months after fledging. During this time, 10 (33%) of 30 young birds vanished at Meandarra. Of these, two were killed by siblings. In 1980, RWR, a male, attacked and killed his female broodmate, OYO, 60 days after both had fledged. RWR viciously pecked OYO and drove her to the ground in 8 min. Several adult males that had provisioned both birds were perched nearby but did not participate in or interfere with the attack. RWR's blows fell mainly on OYO's head and neck. OYO died on the ground while RWR perched and *Bill-wiped* (Dow 1975) on a branch immediately above her.

The second case of killing among broodmates at Meandarra occurred in 1981. Observations of a brood of three provided information about the development of agonistic relations among young miners. Upon fledging, RKO affiliated with YKY, the heaviest member of the brood, rather than WBY, the smallest and last to leave the nest. The proximity of RKO to YKY persisted through week 1, even though aggressive encounters between the two were frequent and sometimes involved physical attacks. WBY spent much of week 1 alone and rarely interacted with its siblings. By week 2, YKY was clearly and consistently dominant over RKO and often perched near WBY. If RKO approached its two siblings, YKY attacked while WBY remained perched. Occasionally, YKY pecked WBY, but never forcefully or persistently. RKO asserted its dominance over WBY in week 3, frequently chasing and pecking it. During these attacks, WBY often moved to YKY's side and perched there. Once, when RKO pecked WBY, YKY immediately pecked RKO, knocking it from its perch. In week 4, YKY still frequently attacked and chased RKO and occasionally supplanted or pecked WBY. RKO still attacked WBY, but WBY often joined YKY in flight chases of RKO. WBY never initiated a solitary attack on RKO. Such behavior continued through week 8, with YKY mounting increasingly strong attacks on RKO. In week 9, 59 days after both had fledged, YKY killed RKO. YKY approached RKO, which was foraging alone, and relentlessly pecked RKO's head and neck. WOG, an adult male that had provisioned both birds from the time they hatched, watched from a nearby perch. The attack lasted 16 min. RKO escaped several times and once managed to reach a nearby tree, but YKY pursued and continued to attack. RKO finally fell to the ground, where YKY delivered five forceful pecks to RKO's neck. RKO, still standing, gasped and vigorously shook its head. YKY flew to a nearby tree and watched RKO stagger 10 m to a clump of bushes, its wings dragging and head hanging. Darkness fell as RKO, still audibly gasping, sat with its eyes closed in a tangle of thorny branches. It was found there, dead, the following morning.

Two of 10 birds that vanished two or three months after fledging were killed by broodmates. Because young miners receive substantial amounts of food from adults for up to three months after they fledge, we attribute the disappearance of the remaining eight birds to mortality rather than dispersal.

A third period of high mortality occurs six to eight months after fledging, after young miners have reached independence. We think that high rates of disappearance in this period reflect dispersal rather than mortality at the natal site.

No banded birds were recovered far from our main study areas. However, two immatures born at Moggill, near Brisbane, and banded in December were later found dead: one 1.3 km north-west in the following February, the other 5.9 km south-west in the following September. A female banded as one of a small group of intruders that appeared for a few days at Laidley was recorded sporadically in the study area in April and July but did not attempt to nest there. In November she was seen in the same colony, 100 m west of the study area. She nested successfully in September 1973 in another colony, 450 m north north-east of the study area. In August 1974 she had moved a further 330 m north-east to nest 770 m north north-east of her original banding site. A male banded as a nestling in 1980 at Meandarra dispersed from the study area in April 1981. He was found in another colony, 600 m south of his natal area, in October. He was still there in December, when the study ended.

Summary and speculation

Many species of Australian honeyeaters are now known to breed cooperatively and are characterized by a high level of aggression – both intraspecific and interspecific. Short-term defense of transitory resources, such as flowers, is probably a common strategy within the family. Miners temporally and spatially extend this strategy and thereby control their food supply by defending their feeding area year-round. Successful indiscriminate interspecific aggression is made possible by the colonial structure and in turn enhances the quality of the colony area by removing potential food competitors. However, poor climatic conditions may reduce available food. Then it may be difficult to sustain the high population density that permits uncontested access to resources through collective interspecific aggression.

The Noisy Miner is an enigma, blending, as it does, extreme aggression with the more cooperative social interactions usually associated with cooperative breeding (Dow 1977). The proximity of individuals and nests seemingly leads to a proliferation of social and sexual bonding, which culminates in cuckoldry and promiscuity. The result is that, at any nest, attendants may include the parents, their close relatives, and a host of other birds that are not related to these other adults nor to the young birds they provision. The level of activity at nests is reflected in the highest feeding rates recorded for birds anywhere, more than 70 per hour to nestlings and 87 per hour to fledglings.

Differences in the breeding strategies of males and females suggest that competition between the sexes for colony resources may have been resolved in different ways. Females compete only with other females for breeding space while attempting to attract as many males as possible by building conspicuous nests and advertising their locations. Space within a colony seemingly does not limit the number of males, because of their extensively overlapping activity spaces. Males, however, presumably compete for food and mates. Competition for males and the food they bring may be manifested as early as the nestling stage through the unusually loud, incessant begging calls of young birds (O'Brien and Dow 1979).

Given the peculiarity of the Noisy Miner's social system, we can only speculate on how some of its features have arisen and how cooperative breeding may have evolved. Our uncertainty about paternal relations in the study populations makes it difficult to assess quantitatively hypotheses such as kin selection. Further, our studies to date have not addressed the topics of habitat saturation (Stacey 1979) or resource localization (Koenig and Pitelka 1981), which are suggested to foster communality in other species. Both of these require data that differ substantially from those we amassed. We can, however, present a plausible scenario for the evolution of cooperative breeding that is consistent with our data. That scenario will undoubtedly be modified as more is learned about Noisy Miners.

Although some authors assumed that monogamy would promote cooperative breeding, the first step toward communality in Noisy Miners could have been promiscuity. Promiscuity among males is not surprising, given the low cost of sperm production (but see Gladstone 1979). Its general occurrence in miners requires that females also benefit (Orians 1969). Through shared paternity, the genetic variability of a brood is increased, and this could be adaptive during environmental fluctuations or in periods of range or colony expansion (Gladstone 1979). The disadvantages of being a cuckolded male are well known (Power *et al.* 1981). But promiscuity among females may under certain circumstances increase a cuckolded male's reproductive success and fitness (Stacey 1982). Two conditions are central to this idea and both seem to apply to Noisy Miners. First, males who copulate with a female must provide parental care. If paternal care is important, males copulating with a female but failing to provision her young – their own offspring – will be at a disadvantage (Emlen 1978). The second condition is that females must be the limiting sex. If mates are difficult to find and to monopolize, males profit most by caring for any offspring that they might possibly have fathered. Coloniality, a biased sex ratio favoring males, and the even dispersion of females throughout our study areas suggest that females are the limiting sex. A male's exclusive defense of his mate would involve aggression and exclusion. Such behavior when directed toward other males might weaken the miners' system of cooperative defense and ultimately cost more than it is worth.

Females that mate with more than one male might experience greater reproductive success because of the greater genetic variability of their broods or because of increased survival of their offspring as a direct result of multi-male care. Because of the increased reproductive success of females assisted by more than one male, competition among females for those males intensifies. Females occupy non-overlapping activity spaces. They resist settlement of their daughters in the natal area, for these represent competitors for males. Thus, dispersal, even with its concomitant risk of mortality, becomes an adaptive strategy for young females.

Increased reproductive success of adults, coupled with an increased mortality rate of females owing to dispersal or exclusion, lowers the probability of a young male finding a mate. The young male remains in his natal area, presumably with the 'permission' of his parents. The retention of the young male could be advantageous to parents and other older auxiliaries because functionally it protects their investment in that male by increasing their inclusive fitness (Rodman 1981), or enhances the survival of individual group members. Alternatively, the retention of a young male could be costly to parents and older auxiliaries and, as payment (*sensu* Gaston 1978; Emlen 1982), this male is manipulated to fill the role of auxiliary. The decrease in fitness suffered by such a male is less than if he had been expelled from his natal area, or left voluntarily.

There may, then, be two origins of auxiliary males. The first to arise serves as an auxiliary because he 'thinks' that he has fathered the young he provisions. The other (the 'manipulated offspring') serves because service is the lesser of two evils. But the second type receives an evolutionary bonus because he often provisions close relatives. In operation, a male may use contextual cues to assess the probability of relatedness between himself and other birds. Such a cue might be the distance between his own activity center and those of other males and females (Hamilton 1964).

As auxiliaries of the second type age and become more experienced, they pose a greater threat to breeding males. If gains in a breeding male's fitness from allowing auxiliary males to copulate with the female outweigh the loss in fitness due to shared paternity, which we suggest occurs in miners, a breeding male should allow copulations by auxiliary males.

As inter-male competition for females increases, social interactions among the males that attend a nest may greatly influence patterns of attendance. Males may follow each other to nearby nests. By attending a nearby nest, they may promote bonding with the female, thereby increasing the probability that they will fertilize her clutches in the future, even though they have not done so in the past.

The crucial factors in this scenario are the probability of breeding in the future and the probability of having bred in the immediate past. The first is related to the biased sex ratio, i.e. to the shortage of females. The second is determined by factors including deceit, supposition, manipulation, and the other more tangible, and hence more measurable, ones such as age and social status.

By studying Noisy Miners in two quite different sites we identified many common features of the ecology and social interactions of the species. We also learned that ecological factors, e.g. habitat, may influence population density and social organization. The ancestral habitat of the Noisy Miner is probably woodland. They now reach their highest densities in woodland with an open canopy and sparse understory. Savannah woodland, which characterizes inland portions of the continent, is typically inhabited by the congeneric Yellow-throated Miner. That both species occurred at Meandarra reflects the extensive clearing of the land for agriculture early in the 1900s (Whitmore *et al.* 1983). The fragmentation of the original woodland has rendered it probably more suitable for Yellow-throated Miners and restricted the distribution of Noisy Miners to denser remnants of vegetation. Hybridization between the two species has been recorded in the area (Dow 1972). Opportunistic infiltration has also been noted in other parts of the Yellow-throated Miner's range, where it has hybridized with the Black-eared Miner (Ford 1981).

Thus, the Meadarra site may have had a lower density of Noisy Miners because of discontinuities in suitable woodland. This would restrict the Miners' ability to establish colonies of the size reported from Laidley, and in turn would limit the concomitant opportunity to exclude other species through numerical force. There were more species of birds at Meandarra, but we do not infer a cause–effect relationship between population density of miners and number of species. Such an inference could only be made from a study that considered the effects of other ecological and environmental variables, such as habitat and climate.

At lower densities, miners tend to appear as members of a more cohesive coterie, without some of the added complexities of interactions seen in larger colonies. Large colonies such as the one studied at Laidley may be much older, having built up in numbers during more salubrious times; Laidley itself may possibly have been in decay during our study.

Group behavior, display and other interactions of miners were qualitatively similar at the two sites. Although birds cooperated in mobbing predators and other species at Meandarra, they were less successful at excluding other birds than were the miners at Laidley. Group activities were more frequently seen at Laidley, presumably because these were, in fact, more common in the population of higher density. Sexual chasing and copulation were noticably more evident among the Laidley birds, and we think that was due to the stimulating effect of a greater number of competing males always being much closer to females.

Clutch size and nesting attempts showed no apparent difference between sites, but nest success was higher at Meandarra, in spite of one year of that study being affected by severe drought. The number of attendants at nests at Laidley was roughly twice that at Meandarra, and the feeding rate produced by birds at Laidley was also roughly double the rate at Meandarra.

Finally, mortality rates (or more correctly mortality and disappearance rates) were higher at Meandarra. This may have resulted from drought conditions as other species of birds showed a marked decline at the same time (Dow, unpublished data).

The only other species in the genus *Manorina* that has been studied in detail is the Bell Miner (Clarke 1984, 1988; Smith and Robertson 1978; Swainson 1970). Studies were all conducted in Victoria, most on the same local population, and essentially on the edge of the species range. Clarke (1988) found Bell Miners to be monogamous, with both male and female auxiliaries attending the nests of breeding pairs. Even breeding pairs fed the nestlings of other pairs. The species bred in every month, and pairs made several successive attempts within a year. After fledging, dependent young were fed by up to 13 birds.

Although Bell Miners share many basic similarities of social behavior and organization with their congener, differences such as strict monogamy while attending the young of other pairs are surprising. Just as we found differences in density to affect certain aspects of social organization in the Noisy Miner, studies of Bell Miners in dense populations, as are more typical nearer the center of their range, may yield other patterns of behavior.

Directions for future research

Before meaningful manipulative experiments can be undertaken, much information about a species must be gathered and synthesized. Although Noisy Miners are among the most common birds in eastern

Australia, biologists knew little of them prior to our studies. Several avenues for further research emerged from our work. First, we must learn more about genetic relationships among adults and the relationships of adults to the young they provision. As we explained, these data are difficult to obtain. We are skeptical of results obtained for other species through destructive sampling and subsequent studies of reproduction, particularly in birds with well-developed social systems. The opportunity to study the patterns of interactions among birds in nature appeals to us.

One way to learn more about the effect of promiscuity on the cooperative behavior of miners is to study copulatory behavior. We think that a female recruits male attendants to her nest by copulating with them. Copulations by miners are frequent and easily observed. In a study of females' copulatory behavior, it would be important to determine the identity of males that copulate with females, the frequency with which they do so, the seasonal distribution of copulations with respect to egg-laying, and the subsequent contribution of each male at the nests of each female. The effects of a male's copulations with any other females would also have to be considered, and could be the subject of a simultaneous study.

Dow (1979a) suggested that visits by males where population densities are high may disturb the female's pattern of incubation, causing reduced hatchability or loss of eggs. At Laidley in 1971, when miners bred most successfully, only 38% of nests known definitely to have received eggs produced at least one nestling. In 1973, the worst year, only 17% of such nests produced at least one nestling. At Meandarra, where the population density was lower, corresponding values were 80% in 1980 and 88% in 1981. A fruitful study could be made of the effect of visits by males on incubation patterns, egg temperatures, and reproductive success in populations of different densities.

A third area deserving attention is the behavioral interactions and social relations among broodmates. Little work on this subject has been done on passerine birds because the young of many species are cryptic or disperse soon after leaving the nest. Social hierarchies among young miners are evident shortly after hatching. Brood members interact frequently in the first few months of life and can be located easily by their vocalizations. Data from about 30 birds at Meandarra suggest that social hierarchies among young birds may affect subsequent dispersal patterns and survival, and thereby influence the structure of the population. At the very least, young miners would be excellent subjects for ontogenetic studies.

Finally, an area of study suggested by O'Brien and Dow (1979) involves manipulating the rate of vocalizations by young miners to discover the role of rate of transmission in communication. Their data suggest that young miners increase their rate of calling as their hunger increases. That the call can be heard far from the nest suggests that it might, in fact, influence the behavior of birds in the area. Calling rates could be manipulated and the behavior of birds nearby could be monitored to see if rates of visiting or feeding increased in response to increased rates of calling.

The Noisy Miner clearly differs in many respects from other cooperatively breeding species. Early reviews of cooperative breeding among birds attempted to categorize species within rigidly defined classifications. Generalized models proved largely unsuccessful in accommodating new facts. As field workers have accumulated details of life history from more and more species, it has become apparent that even among 'typical' cooperative breeders variations abound. That the classification becomes more, not less, difficult with additional information suggests that 'cooperative breeding' is indeed a catch-all for a wide variety of unconventional breeding strategies in which the common feature is provisioning of young by birds other than their parents. Many studies have demonstrated the wide range of variation and flexibility of cooperative breeding systems. It may be better to view the system of the Australian Noisy Miner not as aberrant or atypical but perhaps as a more extreme example of this range, leading us to expect other reports of equally bizarre systems of cooperative breeding among birds.

We are grateful to many friends and colleagues who assisted us in the field. For unrestricted access to properties where our study sites were located, we thank Mr I. G. Burgess of Laidley and Mr and Mrs Ron Jamieson of Meandarra. The drawings of miners are based on published drawings by Peter Slater. Our work was supported financially by grants from the University of Queensland, the Australian Research Grants Committee, the M. A. Ingram Trust, The Rotary Foundation and the Frank M. Chapman Fund of the American Museum of Natural History. This is a contribution of the Meandarra Ornithological Field Study Unit of the University of Queensland.

Bibliography

Clarke, M. F. (1984). Co-operative breeding by the Australian Bell Miner *Manorina melanophrys* Latham: a test of kin selection theory. *Behav. Ecol. Sociobiol.* **14**: 137–46.

Clarke, M. F. (1988). The reproductive behaviour of the Bell Miner, *Manorina melanophrys*. *Emu* **88**: 88–100.

Dow, D. D. (1970). Communal behaviour of nesting Noisy Miners. *Emu* **70**: 131–4.

Dow, D. D. (1972). Hybridization in the avian genus *Myzantha*. *Mem. Qld Mus.* **16**: 265–9.

Dow, D. D. (1973). Flight moult of the Australian honeyeater *Myzantha melanocephala* (Latham). *Aust. J. Zool.* **21**: 519–32.

Dow, D. D. (1975). Displays of the honeyeater *Manorina melanocephala*. *Z. Tierpsychol.* **38**: 70–96.

Dow, D. D. (1977). Indiscriminate interspecific aggression leading to almost sole occupancy of space by a single species of bird. *Emu* **77**: 115–21.

Dow, D. D. (1978a). Breeding biology and development of the young of *Manorina melanocephala*, a communally breeding honeyeater. *Emu* **78**: 207–22.

Dow, D. D. (1978b). Reproductive behavior of the Noisy Miner, a communally breeding honeyeater. *Living Bird* **16**: 163–85.

Dow, D. D. (1979a). The influence of nests on the social behaviour of males in *Manorina melanocephala*, a communally breeding honeyeater. *Emu* **79**: 71–83.

Dow, D. D. (1979b). Agonistic and spacing behaviour of the Noisy Miner *Manorina melanocephala*, a communally breeding honeyeater. *Ibis* **121**: 423–436.

Dow, D. D. (1980). Communally breeding Australian birds with an analysis of distributional and environmental factors. *Emu* **80**: 121–40.

Emlen, S. T. (1978). The evolution of co-operative breeding in birds. In *Behavioural Ecology: An Evolutionary Approach*, ed. J. R. Krebs and N. B. Davies. pp. 254–81. Blackwell: Oxford.

Emlen, S. T. (1982). The evolution of helping. II. The role of behavioral conflict. *Amer. Nat.* **119**: 40–53.

Ford, H. A. (1981). A comment on the relationships between miners *Manorina* spp. in South Australia. *Emu* **81**: 247–50.

Gaston, A. J. (1978). The evolution of group territorial behavior and cooperative breeding. *Amer. Nat.* **112**: 1091–100.

Gladstone, D. E. (1979). Promiscuity in monogamous colonial birds. *Amer. Nat.* **114**: 545–57.

Hamilton, W. D. (1964). The genetical evolution of social behaviour. I and II. *J. Theoret. Biol.* **7**: 1–52.

Kalma, J. D. (1974). Climate of the Balonne–Maranoa area. Part IV. In *Lands of the Balonne–Maranoa Area, Queensland*. CSIRO Land Res. Ser. No. **34**: 1–242.

King, J. A. (1955). Social behavior, social organization and population dynamics in a black-tailed prairie dog town in the Black Hills of South Dakota. *Contr. Lab. Vert. Biol. Univ. Mich.* **67**: 1–123.

Koenig, W. D. and Pitelka, F. A. (1981). Ecological factors and kin selection in the evolution of cooperative breeding in birds. In *Natural Selection and Social Behavior: Recent Research and New Theory*, ed. R. D. Alexander and D. W. Tinkle. pp. 261–79. Chiron Press: New York.

Ligon, J. D. (1980). Communal breeding in birds: an assessment of kinship theory. *Proc. 17th Intern. Ornith. Congr.*: 857–61.

O'Brien, P. H. and Dow, D. D. (1979). Vocalizations of nestling Noisy Miners *Manorina melanocephala*. *Emu* **79**: 63–70.

Orians, G. H. (1969). On the evolution of mating systems in birds and mammals. *Amer. Nat.* **103**: 589–603.

Paton, D. C. and Ford, H. A. (1977). Pollination by birds of native plants in South Australia. *Emu* **77**: 73–85.

Power, D. C., Litovich, E. and Lombardo, M. P. (1981). Male Starlings delay incubation to avoid being cuckolded. *Auk* **98**: 386–9.

Ricklefs, R. E. (1969). An analysis of nesting mortality in birds. *Smiths. Contrib. Zool.* **9**: 1–48.

Rodman, P. S. (1981). Inclusive fitness and group size with a reconsideration of group sizes in lions and wolves. *Amer. Nat.* **118**: 275–83.

Schodde, R., Glover, B., Kinskey, F. C., Marchant, S., McGill, A. R. and Parker, S. A. (1978). Recommended English names for Australian birds. *Emu* **77**: 245–313.

Smith, A. J. and Robertson, B. I. (1978). Social organisation of Bell Miners. *Emu* **78**: 169–78.

Stacey, P. B. (1979). Habitat saturation and communal breeding in the Acorn Woodpecker. *Anim. Behav.* **27**: 1153–66.

Stacey, P. B. (1982). Female promiscuity and male reproductive success in social birds and mammals. *Amer. Nat.* **120**: 51–64.

Swainson, G. W. (1970). Co-operative rearing in the Bell Miner. *Emu* **70**: 183–8.

Walters, M. (1980). *The Complete Birds of the World*. Reed: Sydney.

Whitmore, M. J. (1986). Infanticide of nestling noisy miners, communally breeding honeyeaters. *Anim. Behav.* **34**: 933–5.

Whitmore, M. J. and Dow, D. D. (1982). Noisy Miners. *Austr. Nat. Hist.* **20**: 290–4.

Whitmore, M. J., Dow, D. D., Fisk, P. G. and Moffatt, J. D. (1983). An annotated list of the birds of Meandarra, Queensland. *Emu* **83**: 19–27.

SUMMARY

JAMES N. M. SMITH

Cooperation between the sexes in the rearing of offspring is rare in the animal kingdom, but typical of most species of birds. The usual form of cooperation in birds is for the male and female of a monogamous pair to invest more or less equally in the defense of a breeding area and in the care of young. Thus, most species of birds are cooperative breeders, but this cooperation is limited strictly to the breeding pair, and other conspecifics are vigorously excluded from the vicinity of the nest and young. I therefore use the term 'cooperative breeding' in the restricted sense used elsewhere in this book (see Introduction to this volume).

My perspective in this summary is that of a student of the behavior and population biology of typical birds, i.e. those that cooperate only with a mate while reproducing. Thus, I approach the subject without biases from having worked on a cooperatively breeding species. I first review briefly some questions about cooperative breeders, and the kinds of data that are needed to address them. I then consider how far these questions have been answered by the studies in the book. I conclude by noting that studies of cooperative breeders have enhanced our general understanding of the population biology of birds, and I suggest some directions for future work.

Principal questions

The topic of cooperative breeding has been reviewed very extensively (see e.g. Emlen and Vehrencamp 1983; Emlen 1984; Brown 1987). A historical account is also given in the Introduction to this volume. I pose three questions here: (1) what forms of cooperative breeding occur in nature?; (2) are there particular ecological correlates of cooperative breeding?; (3) is cooperative breeding best explained by kin selection, reciprocity, demographic constraints, or by a simple non-adaptive association between individuals in social groups?

Question 1: what forms of cooperative breeding occur in birds?

This descriptive question has not yet been answered fully, which constrains our ability to address the second and third questions. Brown (1987, Table 2.2) has recently presented a detailed scheme for classifying the known forms of cooperative breeding. Most cooperative breeders reproduce on dispersed territories, while a minority are colonial. A second dichotomy is whether only a single female breeds in each group (singular breeders), or whether two or more females may breed (plural breeders). Brown distinguishes further categories on the basis of sex of helpers and the mating system of the breeders. Even this complex classification fails to include all known cases. Since most recognized species of cooperative breeders live in the tropics, and most tropical birds have yet to be studied in detail, further forms of cooperation are likely to be found in birds (see Chapter 18).

Question 2: what are the ecological correlates of cooperative breeding?

The information needed to address this question is a set of studies that accurately describes the forms of cooperative breeding, and the particular ecological conditions that are faced by members of cooperatively breeding populations. The studies included here are clearly a start in achieving this goal.

Previous failures (see e.g. Dow 1980; but see also below) to find precise geographic or climatologic correlates of cooperative breeding, other than its more frequent occurrence in the tropics, led to the 'habitat saturation' hypothesis (Selander 1964; Brown 1974; Gaston 1978; Stacey 1979; Koenig and Pitelka 1981). This suggests that a principal cause of cooperative breeding is a critical shortage of suitable breeding territories. This forces young birds to remain in their natal groups, or to join other territorial groups, because vacant territories are scarce or absent. This idea was developed by Emlen (1982a, b; 1984), who pointed out that there are ecological constraints other than a shortage of suitable territories, such as risks associated with breeding failure, or dispersal. Stacey and Ligon (1987) have noted that cooperative breeding can be found in the absence of habitat saturation, when variation in habitat quality increases the benefits of philopatry. Young may thus be fitter if they delay dispersal from productive territories, rather than dispersing to empty but poor ones. Thus, habitat saturation is a common, but not necessary, ecological determinant of cooperative breeding.

Question 3: are there adaptive explanations for the various forms of cooperative breeding?

Much of the recent interest in cooperative breeding was sparked by two ideas. The first of these was Hamilton's (1964) theory of inclusive fitness or kin selection (Maynard Smith 1964), described in Stacey and Koenig's Introduction to this volume. Second, Trivers (1971) proposed that reciprocal altruism, usually termed reciprocity in this book, could explain aid-giving between non-relatives. Trivers proposed that selection favors group members that aid others to surive or reproduce, when there is a high probability of the aid being repaid at a later time.

The attractiveness of these two ideas, and the adaptationist zeal of the mid 1970s (e.g. Barash 1975), did not encourage investigation of alternative hypotheses to explain cooperative breeding. Despite this, Zahavi (1974) raised the possibility that helping behavior in birds was actually self-interest, and that apparent helpers might interfere in reproduction to improve their chances of attaining breeding status. This, and earlier non-adaptive explanations for cooperative breeding (see Introduction to this volume) excited little interest in the late 1970s. Only after the challenge to pan-adaptationism by Gould and Lewontin (1979) were non-adaptive proximate explanations of helping reconsidered (see Chapter 13).

The third principal adaptive explanation of cooperative breeding has been most fully stated by Woolfenden and Fitzpatrick (1984). They argued that, while kin selection could lead to the evolution of helping, direct benefits can accrue to a non-dispersing helper if its fitness is enhanced by remaining with the natal group. From the parents' point of view, retention of young is then favored, even if these young do not help in the production of further offspring. Helping behavior, when it occurs, is only one consequence of permanent group-living, which is a more general adaptation. Thus, aid to relatives merely increases the range of demographic conditions under which retention of offspring is favored.

While these three adaptive ideas can be separated logically, they can operate together to explain cooperative breeding, and are difficult to separate in practice.

Scope of the studies in this book

A major feature of this book is that it brings together a set of long-term studies of the demography and social relations of 20 species of birds. In combination with some studies of non-cooperatively breeding birds, larger mammals, insects and fishes (summarized by Clutton-Brock (1988),

Table 1. *Some characteristics of the studies in this book*

Species studied	Duration of the study (years)	No. of study sites	No. of breeding groups studied/year	Seasonal variation in the environment	Annual variation in the environment
Harris' Hawk	7	1	24–29	High	High
Galápagos Hawk	11	2	5–12, 15–31	High	High
Pukeko	3, 3, 7	3	4, 5–10, 13–15	Moderate	Low
Groove-billed Ani	3, 5	1	c. 40	High	Low
Hoatzin	6	1	57–117	High	Moderate
Pied Kingfisher	8	2	9–40	Moderate	Moderate
White-fronted Bee-eater	11	1	15–25	High	High
Green Woodhoopoe	8	1	17–26	High	High
Red-cockaded Woodpecker	7	2	220–234	Moderate	Low
Acorn Woodpecker (HR)	17	1	18–35	Moderate	High
Acorn Woodpecker (WC)	10	1	14–23	Moderate	High
Stripe-backed Wren	11	3	30–85	High	Moderate
Bicolored Wren	11	2	30–87	High	Moderate
Galápagos Mockingbird	10	1, 3	12–38	High	High
Dunnock	5	1	c. 40	Moderate	Low
Arabian Babbler	17	1	15–20	High	High
Splendid Fairy-wren	15	1	8–34	High	Moderate
Noisy Miner	3, 3	2	1, 1	Moderate	Moderate, high
Gray-breasted Jay	17	1	6–7	High	High
Florida Scrub Jay	18	1	28–35	Moderate	Low
Pinyon Jay	16	1	1	Moderate	High

Cases with very small numbers of breeding groups studied are all colonial. Environmental variability is classified on the basis of variation in rainfall, or other factors affecting food supply or survival to a marked extent. Numbers separated by commas refer to different study sites. The long-term studies of Acorn Woodpeckers at Hastings Reservation, California (HR), and Water Canyon, New Mexico (WC), are tabulated separately.

this work is the core of a new subject area in behavioral ecology that can be termed social demography. An increasing overlap between studies of social behavior and population biology was forecast by E. O. Wilson in 1975, but progress towards this meeting of ideas has been most rapid in highly social vertebrates such as cooperatively breeding birds.

Some characteristics of the studies are summarized in Table 1. Several studies have lasted for over 10 years, and some report data on more than one population. The species described here inhabit areas with moderate to high seasonal variability in rainfall and food supply; annual variability in these characteristics ranges from slight to extreme. This suggests that no particular climate or type of environment favors cooperative breeding.

The chapters in this book include a variable amount of empirical data, and take a variety of approaches. Some chapters (e.g. those by Brown and Brown (Chapter 9), Dow and Whitmore (Chapter 18), Zahavi (Chapter 4)) describe only a modest amount of data, but balance this with stimulating ideas, or by describing unusual forms of cooperative breeding. Most other chapters, however, summarize much empirical data. The numbers of breeding groups studied intensively by some investigators (e.g. Walters, see Chapter 3) are truly impressive, as is the rigor with which others (e.g. Woolfenden and Fitzpatrick, see Chapter 8) have measured social and demographic variables over a span of nearly 20 years. Chapter 13 (Craig and Jamieson) is partly a critique of the principal approaches employed to investigate cooperative breeding.

The range of methods employed in most studies is fairly limited (but see Chapter 17). In most cases, the work has been purely descriptive, and the only interventions have been to color-mark individuals. With hindsight, it is a pity that early workers did not attempt more manipulative experiments (e.g. Brown *et al.* 1982), and more studies of the development of helping behavior (e.g. Carlisle and Zahavi 1986; Jamieson 1988). I shall return to this point.

Principal findings of the studies
Question 1: what variety of social systems exists among cooperative breeders?

Table 2 summarizes some social characteristics of the 20 species described in detail in this book. Fifteen of the 20 species defend stable all-purpose group territories, while four are colonial and have no clear breeding territory. Most of the 15 territorial species reside permanently on their group territories, but in three species, the Hoatzin, the Groove-billed Ani and the Pukeko, individuals frequently desert territories to join

Table 2. *Some social characteristics of the species studied in this book*

Species studied	Social dispersion	Mean group size	Principal mating systems	Breeding females per group	Type of nest in plural groups
Harris' Hawk	NT, GH	2.7	M	1	—
Galápagos Hawk	T, D	3.1	PA	1	—
Pukeko	T, D, NBF	2.8, 4.5, 7.0	PG	1–2	Joint
Groove-billed Ani	T, D	4.0	M	1–3	Joint
Hoatzin	T, NBF	3.0	M	1–2	—
Pied Kingfisher	C, NT	2.6, 3.1	M	1	—
White-fronted Bee-eater	C, NT/T	6.4	M	1–3	Single
Green Woodhoopoe	T, D	4.8	M	1	—
Red-cockaded Woodpecker	T, D	2.2	M	1	—
Acorn Woodpecker (HR)	T, D	4.4	M/PG	1–3	Joint
Acorn Woodpecker (WC)	T, D	2.6	M/PA	1	—
Stripe-backed Wren	T, D	4.6	M	1	—
Bicolored Wren	T, D	2.2, 3.0	M	1	—
Galápagos Mockingbird	T, D	4.2	M/PY	1–4	Single/ joint
Dunnock	T, D	3.1	PA/PG/M	1–4	Single
Arabian Babbler	T, D	4–9	PR/PA/M	1–4	Joint
Splendid Fairy-wren	T, D	3.3	M/PR	1–2	Single
Noisy Miner	C, T, ND	7.3	PR	Up to 15[a]	Single
Gray-breasted Jay	T, D	4.5, 8.7	M	1–4	Single
Florida Scrub Jay	T, D	3.0	M	1	—
Pinyon Jay	C, ND	85–112	M	1	—

Categories of social dispersion: C, colonial; D, dispersed; GH, group-hunting; T, territorial; NT, non-territorial; NBF, non-breeding flock; ND, nests dispersed.
Types of mating system: M, monogamy, PA, polyandry, PY, polygyny; PG, polygynandry, PR, promiscuous.
[a] Number of females per colony. (See the text.)

non-breeding flocks. One of the colonial species, the White-fronted Bee-eater, is territorial in the non-breeding season. The last species discussed, the Noisy Miner, is unusual in that it is colonial, yet defends a large group territory in a complex way. Of the 20 species, 19 live in groups with an average size of less than 10. Even in the colonial species, stable subgroups exist within the colony in all but the Pinyon Jay.

By far the most common form of cooperative breeding in birds is the helper-at-the-nest type, where a set of younger non-breeding individuals provides food for the young of a monogamous breeding pair, and also aids in 'sentry duty' or territorial defense. Helpers are usually, but by no means always, the previous offspring of at least one of the breeders. A rarer form, plural breeding, involves more than one breeding female per group; these females may nest separately or jointly, even in the population (Table 2). Some helper species (e.g. Florida Scrub Jay, Galápagos Mockingbird) occasionally exhibit plural breeding. The difference between helping and plural-breeding types is therefore only one of degree. The patterns of parental care in these plural-breeding groups can be very complex, for example in anis (*Crotophaga*), miners (*Manorina*), and Acorn Wood-peckers. The mating system of most species appears to be monogamous (Table 2), but there are cases of polyandry (e.g. in Dunnocks and Galápagos Hawks), and polygynandry (e.g. in the Pukeko, Dunnocks and Acorn Woodpeckers). Polygyny occurs occasionally when adult sex ratios are biased towards females (in Galápagos Mockingbirds, Dunnocks and Hoatzins). One species, the Noisy Miner, may have a promiscuous mating system, where each female copulates indiscriminately with all the males in her part of the colony.

A common characteristic of cooperative breeders (Table 3) is that they avoid inbreeding with close relatives, but there are several apparent exceptions here, notably the Splendid Fairy-wren and the Pukeko.

Several studies in this volume have considered the same species of bird, or two very closely related species, in more than one environment. In most of these cases, interesting variation in the extent and form of cooperative breeding has been noted as a correlate of differences in habitat, food supply or population density (see Chapters 9, 10, 12, 13, 14 and 17). A notable characteristic of such comparative studies is that species exhibiting cooperative breeding do so in almost all populations. The only exceptions to this are the western population of Scrub Jays (see below), and a non-cooperative population of Green Jays at the northern limit of the species' range in Texas (Gayou 1986).

Other species of cooperatively breeding birds (for reviews see Emlen

Table 3. *Some ecological characteristics of cooperative breeders studied in this book*

Species studied	Limiting resources suggested for population	Effects of helpers on reproductive success	Avoidance of inbreeding with close relatives
Harris' Hawk	Need to hunt for large prey	None-slight?	Yes?
Galápagos Hawk	Breeding territories	Marked	No?
Pukeko	Territories in wetland	Slight?	No
Groove-billed Ani	High quality territories	None	No?
Hoatzin	Territories on islands, riverbanks	Moderate	Unknown
Pied Kingfisher	Females limited; food at one site	Moderate– extreme	Likely
White-fronted Bee-eater	Food in some years	Marked	Yes
Green Woodhoopoe	Tree holes for roosting; food	Slight	Yes
Red-cockaded Woodpecker	Groups of tree holes	None	Yes
Acorn Wood-pecker (HR)	Granaries, oaks	Slight	Yes
Acorn Wood-pecker (WC)	Granaries, oaks	Marked	Yes
Stripe-backed Wren	Defense against predators	Extreme	Yes
Bicolored Wren	Defense against predators	Extreme	Yes
Galalápagos Mockingbird	Breeding space	Slight– Moderate	No
Dunnock	Space, females	None–marked	Unknown
Arabian Babbler	Shrub and tree cover	Moderate	Yes
Splendid Fairy-wren	Suitable heathland	Moderate	No
Noisy Miner	Food	Moderate?	Unknown
Gray-breasted Jay	Acorns, space	Moderate?	Unknown
Florida Scrub Jay	Oak scrub habitat	Moderate	Yes
Pinyon Jay	Females	None	Yes

1984; Brown 1987) described to date do not differ greatly from those described here. Also, similar types of cooperative breeding (nest construction and sharing, provisioning and defense of young) are found in a few mammals (for references, see Introduction to this volume), social insects (Brockmann 1984), and even several fishes (Fricke 1979; Taborsky 1984). In some of these organisms, uniparental care, or a limited ability of males to feed young, constrain reproductive cooperation to adults of the same sex, or limit the types of cooperation seen. In birds, juveniles reach adult size within a small fraction of the adult life span, and can then participate to a considerable degree in the reproductive and social activities of a group. Among other vertebrates, perhaps only family groups of cichlid fishes cooperate as fully as birds in the rearing of young (Taborsky, 1984).

Question 2: do ecological constraints explain cooperative breeding?
Some ecological constraints identified by the studies in this book are listed in Table 3. Several species have very specialized requirements for limiting resources. Examples are storage trees for the Acorn Woodpecker, and nest or roosting holes for Green Woodhoopoes and Red-cockaded Woodpeckers. In Acorn Woodpeckers, populations without storage trees live in very small groups or migrate, and hence exhibit little cooperation in breeding (see Chapter 14). Most other populations, e.g. Florida Scrub Jays and Hoatzins, have specialized habitat requirements. The degree of cooperation shown within populations, however, has seldom been related quantitatively to measured habitat characteristics (but see Chapter 14). Also, few studies have contrasted the ecological constraints faced by cooperative breeders to those faced by their non-cooperative relatives. The Florida and western populations of Scrub Jays are an exception here (see Koenig and Mumme 1987, p. 381). A general weakness of the ecological constraints explanation for cooperative breeding is that resource limitation and social constraints also affect the social lives of many non-cooperative breeders in apparently similar ways (e.g. Lack 1968; Patterson 1980; P. B. Stacey and J. D. Ligon, personal communication).

In some species described here, obvious ecological constraints that could inhibit dispersal and promote cooperation are lacking. One such species, the Noisy Miñer, is an obligate cooperative breeder throughout its range. Another, the non-territorial Harris' Hawk has a very wide range, but has only been studied at the northern extreme of this range. Perhaps further studies will identify important ecological constraints in such puzzling cases.

The principal proximate cause of cooperative breeding is delayed juvenile dispersal. This demographic constraint need not map tidily on to a specific ecological requirement. In general, the demographic correlates of cooperative breeding can be summarized more tidily than its ecological correlates (see below). There is, however, one recent and important study of the ecological determinants of cooperative breeding in Australian birds (Ford *et al.* 1988). Ford *et al.* showed that cooperative breeding was most common among insectivorous species inhabiting arid woodlands in Australia, and that insect populations in this habitat fluctuated less during the year than in rainforest habitats. The idea that a limit on cooperative group-living is set by the minimum level of food abundance during the year also applies to the Acorn Woodpecker (see Chapter 14).

In summary, specific ecological factors that promote cooperative group-living can be readily identified in a few species. More general forms of ecological limitation such as specialized habitat requirements apply to most other cases, but also apply to many non-cooperative breeders. The minimum requirement for year-round group territoriality, as seen in most of the studies in this book, is a sufficient food supply during the period of greatest seasonal scarcity.

> *Question 3: are there satisfactory adaptive and evolutionary explanations for cooperative breeding?*
>
> I now consider the three principal explanations in the order in which they were first proposed. I begin with the role of kinship.

Role of kin selection. Most cases of reproductive helping in birds involve close relatives, usually parents and their independent or adult offspring. Cooperation among close relatives is also involved in contests for territories in at least two species, Acorn Woodpeckers and Green Woodhoopoes. In other species, e.g. Splendid Fairy-wrens and the Pukeko, cooperation among relatives arises because individuals often recruit in natal groups and may practice close inbreeding.

Most studies in this book have quantified the contribution of helpers to reproductive success of the group (Table 3). These contributions vary from vital in Stripe-backed Wrens through moderate in Splendid Fairy-wrens to slight in Harris' Hawks, Groove-billed Anis, Red-cockaded Woodpeckers and Pinyon Jays. A detectable enhancement of the reproductive success or survival of relatives is a precondition for kin selection to be a cause of cooperative breeding. In several examples, enhanced reproductive success in groups with helpers is a result of reduced predation on nestlings and dependent fledglings. In other cases, benefits from helpers include more rapid re-nesting (see Chapter 1) and reduced predation on adults (see Chapter 11).

Since the presence of helpers is usually a consequence of previous successful reproduction, the presence and extent of helping is often confounded with differences in parental ability. Therefore, even when pairs without helpers have very poor reproductive success (as in the Stripe-backed Wren), experiments are needed to show that helping *per se* increases reproductive success. The only such experiment published to date (Brown *et al.* 1982) provides some support for a cause and effect relationship, but the experiment was a preliminary one only (Brown 1987). A recent unpublished removal study on Florida Scrub Jays had a similar result (R.L. Mumme, personal communication), but a second such study (M. L. Leonard, A. G. Horn and S. F. Eden, personal communication) failed to find reduced reproductive success after the removal of helpers in Moorhens (*Gallinula chloropus*).

Good evidence for a role of kin selection in directing the target of helping behavior is provided in several chapters of this book. Reyer has shown that adult Pied Kingfishers only accept help from offspring under good feeding conditions, but allow unrelated birds to help when feeding conditions are poor. In White-fronted Bee-eaters and Galápagos Mockingbirds, individuals preferentially aid relatives over non-relatives, when both are simultaneously available to be helped (see Chapters 16 and 10). Similarly, yearling Pinyon Jays leave non-breeding flocks to seek out and aid only close relatives (see Chapter 7). The mechanisms of kin recognition in mammals is usually based on matching with learned or inherited odours of the social group (see e.g. Holmes 1986). In cooperatively breeding birds, presumably visual or vocal cues are used in kin recognition.

The above evidence suggests that kin selection indeed plays a role in cooperative breeding. This role, however, may vary from critical in species such as the Stripe-backed Wren, to modest or absent in the Florida Scrub Jay and Splendid Fairy-wren.

Role of reciprocity. Little evidence is available on the role of reciprocity as a cause of cooperative breeding. Helping behavior is strongly asymmetric in most cases, with young subordinate individuals tending to help older ones, and not vice versa. This greatly reduces the possibility of reciprocal aiding. Reciprocity is most likely to be an important cause of helping in cases where non-related individuals help one another, and/or when helping is not strongly dependent on asymmetries such as age and social dominance. These conditions may apply to cases of cooperation by individuals seeking breeding territories (e.g. Green Woodhoopoes and Galápagos Hawks), to cooperative hunting, as seen in Harris' Hawks (see Chapter 12) and in lions

(Packer and Ruttan 1988), and to colonially breeding species where adults may aid at several nests (e.g. White-fronted Bee-eaters). More detailed lists of possible cases of reciprocity are given by Emlen (1984) and by Brown (1987). In all these cases, non-reciprocal benefits of cooperation may apply, and Occam's razor should perhaps give these precedence over reciprocity.

It may be relevant that one of the most widely quoted examples of reciprocity in mammals, coalition formation in male savannah baboons (Packer 1977), has not been confirmed by more detailed study (Bercovitch 1988). I suspect that reciprocity plays only a minor role in the maintenance of cooperation in most cooperative breeders.

Role of demographic constraints. Woolfenden and Fitzpatrick (1984) proposed the following explanation for cooperative breeding. Offspring that remain in the natal group beyond the minimum breeding age have a higher lifetime reproductive success than those that disperse at this age, because they postpone costs of dispersal and gain breeding opportunities on or near their natal territory. For most studies in this book, data on lifetime breeding success in relation to the time of dispersal are not available. In addition, early and late dispersers may differ in social dominance (e.g. in male Florida Scrub Jays) and may therefore differ in fitness because of this asymmetry alone, regardless of the effect of the timing of dispersal on fitness. Such covariation among variables of interest is a general problem with adaptive explanations of life history phenomena.

We can not therefore easily test the general correctness of Woolfenden and Fitzpatrick's idea at present. We can, however, examine its plausibility, given the demographic results presented in this book. For the idea to apply, the onset of reproduction must be delayed in the helping sex, adult survival must be high, and juvenile survival must be fairly high. Data on three of these points are summarized in Table 4. Few of the species described here breed regularly as yearlings, and in the Gray-Breasted Jay, Arabian Babbler and Pukeko, many individuals do not breed until 3 or more years of age. Annual survival rates of adults range from 0.52 to 0.90 or higher, and most values are over 0.7. Survival from fledging to 1 year ranges from 0.35 to 0.76, with most values reaching to 0.50 or greater. In the Green Woodhoopoe, survival in the first year is actually slightly higher than adult survival. Uneven sex ratios (Table 4) are also an important demographic constraint in several species (e.g. Dunnocks, Galápagos Mockingbirds, Splendid Fairy-wrens). When there is a relative shortage of females in a population, it becomes more advantageous for a male to remain and to help in his natal group, but early dispersal by females is favored.

Table 4. *Demographic characteristics of cooperative breeders described in this book*

Species	Sex ratio of adults (males: females)	Survival from fledging to one year of age	Annual survival of adults	Modal age of first breeding
Harris' Hawk	More males	—	high?	—
Galápagos Hawk	1.68–2.40:1	below 0.50	0.90	—
Pukeko	1.00–1.07:1	low?	c. 0.80	3?
Groove-billed Ani	1:1	c. 0.5	c. 0.72	1
Hoatzin	1.25:1	0.67	c. 0.90	3
Pied Kingfisher	1.58:1	0.42M/?F	0.54–0.74	1F/2M
White-fronted Bee-eater	0.96–1.16:1	0.65–0.70	0.70–0.75	2F/2M
Green Woodhoopoe	0.67–0.94:1	0.50–0.79	0.60–0.70	3F/3M
Red-cockaded Woodpecker	1.6:1	0.42F/0.50M	0.71F/0.77M	1F/2M
Acorn Woodpecker (HR)	1.55:1B	0.57	0.73F/0.83M	2
Acorn Woodpecker (WC)	1.32:1	0.37	0.57	2
Stripe-backed Wren	1.2:1	0.65	0.56–0.72	2
Bicolored Wren	1:1	0.43–0.76	0.68–0.81	2
Galápagos Mockingbird	0.75–2.13:1	0.35	0.41–0.93	1
Dunnock	1.25:1	—	0.60	1
Arabian Babbler	1:1	0.50	0.65–0.85	4
Splendid Fairy-wren	0.96–1.81:1	c. 0.35	0.67–0.72	1
Noisy Miner	2.2–3.3:1	0.50?, 0.60	0.67, 0.63	2
Grey-breasted Jay	—	—	0.81–0.86	3–4?
Florida Scrub Jay	0.76–1.38:1	0.35	0.79	2
Pinyon Jay	0.7–3.2:1	0.41+	0.74	2

Numbers separated by dashes refer both to ranges across different study areas, and to ranges within areas across years; question marks indicate approximate values. Most values are from the chapters in the book, but some are from primary publications cited by the chapter. Dashes indicate absence of the data in the chapter or in primary publications known to me. F = females; M = males; B = breeders only.

Since most cooperative breeders are characterized by delayed onset of breeding and low adult and juvenile mortality, Woolfenden and Fitzpatrick's demographic constraints model is plausible for most species. Very detailed demographic information to test the idea more thoroughly is available for the Florida Scrub Jay (Woolfenden and Fitzpatrick 1984; Chapter 8, this volume; Fitzpatrick and Woolfenden 1988), and the Acorn Woodpecker (Koenig and Mumme 1987; Chapter 14, this volume). In both cases further support has been found for the idea, but new difficulties in testing the idea have arisen because these populations have exhibited unstable demography (see below). In some species (e.g. Harris' Hawks), however, demographic constraints remain to be identified.

Non-adaptive explanations for behaviors involved in cooperative breeding. In Chapter 13, Craig and Jamieson discuss the possibility that behaviors such as helping at the nest require no special adaptive explanation. They suppose in this case that the mere proximity to a begging young is a sufficient stimulus to cause food delivery. This explanation, however, does not explain why some Galápagos Mockingbirds help to feed young in their group, while others do not, and that this is correlated with relative kinship (see Chapter 10), nor does it explain why yearling non-breeders among Pinyon Jays leave the non-breeding flock and seek out their parents to help them to feed dependent young. In these cases, kin selection is a ready explanation. Craig and Jamieson are right to caution against the 'tyranny of bright ideas', but most of the behaviors involved in cooperative breeding can be interpreted in an adaptive framework without *ad hoc* adaptationism.

It is often assumed that proximate adaptive explanations of cooperative breeding can also explain how cooperative breeding evolved. While this may be correct, we have few ways of testing this. Although the common forms of cooperative breeding have evolved in several unrelated groups of birds, phylogenetic analyses (e.g. Sillén-Tullberg 1988) of cooperative breeding may still be worth while. All four species of anis are joint-nesting plural breeders (see Chapter 11), and trends towards cooperative breeding are exhibited by other taxonomic groups such as jays, rails and babblers.

Population stability in cooperative breeders
The studies summarized here include much demographic information for cooperative breeders. Most other detailed demographic observations of birds have been of short-lived species living in temperate environments (see e.g. Clobert *et al.* 1988) or of very long-lived seabirds (see e.g. Ollason and Dunnett 1988). Thus, the studies in this book fill an important

'demographic gap', in that they deal with moderately long-lived species, living predominantly in tropical or warm climates. Further, unlike many studies of temperate birds, most studies here have been conducted in pristine environments.

It is often stated that bird populations are relatively stable in numbers (see e.g. Perrins and Birkhead 1983, p. 136). This is generally true for the cooperative breeders described in the book, and few studies here found marked seasonal changes in group size. Several of the cooperative breeders, however, fluctuated markedly in numbers over longer periods. The failure of acorn crops caused population crashes in Acorn Woodpeckers. Diseases caused crashes in Galápagos Mockingbirds and Florida Scrub Jays, and fires greatly reduced abundances of Splendid Fairy-wrens. Other populations of cooperative breeders underwent steady declines for less obvious reasons (e.g. Stripe-backed Wrens and Pied Kingfishers). This lack of stability in numbers makes it difficult to test the demographic constraints model, and also suggests to me that bird populations are less stable than is commonly supposed.

Future work on cooperative breeding

Future studies of cooperative breeding will certainly delve deeper into the ecological and demographic details of cooperative breeding. These data will be essential to evaluate fully the relative roles of kinship, demographic constraints and other explanations for the existence of cooperation. More studies will be able to present data covering entire life spans of cohorts of individuals, but this alone will not answer all existing problems. In fact, some studies in this book (e.g. Chapters 6 and 17) have obtained critical results without covering entire life spans of individuals. In planning a study, good design and penetrating questions are superior to mere staying power.

New molecular methods of assigning parentage will improve our estimates of kinship and fitness (see e.g. Quinn *et al.* 1987; Burke and Bruford 1987; Wetton *et al.* 1987). This may simply sharpen our traditional focus, if the presumed parents turn out to be the true parents. If, however, helpers are frequently the genetic parents of some of the offspring that they help to rear, a more radical reassessment will be needed. Recent work by T. Burke and N. B. Davies (see Chapter 15) has shown that the extent to which male Dunnocks have access to breeding females, and provision young with food, is indeed related to their actual degree of paternity. Information on paternity may also affect our understanding of close inbreeding. Electrophoretic analysis has shown that the apparently high level of inbreeding in the Splendid Fairy-wren is confounded by a high rate of extra-pair fertilization (I. Rowley and E. Russell, personal communication; see also Chapter 1).

New physiological methods should also sharpen our answers to the traditional questions, and some of these have already been employed in the studies in this book. For example, Reyer measured the energetic costs of helping in the Pied Kingfisher using the doubly labeled water method, and also used endocrinological methods to test whether helpers were physiologically capable of breeding. Future studies should employ such physiological methods with greater frequency.

An area for comparative ecological work is the measurement of food supplies of cooperative and non-cooperative breeders at the season of minimum food availability. It seems highly likely that a richer minimum food supply during times of scarcity is a requirement for the permanent territorial residence seen in most of the species studied in this book. No comparative empirical data, however, exist on this to my knowledge.

Another active area for future work is in mathematical models of cooperation. Hamilton's original models of group-living and kin selection have been developed by Vehrencamp (1983), Brown (1987) and others. The early graphical models of Brown (1969) were developed into more explicit demographic models by Emlen (1982*a,b*), Woolfenden and Fitzpatrick (1984) and Koenig and Mumme (1987). A recent trend has been the development and testing of game theory models of cooperation (see e.g. Axelrod 1984; Craig 1984; Houston and Davies 1985). As in most cases of model-building, further modeling work on cooperative breeding will be most fruitful if it stimulates new empirical work.

A praiseworthy and overdue shift away from purely descriptive work on cooperative breeders is now under way. Following the removal experiment of Brown *et al.* (1982), others have begun experimental studies. Rabenold and his colleagues (see Chapter 6) and Hannon *et al.* (1985) removed individuals from groups to investigate the intensity of competition for breeding spaces. R. L. Mumme and M. L. Leonard and colleagues (see above) have removed helpers in Florida Scrub Jays and Moorhens, respectively, to investigate the effects of helpers on reproductive success. Stacey has begun removals of entire groups to test the importance of habitat saturation in a cooperative (Gray-breasted), and in a non-cooperative (Scrub) jay. This trend to greater reliance on experimental manipulation should continue in non-endangered species. Care, however, will be needed in the selection of questions for experimental testing and in experimental design; bad experiments can interrupt good demographic data sets without yielding clear answers.

A further need in future studies is highlighted by Craig and Jamieson (Chapter 13). Not only can we learn from new approaches to traditional questions, we can profitably address a wider range of questions, such as the development and proximate causation of helping behavior (e.g. Jamieson 1988), and why some species of cooperative breeders avoid close inbreeding while others do not. A parallel with studies of kinship and group-living in primates (see e.g. Smutz *et al.* 1987; Dunbar 1988) is relevant here. Primates exhibit greater social complexity than group-living birds, but many ecological and demographic similarities. Studies of primates have often focused successfully on the development of social relations and on the details of the development of the young (see e.g. Altmann 1980; Dunbar 1988).

Conclusion

The studies in this book have already provided answers to some of the puzzles of cooperative breeding. The principal proximate cause of cooperative breeding is delayed dispersal from the natal group. In a few cases, cooperation is forced on individuals by severe shortages of foods or other limiting resources. High predation rates on dependent offspring, specific habitat requirements and high mortality or early dispersers also are common correlates of cooperative breeding. Kin selection also plays a role in the maintenance of cooperative breeding, and in directing helping behavior towards relatives. No single explanation is likely to account for all cases of cooperative breeding, and many unanswered questions remain.

We have only begun to scratch the surface in experimental, genetic, physiologic and phylogenetic studies of cooperative breeding. The next 25 years of work promise to be as exciting as the 25 years since Hamilton first proposed his genetic theory of social behavior.

Bibliography

Altmann, J. (1980). *Baboon Mothers and Infants*. Harvard University Press: Cambridge.
Axelrod, R. (1984). *The Evolution of Cooperation*. Basic Books: New York.
Barash, D. P. (1975). *Sociobiology and Behavior*. Elsevier: New York.
Bercovitch, F. B. (1988). Coalitions, cooperation and reproductive tactics among adult male baboons. *Anim. Behav.* **36**: 1198–1209.
Brockmann, J. (1984). The evolution of social behaviour insects. In *Behavioural Ecology: An Evolutionary Approach*, 2nd edn, ed. J. R. Krebs and N. B. Davies, pp. 340–61. Blackwell: Oxford.
Brown, J. L. (1969). Territorial behavior and population regulation in birds. *Wilson Bull.* **81**: 293–329.
Brown, J. L. (1974). Alternate routes to sociality in jays – with a theory for the evolution of altruism and communal breeding. *Amer. Zool.* **14**: 63–80.
Brown, J. L. (1987). *Helping and Communal Breeding in Birds: Ecology and Evolution*. Princeton University Press: Princeton, N.J.

Brown, J. L., Brown, E. R., Brown, S. D. and Dow, D. D. (1982). Helpers: effects of experimental removal on reproductive success. *Science* **215**: 421–22.

Burke, T. and Bruford, M. W. (1987). DNA fingerprinting in birds. *Nature (Lond.)* **327**: 149–52.

Carlisle, T. R. and Zahavi, A. (1986). Helping at the nest, allofeeding and social status in immature Arabian babblers. *Behav. Ecol. Sociobiol.* **18**: 339–51.

Clobert, J., Perrins, C. M., McCleery, R. H. and Gosler, A. G. (1988). Survival rate in the great tit *Parus major* in relation to sex, age and immigration status. *J. Anim. Ecol.* **57**: 287–306.

Clutton-Brock, T. H. (1988). *Reproductive Success: Studies of Individual Variation in Contrasting Breeding Systems*. Chicago University Press: Chicago.

Craig, J. L. (1984). Are communal pukeko caught in the prisoner's dilemma? *Behav. Ecol. Sociobiol.* **14**: 147–50.

Dow, D. D. (1980). Communally breeding Australian birds, with an analysis of distributional and environmental factors. *Emu* **80**: 121–40.

Dunbar, R. I. M. (1988). *Primate Social Systems*. Comstock Publishing Associates: Ithaca, NY.

Emlen, S. T. (1982a). The evolution of helping. I. An ecological constraints model. *Amer. Nat.* **119**: 29–39.

Emlen, S. T. (1982b). The evolution of helping. II. The role of behavioral conflict. *Amer. Nat.* **119**: 40–53.

Emlen, S. T. (1984). Cooperative breeding in birds and mammals. In *Behavioural Ecology: An Evolutionary Approach*. 2nd edn, ed. J. R. Krebs and N. B. Davies, pp. 305–39. Blackwell: Oxford.

Emlen, S. T. and Vehrencamp, S. L. (1983). Cooperative breeding strategies among birds. In *Perspectives in Ornithology*, ed. A. H. Brush and G. A. Clark Jr, pp. 93–120. Cambridge University Press: Cambridge.

Fitzpatrick, J. W. and Woolfenden, G. E. (1988). Components of lifetime reproductive success in the Florida scrub jay. In *Reproductive Success: Studies of Individual Variation in Contrasting breeding Systems*, ed. T. H. Clutton-Brock, pp. 305–20. Chicago University Press, Chicago.

Ford, H. A., Bell, H., Nias, R. and Noske, R. (1988). The relationship between ecology and the incidence of cooperative breeding in Australian birds. *Behav. Ecol. Sociobiol.* **22**: 239–49.

Fricke, H. W. (1979). Mating system, resource defence, and sex change in the anemonefish, *Amphiprion akallopisos. Z. Tierpsychol.* **50**: 313–26.

Gaston, A. J. (1978). The evolution of group territorial behavior and cooperative breeding. *Amer. Nat.* **112**: 1091–100.

Gayou, D. C. (1986). The social system of the Texas Green Jay. *Auk* **103**: 540–7.

Gould, S. J. and Lewontin, R. C. (1979). The spandrels of San Marco and the Panglossian paradigm: a critique of the adaptationist programme. *Proc. Roy. Soc. Ser. B* **205**: 581–98.

Hamilton, W. D. (1964). The genetical evolution of social behaviour. I. II. *J. Theoret. Biol.* **7**: 1–52.

Hannon, S. J., Mumme, R. L., Koenig, W. D. and Pitelka, F. A. (1985). Replacement of breeders and within-group conflict in the cooperatively breeding acorn woodpecker. *Behav. Ecol. Sociobiol.* **17**: 303–12.

Holmes, W. G. (1986). Kin recognition by phenotype matching in female Belding's ground squirrels. *Anim. Behav.* **34**: 38–47.

Houston, A. I. and Davies, N. B. (1985). Evolution of cooperation and life history in Dunnocks. In *Behavioural Ecology: the Ecological Consequences of Adaptive Behaviour*, ed. R. Sibly and R. H. Smith, pp. 471–87, British Ecological Society Symposium. Blackwell Scientific Publications: Oxford.

Jamieson, I. G. (1988). Provisioning behaviour in a communal breeder: an epigenetic approach to the study of individual variation in behaviour. *Behaviour* 104: 262–80.

Koenig, W. D. and Mumme, R. L. (1987). *Population Ecology of the Cooperatively Breeding Acorn Woodpecker*. Princeton University Press: Princeton, NJ.

Koenig, W. D. and Pitelka, F. A. (1981). Ecological factors and kin selection in the evolution of cooperative breeding in birds. In *Natural Selection and Social Behavior: Recent Research and New Theory*, ed. R. D. Alexander and D. W. Tinkle, pp. 261–80. Chiron Press: New York.

Lack, D. (1968). *Ecological Adaptations for Breeding in Birds*. Methuen: London.

Maynard Smith, J. (1964). Group selection and kin selection. *Nature* 201: 1145–7.

Ollason, J. C. and Dunnet, G. M. (1988). Variation in breeding success in fulmars. In *Reproductive Success: Studies of Individual Variation in Contrasting Breeding Systems*, ed. T. H. Clutton-Brock, pp. 263–78. Chicago University Press: Chicago.

Packer, C. (1977) Reciprocal altruism in *Papio anubis*. *Nature (Lond.)* 265: 441–3.

Packer, C. and Ruttan, L. (1988). The evolution of cooperative hunting. *Amer. Nat.* 132: 159–98.

Patterson, I. J. (1980). Territorial behaviour and the limitation of population density. *Ardea* 68: 53–62.

Perrins, C. M. and Birkhead, T. R. (1983). *Avian Ecology*. Blackie: Glasgow.

Quinn, T. W., Quinn, J. S., Cooke, F. and White, B. N. (1987). DNA marker analysis detects multiple maternity and paternity in single broods of the lesser snow goose. *Nature (Lond.)* 326: 392–4.

Selander, R. K. (1964). Speciation in wrens of the genus *Campylorynchus*. *Univ. Calif. Publ. Zool.* 74: 1–224.

Sillén-Tullberg, B. (1988). Evolution of gregariousness in aposematic butterfly larvae: a phylogenetic analysis. *Evolution* 42: 293–305.

Smutz, B. B., Cheney, D. L., Seyfarth, R. M., Wrangham, R. W. and Struhsaker, T. T. (eds) (1987). *Primate Societies*. Chicago University Press: Chicago.

Stacey, P. B. (1979). Habitat saturation and communal breeding in the Acorn Woodpecker. *Anim. Behav.* 27: 1153–66.

Stacey, P. B. and Ligon, J. D. (1987). Territory quality and dispersal options in the acorn woodpecker, and a challenge to the habitat-saturation model of cooperative breeding. *Amer. Nat.* 130: 654–76.

Taborsky, M. (1984). Broodcare helpers in the cichlid fish *Lamprologus brichardi*: their costs and benefits. *Anim. Behav.* 32: 1236–52.

Trivers, R. L. (1971). The evolution of reciprocal altruism. *Q. Rev. Biol.* 46: 35–57.

Vehrencamp, S. L. (1983). A model for the evolution of despotic versus egalitarian societies. *Anim. Behav.* 31: 667–82.

Wetton, J. H., Carter, R. E., Parkin, D. T. and Walters, D. (1987). Demographic study of a wild house sparrow population by DNA fingerprinting. *Nature (Lond.)* 327: 147–9.

Wilson, E. O. (1975). *Sociobiology: The New Synthesis*. Harvard University Press: Cambridge, MA.

Woolfenden G. E. and Fitzpatrick, J. W. (1984). *The Florida Scrub Jay: Demography of a Cooperative-breeding Bird*. Princeton University Press: Princeton, NJ.

Zahavi, A. (1974). Communal nesting by the Arabian Babbler: a case of individual selection. *Ibis* 116: 84–7.

INDEX

Acorn Woodpecker, x, xv, 94, 97, 255, 270, 294, 335, 345, 351–2, 401, 403, 415–51, 596, 598–602, 605–7
adoption, 114, 165
age at first breeding, 9, 45–7, 146, 182, 214, 250–1, 275, 305, 387, 533, 569
altruism, x, 21, 125–7, 404, 489
Arabian Babbler, xiv, 105–30, 596, 598, 600, 604–5
asynchronous hatching, 213

Bell Miner, 588
benefits of philopatry model, 57–61, 153, 447–8, 450, 594, 601
Bicolored Wren, 159–95, 596, 598, 600, 605
body temperature, 52
brood parasitism, 353
brood reduction, 83, 432, 568, 581–3
see also egg destruction
Brown Jay, 352
by-product mutualism, 284

cavity nesting and roosting, 51–2, 57–8, 70, 93–5
clutch size, 41, 82–3, 117, 255, 314, 364, 401, 539, 571
coevolution, 199
coloniality, xv, 145, 211–13, 490, 495–7, 503–5, 530–1, 552, 564–7, 580
communal breeding, see cooperative breeding
conservation, 69
cooperative breeding
extent and variability of ix, xiii, 194
in Australia, 25, 602
primitive, 96, 232
cooperative hunting, 378–81

cooperative polyandry, see mating system
copulation, 14, 41, 73, 111, 113, 115, 117, 143, 145–6, 170, 304–5, 314, 365, 375–6, 397–401, 407, 425, 427–8, 468–71, 475–6, 533, 589
coteries, 565–7

Darwin, 291
diet and foraging behavior, ix, 4–5, 33, 70–1, 105–6, 135, 151, 199, 204–5, 213–14, 244, 270–2, 294, 301, 336–7, 360, 378–80, 388, 416, 418–21, 471, 480–1, 492, 497–9, 529–30, 539–40, 564, 566
disease, 258, 497
dispersal, 8–9, 12, 39, 47–9, 79–81, 111, 113–14, 144–6, 168, 172, 180, 183–91, 206–9, 254–5, 278, 301–2, 338–9, 345–7, 395, 397, 436–8, 446, 459–60, 494–5, 501, 503, 531–3, 554–5, 584
distraction display, 4
dominance, 38–41, 73, 112, 115–16, 125–8, 165, 168–9, 205, 252–3, 281, 304, 314–15, 365–6, 392, 397, 399–401, 462, 567
doubly labeled water, 547–8, 608
Dunnock, xv, 327, 457–84, 596, 598–600, 604–5

ecological constraints, 93, 323, 406, 450, 519–24, 551, 593, 601, 604, 606–7
see also habitat saturation
egg destruction, 113, 281, 343, 346, 442–3
electrophoresis, protein, 30, 375–6, 426
El Niño, 291, 295, 313, 319–20
emigration, see dispersal
energetics of feeding, 547–9

fingerprinting, DNA, 195, 428, 450, 463–7, 475, 607
fish, cichlid, xiii
floaters, 59, 77–9, 81, 144–5, 154, 346, 362–4, 367, 390–1, 397
flock size, and composition, 200–3
Florida Scrub Jay, xiv, 80, 124, 194, 230, 241–65, 275, 319, 325–6, 348–51, 425, 490, 596, 598–601, 603–5, 607–8
food storage, 294–5, 294, 416, 418–21, 423
foraging behavior, *see* diet and foraging behavior

Galápagos Hawk, xv, 295, 359–69, 380–1, 596, 598–600, 603, 605
Galápagos mockingbird, xiv, 291–329, 596, 598–600, 603–7
genealogy, of groups, 10, 108–9, 166, 232–4, 280, 303, 374, 396, 502, 532
Gray-breasted Jay, *see* Mexican jay
Gray-crowned Babbler, 352, 484
Green Woodhoopoe, xiv, 33–64, 114, 255, 410, 596, 598, 600–5
Groove-billed Ani, xiv, 335–54, 401, 449, 596, 598, 600, 605
group size and composition, 5, 7–8, 14–16, 26, 38, 75, 110–13, 141–4, 165–7, 245, 248, 272–3, 298–303, 337–9, 363–4, 371–2, 374–5, 390–1, 396, 421–9, 499–502, 505–6
see also helpers and helping; reproductive success

habitat requirements, 3–4, 33, 69–70, 162, 242–4, 336, 388, 415
habitat saturation, 23, 93–6, 123–4, 150, 168, 193–4, 262–5, 282, 323–4, 347–51, 377–8, 392, 394, 445–6, 449, 561, 585, 594, 601
Harris' Hawk, xv, 359–60, 365, 369–81, 596, 598, 600–2, 605–6
hatching failure, 49
Hedge Accentor, *see* Dunnock
Hedge Sparrow, *see* Dunnock
helpers and helping, acceptance of, 230, 549–51, 603
  and feeding rate, 171–3, 216–17, 224, 277, 316–17, 441, 518–19, 537–8, 547–51, 571
  and kinship, 46–7, 76, 144, 171–2, 217–22, 224–5, 251–2, 283, 309–12, 345, 510–15
  and sex ratio, 11, 226–8, 308–9, 533–5, 551
  and social environment, 404–10
  costs and benefits of, 21–4, 152–4, 542–5

effects of, 13, 16–17, 26–7, 61–2, 87, 143, 147–8, 217–24, 233, 433–4
extent of, 73, 306–7, 408–10, 464–8, 505–6, 571
ontogeny of, 173, 408–9, 597, 609
origin of, 16, 76, 144, 217, 224–6, 404–5, 500, 509–9, 533–5, 586
possible non-adaptiveness of, 24, 63, 283, 285, 328, 404–7, 595, 606
see also reproductive success; survivorship
Hoatzin, xiv, 133–54, 161, 596, 598–601, 605
hypothermia, 52

immigration, *see* dispersal
inbreeding and incest avoidance, 12, 16–17, 30, 47–9, 82, 111–12, 123, 188, 250, 305–6, 327, 345, 375, 397, 399, 410–11, 426, 441, 503, 580, 608–9
incubation, 210, 344–6, 350–1, 393, 401–3, 570
and joint-nesting, 352–4
index of kin selection, 439
infanticide, 281, 426, 443–4, 567, 581–3
see also egg destruction

joint-nesting, xiv, 117, 146, 273–4, 305, 309, 335, 343, 350–3, 401–3, 425, 440, 442–3, 449

kin recognition and discrimination, 171, 311–12, 364, 511–15, 602–3
kin selection, x, 23, 151–4, 192–4, 282–3, 345, 409, 438–9, 523–4, 542–5, 579, 593, 595, 603, 608–9

learning skills model, 521–2
lifetime reproductive success, 54–7, 182–3, 260–1, 438–9, 444–6
local resource competition, 88

mammals, xiii
mate-guarding, 400, 427, 462–4, 467–8
mate-sharing, *see* joint-nesting; mating system
mating system, xiv–xv, 7, 14, 30, 39, 41, 73, 112–13, 145–6, 169, 211, 249, 302–9, 327, 337–8, 360, 363–8, 375–6, 401–3, 423, 425–9, 448–50, 458, 460–8, 471–6, 533, 585
membership hypothesis, 283, 285–6
Mexican jay, xiv, 206, 264, 269–87, 307, 401, 596, 598, 600, 604–5, 608

migration, 137–41, 144, 421–3
Moorhen, 352–3, 484, 603, 608
mortality, *see* survivorship

naked mole-rat, xiii
Noisy Miner, xv, 561–90, 596, 598–601,
    605

Occam's razor, 604

parasitism, 14, 22, 170
parental effort, model for, 482–4
    see also 407–10
parental facilitation, 228, 231, 284
physiological constraints, 63
physiology, digestive, 135
Pied Kingfisher, xi, xv, 61, 283, 410,
    529–55, 596, 598, 600, 603, 605, 607–8
Pinyon Jay, xiv–xv, 199–235, 596, 598–600,
    603, 605
play, 121
plural breeding, xiv, 146, 248, 273–4, 303,
    305–6, 325–6, 335, 350, 401–3
polyandry, *see* mating system
polygynandry, *see* mating system
polygyny threshold, 378
population size and stability, 5–6, 44, 91–2,
    120, 148–9, 178–9, 181–2, 200–3,
    246–7, 273, 279, 299–300, 320–2, 339,
    347–8, 418–19, 440, 459, 493, 496–7,
    531, 564, 606–7
predators, 45, 83, 117, 147, 175–6, 244, 295,
    344, 351, 533, 581
primates, ix, 609
prisoner's dilemma, 284, 368, 394
promiscuity, and cooperative breeding, 585
Pukeko, xv, 335, 351–2, 387–411, 596,
    598–600, 602, 604–5

reciprocity, 58, 60–2, 125, 153, 171, 194,
    227, 277, 281–2, 319, 368, 409, 593,
    595, 603–4
Red-cockaded Woodpecker, xiv, 69–98,
    415, 596, 598, 600–1, 605
relatedness, 16, 46–7, 60, 171, 280, 345,
    364, 425, 428–9, 441, 510–15, 535–6,
    576, 579
repayment model, 88, 553
reproductive behavior, 13–16, 41–2, 82–4,
    117, 170–1, 210–15, 255, 342–5, 356–66,
    371–5, 400–3, 429–35, 461–3, 504–6,
    533–4, 569–80

reproductive competition, 123–4, 183–6,
    314–15, 403, 441–3
reproductive success and group size, 18–19,
    22, 26, 43–5, 50–1, 84–7, 117–19,
    146–7, 173–8, 222–4, 255–8, 314–18,
    340–5, 366–7, 376–7, 394–5, 429–34,
    472–5, 515–18, 538–9, 572–3, 581
    see also helpers and helping
resource localization, 585
    see also habitat saturation

Scrub Jay, *see* Florida Scrub Jay
senescence, 259
sentinel behavior, 122, 127, 205–6, 253, 375
sex ratio, 11, 26, 39, 88, 144, 179, 208, 227,
    308–9, 322, 372, 390–2, 440, 477, 493,
    509–10, 531, 535, 551–4, 568, 585
sexual dimorphism, 3, 33, 477, 530, 561–2
    in foraging, 33, 71
siblicide, 582–3
skill hypothesis, 286, 521–2
social insects, x
social status hypothesis, 125–9
Splendid Fairy Wren, xiv, 3–28, 327, 596,
    598, 600, 602–5, 607–8
Stripe-backed Wren, 159–95, 596, 598, 600,
    602–3, 605, 607
Superb Blue Wren, 411
survivorship, 19–20, 25–6, 51, 88–9, 92,
    119, 148–9, 179–82, 207, 209–10,
    255–6, 258–9, 278–9, 318–20, 338,
    341–2, 367, 395, 435–7, 459, 531,
    582–3
Swamphen, *see* Pukeko

territorial budding, 90–2, 113, 144, 146,
    148, 153, 247
territorial inheritance, 12, 61, 77, 90, 114,
    278, 283, 306, 345, 437
territoriality, 5, 36–7, 73–5, 90–1, 106, 110,
    137, 140–2, 167, 245–6, 297–300, 339,
    351–2, 361–3, 376, 389–94, 416, 418,
    461–3, 477–80, 493, 497–9, 529, 564
territory size and quality, 36–7, 53–8, 74–5,
    140–1, 148, 150, 176–7, 195, 264,
    229–30, 328, 339–40, 348, 364, 392–3,
    432, 444–5, 477–80, 499

variance enhancement and utilization, 284

White-fronted Bee-eater, xi, xv, 61, 73, 283,
    489–524, 568, 596, 598–600, 603–5

Printed in the United States
by Bookmasters

Printed in the United States
By Bookmasters